2nd International Planetary Probe Workshop

August 23-26, 2004

NASA Ames Conference Center
Moffett Field, California USA

Sponsored by:
The Aerospace Corporation
Applied Research Associates, Inc.
Ball Aerospace
The Boeing Company
ELORET Corporation
European Space Agency
Idaho NASA Space Grant Consortium
Jet Propulsion Laboratory
Lockheed Martin
NASA
NASA Ames Research Center
NASA Langley Research Center
NASA Marshall Space Flight Center
SJSU, Mechanical Aerospace Engineering Department
University Affiliated Research Center
University of Idaho College of Engineering

Acknowledgement and Appreciation

It is with great appreciation that we acknowledge the hard work and support that was supplied by the International Steering Committee and the Local Organizing Committee. The success of the 2nd International Planetary Probe Workshop was primarily due to the enthusiasm and encouragement of them all. Many thanks!

Publication   Proceedings of the 2nd International Planetary Probe Workshop
                      NASA Ames Research Center, Moffett Field CA, USA
                      NASA/CP-2004-213456

Assembly Editor   Marla Arcadi
                      ELORET Corporation

Graphics   Jay Nuez
                      ELORET Corporation

Available from:

NASA Center for AeroSpace Information
7121 Standard Drive
Hanover, MD 21076-1320
(301) 621-0390

National Technical Information Service
5285 Port Royal Road
Springfield, VA 22161
(703) 487-4650

International Steering Committee

Ethiraj Venkatapathy, Chair (NASA Ames Research Center, USA)
David Atkinson, Co-Chair (University of Idaho, USA)
Jean-Pierre Lebreton, Co-Chair (ESA / European Space Research and Technology Center, The Netherlands)
Jim Arnold (University Affiliated Research Center, UC Santa Cruz, USA)
Sushil Atreya (University of Michigan, USA)
Reta Beebe (New Mexico State University, USA)
Bernie Bienstock (Boeing, USA)
Bobby Braun (Georgia Institute of Technology, USA)
Rodolphe Cledassou (Natinal Center for Space Studies, France)
Marcello Coradini (ESA Headquarters, France)
Jim Cutts (Jet Propulsion Laboratory, USA)
Prasun Desai (NASA Langley Research Center, USA)
Larry Esposito (University of Colorado, USA)
Francesca Ferri (University of Padova, Italy)
Ralph Lorenz (Lunar and Planetary Laboratory / University of Arizona AND
Space Sciences Research Institute, The Open University, UK)
Tobias Owen (University of Hawaii, USA)
Jim Robinson (NASA Headquarters, USA)
Maarten Roos-Serote (Lisbon University - Center for Astronomy and Astrophysics / Lisbon Astronomical Observatory, Portugal)
Tom Spilker (Jet Propulsion Laboratory, USA)
Bill Willcockson (Lockheed Martin, USA)
Rich Young (NASA Ames Research Center, USA)
John Zarnecki (Planetary and Space Sciences Research Institute, The Open University, UK)

Local Organizing Committee

Ed Martinez, Chair (NASA ARC)
Marla Arcadi, Deputy Chair (ELORET Corporation)
Charles A. Smith (NASA ARC)
Jim Arnold (UARC, UC Santa Cruz)
Terrill Buffum (ELORET)
Rebecca Allen Diamond (ELORET)
Peter Gage (ELORET)
Periklis Papadopoulos (SJSU)
Derrick Thomas (NASA ARC)

Table of Contents

	Page	DVD*
Foreword E. Venkatapathy, D. Atkinson	3	

Outlook for Probe Missions
Chair: Dave Atkinson   Co-Chair: Rich Young

	Page	DVD*
Welcome and Introduction E. Venkatapathy		1.0 v, p
Now, The Time for Probes and In-Situ Science G. Scott Hubbard		1.1 v, p
NASA's Solar System Exploration Program James Robinson		1.2 v, p
Planetary Exploration in ESA Gerhard H. Schwehm		1.3 v, p
Basic Questions About the Solar System: The Need for Probes Andrew P. Ingersoll	7	1.4 v, p
Technology for Entry Probes James A. Cutts, J. Arnold, E. Venkatapathy, E. Kolawa, M. Munk, P. Wercinski, B. Laub		1.5 v, p

Mars
Chair: Jeff Umland   Co-Chair: Prasun Desai

	DVD*
Technology Development for NASA Mars Missions Samad Hayati	2.1 v, p
ESA's Mars Program: European Plans for Mars Exploration Francois Forget	2.2 p
MER EDL: Overview and Reconstruction Status Prasun N. Desai, Wayne J. Lee	2.3 p
Mars Science Laboratory Overview & MSL EDL Challenges Jeffrey W. Umland	2.4 p
MSL Entry, Descent and Landing Performance and Environments Mary Kae Lockwood, A. Dwyer-Cianciola, A. Dyakonov, K. Edquist, D. Powell, S. Stripe, et al	2.5 p
A Model Based Mars Climate Database for the Mission Design Francois Forget	2.6 p

__Note:__ A DVD is included in the back of this CP with referenced video (v) and PowerPoint (p) presentations

	Page	DVD

Pascal: A Mars Climate Network Mission *2.7 p*
Anthony Colaprete, Bob Haberle, The Pascal Team

Overview of the Phoenix Entry, Descent and Landing System *2.8 p*
Rob Grover

Should We Believe Atmospheric Temperatures Measured by Entry Accelerometers Traveling at "Slow" Near-Sonic Speeds? 13
Paul Withers

The Next Generation of Planetary Atmospheric Probes 21
Howard Houben

Gas Giants
Chair: Bernie Bienstock Co-Chair: Jim Robinson

NASA Outer Solar System Exploration *3.1 p*
Jay T. Bergstralh

Project Prometheus and Future Entry Probe Missions *3.2 p*
Thomas R. Spilker

Neptune Polar Orbiter with Probes 29
Bernard Bienstock, D. Atkinson, K. Baines, P. Mahaffy, P. Steffes, S. Atreya, A. Stern, M. Wright, et al

Atmospheric Models for Aeroentry and Aeroassist 41
C.G. Justus, Aleta Duvall, Vernon W. Keller

An Investigation of Aerogravity Assist at Titan and Triton for Capture into Orbit About Saturn and Neptune 49
Philip Ramsey, James Evans Lyne

Software Risk Identification for Interplanetary Probes 59
Robert J. Dougherty, Periklis E. Papadopoulos

Aerothermodynamic Testing of Aerocapture and Planetary Probe Geometries in Hypersonic Ballistic-Range Environments *3.7 p*
M.C. Wilder, D.C. Reda, D.W. Bogdanoff, J. Olejniczak

Stagnation Point Heat Transfer with Gas Injection Cooling 69
B. Vancrayenest, M. D. Tran, D. G. Fletcher

Chemistry Modeling for Aerothermodynamics and TPS 75
Dunyou Wang, James Stallcop, Christopher Dateo, David Schwenke, Timur Halicioglu, Winifred Huo

Direct Communication to Earth from Probes *3.10 p*
Scott J. Bolton, William M. Folkner, Douglas S. Abraham

Benefits of Application of Advanced Technologies for a Neptune Orbiter, Atmospheric Probes and Triton Lander 81
Alan Somers, Luigi Celano, Jeffrey Kauffman, Laura Rogers, Craig Peterson

	Page	DVD

On Nonequilibrium Radiation in Hydrogen Shock Layers — 91
Chul Park

Calculation of H_2-He Flow with Nonequilibrium Ionization and Radiation: an Interim Report — 99
Michiko Furudate, Keun-Shik Chang

Clouds of Neptune and Uranus — 107
Sushil K. Atreya, Ah-San Wong

Titan
Chair: Ralph Lorenz   Co-Chair: Jean-Pierre Lebreton

The Huygens Mission to Titan: Overview and Status — *4.1 v, p*
Jean-Pierre Lebreton, Dennis Matson

Thermal Protection of the Huygens Probe During Titan Entry: Last Questions — 113
Jean-Marc Bouilly

Revalidation of the Huygens Descent Control Sub-System — 121
J.C. Underwood, J.S. Lingard, M.G. Darley

The Huygens Atmospheric Structure Instrument (HASI): Expected Results at Titan and Performance Verification in Terrestrial Atmosphere — 129
F. Ferri, M. Fulchignoni, G. Colombatti, P.F. Lion Stoppato, J.C. Zarnecki, A.M. Harri, et al

A Surface Science Paradigm for a Post-Huygens Titan Mission — 137
Wayne Zimmerman, Jonathan Lunine, Ralph Lorenz

Attitude Issues on the Huygens Probe: Balloon Dropped Mock up Role in Determining Reconstruction Strategies During Descent in Lower Atmosphere — 147
C. Bettanini, F. Angrilli

Atmospheric Stability & Turbulence from Temperature Profiles over Sicily During Summer 2002 & 2003 HASI Balloon Campaigns — 153
G. Colombatti, F. Ferri, F. Angrilli, M. Fulchignoni, HASI Balloon Team

Parachute Dynamics Investigations Using a Sensor Package Airdropped from a Small-Scale Airplane — 163
Jessica Dooley, Ralph D. Lorenz

Simulation Results of the Huygens Probe Entry and Descent Trajectory Reconstruction Algorithm — 171
B. Kazeminejad, D.H. Atkinson, M. Perez-Ayucar

Study of Some Planetary Atmospheres Features by Probe Entry and Descent Simulations — 181
P.J.S. Gil, P.M.B. Rosa

	Page	DVD

Venus and Special Topics
Chair: Larry Esposito Co-Chair: Eric Chassefiere

Future Venus Probe Missions — *5.1 p*
Larry Esposito

Lavoisier: A Low Altitude Balloon Network for Probing the Deep Atmosphere and Surface of Venus — 189
E. Chassefiere, J.J. Berthelier, J.-L. Bertaux, E. Quemerais, J.-P. Pommereau, P. Rannou, et al

ESA Venus Entry Probe Study — 201
M.L. van den Berg, P. Falkner, A. Phipps, J.C. Underwood, J.S. Lingard, J. Moorhouse, et al

Rotary-Wing Decelerators for Probe Descent Through the Atmosphere of Venus — 209
Larry A. Young, Geoffrey Briggs, Edwin Aiken, Greg Pisanich

A Search for Viable Venus and Jupiter Sample Return Mission Trajectories for the Next Decade — 217
Jason N. Leong, Periklis Papadopoulos

Helium-3 Mining Aerostats in the Atmospheres of the Outer Planets — *5.6 p*
Jeffrey E. Van Cleve, Carl Grillmair, Mark Hanna, Rich Reinert

Re-Entry Simulation and Landing Area for YES2 — *5.7 p*
Silvia Calzada

Neutral Mass Spectrometry for Venus Atmosphere and Surface — *5.8 p*
Paul Mahaffy

The Venus SAGE Atmospheric Structure Investigation — *5.9 p*
Anthony Colaprete, David Crisp, Clayton La Baw, Stephanie Morse

Synergy Between Entry Probes and Orbiters — 225
Richard E. Young

Cross-Cutting Topics
Chair: Michelle Munk Co-Chair: Neil Cheatwood
Panel Moderator: Dan Dumbacher

In-Space Propulsion (ISP) Aerocapture Technology — *6.1 p*
Michelle M, Munk, Bonnie F. James, Steve Moon

Family System of Advanced Charring Ablators for Planetary Exploration Missions — *6.2 p*
William M.Congdon, Donald M. Curry

Development of Solid State Thermal Sensors for Aeroshell TPS Flight Applications — 235
Ed Martinez, Tomo Oishi, Sergey Gorbonov

Ultralightweight Ballute Technology Advances — 239
Jim Masciarelli, Kevin Miller

	Page	DVD

Entry, Descent and Landing Using Ballutes — *6.5 p*
Daniel T. Lyons, Angus McRonald

New Approach for Thermal Protection System of a Probe During Entry — 245
Boris Yendler, Nathan Poffenbarger, Amisha Patel, Ninad Bhave, Periklis Papadopoulos

HyperPASS, a New Aeroassist Tool — 251
Kristin Gates, Angus McRonald, Kerry Nock

Genesis Sample Return Capsule Overview — *6.8 p*
Bill Willcockson

Performance of a Light-Weight Ablative Thermal Protection Material for the Stardust Mission Sample Return Capsule — 257
M.A. Covington

Development and Test Plans for the MSR EEV — 269
Robert Dillman, Bernard Laub, Sotiris Kellas, Mark Schoenenberger

Validation of Afterbody Aeroheating Predictions for Planetary Probes: Status and Future Work — 275
Michael J. Wright, J.L. Brown, K. Sinha, G.V. Chandler, F.S. Milos, D.K. Prabhu

Emerging Technologies
Chair: Jim Arnold   Co-Chair: Raj Venkatapathy

A Survey of the Rapidly Emerging Field of Nanotechnology: Potential Applications for Scientific Instruments and Technologies for Atmospheric Entry Probes — 289
M. Meyyappan, J.O. Arnold

Pico Reentry Probes: Affordable Options for Reentry Measurements and Testing — 291
William H. Ailor, V.B. Kapoor, G.A. Allen Jr, E. Venkatapathy, J.O. Arnold, D.J. Rasky

Nanostructured Thermal Protection Systems for Space Exploration Missions — 301
J.O. Arnold, Y.K. Chen, T. Squire, D. Srivastava, G. Allen Jr, M. Stackpoole, H.E. Goldstein, et al

NASA Ames Arc Jets and Range, Capabilities for Planetary Entry — 313
Ernest F. Fretter

Could Nano-Structured Materials Enable the Improved Pressure Vessels for Deep Atmospheric Probes? — *7.6 p*
D. Srivastava, A. Fuentes, B. Bienstock, J.O. Arnold

The Instrumented Frisbee® as a Prototype for Planetary Entry Probes — 317
Ralph D. Lorenz

Thermal, Radiation and Impact Protective Shields (TRIPS) for Robotic and Human Space Exploration Missions — 325
M.P. Loomis, J.O. Arnold

Deep Space Network Capabilities for Receiving Weak Probe Signals — *7.9 p*
Sami Asmar, Doug Johnston, Robert Preston

Page *DVD*

Honored Historians

From H.G. Wells to Unmanned Planetary Exploration   **337**
John W. Boyd, NASA Ames Research Center

Lessons From the Pioneer Venus Program   **343**
Steven D. Dorfman, Boeing Satellite Systems

List of Attendees   **349**

Foreword

For probe missions, 2004 was a bumper year, beginning with two successful Mars missions and concluding with the descent of Huygens to the surface of Titan. In August, an international community of scientists, engineers, mission designers, instrument/sensor designers, program and agency leaders gathered in California to discuss planetary probe science and technology, both for missions in progress, such as Genesis and Huygens, and for future missions, including Stardust, Phoenix, MSL, and proposed missions to the Gas Giants and Venus.

This international workshop grew out of the need to create a forum for communication between the leaders in science and enablers of technology, to allow the leadership to chart and market missions of the future. This year's meeting was a great success in gathering both the different disciplinary experts and a large number of students, whose attendance and engagement in the workshop, it is hoped, will excite them about the future and allow them to imagine their own possibilities.

The workshop was organized in various themes, and the Proceedings capture it in its entirety, through the papers submitted in time for this publication and through the DVD, where the remaining presentations are to be found. At this workshop, an expert panel discussion was introduced at each session. Although these discussions were not captured, they allowed an exchange of ideas across a broad range of related topics, including budget issues and how to enable future missions.

Emphasis was not given to any one destination or mission. Instead, we were able to look at each in turn and ask, in the context of probes, how mission experiences at one destination might help us plan others. How, for example, can the challenges and successes of technology investment for the Mars missions serve as a model for SSE or other destinations? Papers and panel discussions on the entry trajectory reconstruction of the Mars Exploration Rover (MER), for example, allowed us to consider its improvement for other missions, such as Huygens and Stardust. We were also concerned with the issue of cross cutting and emerging technologies in the broadest context, with a view to the probe community at large, rather than focused on any one project.

The success of the workshop cannot be measured purely by attendance and participation, which were excellent, but by activities that will blossom in later years, when we truly are sending 'multiple probes to multiple worlds', as challenged by Dr. Toby Owen and Dr. Sushil Atreya at the 1st International Probe Workshop in Lisbon, Portugal in 2003. As we look to the future, we see an increasing need for in-situ and sample return missions, because so much has been learned from orbiters and distant optical observations. Although the probe community is relatively small and missions take decades from concept to full realization of the science data, the diversity (both in age and discipline) of those attending the workshop gives us the best hope for the future.

The community decided to hold the next meeting in Greece, to celebrate the success of all the '05 missions. We hope to see you there, in June.

Ethiraj Venkatapathy David H. Atkinson
ISC Chair *ISC Co-Chair*

Outlook for Probe Missions

BASIC QUESTIONS ABOUT THE SOLAR SYSTEM: THE NEED FOR PROBES

Andrew P. Ingersoll[1]

[1]*Division of Geological and Planetary Sciences, California Institute of Technology
Pasadena, CA 91125, USA, api@gps.caltech.edu*

ABSTRACT

Probes are an essential element in the scientific study of planets with atmospheres. In-situ measurements provide the most accurate determination of composition, winds, temperatures, clouds, and radiative fluxes. They address fundamental NASA objectives concerning volatile compounds, climate, and the origin of life. Probes also deliver landers and aerobots that help in the study of planetary surfaces. This talk focuses on Venus, Titan, and the giant planets. I review the basic science questions and discuss the recommended missions. I stress the need for a balanced program that includes an array of missions that increase in size by factors of two. Gaps in this array lead to failures and cancellations that are harmful to the program and to scientific exploration.

1. INTRODUCTION

1.1 Relevance to NASA/OSS Objectives

- Origin and evolution of our solar system (SS)
- What our SS tells about extrasolar systems
- Distribution of volatile compounds in SS
- Differences among terrestrial planets
- SS characteristics that led to origin of life
- Sources of prebiotic compounds
- Habitable zones in the SS

These objectives are from the SSE Roadmap. I was on the Decadal Survey Committee and served as vice chair of the Giant Planets Panel. Reta Beebe was the chair of that panel, and Mike Belton was chair of the committee. I'll tell you what the Decadal Survey has recommended in the way of probes and missions to planets with atmospheres. Venus is relevant to differences among the terrestrial planets. Titan may have an ocean that has gotten mixed up with the organics on the surface, so Titan is relevant to origin of life, prebiotic compounds, and habitable zones. Giant planets are relevant to the origin of the solar system and distribution of volatiles in the solar system,

1.2 Scope of this Talk and Past Probe Missions

- Venus - Venera probes (Soviet Union) and Pioneer Venus Probes (US)
- Titan - Huygens probe (ESA) and Cassini spacecraft (US)
- Giant Planets - Galileo probe (US)

I'll be talking about Venus, Titan, and the giant planets. These are the objects with thick atmospheres, where probes are the only way to make in-situ measurements. There have been ~ 10 probes into the Venus atmosphere, one to Titan, and one to Jupiter. There were ~ 10 missions involving probes that were recommended during the Decadal Survey.

1.3 Documents and Sources

- Decadal Survey (DS) - *New Frontiers in the Solar System,* Michael J. S. Belton, chair (National Academy of Sciences, 2003)

- Community White Papers (CWP) - *The Future of Solar System Exploration, 2003-2013,* Mark V. Sykes, editor (Astronomical Society of the Pacific, 2002)

The Decadal Survey was published in 2003, although most of the work was done in 2002. The Community White Papers were published in 2002 but they were part of the whole decadal study of what does the scientific community feel is the next step in the exploration of these objects.

1.4 What Probes Measure

- Atmospheric chemical composition - major and minor constituents including noble gases and isotopes (gas chromatographs, mass spectrometers, optical spectrometers, chemical sensors)

- Atmospheric physical structure - T, P, ρ, turbulence, waves - all as functions of depth

- Clouds – particle properties (composition, size, shape, optical constants) and vertical distribution

Probes are the gold standard for measuring the composition of an atmosphere. Remote sensing gives you pieces of the composition, but there's no way you could get the isotopes and noble gases by remote sensing to the accuracy that you get by in-situ measurements. The composition is relevant to the origin of the solar system, the distribution of volatiles, and the history of surface-atmosphere interactions. Probes are also the gold standard for the physical structure of the atmosphere, which is relevant to the dynamics and sometimes the surface-atmosphere interaction. And to do sampling of clouds you need probes.

- Large-scale winds from Doppler tracking and VLBI, turbulence and waves from accelerometers and on-board wind sensors
- Radiative heating from optical and IR sensors (net flux radiometer)
- Miscellaneous – lightning, He/H2, radio opacity (NH3), exosphere composition

Probes can measure winds from accelerometers and onboard wind sensors. And again, probes are the gold standard, although you can do things from orbit on all of these objects. For Jupiter, we have a lot of cloud top winds from Voyager and Galileo and Cassini. But if you want to understand the dynamics, you need to know what the winds are doing below the clouds. You also need the radiative energy sources. And, of course, on the Galileo probe there was a lightening detector and a helium-to-hydrogen detector. The Galileo probe's radio signal passing through the atmosphere was really the best estimate of the ammonia abundance, although the mass spectrometer provided another estimate.

- Probes deliver landers to the surface of Venus and Titan, and with difficulty, Mars
- Observe surface morphology, mineralogy, elemental composition, subsurface structure, seismology, surface-atmosphere interaction
- Probes deliver aerobots (balloons, airplanes, helicopters) that survey the surface remotely at altitudes 1/100 of orbital altitude

Probes are delivery vehicles because if you have a thick atmosphere, you really have to get to the surface by means of probes. Probes also can provide a platform for launching balloons, airplanes, and helicopters that allow you to travel around and look at many places on the surface and look at it from much closer distances and with higher resolution than you get from orbit.

2. VENUS

2.1 Venus Science Questions

- Surface history - impacts, tectonics, volcanism, erosion; global resurfacing "event" 500 Myr BP?
- History of water (loss of an ocean) and evolution of the atmosphere - noble gas composition, isotopes of O & H, outgassing, reactions with surface (C, S, O, H, Cl)
- Greenhouse effect and climate - trace gases, clouds, penetration of sunlight, IR opacity
- Super-rotation of the atmosphere

From the number of impact craters on the surface of Venus you estimate that the surface is approximately 500 million years old, which is still only one-tenth of the age of the solar system. So something's been going on, and it's not erosion. There's not a lot of evidence of degradation of surface features by erosion. The surface features are rather pristine. The other source of resurfacing is internal processes, volcanoes or something, and that whole resurfacing of Venus takes place every 500 million years. It is a principal mystery, and getting down to the surface, measuring the surface composition, is one of the principal goals for Venus.

Another important goal is what happened to the water? If Venus is the Earth's sister planet, where is the ocean? It's quite possible that there was an ocean on Venus. It might have been a steam ocean because Venus is that close to the sun, but what happened to all that water vapor, those 300 bars of water vapor? The clue may be in the isotopes of oxygen and hydrogen, which you can best measure with probes.

Venus has probably the hottest surface in the solar system. It gets hotter than Mercury. And this greenhouse effect is there because trace gases and many other things plug up the IR windows. Understanding that is very important. The atmosphere rotates 50 times faster than the solid planet, which is its own mystery.

2.2 Recommended Venus Missions

- VISE (DS) - Compositional and isotopic analysis of atmosphere, core sample of the surface lofted to balloon altitude and analyzed there, winds and radiometry during descent and at balloon station
- Noble gas and trace gas explorer (CWP) - single probe to the surface

- Dynamics explorer (CWP) - Four to eight probes
- Landers (months to year), sample return (CWP)

The Decadal Survey recommended a mission called VICE, as part of the New Frontiers line. It involves a probe, a main balloon, and a sub-balloon that drops to the surface, grabs a piece of the surface and then quickly gets back up for a rendezvous with the main balloon. The balloon carries instruments that work at room temperature but not at the surface. The balloon also does chemical analyses of the atmosphere and, of course, the little sub-balloon can do analysis on the way up and down. It's a rather ambitious concept, and there are technological hurdles that have to be overcome.

The Community White Paper said, make it simpler. Let's just have a balloon that does a good job on the isotopes and trace gases—a single probe down to the surface and no rendezvous with a main balloon.

For the dynamics, to understand the super-rotation of the atmosphere you really need a lot of probes to sample space and time, and so the Community White Paper has recommended 4 to 8 probes. And ultimately we would like to learn how to go down to the surface of Venus and operate there with instruments for a long time at temperatures of 730 K.

3. TITAN

3.1 Titan Science Questions

- Composition and extent of surface organics
- Subsurface ocean - composition, depth
- Evidence of episodic heating and exposure of organics to aqueous solutions
- Atmospheric dynamics - winds, clouds, precipitation, radiative heating, atmosphere-surface interaction

Titan doesn't have enough gravity to hold on to the lightest gases like hydrogen, so it is constantly increasing the carbon to hydrogen ratio. This means the number of multi-carbon molecules is increasing, so it's a very interesting place for organic chemistry. There's a possibility that these organics form liquid lakes or oceans. Also there's nitrogen present. Titan gives you a chance to watch an atmosphere that's evolving, as the Earth may have evolved billions of years ago.

There's also a possibility of a subsurface ocean sort of like the subsurface ocean on the Galilean satellites. And there's even the possibility that the ocean and the surface organics have gotten in contact at times, and this becomes an interesting habitability question. The atmosphere dynamics could be like Venus with a strong super-rotation. The winds appear to be large even though the amount of sunlight is small.

3.2 Recommended Titan Missions

- Titan Explorer (DS) - orbiter (for relay) and probe (aerobot); use atmosphere for mobility; descend repeatedly to the surface; make high-resolution remote observations and repeated atmospheric measurements
- Airship and mobile lander (CWP) - balloons, airplane, or helicopter.
- Radioisotope power source is critical (DS)

Cassini carries the Huygens probe, which will enter Titan's atmosphere in January 2005. After the Huygens probe, there will still be the need for atmospheric mobility to descend rapidly to the surface many times and also make atmospheric measurements in many places. This would be a more ambitious device than Huygens, which just goes down in one place. So mobility is the key to the future, according to the Community White Papers and the Decadal Survey. But if the probes are to last for a long time they will need power, and solar power is not going to work. The outer solar system missions need radioisotope power, and that's an area where development is needed.

4. GIANT PLANETS

4.1 Giant Planet Science Questions

- Composition of their atmospheres and interiors - water is key (O/H ratio); major elements, noble gases, isotopes; cloud base may exist only for Jupiter and Saturn
- Liquid oceans on Uranus and Neptune?
- Deep winds, temperature structure, clouds, radiation, convection, lightning - relation to meteorology at cloud top level

Now let's talk about giant planets. The Galileo Probe did a wonderful job, but it really did not settle the issue of the oxygen to hydrogen ratio on Jupiter. That is a crucial issue because we are talking about the most abundant element and the third most abundant element, with helium in between, so water was probably the most abundant compound in the early solar system. And yet we don't know how much water is on Jupiter. To find out, we have to get down to where water is well mixed.

At least at the Galileo Probe site, we didn't do it. We need to get down to 100 bars on Jupiter and all of the giant planets. Having probes in all the giant planets makes a great deal of sense. Having deep probes to Jupiter at a number of different latitudes makes a great deal of sense too.

4.2 Recommended Giant Planet Missions

- JPOP (DS) - Polar orbiter with probes at 3 different latitudes (≤ 30°) down to 100 bars
- Neptune Orbiter with Probes (DS) - measure planet's C, S, noble gases, isotopes
- Jupiter Microwave Sounder (CWP) - Either orbiter or flyby; water & ammonia to 100 bars and below, no noble gases or isotopes
- Deep probes to all giant planets (CWP); can we detect probe signals directly at Earth?

The Decadal Survey recommended JPOP, a Jupiter Polar Orbiter with Probes, as part of the New Frontiers line. JPOP involves probes to 100 bars at three different latitudes. That can be done. What can't be done is to guarantee that we will meet an ambitious set of scientific goals and do it within a specified cost cap. There's too much uncertainty. Something's got to give.

The Community White Papers recommended something complimentary to probes—an orbiter or a fly-by that has a microwave sounder that can peer into the planet down to 100 bars. It measures the water and ammonia abundance but not as accurately as the instruments on a probe. It can't do everything probes can do, but it can get a global picture. Probes are limited in this respect, because the number of probes is limited. The microwave sounder is the Juno mission, which has Scott Bolton as PI and is working its way through the system right now. I feel these missions are complimentary, and I'd love to see both probes and a microwave sounder.

It would be nice if we could detect probes directly at Earth because communication between a probe and the fly-by or orbiter that delivered the probe is a difficult constraint on the probe. If we could cut that constraint and listen to the probe directly, it would be wonderful.

5. GENERIC RECOMMENDATIONS (DS)

- Thermal protection system for probes and aerocapture - Jupiter is the driver
- Radioisotope power sources
- Nuclear-powered electric propulsion
- Advanced telecommunications
- Balanced program - factors of 2 in cost - Discovery, New Frontiers, Flagship
- 2xFlagship, Prometheus, humans on Moon/Mars

All but the last bullet is right out of the Decadal Survey. The last bullet is my own interpretation of where we stand. The Decadal Survey said, get to work on thermal protection systems for probes. Using atmospheric drag to go into orbit is called aerocapture. The technology is similar to that of an entry probe and involves high-speed entry and thermal protection. I'm the PI on a Neptune vision mission study, which is to think about the decade after the one we are in. We are studying a mission that does not use nuclear electric propulsion—the Prometheus-type technology. We can use ordinary chemical or maybe solar electric propulsion to get out to Neptune, but then we need aerocapture to get into orbit. With aerocapture we can do a Cassini class mission for Cassini class dollars and that's what the Decadal Survey calls a Flagship mission. But, of course, the Decadal Survey also says nuclear power for propulsion is a wonderful thing and perhaps that's true. I'm concerned that it's a long way off in the future and we have to have something to keep us going in between.

The Decadal Survey says we need a balanced program, and that means an array of missions differing in cost by factors of 2. An example would the Discovery class, 2 times Discovery, which is the New Frontiers line, 4 times Discovery, which is a Flagship, and so on. Now here I'm going to add my own controversial thing. I think we need 2 and possibly 4 times Flagship, which is 8-16 times Discovery. Right now we don't have any Flagships to any of these objects, and we aren't even thinking about anything in the 2 times Flagship category. Prometheus is 4-8 times Flagship. We sort of leapt ahead to Prometheus and leapt even beyond that in talking about humans on Moon and Mars. I think it's a very dangerous situation when we have this gap between things that we can do and the grand things that we would like to do some day. There are wonderful things we can do in this gap and there's a danger that they won't ever get done. We're relying on this pie-in-the-sky of Prometheus and beyond, which is not a good situation to be in.

Mars

SHOULD WE BELIEVE ATMOSPHERIC TEMPERATURES MEASURED BY ENTRY ACCELEROMETERS TRAVELLING AT "SLOW" NEAR-SONIC SPEEDS?

Paul Withers[1,2]

[1] *Center for Space Physics, Boston University, 725 Commonwealth Avenue, Boston, MA 02215, USA (email: withers@bu.edu)*
[2] *Planetary and Space Science Research Institute, Open University, Walton Hall, Milton Keynes, MK7 6AA, Great Britain*

ABSTRACT

Mars Pathfinder's Accelerometer instrument measured an unexpected and large temperature inversion between 10 and 20 km altitude. Other instruments have failed to detect similar temperature inversions. I test whether this inversion is real or not by examining what changes have to be made to the assumptions in the accelerometer data processing to obtain a more "expected" temperature profile. Changes in derived temperature of up to 30K, or 15%, are necessary, which correspond to changes in derived density of up to 25% and changes in derived pressure of up to 10%. If the drag coefficient is changed to satisfy this, then instead of decreasing from 1.6 to 1.4 from 20 km to 10 km, the drag coefficient must increase from 1.6 to 1.8 instead. If winds are invoked, then speeds of 60 m s^{-1} are necessary, four times greater than those predicted. Refinements to the equation of hydrostatic equilibrium modify the temperature profile by an order of magnitude less than the desired amount. Unrealistically large instrument drifts of 0.5 – 1.0 m s^{-2} are needed to adjust the temperature profile as desired. However, rotational contributions to the accelerations may have the necessary magnitude and direction to make this correction. Determining whether this hypothesis is true will require further study of the rigid body equations of motion, with detailed knowledge of the positions of all six accelerometers.

The paradox concerning this inversion is not yet resolved. It is important to resolve it because the paradox has some startling implications. At one extreme, are temperature profiles derived from accelerometers inherently inaccurate by 20 K or more? At the other extreme, are RS temperature profiles inaccurate by this same amount?

1 INTRODUCTION

The aim of this paper is to investigate an unusual temperature inversion measured by the accelerometer (ACC) instrument on Mars Pathfinder (MPF) during its descent through the atmosphere of Mars. A temperature inversion occurs when temperature increases as altitude increases within an atmosphere. Temperatures usually decrease as altitude increases. Neither models nor other observations have shown similar inversions. Is the inversion seen by MPF real? I will attempt to remove this temperature inversion from the MPF data by testing the limits of the assumptions underlying the data processing. If I suceed, then the temperature inversion may be an artifact of the data acquisition and processing. If I do not succeed, then the temperature inversion is robust and real, and its presence in this dataset and absence from other datasets must be explained.

Fig. 1 shows the vertical temperature profile derived from MPF's ACC data using the techniques of Withers et al. [10] and Withers [11]. It is very similar, but not identical, to the profiles derived by other workers [6, 9]. Uncertainties are not shown because my data processing software has not yet been extended to include a thorough error analysis. Errors are shown in Magalhães et al. [6]. The MPF ACC Science Team has archived its reconstructed trajectory and atmospheric structure, with extensive documentation, at the Planetary Data System (PDS) [8]. A noteworthy feature of this temperature profiles is the large temperature inversion below 20 km. Some workers have suggested that it is due to radiative cooling from a water ice cloud [4, 1, 2]. However, repeated Mars Global Surveyor (MGS) Radio Science (RS) occultation measurements of temperature and pressure in the lower atmosphere at similar latitudes, longitudes, and local solar times (LSTs) one Mars year later do not reveal any trace of an inversion [5]. Hinson and Wilson [5] do see inversions at other latitudes and longitudes at this LST and season (Ls),

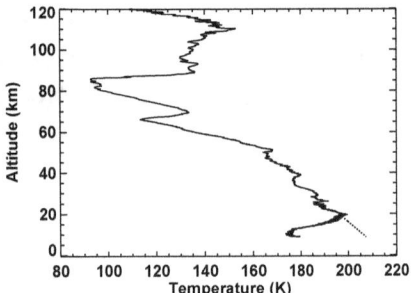

Figure 1: Temperature as a function of altitude for MPF (dashed line) and a model (solid line). The vertical axis has a linear scale from 0 to 50 km. Figure from Colaprete and Toon [2].

but they state that the RS inversions have distinctly different structures to the MPF inversion. Based on visual inspection of some RS inversions shown in Fig. 2 of Hinson and Wilson [5], I think that the RS inversions are narrower in width than the MPF inversion and have an asymmetric vertical profile, whereas the MPF inversion has a more symmetric profile. Colaprete and Toon [2] show (their Fig. 5, reproduced in Fig. 2) a qualitative comparison between the MPF inversion and a model inversion. A quantitative comparison is not discussed. The model inversion appears to have the same discrepancies with respect to the MPF inversion as the observed RS inversions in Hinson and Wilson [5]. Comparing static stabilities might be the best way to quantify the differences between RS inversions, the MPF inversion, and model inversions [6, 5].

2 TEMPERATURE PROFILES

Fig. 3, courtesy of John Wilson, shows an MGS Thermal Emission Spectrometer (TES) profile, which has poor vertical resolution, several MGS RS profiles, which have good vertical resolution, and the MPF ACC profile, which also has good vertical resolution. Nothing like the MPF inversion is seen in the TES or RS data. The MPF temperature maximum occurs near 20 km. The MPF temperature inversion can be removed, and the profile made more consistent with the TES and RS data, if temperatures, T, below 20 km altitude are replaced by the following equation:

$$T/K = 216 - z/km \qquad (1)$$

where z is altitude. Throughout this paper, "altitude" means radial distance above the MPF landing site, which 3389.715 km from the centre of mass of

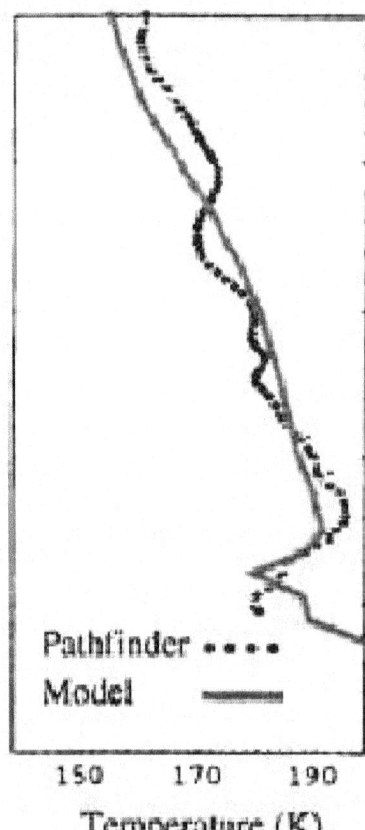

Figure 2: Original temperature profile from MPF (solid line) and desired temperature profile (dotted line).

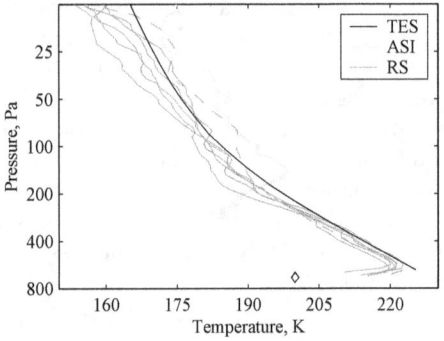

Figure 3: One TES temperature profile (smooth line), several RS profiles (extending to the base of the figure), and the MPF ACC temperature profile (with large temperature inversion). Figure provided by John Wilson.

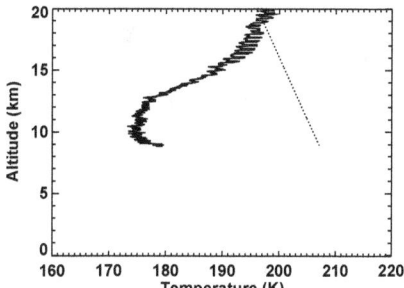

Figure 4: Original temperature profile from MPF (solid line) and desired temperature profile (dotted line).

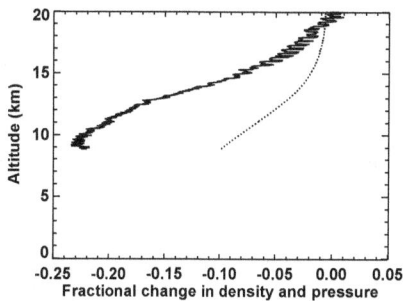

Figure 5: Fractional change in density (solid line) and pressure (dotted line) necessary to obtain desired temperature profile.

Mars. This "desired" temperature profile is also plotted on Fig. 1. Fig 4 shows the original and desired temperatures profiles below 20 km altitude. A 30 K shift in temperature will not be easy to obtain.

3 DENSITY AND PRESSURE

Density, ρ, pressure, p, and temperature are related by the ideal gas law:

$$T = \frac{m_{mean}}{k_{Boltzmann}} \times \frac{p}{\rho} \qquad (2)$$

where m_{mean} is the mean mass of an atmospheric molecule and $k_{Boltzmann}$ is Boltzmann's constant. m_{mean} may vary slightly in the lower atmosphere of Mars, but not by enough to change the temperature profile as desired. Hence atmospheric density or pressure or both must be altered in Eqn. 2 to alter the temperature. Atmospheric pressure is related to atmospheric density by the equation of hydrostatic equilibrium, Eqn. 3.

$$p(z) = p(z_0) - \int_{z_0}^{z} \rho g dz \qquad (3)$$

where g is the magnitude of the acceleration due to gravity. How must density, and pressure change to be consistent with this desired temperature profile? Since pressure is effectively vertically-integrated density, its value just below 20 km is strongly influenced by the densities above 20 km. So density just below 20 km must decrease from its original value to shift the temperature from its original to its desired value. As we progress downwards, density must continue to decrease from its original value. Pressures must decrease also, since they are vertically-integrated densities, but not by as much. Fractional changes in density and pressure required to give the desired temperature profile are shown in Fig. 5. The fractional change in density cannot stabilize, because then the fractional change in pressure would also stabilize (at a lower altitude) and the temperature would be back to its original value. This is quite a strong constraint, which also applies to anything else that affects the density profile.

The MPF ACC measured accelerations and then converted them into densities. Pressures and temperatures were derived from the density profile using Eqns. 3 and 2. Accelerations were converted into densities using the drag equation:

$$\rho = \frac{-2m}{C_A A} \times \frac{a_z}{v_R^2} \qquad (4)$$

where m is the mass of the spacecraft, A is the reference area of the spacecraft, C_A is the axial force coefficient appropriate to the present angle of attack, atmospheric composition, density, and temperature, a_z is the acceleration along the spacecraft's z-axis, and v_R is the speed of the spacecraft relative to the atmosphere. I used aerodynamic databases from Moss et al. [7] and Gnoffo et al. [3], generally assuming an angle of attack of zero.

If the desired changes in ρ were due solely to changes in m, m would have to decrease by 20%, or over 100 kg. This is unrealistic. If the desired changes in ρ were due solely to changes in A, the spacecraft radius would have to increase by 10%, or 13 cm. This is also unrealistic. Even if the magnitude of the changes were reasonable, it would still be challenging to get the desired change as a function of altitude and to explain why the change starts below 20 km.

4 AERODYNAMICS

If the desired changes in ρ were due solely to changes in C_A, then the vertical profile of C_A, shown in Fig. 6, would change dramatically. Is this realistic? The MPF aerodynamic database was generated

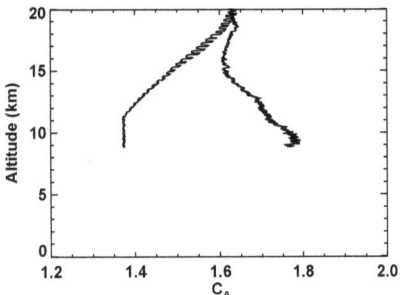

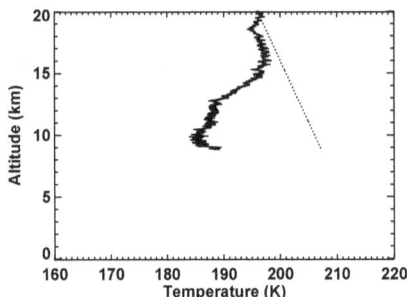

Figure 6: Original vertical profile of C_A for MPF reconstruction (values 1.4 – 1.6) and profile of C_A required to obtain desired temperature profile (values 1.6 – 1.8).

Figure 7: Temperature profile from MPF, assuming that values of $C_A = 1.3$ in the aerodynamic database should actually be 1.5, (solid line) and desired temperature profile (dotted line).

by running computationally-intensive numerical simultations under a discrete set of flow conditions. One value of C_A was extracted from each simulation, yet MPF's actual trajectory encompassed an infinite continuum of flow conditions. Appropriate values of C_A are formed by interpolation within the finite set of simulations. The interpolation might not be accurate and the simulations themselves might not be accurate. It has been stated that uncertainties in C_A for MPF's atmospheric entry are from 1 to 3% [6]. 5 to 10% seems more realistic to me, with greatest uncertainties at low altitudes. Below 20 km altitude, flow conditions for which there are numerical simulations are at 18 km and at 9 km. The shape of the large decrease in C_A between these altitudes is unconstrained by simulations or by observational data. This will make interpolation of C_A in this region very difficult. It does not matter much if the simulated value of C_A is wrong when C_A is not changing much. Derived temperatures are quite insensitive to this. However, when C_A is changing rapidly, derived temperatures are much more sensitive. Even allowing for reasonable errors in simulated values of C_A and in interpolating between simulations, the changes in C_A necessary to obtain the desired temperature profile seem unreasonably large.

Suppose that the simulations of Gnoffo et al. [3] at Mach number (Ma) 1.9 and 2.0 are somehow systematically flawed and they should have generated $C_A \approx 1.5$ instead of the actual values of 1.3. Rederiving the temperature profile with this updated aerodynamic database *still* produces a large temperature inversion, Fig. 7. It appears that C_A must increase by 15% as altitude decreases from 20 km to 8 km to obtain the desired temperature profile, though the numerical simulations predict that it decreases by 15%.

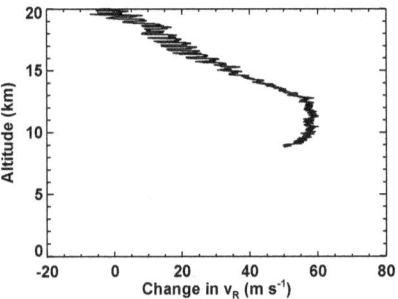

Figure 8: Change in v_R necessary to obtain the desired temperature profile.

5 WIND

Atmospheric dynamics can affect the results of Eqn. 4. Any wind in the atmosphere affects v_R. Fig. 8 shows the difference between the magnitude of v_R necessary to obtain the desired temperature profile and the magnitude of v_R in the original reconstruction. The horizontal wind necessary to account for this change, assuming that it is blowing in the most favourable direction, is 10 – 20% greater than this difference. These wind speeds are significantly greater than the 15 m s^{-1} speeds predicted by general circulation models for these conditions [6].

We can combine considerations of winds and errors in C_A. Suppose $C_A \approx 1.5$ at Ma=1.9 and Ma=2.0 instead of the actual 1.3 — what horizontal winds are needed in this case to obtain the desired temperature profile? Fig. 9 shows the difference in magnitudes of v_R for this case. The change to C_A is about twice as much as is reasonable and the derived wind speed is about twice as much as is reasonable as well — even if it is blowing in the most favourable direction. These techniques have not been successful at removing the

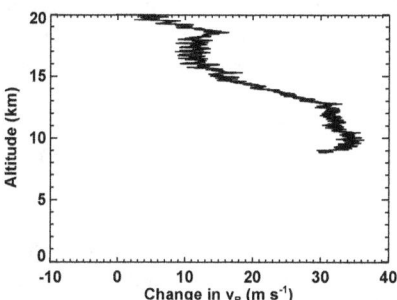

Figure 9: Change in v_R necessary to obtain the desired temperature profile, assuming that values of C_A = 1.3 in the aerodynamic database should actually be 1.5.

temperature inversion.

6 REFINED HYDROSTATIC EQUILIBRIUM

The equation of hydrostatic equilibrium (Eqn. 3) is not completely accurate. It has neglected higher order gravitational terms, terms due to planetary rotation, and the effects of atmospheric dynamics. Correcting for the first two categories gives:

$$\Delta p = \rho \left(g_{eff,r} \Delta r + g_{eff,\theta} r \Delta \theta + g_{eff,\phi} r \sin\theta \Delta \phi \right) \quad (5)$$

$$g_{eff,r} = \frac{-GM}{r^2} - \cdots \quad (6)$$

$$\frac{9}{2} \frac{GM}{r^2} \left(\frac{r_{ref}}{r} \right)^2 (\cos\theta \cos\theta - 1) C_{20} + \cdots \quad (7)$$

$$r\Omega^2 \sin\theta \sin\theta \quad (8)$$

$$g_{eff,\theta} = \frac{-3GM}{r^2} \left(\frac{r_{ref}}{r} \right)^2 \sqrt{5} \sin\theta \cos\theta C_{20} + \cdots \quad (9)$$

$$r\Omega^2 \sin\theta \cos\theta \quad (10)$$

$$g_{eff,\phi} = 0 \quad (11)$$

where the terms have their usual meanings. A latitude of 20° corresponds to $\sin\theta\cos\theta = 0.3$. $C_{20} \sim -10^{-3}$, $GM/r^2 = 3.7$ m s^{-2}, and $r\Omega^2 = 0.01$ m s^{-2}. The effect of these corrections is critically dependent on the trajectory. A shallow entry trajectory along a meridian of constant longitude will have a large correction. A steep entry trajectory along a circle of constant latitude will have a small correction. I have rederived pressure and temperature profiles for MPF using this correction; both change by less then 1% at all altitudes. Further corrections to hydrostatic equilibrium due to atmospheric dynamics are about an order of magnitude smaller than these corrections.

Horizontal gradients in the atmosphere, regardless of their cause, can be investigated. If $\partial \ln \rho / \partial \theta = k$ and $\partial \ln \rho / \partial r = H$, if MPF's atmospheric trajectory can be modelled as $r + \alpha\theta =$ constant, and if the atmosphere is isothermal with temperature T_{iso}, then temperatures derived from entry accelerometer data using Eqn. 3 will be given by $T = T_{iso}(1 - kH/\alpha)$. For MPF, $\alpha \sim 40$ km deg^{-1} and $H \sim 10$ km. A reasonable value of k is 0.01 deg^{-1}, which alters temperatures by a few tenths of one percent.

7 ANGLE OF ATTACK

Eqn. 4 conceals a dependence upon angle of attack, α. α is not measured directly during the entry, it is inferred indirectly from the ratio of axial to normal accelerations using the ratio of axial to normal force coefficients. If force coefficients in the aerodynamic database are incorrect by a few percent, then α could also be inaccurate. Does allowing α to change from its nominal value help obtain the derived temperature profile? Nominal values of α are so close to zero that the only way changes in α can significantly change C_A is for α to increase to some relatively large value. Recall that I need C_A to increase at low altitudes to obtain the desired temperature profile. However, C_A always decreases as α increases. This approach will not help remove the temperature inversion.

8 ANGULAR ACCELERATIONS

Measured accelerations may be corrupted due to rotation of MPF, a rigid body, about its centre of mass. Accelerations along MPF's axis of symmetry are measured by its z-axis accelerometer. This accelerometer is about 5 cm away from the centre of mass along the z-axis. It is much closer to the z-axis than this. Rotational contributions to the measured axial acceleration, such as $\underline{\Omega} \times (\underline{\Omega} \times \underline{r})$ and $\underline{\dot{\Omega}} \times \underline{r}$, might be important. I have not looked at the full rigid body equations of motion to determine how the rotation affects all six (three science and three engineering) accelerometer measurements, but I have made a rough estimate of its importance.

Axial accelerations need to decrease in magnitude by 0.5 – 1.0 m s^{-2} at all altitudes below 20 km to obtain the desired temperature profile. The nominal measurements have magnitudes of 8 m s^{-2} at 9 km and 30 m s^{-2} at 18 km. The instrument resolution at this time is 0.05 m s^{-2}, which is effectively the measurement uncertainty. Such a large instrumental drift, or similar error, can be ruled out. The normal acceleration measurements, on the order of 0.05 m s^{-2}, are so small that they cannot possibly be altered by rotational effects to obtain the desired temperature profile.

Ω was not measured directly by MPF, which did not carry any gyroscopes. MPF had a pre-entry roll about its symmetry axis of 0.06 rad s^{-1}. The time series of measured x-axis and y-axis accelerations show interesting oscillations with an angular frequency $\sim$ 4.5 rad s^{-1} at 20 km and $\sim$ 2.5 rad s^{-1} at 10 km. Overtones are also present. I estimate the maximum angular acceleration contribution to the measured z-axis accelerations to be 4.5 rad s^{-1} × 4.5 rad s^{-1} × 50 mm = 1.0 m s^{-2} at 20 km and 2.5 rad s^{-1} × 2.5 rad s^{-1} × 50 mm = 0.3 m s^{-2} at 10 km. These are of the desired magnitude. However, until a proper study of the full rigid body equations of motion has been made it would be premature to conclude that this effect is responsible for all of the temperature inversion.

9 CONCLUSIONS

The only effect that seems to offer any hope for removing the temperature inversion from the MPF ACC dataset is due to rotation, possibly assisted by $\sim$ 5% uncertainties in C_A at low Ma. MPF's moments of inertia are known. The positions and orientations of its six accelerometers are presumably also known, though I have not seen them. Measured accelerations can be divided into two contributions: oscillating and non-oscillating. Oscillating contributions are presumably due to rotation, so it might be possible to identify these terms separately within the measured accelerations. Is it possible to deduce MPF's angular velocity vector from these data? Maybe.

Further study of the six-degrees-of-freedom behaviour of MPF during its entry would be helped by analysis of data from Galileo and Pioneer Venus, which measured atmospheric temperatures and pressures directly immediately after parachute deployment, and from MER, which carried 3-axis accelerometers and gyroscopes on both the rover and the backshell.

The paradox concerning this inversion is not yet resolved. It is important to resolve it because the paradox has some startling implications. At one extreme, are temperature profiles derived from accelerometers inherently inaccurate by 20 K or more? At the other extreme, are RS temperature profiles inaccurate by this same amount?

Does this work have any impact on Huygens? I recommend that the Huygens Science Teams investigate how rotation affects each accelerometer instrument onboard and consider how these effects could be mitigated.

More general questions raised by this work concern uncertainties in the aerodynamic database. Should uncertainties in simulated values be represented by a normal distribution, some other distribution, or a possible systematic error? How should uncertainties be assigned to interpolated values of drag coefficients when the way in which the actual drag coefficient changes between simulation points is not known and probably not linear? How do these uncertainties map to uncertainties in derived atmospheric properties? How do they affect derived angles of attack, which in turn affect drag coefficients, and so on?

References

[1] Colaprete A., et al. Cloud formation under Mars Pathfinder conditions, *J. Geophys. Res.*, Vol. 104(E4), 9043-9053, 1999.

[2] Colaprete A. and Toon O. B. The radiative effects of Martian water ice clouds on the local atmospheric temperature profile, *Icarus*, Vol. 145, 524-532, 2000.

[3] Gnoffo P. A., et al. Influence of Sonic-Line Location on Mars Pathfinder Probe Aerothermodynamics, *J. Spacecraft and Rockets*, Vol. 33(2), 169-177, 1996.

[4] Haberle R. M., et al. General circulation model simulations of the Mars Pathfinder atmospheric structure investigation/meteorology data, *J. Geophys. Res.*, Vol. 104(E4), 8957-8974, 1999.

[5] Hinson D. P. and Wilson R. J. Temperature inversions, thermal tides, and water ice clouds in the Martian tropics, *J. Geophys. Res.*, Vol. 109, E01002, doi:10.1029/2003JE002129, 2004.

[6] Magalhães J. A. et al., Results of the Mars Pathfinder atmospheric structure investigation, *J. Geophys Res.*, Vol. 104(E4), 8943-8955, 1999.

[7] Moss J. N., et al. Mars Pathfinder Rarefied Aerodynamics: Computations and Measurements, *36th AIAA Aerospace Science Meeting, January 12-15, Reno, Nevada, USA*, AIAA 98-0298, online at http://techreports.larc.nasa.gov/ltrs/PDF/1998/aiaa/NASA-aiaa-98-0298.pdf, 1998.

[8] PDS — http://atmos.nmsu.edu/PDS/data/mpam_0001/

[9] Spencer D. A., et al. Mars Pathfinder Entry, Descent, and Landing Reconstruction, *J. Spacecraft and Rockets*, Vol. 36(3), 357-366, 1999.

[10] Withers P., et al. Analysis of entry accelerometer data: A case study of Mars Pathfinder, *Planetary and Space Sci.*, Vol. 51, 541-561, 2003.

[11] Withers P. Tides in the Martian Atmosphere - And Other Topics, PhD dissertation, *University of Arizona*, 2003.

THE NEXT GENERATION OF PLANETARY ATMOSPHERIC PROBES

Howard Houben

Bay Area Environmental Research Institute
MS 245-3, Ames Research Center, Moffett Field, CA, 94035-1000, USA
houben@humbabe.arc.nasa.gov

ABSTRACT

Entry probes provide useful insights into the structures of planetary atmospheres, but give only one-dimensional pictures of complex four-dimensional systems that vary on all temporal and spatial scales. This makes the interpretation of the results quite challenging, especially as regards atmospheric dynamics. Here is a planetary meteorologist's vision of what the next generation of atmospheric entry probe missions should be: Dedicated sounding instruments get most of the required data from orbit. Relatively simple—and inexpensive—entry probes are released from the orbiter, with low entry velocities, to establish ground truth, to clarify the vertical structure, and for adaptive observations to enhance the dataset in preparation for sensitive operations. The data are assimilated onboard in real time. The products, being immediately available, are of immense benefit for scientific and operational purposes (aerobraking, aerocapture, accurate payload delivery via glider, ballooning missions, weather forecasts, etc.).

Key words: atmospheric probes, data assimilation.

1. INTRODUCTION

There are eight planetary atmospheres in the solar system exhibiting a wide range of behaviors, with variations on many spatial and temporal scales. The planets are conveniently grouped into pairs, making the solar system an excellent laboratory for the study of comparative atmospheres and associated scientific questions dealing with fluid dynamics, planetary evolution, climate change and predictability, and habitability. The challenge is to find methods to address the wide range of questions involved in a useful scientific manner. Entry probes working in co-ordination with orbital sounders can be an important part of this process.

Earth and Mars both have optically thin atmospheres that are primarily forced from below by the absorption of solar radiation at their solid surfaces. It is not surprising that these are the best studied atmospheres in the solar system. But many questions remain unanswered, particularly related to climate change and predictability. In the case of the Earth, there is a major international effort (with political ramifications) to predict changes in climate over the next hundred years. While the martian meteorology appears to be quite repeatable during the aphelion season [1], major planetwide dust storms occur on an irregular basis during some perihelion seasons [2]. Mars has received the most exploration attention of any planet besides the Earth, of course. But, the PMIRR instrument that was intended to be a dedicated atmospheric sounder was flown on two spacecraft that failed to achieve orbit (Mars Observer and Mars Climate Orbiter). A French mission with a microwave sounder (MAMBO) was canceled.

Venus and Titan have optically thick atmospheres, forced from above. Both appear to super-rotate with zonal wind speeds much greater than the rotation speed of the underlying solid surface. The source and mechanism of angular momentum transfer for these super-rotations are still uncertain. Both atmospheres are to be studied in great detail in the near future, by the Venus Express and Cassini-Huygens Probe missions, respectively.

Jupiter and Saturn have deep atmospheres (with no solid underlying surface). They have internal heat sources greater than the solar heating, and no large equator-to-pole temperature gradient. Yet, small scale meridional gradients in temperature are consistent with a prominent structure of (relatively clear) belts and cloudy zones with alternating easterly and westerly winds. (Whether these are related to a deep interior circulation is unknown.) Both planets have strong prograde equatorial jets. Large vortices like Jupiter's Great Red Spot have fascinated observers for as long as telescopes have been available, yet the nature of the red chromophore is still unknown. The bulk compositions of Jupiter and Saturn (e.g., helium to hydrogen ratios) are of great astrophysical interest [3]. How these relate to the atmospheric values may provide a key to their interior structures and evolution. The Galileo orbiter's atmospheric observations of Jupiter were curtailed because of the loss of the high gain antenna. Cassini made some observations of the jovian atmosphere during its flyby and will have the opportunity to make many observations of the saturnian atmosphere.

At first glance, the ice giants Uranus and Neptune appear to be smaller versions of Jupiter and Saturn. But they are characterized by relatively small internal heat fluxes and external forcing. That Neptune's zonal jets are the largest measured on any planet is another mystery.

Although there have been a number of spacecraft missions that have made observations of these atmospheres, none (besides the Earth) has been studied systematically from the point of view of a meteorologist. Such systematic study would require global coverage over a radiative time scale with an effective spatial resolution adequate to resolve the radius of deformation. (See the discussion and table in the next section.) We are receiving a large number of atmospheric measurements from the Mars Global Surveyor (mostly from the Thermal Emission Spectrometer, TES) and several instruments on the European Mars Express. It should be noted, however, that TES was primarily designed to look at martian surface composition. Its coarse resolution in the 15 micrometer carbon dioxide band allows the temperature structure of the atmosphere to be determined with a broad weighting function. (Somewhat better vertical resolution is obtained from the less frequent limb scans that sacrifice horizontal resolution.) On the other hand, the highly elliptical orbit of Mars Express gives only limited coverage of the planet over the course of a day. Both missions have taught (and are teaching us) a lot about Mars and about how best to study planetary atmospheres. No Mars Express data are in the public domain, however, and it is frustrating to think of doing meteorology (the value of which clearly degrades with time) at a remove of many months or a year. These missions also serve as excellent testbeds for the Mars Climate Sounder—which should be the first dedicated planetary atmospheric sounder in a suitable orbit for specifying the global meteorology—that is scheduled to be launched at the next opportunity (August 2005).

Of course, there have been a number of exciting probe missions to the planets. The Viking, Pathfinder, and Mars Exploration Rover missions all did entry, descent, and landing science. In spite of the quality of the data, these entry profiles are swamped in number by radio occultation temperature profiles from Viking and Mars Global Surveyor. That the Viking profile was relatively warm compared to Earth-based measurements and subsequent entry profiles has been remarked, but the implications are not clear [4]. The Pathfinder profile was measured at night and how much of its colder temperatures should be attributed to day-night differences is an open question. (Near-surface wind measurements deduced from the Viking entries show a spiral opposite in sense to what would be expected in an Ekman boundary layer, but this issue has also received little attention [5].) In general, the interpretation of probe measurements must deal with representativeness questions of this kind. They offer at best a one-dimensional picture of a highly variable four-dimensional dynamical system.

Similar problems plague the interpretation of the Galileo Probe results, at least as pertains to composition of the jovian atmosphere (a key goal for in situ science was the determination of the helium, nitrogen, and oxygen or water mixing ratios). The probe only detected water in small amounts at great depth. But it had entered a 5-micron hotspot (a gap in the normal planetary cloud cover where thermal radiation is much higher than for the bulk of the planet) [6]. How the special thermal environment of the hotspot effects the representativeness of its volatile components is still something of an open question. It depends crucially on the meteorological interpretation of these features [7]. (Interestingly, a similar question — of compositional and isotopic variations in space and time — has recently arisen in the Mars context where the partial condensation of carbon dioxide, the principal atmospheric component, has been seen to modify the winter polar composition [8].)

Venus has also been visited by a number of probes. The Pioneer Venus probes failed to make measurements all the way down to the surface and, as a result, were unable to shed light on the direction of the planet-atmosphere torque. The Vega Balloons were a bit of a departure, allowing direct (vertical and horizontal) wind measurements over a longer than normal probe lifetime, albeit at a single altitude [9].

The Huygens Probe is now poised to enter Titan's atmosphere. Will it resolve questions of the origin and nature of the atmospheric super-rotation? Will it settle the issue of the existence of hydrocarbon lakes? Or of methane precipitation? A strict reading of past experience makes it unlikely that this will be the case. Great science (and more new ideas) will undoubtedly result from this mission, but a comprehensive meteorological picture is not going to emerge from the probe measurements alone. Undoubtedly, global observations from Cassini will be of great importance in the interpretation of the probe results. [A paper at this meeting by R. Young further addresses the need for synergy between probe and orbiter measurements.]

This is an opportunity to re-evaluate how to study (and observe) planetary atmospheres. Now that an initial reconnaissance of the atmospheres in the solar system has taken place, and broad experience with entry probes has been obtained, it should be possible to design a mission architecture that will address and find answers to the outstanding scientific and operational questions associated with these bodies.

2. PLANETARY METEOROLOGY MISSION CONCEPT

For all their diversity, we believe that all of the planetary atmospheres obey the same fluid dynamic laws (as embodied in the Navier-Stokes equations, for example). Meteorology, then, is essentially a problem of determining and providing the correct global (planetary radius, rotation rate, gravity, composition, mass of atmosphere), boundary (topography, insolation) and initial conditions (i.e., the current state of the atmosphere) for the solution of these equations. However, we believe that the governing laws are chaotic (i.e., extremely sensitive to initial conditions) and, therefore, that in order to maintain an accurate description of an atmospheric state it is necessary to continually update our knowledge based on new observations. For this reason, the concept of sending a single probe into an atmosphere to make measurements over a very short period of time will always be flawed. It is impossible to know whether the measurements made at that time are, in fact, representative of any other time and place on the planet. Rather, we want sustained time coverage of the entire planet. This provides a meteorological context in which to interpret other detailed measurements that can be made in situ (and which are then quite useful in improving our knowledge of the atmospheric state). In addition, since the different meteorological variables are (frequently) related by balance relations which involve their gradients, we want a synoptic view of the planet (in which quantities at different locations are determined at the same time).

Our requirements are a bit daunting: four-dimensional coverage compared to the one-dimensional observations from traditional probe missions. However, there is a huge payoff, scientific and practical, from the more comprehensive approach. The goal of exploration, after all, is to discover things that are previously unknown. This requires a superabundance of observations, so that new parameters can be fit. But if we are able to gather all of this data — and process it in time — we will have a very accurate analysis of the weather system under investigation and many operational uses of this knowledge will be available.

Based on the equations of motion, we can estimate the time and length scales over which significant changes in the general circulation of a planet take place. Temporal changes are usually controlled by diabatic heating (from the absorption of solar radiation), and so the crucial time interval is the radiative time constant, the e-folding time for atmospheric temperature changes by radiation: $\tau = \rho H c_p / 4 \epsilon \sigma T^3$. Values for all of the planets are given in Table 1 (as derived from the Planetary Data System's Planetary Atmospheres Node). Because of the density factor in this relation, the values of τ can vary widely through the altitude range of interest. We have chosen "tropopause" values as being typical of the region that is easily observed by remote sensing and by entry probes. The important length scale is the Rossby radius of deformation [10] that gives the scale of baroclinic waves in the atmosphere (i.e., of the storm systems that constitute the primary deviation from zonal symmetry). $L = (gH)^{\frac{1}{2}}/f$. There is relatively little variation of this length scale from planet to planet. (See Table 1. For Venus and Titan, we choose values of f that correspond to the super-rotation rather than the solid body rotation rate.) The (polar) orbital period of a remote sensing instrument should be short enough so that it can see the whole planet with spatial resolution L within timescale τ, or $P \sim \tau L / 2R$.

Table 1. Required periods for atmospheric sounders

Planet	R, km	τ, days	L, km	P, days
Earth	6378	48	1862	7.0
Mars	3395	2	1391	0.4
Venus	6070	3	5631	1.4
Titan	2440	19	1145	4.5
Jupiter	71300	1650	1950	22.5
Saturn	60100	7590	1780	112.0
Uranus	24500	48000	1380	1350.0
Neptune	25100	44000	1930	1700.0

In fact, accumulating a lot of information from orbit is not difficult. TES has returned on the order of 200,000,000 infrared spectra over 3 Mars years. And only about 10% of the bandwidth is devoted to the atmospheric component of the observation. If the data can be made usable, there is a long list of applications that will benefit. Already, aerobraking has become an important part of the Mars payload delivery system. With increased confidence confidence in the forecasts of upper atmosphere densities, we can expect aerocapture missions and/or glider systems for accurate payload deliveries. The challenges of exploring a planet with as much surface area as the Earth will eventually lead to airplane or balloon missions that will require meteorological analyses and forecasts. And, of course, human exploration will depend on dust storm warning systems and the like. In order that this information be usable at Mars, it is highly desirable that not only data gathering, but routine data processing take place there. Already, the burden of aerobraking requires that a large Earth-based team make decisions about upper atmospheric densities. This would be unnecessary if the spacecraft could collect and evaluate the data autonomously. The case becomes stronger as we move to more distant planets where the two-way communication times are much too long to allow for continuous mission management from Earth. On the other hand, increased autonomy will allow for improved data products. Adaptive observations can be made to improve analyses in critical areas or where errors are large. Certainly, one can envision a small radiosonde network in support of human exploration, for example.

Clearly, then there is still a role for probes in providing the ground truth (from in situ measurements) that calibrates the more prolific orbital observations. Small mass (to increase the number of probes that can be utilized for a variety of purposes) and long lifetimes (to increase the data return) are clearly desirable. Because winds are more diagnostic than temperatures in determining the meteorology, a possible approach is the deployment of small floaters that become entrained in the wind and can be tracked from the primary spacecraft. (There is clearly a challenge in this type of design as most probes would drift in the zonal direction, while the orbiters would presumably be highly inclined in order to obtain global coverage. So the tracking problem is operationally difficult, at least at the terrestrial planets. On the other hand, it may be possible to recover the probes after they become occulted by the planet, and to download stored data. Alternatively, perhaps a system of weather balloons would act as semi-permanent wind probes. If these include lidars for measuring atmospheric vertical structure, a very rich dataset could be obtained.)

The most important point to make is that the quantity of data is very important. Four-dimensional systems have many degrees of freedom and it is impossible to constrain them without adequate data. On the other hand, data quality is less important. Various measurements are physically linked by the equations of motion. So given enough observational data, one can add many constraints to that dataset. In practice, for example, most operational weather analyses and forecasts use initialized data (i.e., data that has been filtered and modified to give better operational results). It doesn't matter very much if this data is noisy, as long as the statistics of that noise are known. It is then possible to modify it into the best form for operational use. Very accurate measurements of a few quantities (which are plagued in any case with the representativeness issues discussed above) are not nearly as useful. The trend in planetary probes — towards greater sophistication and expense to make a few localized measurements — flies in the face of this practical reality. A move towards cheaper, simpler probes is clearly in the cards.

3. NEW SOFTWARE TOOLS FOR PLANETARY ATMOSPHERES OBSERVATIONS

Based on a lot of technology transfer from the terrestrial meteorology community, the software tools for a new generation of planetary atmospheres observation missions are available. In fact, thanks to the limitation on resources that can be devoted to planetary exploration (as opposed to the high economic value of terrestrial numerical weather prediction), many of the planetary atmospheres tools are simpler to implement (and thus are better suited for the real time onboard processing possibilities that may be vital to remote operations). The tools are essentially 1) a versatile predictive modeling capability; 2) robust data assimilation techniques; and 3) adequate filtering methodologies.

The EPIC model [11] is a hybrid coordinate model that has been used to model a large number of planetary atmospheres [12], including the 5-micron hotspots that are so crucial to the interpretation of Galileo data [7]. The choice of vertical coordinate is suitable for modeling the deep atmospheres of the giant planets (for which the model was designed), but can also be used to simulate an atmosphere with a solid lower boundary. EPIC incorporates all of the physics that is needed for a weather forecasting model. And the code has been made available for all interested users via the Planetary Data System. This is just one example of a general circulation model that is well-tested and readily available to be applied to the analysis of planetary atmospheric data.

Data assimilation combines the forecast model and observations to produce an analysis of the state of the atmosphere that is consistent with both the governing equations of the model and the data [13]. The analysis can be thought of as a weighted average of the model predicted state and the measured state [14]. Determining the optimal average requires that both the model forecast errors and the observational errors (which include not only instrumental noise but representativeness errors) be known. The latter can clearly be inferred from the statistical properties of the data themselves. Model errors are another matter. While there are a number of approaches to determining forecast errors for well-tested terrestrial numerical weather prediction models (e.g., by making ensemble forecasts or from knowledge of the intrinsic co-variability in the atmosphere) [15, 16], planetary atmospheres and general circulation models are not so well known. Therefore it is preferable to deal with the the statistics of the residuals of predicted observables (like infrared radiances), rather than the presumed model errors in wind forecasts, say. We have shown that use of the resulting innovation covariance matrix [17, 18] leads to a viable and efficient formulation of the data assimilation problem.

The resulting observation space assimilation procedure greatly reduces the calculations required to make meteorological analyses, while putting the emphasis on producing a quality controlled version of the original data. The algorithm cycles (i.e., each new observation leads to an updated estimate of the atmospheric state which is used for the analysis of the following observation) and so is suitable for real time operation.

Given the estimate of the true value of the observed quantities, it is still necessary to find the atmospheric state (temperatures, winds, etc.) consistent with these measurements. This is best done by the four-dimensional variational (4DVAR) technique if an adjoint form of the predictive model is available. (The adjoint determines the sensitivity of the model's final state to the initial state and therefore gives guidance as to how to adjust the initial state in order to arrive at a desired final state.) For example, to test the viability of 4DVAR in the martian context, we have formulated a Martian general circulation model with an adjoint version [19]. The dynamical core of this model is based on the baroclinic spectral formulation that has long been used in terrestrial numerical weather prediction [20-22]. The model includes realistic topography and a diurnal cycle, but is treated as imperfect when doing assimilations. In practice, this means that the diabatic forcing (a complicated function of the highly variable atmospheric dustiness on Mars) is assimilated rather than predicted by the model.

Will the answer to the assimilation problem derived by this technique be unique? Not necessarily, as a primitive equation model has a large number of free modes which are useful for matching observed conditions, but which do not correspond to real motions in the atmosphere. These rapidly varying modes, usually called gravity waves in the atmospheric modeling community, must be filtered out of the solutions in order to obtain reliable forecasts. This initialization process has been the subject of much study in numerical weather prediction [23]. It is frequently achieved in terrestrial modeling by imposing a balance condition between the model wind and mass (temperature) fields. However, it is difficult to impose such a condition on the Martian atmosphere as that would filter out atmospheric tides which are known to be important and which do not satisfy a balance relation. Instead, we have recently developed a digital filtering technique [16] that minimizes the ill effects of fast gravity waves.

Digital filtering of gravity waves sidesteps the issues of evaluating in advance what the balance relations for a given planetary atmosphere should be. The modes that are filtered do not represent physical reality in any case, and so manipulating their amplitudes has a small effect on the ultimate quality of the results. What is most important in all of these procedures is the minimum foreknowledge of the given system that is required to produce high level atmospheric dynamics products. Since our goal is exploration, it is important that we do not assume that we have good preknowledge of the conditions in a planetary atmosphere.

To implement the onboard processing of data for a planetary mission, it is important that observational teams produce forward models for their instruments. (In any case, this would be a desirable part of the instrument design process.) Such models need to be constructed properly so that linearization and adjointing can be implemented and lead to robust codes in preparation for the 4DVAR analyses. Only an instrument whose signal is, in principle, invertible is going to be a reasonable flight prospect, so it is to be expected that such robust codes can be produced. Once available, the codes and the onboard processing relieve the instrument teams of the responsibility for producing standard data products, so they can devote their efforts to higher level scientific concerns.

The upcoming Mars Climate Sounder (MCS) will provide an opportunity to test this software and this modeling approach (though not yet the onboard processing aspects of the proposal). It is likely that at least some of the other atmosphere-observing instruments orbiting the planet at this time will overlap with MCS and provide an excellent opportunity for validating its products.

4. NEW HARDWARE TOOLS FOR PLANETARY ATMOSPHERES OBSERVATIONS

The goal of this paper is to challenge the experienced probe community to come forward with a new set of hardware components that will match the available software techniques in providing all elements needed for future planetary atmospheres missions.

Clearly, a new generation of sounders will be needed. Most planetary spacecraft to date have depended on infrared spectrometers to determine vertical structure. But the wave of the future seems to be moving towards microwave sensors. The ability to measure individual lines in the microwave region leads to greatly enhanced sounder capabilities. Among these are the ability to penetrate clouds and dust, the ability to measure isotopes and tracers, and probably the ability to determine horizontal winds from Doppler shifts.

The new entry probes will provide crucial observations, but many fewer measurements than the sounders. So it is appropriate that they be less expensive and significantly lighter. It might be possible for very light probes to float in the atmosphere, significantly increasing the data return and the decreasing the cost/benefit ratio for these mission components. The exciting new role of the entry probe will be to make targeted observations. Thus, it is desirable that the orbiter be capable of carrying a large number of entry probes with an accurate release mechanism.

5. CONCLUSIONS

The World Meteorological Organization has recently begun the THORPEX program to improve the accuracy of terrestrial weather forecasts. Among the objectives of this program are [24]

- the incorporation of model uncertainty into data-assimilation systems;
- developing adaptive data-assimilation and target-observing strategies;
- improving the assimilation of observations of physical processes and atmospheric composition; and
- the introduction of interactive procedures that make the forecast system more responsive to user needs.

The planetary exploration program is now in a position to take advantage of the terrestrial weather forecasting experience in the design of planetary atmospheres missions with these valuable capabilities. The tools that are becoming of greater importance to weather forecasters can be used as readily by spacecraft at other planets. They will lead to greater autonomy and adaptability of missions. This will in turn lead to better quality high-level scientific products and real-time availability of data for operational purposes.

The Mars program will be the testbed for these technologies. We have already learned that most of the required information to determine the state of the atmosphere can be obtained from low orbit, even with instruments that are not specifically designed for atmospheric observations. However, to make the most of the data, assimilation techniques must be used to assure that the retrievals of structure and winds are physically consistent. This same assimilation allows supplementary data to be used for calibration and validation. They also allow targeted in situ observations to provide the crucial ground truth for the remote sensing instruments.

The design of a new generation of small entry probes that can be released by the orbiters to provide ground truth measurements is the crucial next step in the development of this exploration program.

REFERENCES

1. Liu, J., M. I. Richardson, and R. J. Wilson, An assessment of the global, seasonal, and interannual spacecraft record of Martian climate in the thermal infrared, *J. Geophys. Res.*, **108 (E8)**, 10.1029/2002JE001921, 2003.

2. Zurek, R. W,. and L. J. Martin, Interannual variability of planet-encircling dust storms on Mars, *J. Geophys. Res.*, **98**, 3247–3259, 1993.

3. Owen, T. C., P. Mahaffy, H. B. Niemann, S. K. Atreya, T, M, Donohue, A. Bar-Nun, and I. de Pater, A new constraint on the formation of giant planets, *Nature*, **402**, 269–270, 1999.

4. Magalhaes, J. A., J. T. Schofield, and A. Seiff, Results of the Mars Pathfinder atmospheric structure investigation, *J. Geophys. Res.*, **104**, 8943–8955, 1999.

5. Seiff, A., Mars atmospheric winds indicated by motion of the Viking landers during parachute descent, *J. Geophys. Res.*, **98**, 7461–7474, 1993.

6. Young, R. E., The Galileo probe mission to Jupiter: Science overview, *J. Geophys. Res.*, **103**, 22775–22790, 1998.

7. Showman, A.P. and T. E. Dowling, 2000. Nonlinear simulations of Jupiter's 5-micron hot spots, *Science*, **289**, 1737–1740, 2000.

8. Feldman, W. C., T. H. Prettyman, W. V. Boynton, J. R. Murphy, S. Squyres, S. Karunatillake, S. Maurice, R. L. Tokar, G. W. McKinney, D. K. Hamara, N. Kelly, and K. Kerry, CO2 frost cap thickness on Mars during northern winter and spring, *J. Geophys. Res.*, **108 (E9)**, 10.1029/2003JE002101, 2003.

9. Blamont, J.E., R.E. Young, A. Seiff, B. Ragent, R.Z. Sagdeev, V.M. Linkin, V.V. Kerzhanovich, A.P. Ingersoll, D. Crisp, L.S. Elson, R.A. Preston, G.S. Golitsyn, and V.N. Ivanov. Implications of the VEGA balloon results for Venus atmospheric dynamics. Science 231, 1422-1425, 1986.

10. Holton, J. R., *An Introduction to Dynamic Meteorology*, Academic Press, New York, 319 pp., 1972.

11. Dowling, T.E., A. S. Fischer, P. J. Gierasch, J. Harrington, R. P. LeBeau, and C. M. Santori, The Explicit Planetary Isentropic-Coordinate atmospheric model, *Icarus*, **132**, 221–238, 1998.

12. Stratman, P. W., A. P. Showman, T. E. Dowling, and L. A. Sromovsky, EPIC simulations of bright companions to Neptune's Great Dark Spots, *Icarus*, **151**, 275–285, 2001.

13. Ghil, M., and P. Malanotte-Rizzoli, Data assimilation in meteorology and oceanography, *Adv. Geophys.*, **33**, 141–246, 1991.

14. Talagrand, O., Assimilation of observations, an introduction, *J. Meteor. Soc. Japan*, **75**, 191–209, 1997.

15. Daley, R., and E. Barker, *NAVDAS Source Book*, NRL Publication NRL/PU/7530–01-441, 161pp., 2001.

16. Kalnay, E., *Atmospheric Modeling, Data Assimilation and Predictability*, Cambridge U. Press, 341pp., 2003.

17. Menard, R., S. E. Cohn, L.-P. Chang, and P. M. Lyster, Stratospheric assimilation of chemical tracer observations using a Kalman filter, part I, Formulation, *Mon. Wea. Rev.*, **128**, 2654–2671, 2000.

18. Houben, H., Planetary meteorology by onboard data assimilation, Roger Daley Memorial Symposium, Montreal, Canada, 2003.

19. Houben, H., Assimilation of Mars Global Surveyor meteorological data, *Adv. Spa. Res.*, **23**, 1899–1902., 1999.

20. Bourke, W., B. McAvaney, K. Puri, and R. Thurling, Global modeling of atmospheric flow by spectral methods, in *General Circulation Models of the Atmosphere*, ed. J. Chang, *Methods in Computational Physics*, **17**, pp. 267–324, Academic Press, New York, 1977.

21. Haltiner, G. J., and R. T. Williams, *Numerical Prediction and Dynamic Meteorology*, Second Edition, John Wiley & Sons, New York, 477 pp., 1980.

22. Krishnamurti, T. N., H. S. Bedi, and V. M. Hardiker, *An Introduction to Global Spectral Modeling*, Oxford U. Press, 1998.

23. Daley, R., *Atmospheric Data Analysis*, Cambridge University Press, 457 pp., 1991.

24. Shapiro, M. A., and A. J. Thorpe, *THORPEX, A Global Atmospheric Research Programme, International Science Plan*, World Meteorological Organization, 2004.

Gas Giants

NEPTUNE POLAR ORBITER WITH PROBES*

2nd INTERNATIONAL PLANETARY PROBE WORKSHOP, AUGUST 2004, USA

Bernard Bienstock[1], David Atkinson[2], Kevin Baines[3], Paul Mahaffy[4], Paul Steffes[5], Sushil Atreya[6], Alan Stern[7], Michael Wright[8], Harvey Willenberg[9], David Smith[10], Robert Frampton[11], Steve Sichi[12], Leora Peltz[13], James Masciarelli[14], Jeffrey Van Cleve[15]

[1]Boeing Satellite Systems, MC W-S50-X382, P.O. Box 92919, Los Angeles, CA 90009-2919, bernard.bienstock@boeing.com
[2]University of Idaho, PO Box 441023, Moscow, ID 83844-1023, atkinson@ece.uidaho.edu
[3]JPL, 4800 Oak Grove Blvd., Pasadena, CA 91109-8099, kbaines@pop.jpl.nasa.gov
[4]NASA Goddard Space Flight Center, Greenbelt, MD 20771, paul.r.mahaffy@nasa.gov
[5]Georgia Institute of Technology, 320 Parian Run, Duluth, GA 30097-2417, ps11@mail.gatech.edu
[6]University of Michigan, Space Research Building, 2455 Haward St., Ann Arbor, MI 48109-2143, atreya@umich.edu
[7]Southwest Research Institute, Department of Space Studies, 1050 Walnut St., Suite 400, Boulder, CO 80302, astern@boulder.swri.org
[8]NASA Ames Research Center, Moffett Field, CA 94035-1000, mjwright@mail.arc.nasa.gov
[9]4723 Slalom Run SE, Owens Cross Roads, AL 35763, harvey@willenbergs.com
[10] Boeing NASA Systems, MC H013-A318, 5301 Bolsa Ave., Huntington Beach, CA 92647-2099, david.b.smith8@boeing.com
[11]Boeing NASA Systems, MC H012-C349, 5301 Bolsa Ave., Huntington Beach, CA 92647-2099 robert.v.frampton@boeing.com
[12]Boeing Satellite Systems, MC W-S50-X382, P.O. Box 92919, Los Angeles, CA 90009-2919, stephen.f.sichi@boeing.com
[13]Boeing NASA Systems, MC H013-C320, 5301 Bolsa Ave., Huntington Beach, CA 92647-2099, leora.peltz@boeing.com
[14]Ball Aerospace & Technologies Corp., P.O. Box 1062, Boulder, CO 80306-1062 jmasciar@ball.com
[15]Ball Aerospace & Technologies Corp., P.O. Box 1062, Boulder, CO 80306-1062 jvanclev@ball.com

ABSTRACT

The giant planets of the outer solar system divide into two distinct classes: the 'gas giants' Jupiter and Saturn, which consist mainly of hydrogen and helium; and the 'ice giants' Uranus and Neptune, which are believed to contain significant amounts of the heavier elements oxygen, nitrogen, and carbon and sulfur. Detailed comparisons of the internal structures and compositions of the gas giants with those of the ice giants will yield valuable insights into the processes that formed the solar system and, perhaps, other planetary systems. By 2012, Galileo, Cassini and possibly a Jupiter Orbiter mission with microwave radiometers, Juno, in the New Frontiers program, will have yielded significant information on the chemical and physical properties of Jupiter and Saturn. A Neptune Orbiter with Probes (NOP) mission would deliver the corresponding key data for an ice giant planet.

---

* This work discussed in this paper was funded under a NASA Visions Contract, NNH04CC41C.

Such a mission would ideally study the deep Neptune atmosphere to pressures approaching and possibly exceeding 1000 bars, as well as the rings, Triton, Nereid, and Neptune's other icy satellites. A potential source of power would be nuclear electric propulsion (NEP). Such an ambitious mission requires that a number of technical issues be investigated, however, including: (1) atmospheric entry probe thermal protection system (TPS) design, (2) probe structural design including seals, windows, penetrations and pressure vessel, (3) digital, RF subsystem, and overall communication link design for long term operation in the very extreme environment of Neptune's deep atmosphere, (4) trajectory design allowing probe release on a trajectory to impact Neptune while allowing the spacecraft to achieve a polar orbit of Neptune, (5) and finally the suite of science instruments enabled by the probe technology to explore the depths of the Neptune atmosphere. Another driving factor in the design of the Orbiter and Probes is the necessity to maintain a fully operational flight system during the lengthy transit time from launch through Neptune encounter, and throughout the mission.

Following our response to the recent NASA Research Announcement (NRA) for Space Science Vision Missions for mission studies by NASA for implementation in the 2013 or later time frame, our team has been selected to explore the feasibility of such a Neptune mission.

SECTION 1: INTRODUCTION, OUTLINE/OVERVIEW, BACKGROUND

Solar system exploration has historically been divided into three overlapping stages – Reconnaissance, Exploration, and In-Depth Study [1]. Since the advent of outer planetary exploration in the 1970s, an initial reconnaissance of the gas giants Jupiter and Saturn has been completed by the Pioneers 10 and 11, Voyagers 1 and 2, and Ulysses spacecraft. Exploration of the Jupiter and Saturn Systems is in its early stages, initiated by the multiyear encounters of the Galileo spacecraft and the Cassini / Huygens spacecraft, respectively. However, the Ice Giant planets Uranus and Neptune have only been visited once each, by Voyager 2 in 1986 and 1989, respectively. Comparative exploration of one or both of the Ice Giants is the natural next step in the continuing progression of outer solar system exploration.

Although sharing a number of characteristics, each of the Gas and Ice Giant Planets is a unique system. Not only is each a miniature planetary system in its own right, with moons and rings, dynamic atmospheres and magnetospheres, but the outer planets contain a physical and chemical record of conditions at the time of solar system formation that is complementary to but different from the record encoded in the terrestrial planets.

Jupiter and Saturn are dominated by H and He. However, conditions change markedly as we move beyond the inner outer solar system to the regions where the Ice Giants Uranus and Neptune reside. Both Uranus and Neptune contain much higher levels of condensed refractories and volatile ices such as nitrogen, oxygen, sulfur, and carbon [2]. Careful examination and comparison of the compositions and internal structures of the Ice and Gas Giants will provide important clues regarding the mechanisms by which the solar system and, by extension, extra-solar planetary systems formed.

Hammel (2001) [3] points out that previous studies of planetary atmospheres have stressed the identification of physical and chemical processes underlying many of the phenomena observed in planetary atmospheres. However, now that preliminary studies of individual atmospheres in the outer solar system have been completed, more detailed comparative studies of the atmospheres, satellites, rings, and magnetospheres of the Gas and Ice Giants are needed.

Although quite diverse, the terrestrial planets and the Gas Giants share a number of characteristics. The similarities and differences between the moons, atmospheres, geology, chemistry, and magnetospheres of the inner and outer solar system provide a natural laboratory for identifying and understanding conditions favorable for enabling and supporting biological activity, understanding and controlling the effects on the Earth's atmosphere of human activity, as well as interpreting observations of extra-solar planetary systems.

The overall justification for the Neptune Orbiter with Probes (NOP) mission therefore derives from the importance of continuing comparative studies of the Gas and Ice Giants, as well as between bodies in the inner and outer solar system, to address questions of planetary origins, and to help discriminate between possible theories of solar system formation and evolution. The NOP mission provides an opportunity to contribute to these goals by exploring a member of the family of Ice Giants.

The proposed NOP mission directly addresses a number of common goals, objectives, and themes in the National Academy of Science Decadal Survey, NASA's Solar System Exploration theme, and the Solar System Exploration Roadmap.

Solar System Exploration Roadmap, 2003

The Solar System Exploration (SSE) Roadmap, 2003 [4] lists possible mid-term and long-term flagship missions that should be selected to build on results of earlier investigations. One of the high priority missions listed is the NOP mission. Of the 8 primary objectives enumerated in the Roadmap, the first two contain elements that are directly addressed by the Neptune Orbiter with Probes mission:

How did planets/minor bodies originate. Understand the initial stages of planet and satellite formation, and study the processes that determined the original characteristics of the bodies in our solar system

How solar system evolved to current state. Determine how the processes that shape planetary bodies operate and interact, understand why the terrestrial planets are so different from one another, and learn what our solar system can tell us about extra solar planetary systems.

Furthermore, the SSE Roadmap indicates that "comprehensive exploration of the Ice Giant Neptune will permit direct comparison with Jupiter and more complete modeling of giant planet formation and its effect on the inner solar system." The Roadmap also provides the Neptune Orbiter with Probes mission as an example of a high priority Flagship mission that would provide major scientific advances.

National Academy of Science Decadal Survey for Solar System Exploration

The National Academy of Science Decadal Survey for Solar System Exploration [5] has recommended that in-depth studies of the Neptune system be given high priority. Additionally, the Primitive Bodies Panel lists a Neptune/Triton mission among its highest priorities for Medium Class, and the Giant Planets Panel lists a Neptune Orbiter with multiple entry Probes as its highest priority in the next decade. The Decadal Study emphasizes that it is only through a comparison of composition and interior structure of the giant planets in our solar system that we can advance our understanding of how our planetary system formed. Additionally, detailed study of the giant planets can help us extrapolate to planetary systems around other stars. All three of the themes developed in the Decadal Study Report: 1) Origin and evolution; 2) Interiors and atmospheres; and 3) Rings and plasmas are addressed by a Neptune Orbiter with Probes mission.

"The primary probe science goal is the use of composition and temperature data in the Neptune atmosphere from the stratosphere to hundred/kilobar pressures to advance the understanding of solar system formation. Complementary probe measurements of winds, structure, composition and cloud particle size and lightning are also suggested. Critical measurements are CH_4, NH_3, H_2S, H_2O, PH_3, and the noble gases He, Ne, Ar, Kr, and Xe. Although the average atmospheric O abundance is not likely to be measured by 100 bar, C in methane and the noble gases will reveal the elemental abundance that can constrain models of Neptune's formation when analyzed in the context of data from other giant planets such as Jupiter and Saturn." 2003 NASA Strategic Plan and the more recent report of the President's Commission on Implementation of United States Exploration Policy provides the broad motivation for a Neptune mission. The Strategic Plan places the outer planet exploration program in the context of the study of the origin of the solar system and the building blocks of life. The President's Commission Report describes the same themes in its National Science Research Agenda organized around the themes Origins, Evolution, and Fate. Neptune exploration is further motivated by sub-themes described in the NASA Strategic Plan, including formation of the solar system, comparative planetology, and solar controls on climate.

Solar System Exploration Theme

NASA's Solar System Exploration theme also lists a Neptune mission as one of its top priorities for the mid-term (2008-2013) [6,7]. In a recent NASA study, a Neptune mission was highly ranked for its connections to astrophysical problems beyond the Solar System, including geology, ring systems and ring dynamics, atmospheric dynamics and structure, magnetic field structure and generation, Triton pre-biotic chemistry, and as an analog for local extrasolar planets. The Neptune mission is described as "almost Cassini-like in scope, near Discovery-like in cost" [8].

SECTION 2: SCIENCE

Very little is known about the overall composition, structure, and dynamics of Neptune's deep atmosphere. It is proposed that multiple entry Probes be used to sample the composition, cloud and energy structure, and atmospheric dynamics of the Neptune atmosphere at several latitudes. Voyager and ground-based observations have revealed in Neptune's atmosphere the presence of hydrogen, helium (indirectly), methane (and only two of its photochemical products, acetylene and ethane), hydrogen cyanide, carbon monoxide, and $H3+$. This list is sparse when compared to the 28 (neutral) molecules, one ion, and multiple important isotopes that have been measured in the atmosphere of Jupiter [9]. Moreover, even for the species detected in Neptune's atmosphere, the actual mixing ratio measurement is either highly uncertain or just not available.

To understand the formation of Neptune and Neptune's atmosphere, detailed knowledge of atmospheric composition is crucial. Composition measurements in the region of 10-1000 mbar (lower stratosphere to upper troposphere) can offer valuable information on dynamical and photochemical processes in these regions. Elemental abundances of the heavy elements, at least C and S, as well as helium and the other noble gases, Ne, Ar, Kr, Xe, in the well-mixed atmosphere below the cloud layers, are needed to constrain the formation models of Neptune, as well as the origin and evolution of its atmosphere. The O/H and N/H ratios in the well-mixed atmosphere are desirable but not required. Supporting composition measurements on disequilibrium species, PH_3, GeH_4, AsH_3; isotope ratios, $^{15}N/^{14}N$, and D/H, primordial molecules, N_2 together with CO and HCN, are also highly desirable.

Based on the known composition, the deepest (probe accessible) cloud on Jupiter is expected to be water (ice and droplets) at approximately 5 bars for 3 x solar O/H and at 10 bars for 10 x solar O/H (10 times the solar value of O/H) [10,11]. Similarly, the deepest cloud on Neptune is also expected to be water. Thermochemical equilibrium calculations predict a water "ice" cloud base at approximately 100-bar level (273 K), with a cloud of water/ammonia droplets – aqueous solution – forming below this level. The base of the water cloud could therefore be as deep as approximately 370 bar (460 K) for 30x solar O/H or 500 bar (500 K) for 50x solar[11,12].

Models of formation of Jupiter and the other giant planets predict that heavy elements become increasingly enriched from Jupiter to Neptune [9]. This indeed appears to be the case, as the C/H ratio, 3x solar at Jupiter, is found to increase to 20-30x solar at Uranus and 30-50x solar at Neptune. Icy planetesimal models [9] predict that other heavy elements including O (as in water) would also be similarly enhanced in the atmosphere of Neptune [11]. This implies that to ensure the measurement of O/H on Neptune, a Probe would have to make measurement of water vapor to depths well below the water cloud, i.e. to >500 bar level.

However, a theorized deep water-ammonia ionic ocean would prevent water and ammonia from being well-mixed at pressures less than 10-100 kilobars [11]. Although no Probe in the foreseeable future can access these pressure levels, the elements C, S, He, Ne, Ar, Kr, and Xe (same as Jupiter except for O and N) can easily be reached at pressures of 50-100 bar. Combining mixing ratios of these elements with the isotopic data on D/H and $^{15}N/^{14}N$, along with the available elemental information on Jupiter from Galileo and Juno (if selected following the on-going Phase A study), along with C, $^{15}N/^{14}N$, and He at Saturn by the Cassini orbiter, will be adequate for constraining the models of formation of Neptune and its atmosphere. It is also important to recognize that even though the O/H and N/H will not be accessible at Neptune, measurement of the NH_3 and H_2O profiles to the maximum attainable depths are still valuable for gaining insight into the interior processes including the existence of the purported ionic ocean.

Neptune does possess an internal heat source, and, similar to Jupiter, we would expect to find that this is variable from equator to pole, resulting in variable convective processes and latitude-dependent winds with depth. For purposes of studying global atmospheric dynamics, as well as possible variability in composition, the proposed NOP mission therefore targets entry Probes to multiple latitudes.

TRITON, RINGS, MAGNETOSPHERE AND ICY SATELLITES

A mission to the Neptune system has strong connections to astrophysical problems beyond the Solar System, including geology, ring systems and ring dynamics, atmospheric dynamics and structure, magnetic field structure and generation, possible pre-biotic chemistries, and as an analog for local extrasolar planets.

Triton

Triton is perhaps the key element to selecting Neptune over Uranus as the prime representative of the Ice Giants for exploration. The largest satellite of Neptune, Triton is in a high inclination, retrograde orbit, and appears to be a captured object. Imaging from Voyager 2 showed a marked disparity between different regions on Triton's surface suggesting significant differences in the dynamic and impact history of these regions. Additionally, it is suspected that Triton may be related to Charon and Pluto, and possibly to comets and Kuiper Belt Objects (KBOs), although it is unknown how the composition and inventory of volatiles of Triton compares with these objects. Key science questions include the composition of the surface ice, the abundance of N_2, CO, hydrocarbons, nitriles, and noble gases on Triton's surface and in Triton's atmosphere, and the distribution and sources of aerosols.

Satellites and Rings

The properties of Neptune's system of satellites are largely unknown, including overall densities and composition, whether they are mostly silicate or icy. Are the dark surfaces siliceous or carbonaceous? What is the dynamic / collision history of these objects? The relationship between the satellites and Neptune's rings is also unknown. Do the satellites contribute to the generation and maintenance of the rings, and if so – how? What is the composition of the rings, what are their ages and dynamical evolutional history? Can the overall structure of the rings, including ring arcs be explained?

Magnetic Field

Neptune's magnetic field is unique in the solar system; it mainly consists of a dipole field that is offset and highly inclined with respect to Neptune's center. What is the mechanism by which the magnetic field is generated? How does the structure of the field affect interactions with the solar wind, and the structure and properties of the magnetosphere? Does the magnetosphere change over a Neptune year? Additionally, Neptune's aurora can be used as a partial diagnostic and probe of the magnetic field. What are the processes of the auroral emission, and how do these compare to those on Saturn and Jupiter, as well as the Earth?

The complexity and scientific richness of the Neptune systems requires a well-defined list of science and measurement goals and objectives, and a highly integrated suite of remote sensing and in situ instruments. A detailed discussion of the specific goals and required instrumentation is beyond the scope of this paper. However, a listing of important science goals and issues in the Neptune system can be found in [2]. The NOP Science Goals and Objectives are listed in Table 1a, and specific measurement objectives are provided in Table 1b.

Table 1a. Science Goals and Objectives

1. Origin and evolution of Ice Giants – Neptune atmospheric elemental ratios relative to Hydrogen (C, S, He, Ne, Ar, Kr, Xe) and key isotopic ratios (e.g., D/H, $^{15}N/^{14}N$), gravity and magnetic fields.
2. Planetary Processes – Global circulation, dynamics, meteorology, and chemistry. Winds (Doppler and cloud track), cloud structure, microphysics, and evolution; ortho/para hydrogen ratio; Photochemical species (C-H hydrocarbons, HCN); tracers of interior processes such as N_2, and disequilibrium species, CO, PH_3, GeH_4, AsH_3).
3. Triton – Origin, Plumes, Atmospheric composition and structure, surface composition, internal structure, and geological processes
4. Rings – Origin and evolution, structure (waves, microphysical, composition, etc.)
5. Magnetospheric and Plasma Processes
6. Icy Satellites – Origin, evolution, surface composition and geology Neptunian magnetosphere.

Table 1b. Measurement Goals and Objectives

• Neptune – Measurement of profiles of N_2, HCN, H_2S, NH_3, CH_4, H_2O, PH_3 and other disequilibrium species, hydrocarbons, noble gases and their isotopic ratios, D/H, $^{15}N/^{14}N$; Profiles of atmospheric temperature, pressure, and density; atmospheric dynamics; radiative balance and internal heat; Cloud particle size/density, microphysical properties; storm evolution, lightning, stratospheric emissions, nightside thermal imaging; Interior: Gravitational field measurements
• Triton – Geological mapping, surface composition/roughness and thermal mapping, topography, subsurface mapping and interior/seismometry
• Rings – Composition, waves, dynamics
• Magnetosphere – Magnetic field; Plasma composition and electric fields

In the following tables of instruments (Tables 2a and 2b) the numbers in parentheses following each Measurement Goal refer to a Science Goal in Table 1a.

Table 2a. Probe Instruments

Instrument	Measurements*	Heritage
Gas Chromatograph Mass Spectrometer (GCMS)	• Profiles of N_2, CO, HCN, H_2S, NH_3, CH_4, H_2O, etc.: Stratosphere to deep atmosphere (1,2) • D/H (1) • $^{15}N/^{14}N$ (1) • Disequilibrium species (2, 1) • Hydrocarbons (1,2) • Noble gases (He, Ne, Ar, Kr, Xe) (1) • Isotopic ratios (1)	Stand-alone GC and MS experiments have been flown on the Pioneer Venus Probes. An MS was flown on the Galileo Jupiter Probe. A GCMS is now flying on the Huygens Probe as part of the Cassini-Huygens mission.
Atmospheric Structure Instrument (ASI), including 3-axis accelerometers: x, y, z and redundant z	• Density (2) • Temp/pressure profile (2) • Wind dynamics (2)	Used in many planetary Probe missions to Venus, Mars, and Jupiter
Net Flux Radiometer (NFR)	• Radiative balance and internal heat (1,2)	Used on Jupiter and Venus Probes
Nephelometer	• Cloud particle size/density, microphysical properties (2)	Flown on Pioneer Venus Probes.
Helium Abundance Detector (HAD)	• Detailed helium measurements (1)	Some redundancy with the GCMS. Flown on Galileo Probe.
Ortho/Para H2 Experiment	• Vertical atmospheric transport (2)	First flight for this instrument.
Lightning Detector	• Lightning (2)	Flown on Galileo Probe.
Doppler Wind Experiment (DWE)	• Vertical Profile of zonal winds, atmospheric waves (2)	Flown on Galileo and Huygens Probes
ARAD (Analog Resistance Ablation Detector)	• TPS recession as a function of time, allows for determination of flight aerodynamics and aerothermal loads (1,2)	Provides science and engineering data. Flown on the Galileo Probe. A must for planetary entry Probes.

* Numbers in parentheses refer to a Science Goal in Table 1a

Table 2b. Orbiter Instruments

Instrument	Measurements*	Heritage
High Resolution UV Spectrometer	• Neptune thermospheric and auroral emissions, occultation number density profiles (2) • Triton: atmospheric emissions, occultation number density profiles, surface composition (3) • Rings: composition (4)	Galileo UVS Cassini UVIS New Horizons/ ALICE
High Resolution IR Spectrometer	• Atmospheric composition (1, 2) • Triton and icy satellite surface composition/roughness and temperature (3, 6)	Imaging experiment included as part of Galileo (NIMS) and Cassini (VIMS) orbiter payloads.
High Resolution Camera	• Triton Surface, geological mapping (3) • Rings: waves, structure and dynamics (4) • Neptune Atmosphere, meteorology, dynamics, storm evolution, and lightning (2) • Icy Satellites (6)	Voyager/ ISS Galileo/SSI Cassini/IS
Mid and far IR spectrometer	• Neptune: detailed atmospheric composition, thermal mapping (3-D wind fields) (1,2) • Triton: surface thermal mapping (3) • Rings: particle size and thickness (4)	Voyager/IRIS Cassini/CIRS
Plasma wave instrument	• Plasma composition and electric fields (5)	Galileo/PWS, Cassini/RPWS
Ion / neutral mass spectrometer (INMS)	• Protons, heavier ions, neutral particles/atoms (3,5)	Cassini/INMS

Instrument	Measurements*	Heritage
Ka/X/S-band radio science	• Atmospheric pressure, temperature profile, density (2) • Abundance profiles of PH_3, H_2S, and NH_3 (1,2) • Gravitational field measurements (interior structure) (1,2) • Ring occultations for particle size and ring thickness (4)	Flown on Cassini Orbiter for studies of Saturn/Titan atmospheres. Earlier Voyager heritage.
Uplink radio science	• Neptune and Triton atmospheric pressure, temperature profiles, density (2)	Developed and ready for launch as part of New Horizons mission to Pluto.
Bistatic radar	• Triton and possibly other satellite surface texture, mapping (3)	Uses incumbent radio science system. Demonstrated on Mars Global Surveyor Mission
Magnetometer	• Magnetic fields (1,5)	Galileo/Magnetometer, Cassini/Magnetometer
Laser altimeter	• Triton topography (3)	Mars MOLA, NEAR
Microwave radiometer	• Neptune deep atmosphere composition (1,2) • Triton composition (3) • Neptune, Triton, icy satellite brightness temperatures (1, 2, 3, 6)	Demonstrated with both Magellan and Cassini RADAR systems operating in passive modes
Bolometer Array	• Triton, icy satellite, and possibly ring surface temperature distribution (3, 4, 6)	
Penetrating radar	• Triton subsurface mapping, altimetry, surface emmissivity/ roughness (3)	Magellan/ RADAR Cassini/RADAR under development for JIMO.

* Numbers in parentheses refer to a Science Goal in Table 1a

Candidate Instrument Payloads

As illustrated in Table 2b, the Orbiter is the core of the Neptune mission, providing a remote sensing platform and in-situ instruments for the study of Neptune's magnetic field, and primary data links. A key element of the Orbiter instrument payload would be an integrated imaging package comprising multiwavelength imagers and spectrometers and a microwave radiometer. Space physics detectors might include a magnetometer and a plasma wave detector. An Ion and Neutral Mass Spectrometer could obtain chemical and isotopic measurements from the atmosphere of Triton. Radio science investigations would be enhanced by including an uplink capability enabled by ultrastable oscillators. Multiple entry Probes and Triton Lander(s) are also an essential part of the atmospheric and surface structure and chemistry on Neptune and Triton, respectively. Probe instrumentation, listed in Table 1a, is similar to that flown on Galileo and Huygens, including a Gas Chromatograph/Mass Spectrometer (GCMS), sensors for measuring temperature, pressure and acceleration, solar and IR radiometers, and a nephelometer.

SECTION 3: PROMETHEUS ARCHITECTURE; CAPABILITIES, ADVANTAGES, BENEFITS

Use of Prometheus technology, while providing great flexibility in mission planning, has several liabilities, including long mission duration and the necessity for extremely long burns from a 2-3 N thruster system to accomplish the required ΔV. A comparison of typical chemical operating parameters with the nuclear electric propulsion (NEP) performance is provided in Table 3. At present the Jupiter Icy Moons Orbiter (JIMO) NEP system is not qualified for a Neptune mission, although this technology is expected to be qualified in time for incorporation into a Neptune Orbiter mission.

Table 3. Thrust Characteristics of NEP and Chemical Propulsion Systems

Parameter	Chemical Propulsion	Nuclear Electric Propulsion
Isp (sec)*	320	7000
Thrust (N)	100	2-3

* Isp, or specific impulse, is a measure of fuel efficiency

Using thrust constraints imposed by NEPP, the transit time for a Neptune mission will be approximately 16 years. This extremely long time can be shortened somewhat by using higher thrust NEP, which should be available in the 2015 to 2018 time frame. In addition, a Jupiter gravity assist will also shorten the mission, although this constraint affects the launch period and in fact limits the launch opportunities to once every 12 years. This fact alone may negate the advantages of planning a Neptune mission that requires a Jupiter gravity assist.

Additional advantages of NEP result from the nature of a low thrust, highly efficient system. Although NEP requires that thrusters be fired for long periods of time (on the order of months), this continuous thrust provides the ideal situation for on-board navigation. Midcourse corrections are not required since they can be incorporated during the long thrust periods. In addition, the high Isp affords great flexibility in mission planning at Neptune, although careful planning is required to begin the thrusting at the proper time to execute the desired ΔV maneuver.

The use of the highly efficient NEP on the Prometheus platform also results in considerable mass available for the science payload. On JIMO, that mass is specified at 1500 kg. But the mass estimate for shielding of electronic components from the intense radiation at Europa is on the order of 1000 kg. The radiation fields at Neptune, and its moon Triton, are significantly lower than those at Jupiter. Thus, it is expected that the Neptune Orbiter could support a payload mass of nearly 2500 kg.

Since the Neptune Orbiter payload consists of three planetary entry Probes, with a relatively low mass allocated for Orbiter science payload, the mass allocation for the Neptune Probes will be significant. Assuming a Probe design goal of approximately 300 kg, or close to the Galileo Probe mass of 339 kg, the 3 Neptune Probes consume only 36%, or 900 kg of the available 2500 kg payload mass. Allocating 100 kg for the Neptune Orbiter science payload still leaves 1500 kg of mass available for other science within the Neptune planetary system. A tantalizing use for this mass is to fly two Triton Landers, at 750 kg each, for a detailed evaluation of a moon that is nearly the scientific equal of Titan and Europa. A summary of the payload mass allocations for the Neptune Orbiter with Probes mission is given in Table 4.

Table 4. Payload Mass Allocations

Element	Unit Mass (kg)	Count	Total Mass (kg)
Probe	300	3	900
Orbiter	100	1	100
Triton Lander	750	2	1500
Total			2500

The use of nuclear electric power, the second "P" in NEPP, also enhances the NOP mission. The continuously available, high level of electrical power permits simultaneous operation of all science instruments on the Orbiter and a high-power transmitter for high data rate transmissions to Earth. Thus, high power science, such as a Triton radar imager, can operate simultaneously with other instruments. This is in marked contrast to previous outer planet missions, where the relatively low power generated by Radioisotope Thermoelectric Generators (RTGs) necessitated cycling of the science instruments to remain within the total science power allocation.

SECTION 4: MISSION DESIGN ISSUES

Defining a robust strategy for multiple Probe release, Lander operations, and the Orbiter mission (over and above serving as the communications link for the Probes and Landers) at Neptune and Triton is a challenging task. As is common for complex missions of this type, optimization of each mission element must be balanced with the science requirements and mission design constraints of the other elements. Ideally, the selected design will provide a balanced mission that achieves all the science goals for each element. A discussion of the considerations for each element is presented below.

Probe Mission

The current concept includes three identical Probes that will sequentially enter the Neptune atmosphere at three different latitudes. The well-mixed nature of the Neptune atmosphere and the desire to study the atmospheric dynamics drives the Probe entry to three widely spaced latitudes. In addition, other than possible slight seasonal effects, the Neptune atmosphere is postulated to be symmetric about the equator so that a Probe entering at 15° north latitude would sense the same atmosphere as a Probe entering at 15° south latitude. For planning purposes, the three Probes should be targeted to high (60° - 90°), mid (30° - 60°) and low (0° - 30°) latitudes. Entry at 90° and 0° latitude should be avoided, since the atmosphere at these locations will undoubtedly have singularities not representative of low and high latitude atmospheric dynamics.

Probe depth is another mission design driver. For a Neptune atmospheric entry Probe, it is desirable to descend to a depth of 200 bars. Descent to this depth may require up to 4 hours assuming reasonable ballistic coefficients and the current model of the Neptune atmosphere. Thus a constraint on the Orbiter trajectory is to maintain the Probe in radio view for at least 5 hours, a duration that allows margin should a Probe descend deeper than the baseline depth by operating beyond its design life.

An additional science consideration for the Orbiter is dictated by science mission requirements for long-term observations of Neptune, and Neptune's rings and icy satellites. Although Orbiter science will be conducted throughout the Probe and Lander segments of the Neptune encounter, it is expected that long-term observations of Neptune, the rings, and the icy satellites will continue during a post-Probe, post-Lander extended mission.

The Orbiter must allocate sufficient resources to support the three Probe missions, including payload mass, ΔV (fuel), power, and relay link communication system design. Probe release strategy is also a significant mission driver for the Orbiter. Figure 1 indicates two options for Probe release. Separation of the first Probe as the Orbiter approaches Neptune is a strategy that optimizes the first Probe mission only. The subsequent two Probes will be released from an elliptical orbit, at apoapsis, in order to allow the Orbiter to perform a deflection mission to avoid following the same planetary entry trajectory as the Probe.

A possible Probe release strategy is to complete the first Probe mission, complete a preliminary analysis of the Probe data, and to then have the option of modifying the second (and third) Probe missions based on the initial data reduction/analysis from the first (and second) Probes. Following each release, the Orbiter must execute an inclination change to allow the next vehicle to enter at a different latitude. Thus a possible strategy is to select a sequence with a low latitude change requirement. Finally, long term observations of Triton are an essential component of the NOP mission and must also be factored into the mission design. The phasing of the Orbiter Triton mission must be factored into the Orbiter mission planning. A long-term observation of Triton is required, that in turn drives the Orbiter to continue to orbit Triton after completion of the Probes mission support.

Figure 1. Neptune Mission Design

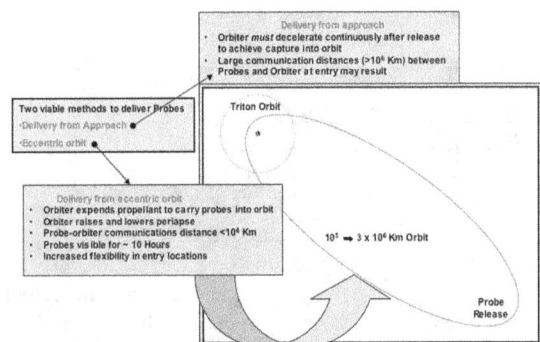

Technology and Engineering

A host of technology issues exist in the Neptune Orbiter with Probes mission design. These include the following:

Pressure Vessel Design. Even at 200 bars, the pressure vessel design is a challenge. Remember that the Pioneer Venus Probes were designed for 100 bars [13]. The question here is what type of technologies enable the fabrication of a 200 bar pressure vessel and penetrations (windows, inlets, feedthroughs, etc.). At the conference, a titanium alloy with carbon nanotubes was discussed as a potential light-weight pressure vessel material. This would, of course, save tremendous mass.

The Probe thermal design is also a challenge. Heaters will likely be required to maintain Probe temperatures above minimum limits from Orbiter separation to atmospheric entry. After entry, the thermal design must maintain electronics and science instrument temperatures within their operating ranges as thermal input increases due to atmosphere heating and energy dissipation within the Probe. The use of carbon nanotube passive heat pipes might prove useful since conventional heat pipes are not viable in a high-G Probe mission.

Deceleration Module. Significant work is required to define, design, develop and test materials appropriate for a high speed Neptune entry. High latitude entries will result in increased atmosphere-relative entry speeds, requiring high capability thermal protection system (TPS) materials.

Staging Systems. Although parachutes have been used successfully on the previous Probe and Probe/Lander missions (Pioneer Venus, Viking, Galileo, MER, etc.), they have required extensive development. Due to the inherent unreliability of parachutes, other staging techniques (separation of the Probe from the deceleration module) should be considered.

RF Link Design. The Radio Frequency (RF) link design, including frequency, power, and data rate, is dependent on the atmosphere model and the depth. Once the link is designed, there will be a need to design RF electronics with high efficiencies to limit thermal dissipation and provide the necessary power to allow reception of the Probe signal by the Orbiter.

Battery Design. The Probe battery design is of great concern in determining the total Probe electrical power budget. A primary consideration includes the power required during Probe coast (the period between Probe separation from the Orbiter and Probe pre-entry warm-up). The need for heaters may require batteries with high energy densities. The use of RHUs (radioisotope heater units), as were used on the Galileo and Huygens Probes, would help alleviate the electrical power requirement for heaters prior to Probe entry. The Probe battery requirements will also be driven by the depth to which the Probe must operate, the time to depth as dictated by the descent system, and the mass and power requirements of the Probe science and engineering payload.

Electronic Design. The use of low power electronic devices that can operate over a wide temperatures range should be considered. These devices should be used throughout the Probe, within both housekeeping and science electronics, to ease the thermal control requirements.

Probe Miniaturization. The deployment of multiple Probes into the Neptune atmosphere is likely to be realized only with considerable Probe miniaturization. This can be accomplished not only through the continued miniaturization of the required instruments and sensors, but also by integrating all Probe elements. Power, mass and volume efficiencies can be achieved by, for example, using a single processor for all Probe elements and by mounting most of the Probe subsystems in a common vacuum or pressurized vessel. Of course, multiple processors, with appropriate, autonomous switching logic, must be included in the design to guarantee the highest probability of mission success. An aggressive program to develop and test highly integrated miniature Probes capable of obtaining the required measurements is essential for this mission.

Entry Environment and TPS Testing

The Galileo Probe entered Jupiter at a velocity of about 47 km/s, and experienced a peak net heat flux (including the effects of ablation) of nearly 30 kW/cm^2. This was by far the most energetic planetary entry attempted in the history of spaceflight. Direct entry at Neptune will not be as stressing, but entry velocities will be in the range of 23-30 km/s, depending on the interplanetary trajectory and desired entry latitude. At these conditions peak heat fluxes of several thousand W/cm^2 will be encountered. By comparison the Shuttle Orbiter experiences heat fluxes on the order of 40 W/cm^2.

Barring significant advances in TPS technology, there are few materials that can withstand these heat fluxes effectively. The only material ever used in such an environment is fully dense carbon phenolic of the kind used on the Pioneer Venus and Galileo Probes. Unfortunately, heritage carbon phenolic can no longer be produced, since the heritage Rayon fabric is no longer manufactured. The Air Force is currently qualifying a new carbon phenolic material for ballistic missile applications. Once this qualification is complete, the new material can readily be adapted to planetary entry Probes.

However, even when a suitable material is created, the performance characteristics of the material will be different and must be characterized *in a relevant environment* to ensure that the TPS system will perform as expected. Act jet facilities provide the best test environment for planetary TPS, and such facilities have been used to flight-qualify the thermal protection materials for every NASA planetary Probe to date, including Mars Pathfinder, MER, Pioneer Venus, and Galileo. The environment experienced by a giant planet entry Probe is very different from that seen by Probes entering the inner planets. For example, 50% or more of the aerothermal heating will be due to radiation produced in the hot shock layer. Furthermore, the atmosphere of all giant planets consists primarily of hydrogen and helium, much different than the $N_2/O_2/CO_2$ atmospheres of the inner planets. Therefore, in order to test materials in flight relevant conditions, a facility is required that can provide a combination of convective and radiative heating in a H_2/He environment.

During the Galileo program an arc jet facility known as the Giant Planet Facility (GPF) was constructed at NASA Ames Research Center specifically for material testing. However, this facility was shut down and dismantled soon after the conclusion of Galileo testing. No facility currently exists that can meet the requirements for giant planet TPS qualification. But, since the net heat fluxes encountered at Neptune are likely much lower than those seen by Galileo, the full capabilities of the GPF may not be required. One alternative is to conduct much of the testing in existing facilities using air as the test gas, and supplement this with limited testing in the correct gas in the presence of radiative heating. This could be accomplished either via a reconstruction of the GPF, or with a (lower cost) subscale option, such as the Developmental Arc Facility (DAF) currently under construction at NASA Ames. Both options will be explored during the development of this mission concept. In any case, development of a suitable facility will take some time and must be accommodated in any giant planet entry Probe mission timeline.

Parachuteless Entry Option

Parachutes are used during atmospheric Probe entries for three main reasons: (1) to separate the hot aeroshell from the payload, (2) to provide stability in the subsonic regime, and (3) to slow descent through the atmosphere to enable science objectives. For this mission, it is desired to reach atmospheric pressures of 100 bars or greater. Preliminary studies have shown that it will take many hours to reach this depth with a parachute, which is undoubtedly too long to maintain radio contact with the Orbiter. For this reason it would be desirable to eliminate the parachute if possible. In order to do this, a design must be chosen that is unconditionally stable from the hypersonic to subsonic flow regimes. This can be done; for example, by a sphere-cone with a spherical base and a cone angle of 45° or less.

Another concern is separation of the aeroshell from the payload. This is important both to permit scientific instrument access to the atmosphere and to eliminate the aeroshell before the heat absorbed during the hypersonic entry is conducted into the payload. There have been other methods of aeroshell separation proposed, including shaped charges that split the aeroshell into "petals" that are then shed, but a high drag device is by far the best option. This study will look at several descent options, but at this point the most attractive option is to use a parachute only for aeroshell separation. Once separation is achieved the parachute would be cut free and the payload would continue its descent.

CONCLUSIONS

A complex Neptune Orbiter with Probes mission, that includes Triton Landers, will benefit considerably from development now under way to develop a nuclear electric power and propulsion capability for solar system exploration. Challenges abound in the development of the technologies necessary to support this ambitious mission. In addition, the extremely long mission duration, while daunting, must be viewed as path to completing a comprehensive study of Neptune, Triton and the other Neptunian moons. The time required to transverse nearly the entire solar system in order to complete a comprehensive Neptune study should be considered as an investment in a mission that will undoubtedly be conducted only once in this century or perhaps in the centuries to come.

REFERENCES

1. National Research Council, *An Integrated Strategy for Planetary Sciences 1995-2010*, National Academy Press, Washington, D.C. 1994.

2. Stern, S.A., et al., "Future Neptune and Triton Missions," Chapter in *Neptune and Triton*, Dale Cruikshank, editor, University of Arizona Press 1997.

3. Hammel, H.B., C.C. Porco, and K. Rages, "The Case for a Neptune Orbiter/Multi-Probe Mission", Innovative Approaches to Outer Planetary Exploration 2001-2020, Lunar and Planetary Institute, Houston, Texas, 21-22 February, 2001.

4. Solar System Exploration Roadmap, JPL Publication 400-1077, May, 2003. Available at http://spacescience.nasa.gov/admin/pubs/strategy/2003/index.html

5. National Academy of Science Decadal Survey for Solar System Exploration: An Integrated Exploration Strategy, Space Studies Board of the National Research Council (2003). Also available online at http://books.nap.edu/html/newfrontiers/0309084954.pdf.

6. Solar System Exploration Roadmap, 2003. Available at http://spacescience.nasa.gov/admin/divisions/se/SSE_Roadmap.pdf.

7. NASA (1999) Exploration of the Solar System: Science and Mission Strategy.

8. Porco, C. C. (1998) Report of the NASA SSES Astrophysical Analogs Campaign Strategy Working Group.

9. Atreya, S.K., Mahaffy, P.R., Niemann, H.B., Wong, M.H., Owen, T.C., Composition and Origin of the Atmosphere of Jupiter – An Update and Implications for the Extrasolar Giant Planets, *Planetary and Space Science*, 51, 105-112, 2003. (This paper is an update of an earlier comprehensive publication, "A Comparison of the Atmospheres of Jupiter and Saturn: Deep Atmospheric Composition, Cloud Structure, Vertical Mixing, and Origin," S. K. Atreya, M. H. Wong, T. C. Owen, P. R. Mahaffy, H. B. Niemann, I. de Pater, Th. Encrenaz, and P. Drossart, *Planet Space Sci., 47*, 1243, 1999.)

10. Atreya, S.K., *Atmospheres and Ionospheres of the Outer Planets and their Satellites*, Chapter 3, Springer-Verlag, New York-Berlin (1986).

11. Atreya, S.K., Wong, A.S., Coupled Clouds and Chemistry of the Giant Planets – A Case for Multiprobes, a chapter in *Outer Planets* (R. Kallenbach, T. Owen, T. Encrenaz, eds.), Kluwer Academic, (in press, 2004).

12. Wong, A.S., and S.K. Atreya, "Neptune's Cloud Structure," 2nd International Planetary Probe Workshop, NASA Ames Research Center, 23–26 Aug 2004.

13. Bienstock, B.J., Pioneer Venus and Galileo Probe Heritage, *Proceedings of the International Workshop Planetary Atmospheric Entry and Descent Trajectory Analysis and Science*, ESA SP-544, 37-45, 2004.

ATMOSPHERIC MODELS FOR AEROENTRY AND AEROASSIST

C. G. Justus[1], Aleta Duvall[1], Vernon W. Keller[2]

[1]NASA MSFC ED44/Morgan Research, Marshall Space Flight Center, AL 35812 (USA), Email: Jere.Justus@msfc.nasa.gov, Aleta.Duvall@msfc.nasa.gov
[2]NASA Marshall Space Flight Center, ED44, Marshall Space Flight Center, AL 35812 (USA), Email: Vernon.Keller@nasa.gov

ABSTRACT

Eight destinations in the Solar System have sufficient atmosphere for aeroentry, aeroassist, or aerobraking/aerocapture: Venus, Earth, Mars, Jupiter, Saturn, Uranus, and Neptune, plus Saturn's moon Titan. Engineering-level atmospheric models for Earth, Mars, Titan, and Neptune have been developed for use in NASA's systems analysis studies of aerocapture applications. Development has begun on a similar atmospheric model for Venus. An important capability of these models is simulation of quasi-random perturbations for Monte Carlo analyses in developing guidance, navigation and control algorithms, and for thermal systems design. Characteristics of these atmospheric models are compared, and example applications for aerocapture are presented. Recent Titan atmospheric model updates are discussed, in anticipation of applications for trajectory and atmospheric reconstruct of Huygens Probe entry at Titan. Recent and planned updates to the Mars atmospheric model, in support of future Mars aerocapture systems analysis studies, are also presented.

1. INTRODUCTION

Engineering-level atmospheric models have been developed, or are under development, for five of the eight possible Solar System destinations where aerocapture could be used. These include Global Reference Atmospheric Models (GRAMs) for Earth (GRAM-99) [1, 2], Mars (Mars-GRAM 2001) [3-6], Titan (Titan-GRAM) [7], Neptune (Neptune-GRAM) [8], and Venus-GRAM (under development). Physical characteristics of the various planetary atmospheres vary significantly. Likewise, significant variation is found in the amount of available data on which to base the respective engineering-level atmospheric models. The detailed characteristics of these models differ accordingly.

Earth-GRAM is based on climatology assembled from extensive observations by balloon, aircraft, ground-based remote sensing, sounding rockets, and satellite remote sensing. Details are provided in the GRAM User's Guide [1]. Mars-GRAM is based on climatologies of General Circulation Model (GCM) output, with details given in the Mars-GRAM User's Guide [3]. Mars-GRAM has been validated [4-6] by comparisons against observations made by Mars Global Surveyor, and against output from a separate Mars GCM. In contrast, data used to build Titan-GRAM and Neptune-GRAM are more limited, deriving primarily from Voyager observations and limited ground-based stellar occultation measurements. Titan-GRAM is based on data summarized in [9], while Neptune-GRAM was built from summaries of data contained in [10]. For Venus, a substantial amount of data has been collected from orbiter and entry probe observations. These have been summarized in the Venus International Reference Atmosphere (VIRA) [11], which forms the basis for Venus-GRAM (under development).

Fig. 1 shows the wide variety of temperature profiles encountered among the planets and Titan. For Earth, Venus, Mars, and Titan, height is measured from a reference surface (mean sea level on Earth). On Neptune, height is measured above the level at which

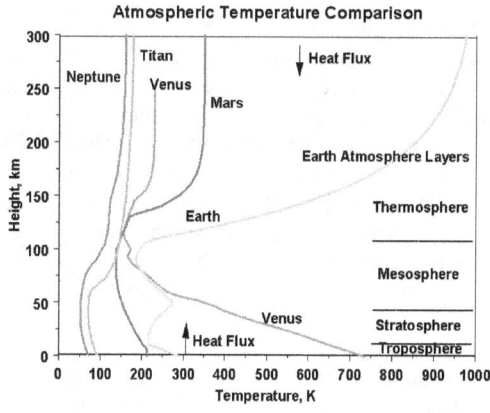

Fig. 1. Comparison of temperature profiles among the planets and Titan.

pressure is one bar (Earth normal sea-level pressure). All of the planets exhibit a troposphere region, where temperature decreases with altitude, indicative of heat flow upward from the surface (on average). All of the planets exhibit a thermosphere region, where (on average) temperature increases with altitude, because of absorption of heat flux from the Sun as it penetrates into the atmosphere. All of the planets have stratospheres, where temperature decrease above the surface diminishes, and remains relatively constant until the base of the thermosphere (Earth being the exception to this, where the presence of ozone and resultant atmospheric heating produces a local temperature maximum in Earth's stratosphere-mesosphere region).

For interest in aerocapture or aerobraking, atmospheric density is the most important parameter. Fig. 2 compares density profiles on the planets and Titan.

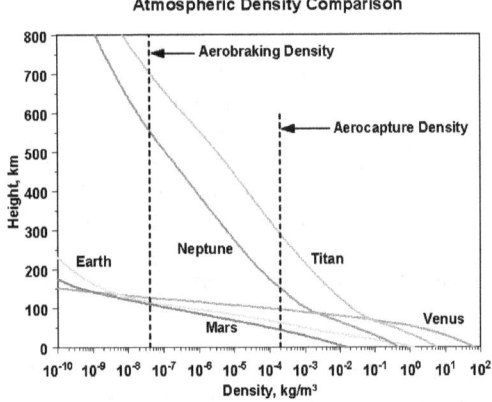

Fig. 2. Comparison of density profiles among the planets and Titan.

Vertical dashed lines in Fig. 2 indicate typical density values at which aerocapture or aerobraking operations would occur. Intersections of the aerocapture dashed line with various density curves shows that aerocapture would occur at a wide range of altitudes at the various destinations, varying from about 50 km at Mars to about 300 km at Titan. Aerobraking at Earth, Mars, and Venus would take place near, and just above, the 100 km level. At Neptune and Titan, aerobraking would be implemented near 550 km and 750 km, respectively.

Fig. 2 shows that density decreases fairly rapidly with altitude for the terrestrial planets (Venus, Earth, Mars), while it decreases rather slowly for Neptune and Titan. This effect is explained by differences in density scale height, H, for the various planets and Titan. Density decreases rapidly with altitude if H is small, while it decreases slowly if H is large. H is proportional to pressure scale height $[\,R\,T\,/\,(\,M\,g\,)\,]$. For the terrestrial planets, molecular mass M is large ($M \approx$ 29-44), so H is small. On Neptune, H is large because M is small for Neptune's hydrogen-helium atmosphere ($M \approx 2$). For Titan, H is large despite the high molecular mass of its atmosphere ($M \approx 29$), because its gravity is low.

2. BASIS FOR THE ATMOSPHERIC MODELS

In Earth-GRAM, Mars-GRAM, and Venus-GRAM, input values for date, time, latitude, longitude, etc. are used to calculate planetary position and solar position. In this manner, effects of latitude variation and seasonal and time-of-day variations can be computed explicitly. A simplified approach is adopted in Titan-GRAM and Neptune-GRAM, whereby these effects (as well as effects of relatively large measurement uncertainties for these planets) are represented within a prescribed envelope of minimum-average-maximum density versus altitude. Fig. 3 shows this envelope for Titan.

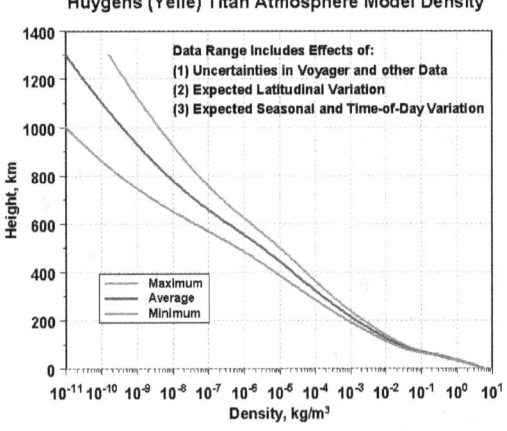

Fig. 3. Minimum, average, and maximum density profiles for Titan [9].

Engineering atmospheric model data developed for the Huygens entry probe [9] are used to define the Titan envelope. For Neptune, data from [10] are employed to generate a comparable minimum-maximum envelope, as shown in Fig. 4. A single model input parameter, Fminmax, allows the user of Titan-GRAM or Neptune-GRAM to select where within the min-max envelope a particular simulation will fall. Fminmax = -1, 0, or 1 selects minimum, average, or maximum conditions, respectively, with intermediate values determined by interpolation; i.e., Fminmax between 0 and 1 produces values between average and maximum. Effects such as variation with latitude along a given trajectory path can be computed using the appropriate representation of Fminmax variation with latitude.

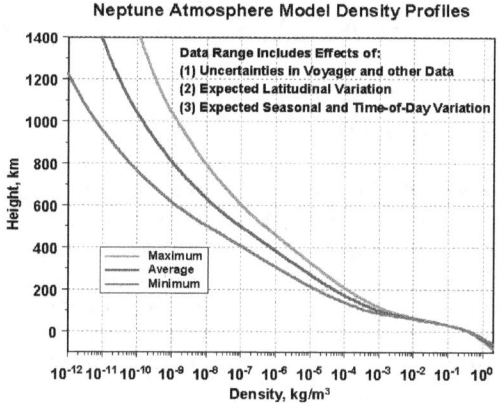

Fig. 4. Minimum, average, and maximum density profiles for Neptune from data in [10].

Since drag is proportional to density, density is the most important atmospheric parameter for aerocapture. Next most important is height variation of density (as characterized by density scale height). Density scale height is important in determining aerocapture corridor width, or entry angle range that allows the vehicle to achieve capture orbit without "skipping out" or "burning in". As discussed above, small density scale height means rapid change of density with altitude, which results in low corridor width. Large density scale height implies slow density change with altitude, and large corridor width.

Fig. 5 compares height profiles of density scale height among the planets and Titan. Aerocapture altitude (c.f. discussion of Fig. 2) is indicated by letter A in Fig. 5. This figure shows low density scale heights (4-8 km) at aerocapture altitudes for the terrestrial planets. Larger scale heights ($\approx$ 30-50 km) occur at aerocapture altitudes on Neptune and Titan.

3. TITAN-GRAM GCM OPTION

An option has recently been added for using Titan General Circulation Model (GCM) data as input for Titan-GRAM. The Titan GCM data used are from graphs in [12]. Upper altitudes for the Titan GCM option are computed using a parameterized fit to Titan exospheric temperatures, taken from graphs in [13].

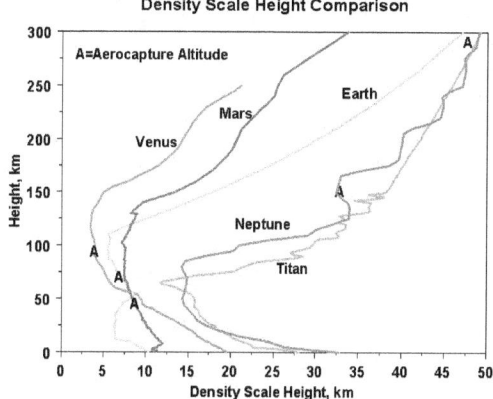

Fig. 5. Comparison of atmospheric density scale height among the planets and Titan.

Fig. 6 shows a height-latitude cross section of density, expressed as percent deviation from the mean, for Voyager encounter date November 12, 1980 (planetocentric longitude of Sun $Ls = 8.8°$), 00:00 GMT, longitude zero, local solar time 0.7 Titan hours. Fig. 7 compares vertical density profiles at latitude zero, local solar time 1 hour and 13 hours on the Voyager encounter date, with the Huygens Yelle [9] minimum-maximum density envelope from Fig. 3. This figure shows that the Titan GCM results correspond fairly closely with Yelle maximum conditions up to about 300 km altitude, and agree quite closely with Yelle average conditions (vertical line at 0 in Fig. 7) above about 500 km.

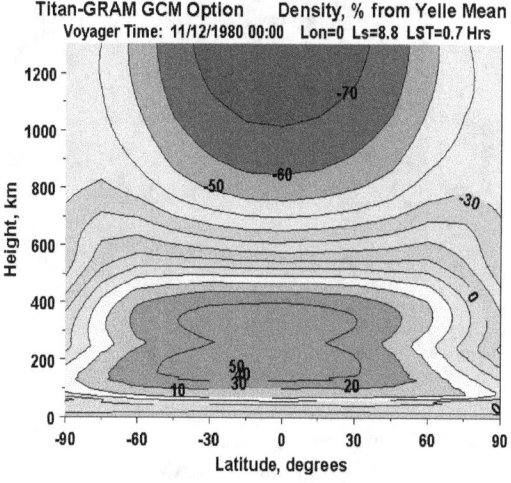

Fig. 6. Density (percent deviation from mean) versus height and latitude, using Titan-GRAM GCM option.

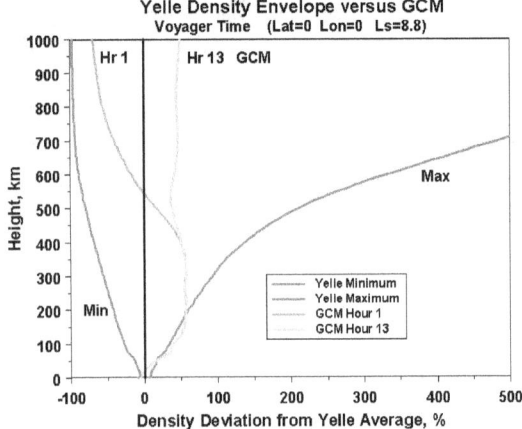

Fig. 7. Comparison of two selected Titan-GRAM density profiles (GCM option) with minimum-maximum envelope from Huygens Yelle model [9].

4. VENUS-GRAM DEVELOPMENT

Based on the Venus International Reference Atmosphere (VIRA) [11], Venus-GRAM is being developed and applied in ongoing Venus aerocapture performance analyses. Fig. 8 shows a plot of density (percent deviation from the mean) versus height and latitude from Venus-GRAM. Conditions in Fig. 8 are for $Ls = 90°$ and local solar time = 12 Venus hours.

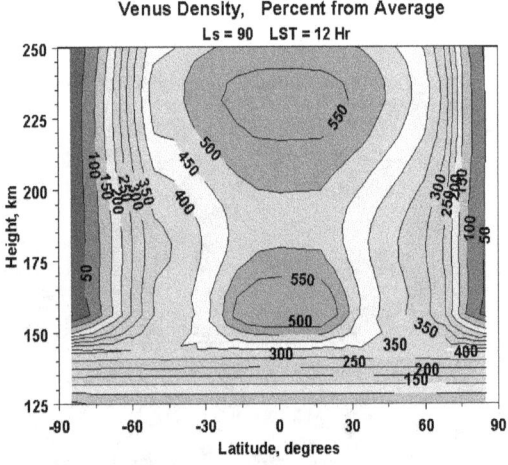

Fig. 8. Example height-latitude density cross section from Venus-GRAM.

Below about 100 km altitude on Venus, we find that temperature, density, and density scale height conditions are very uniform with both latitude and time of day. VIRA data below 100 km altitude vary only slightly with latitude and have no dependence on local solar time. Between 100 km and 150 km, VIRA data depend on local solar time, but not latitude. From 150 km to its top at 250 km, VIRA depends on solar zenith angle, which is affected by both latitude and local solar time.

5. NEW MARS-GRAM FEATURES

During Mars Global Surveyor (MGS) aerobraking operations, large density variations were observed between successive periapsis passes [14]. These appeared to be longitude-fixed or terrain-fixed waves, usually dominated by wave-2 or wave-3 components (wave-n meaning that n wavelengths fit around a 360° longitude circle). During Mars Odyssey aerobraking, similar large-amplitude density variations were observed. However, during some periods, Odyssey-observed density variations appeared to be traveling waves whose phase speed relative to a fixed longitude seemed to remain constant for a matter of a few days. Mars-GRAM 2001 has an option to represent terrain-fixed waves of the type observed by MGS. Work is underway to develop a new version of Mars-GRAM that will (among other features) include the option to allow user input values for phase speed of traveling wave components, of the type observed by Mars Odyssey.

Also during MGS and Odyssey aerobraking, it was observed that Mars-GRAM produced better correspondence with observed atmospheric density if the altitude scale of its input Mars Thermospheric General Circulation Model (MTGCM) data base was shifted (described as a "height offset"). New sets of Mars General Circulation Model (MGCM) and MTGCM data are being produced for use as input in the next Mars-GRAM update. These GCM model runs include better treatment of the matchup conditions (both mean conditions and upward wave fluxes) between the upper boundary of MGCM and lower boundary of MTGCM (at the 1.32 μbar level, near 80 km). A new non-local thermodynamic equilibrium (non-LTE) method for treating near-infrared heating and CO_2 15-micron cooling will also be employed in the MTGCM model runs. This methodology is based on a non-LTE model of López-Valverde and López-Puertas [15]. More realistic dynamics in both MGCM and MTGCM data sets is also anticipated from the use of latitude and seasonal variations of dust optical depth observed by MGS Thermal Emission Spectrometer (TES) in its mapping years 1 and 2. It is hoped that these new MGCM/MTGCM input data sets for Mars-GRAM will significantly lessen the need for height offset, and significantly improve the correspondence with observed densities during Mars-GRAM use in support of aerobraking operations for Mars Reconnaissance Orbiter.

6. NEW MARS-GRAM SLOPE-WIND FEATURE

For potential applications in preliminary site screening for Mars landers, a new slope wind feature is being developed for Mars-GRAM. Slope winds are computed in Mars-GRAM from a diagnostic (algebraic) relationship based on [16]. This approach differs from mesoscale models, such as Mars Regional Atmospheric Model System (MRAMS) [17], and Mars Mesoscale Model version 5 (MMM5) [18], which use prognostic, full-physics solutions to the time- and space-dependent differential equations of motion. As such, slope winds in Mars-GRAM will be consistent with its "engineering-level" approach, and will be extremely easy and fast to evaluate, compared with mesoscale model solutions. Mars-GRAM slope winds are not being suggested as a replacement for more sophisticated, full-physics mesoscale models, but may have value, particularly for preliminary screening of large numbers of candidate landing sites for future Mars missions.

Terrain slopes used in the slope wind model are computed from 0.5° _ 0.5° Mars Orbiter Laser Altimeter (MOLA) topography. Mars-GRAM slope winds will be added to winds from MGCM, which have a resolution of 7.5° _ 9° in latitude and longitude. The Mars-GRAM slope wind model will thus add significantly higher resolution information about possible near-surface winds than is provided by MGCM.

Fig. 9 shows Mars-GRAM slope winds, evaluated at a level 2 km above local terrain height for the Gusev Crater area, at the date and time of Rover Spirit landing. If this wind field is valid, then Spirit would have experienced up to ~ 25 m/s winds "opposing" its entry into Gusev Crater near an altitude of 1-2 km above surface level. Spirit experienced significant turbulence or winds during its descent (Prasun Desai, private communication), causing it to fire its Transverse Impulse Rocket System to correct for off-vertical firing of its main retrorockets, and to reduce its lateral impact speed.

Fig. 10 shows MOLA terrain heights in a portion of the eastern end of Valles Marineris, used in these preliminary tests of the slope wind model.

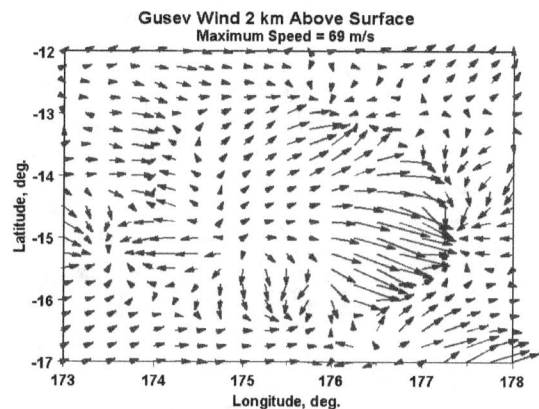

Fig. 9. Slope wind vectors at Gusev Crater, 2 km above surface, for date and time of Spirit landing.

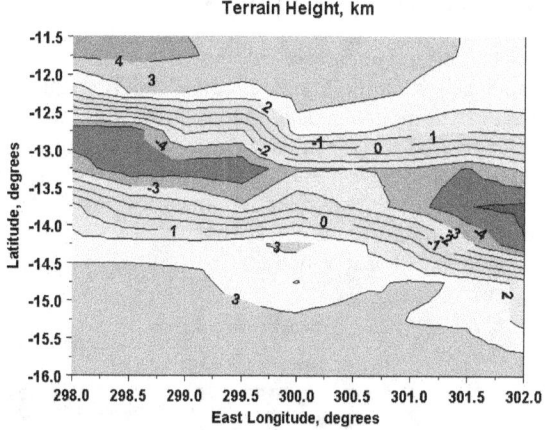

Fig. 10. Terrain Heights in portion of Valles Marineris region.

Fig. 11 shows northward component of Mars-GRAM slope winds, evaluated at a level 1 km above local terrain height for the study area shown in Fig. 10. The season assumed is Ls = 0° (northern spring equinox) at local time 14 hours. Comparison of Fig. 11 with Fig. 10 shows that the major pattern for slope winds at this time is northward and upward along the north wall of the valley and southward and upward along the south wall (i.e. upslope flow on both valley walls), a reasonable situation for early afternoon local time. These examples of test output from the new Mars-GRAM slope wind model may be compared with wind simulations from Mars mesoscale models, presented by Rafkin and Michaels [19] and Kass, et al. [20].

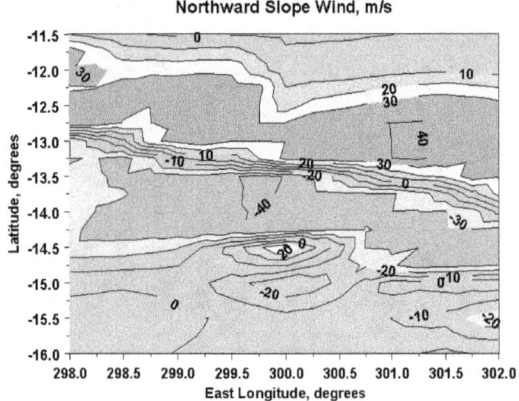

Fig. 11. Northward slope winds at Ls = 0° and LST = 14 hours, 1 km above terrain surface.

7. PERTURBATION MODELS

An important feature of all the GRAM atmospheric models is their ability to simulate "high frequency" perturbations in density and winds, due to such phenomena as turbulence and various kinds of atmospheric waves. As illustrated in Fig. 12, Earth-GRAM altitude, latitude, and monthly variations of perturbation standard deviations are based on a large climatology of observations.

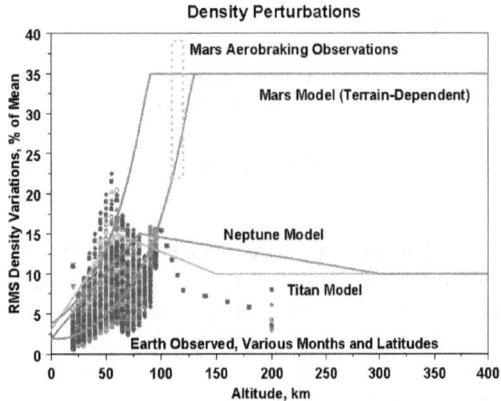

Fig. 12. Height variation of density perturbation model standard deviations for Earth, Mars, Titan, and Neptune.

For Titan-GRAM and Neptune-GRAM, perturbation standard deviations are computed from an analytical expression for gravity wave saturation conditions, explained more fully in [7]. As shown in Fig. 12, the resulting vertical profiles of standard deviations for Titan and Neptune are not dissimilar to Earth observations, when expressed as percent of mean density. For Mars-GRAM, a similar gravity wave saturation relation is used to estimate density perturbation standard deviations, except that effects of significant topographic variation on Mars are also taken into account. Up to about 75 km altitude, the Mars model density standard deviations are also fairly consistent with Earth observations. By about 100 km to 130 km altitude, Mars model density standard deviations increase to about 20% to 35% of mean value, consistent with observed orbit-to-orbit density variations observed by Mars Global Surveyor and Mars Odyssey.

A typical application of the Neptune-GRAM perturbation model is shown in Fig. 13.

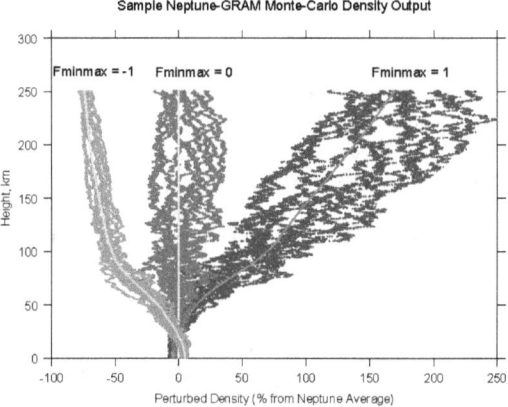

Fig. 13. Sample Monte Carlo density perturbations from Neptune-GRAM, expressed as percent deviation from Neptune mean value.

Neptune-GRAM was recently utilized in Neptune aerocapture systems analysis studies. The chosen aerocapture design reference mission included simulations which involved capture into a highly eccentric orbit, to allow the orbiter to periodically visit Triton for scientific observations. The ability to successfully aerocapture into such an eccentric orbit depends very significantly on details of Monte Carlo trajectory simulations, particularly on atmospheric density variations such as illustrated in Fig 13. For such an eccentric orbit, there is relatively little margin for error between a captured orbit and one which exceeds escape velocity upon atmospheric exit, a result which could ultimately lead to mission failure. Neptune-GRAM was used to define an aerocapture corridor width consistent with mission success.

8. CONCLUSIONS

The engineering-level atmospheric models presented here are suitable for a wide range of mission design, systems analysis, and operations tasks. For orbiter missions, applications include analysis for aerocapture or aerobraking operations, analysis of station-keeping issues for science orbits, analysis of orbital lifetimes for end-of-mission planetary protection orbits, and atmospheric entry issues for accidental break-up and burn-up scenarios. For lander missions to Venus, Mars and Titan, and for Earth-return, applications include analysis for entry, descent and landing (EDL), and guidance, navigation and control analysis for precision landing and hazard avoidance. Perturbation simulation capabilities of these models make them especially useful in Monte Carlo analyses for design and testing of guidance, navigation, and control algorithms, and for heat loads analysis of thermal protection systems.

9. ACKNOWLEDGMENTS

The authors gratefully acknowledge support from the NASA Marshall Space Flight Center In-Space Propulsion Program. Particular thanks go to Bonnie James (MSFC), Manager of the Aerocapture Technology Development Project, to Michelle M. Munk (LaRC/MSFC), Lead Systems Engineer for Aerocapture, and to Melody Herrmann (MSFC), team lead and Mary Kae Lockwood (LaRC), technical lead for the Titan/Neptune Systems Analysis study. Model user feedback and suggestions from the following individuals are also greatly appreciated: Dick Powell, Brett Starr, and David Way (NASA LaRC), and Claude Graves, Jim Masciarelli, Lee Bryant, Tim Crull, and Tom Smith (NASA JSC). External review comments from Prof. Darrell Strobel (Johns Hopkins University) were especially helpful.

10. REFERENCES

1. Justus, C. G., and Johnson, D. L., "The NASA/MSFC Global Reference Atmospheric Model - 1999 Version (GRAM-99)", NASA/TM-1999-209630, 1999.
2. Justus, C.G., Duvall, A. L., and Johnson, D. L., "Earth Global Reference Atmospheric Model and Trace Constituents", *Advances in Space Research*, in press.
3. Justus, C. G., and Johnson, D. L., "Mars Global Reference Atmospheric Model 2001 Version (Mars-GRAM 2001) Users Guide", NASA/TM-2001-210961, April, 2001.
4. Justus, C. G., Duvall, A. L., and Johnson, D. L., "Mars-GRAM Validation with Mars Global Surveyor Data", *34th COSPAR Scientific Assembly*, Houston, Texas, Paper C3.3-0029-02, October, 2002.
5. Justus, C. G., Duvall, A. L., and Johnson, D. L., "Global MGS TES Data and Mars-GRAM Validation", *Advances in Space Research*, in press.
6. Justus, C. G., Duvall, A. L. and Johnson, D. L., "Mars Global Reference Atmospheric Model (Mars-GRAM) and Database for Mission Design", *International Workshop on Mars Atmosphere Modeling and Observations*, Granada, Spain, January, 2003.
7. Justus, C.G., Duvall, A. L., and Johnson, D. L., "Engineering-level model atmospheres for Titan and Neptune", *39th AIAA/ASME/SAE/ASEE Joint Propulsion Conference*, Huntsville, Alabama, Paper AIAA-2003-4803, July, 2003.
8. Justus, C.G., Duvall, A. L., and Keller, V. W., "Engineering-level model atmospheres for Titan and Mars", *International Workshop on Planetary Probe Atmospheric Entry and Descent Trajectory Analysis and Science*, Lisbon, Portugal. October, 2003.
9. Yelle, R.V., Strobel, D. F., Lellouch, E., and Gautier, D., " Engineering Models for Titan's Atmosphere", in *Huygens Science, Payload and Mission*, ESA SP-1177, August, 1997.
10. Cruikshank, D.P. (ed.), *Neptune and Triton*, University of Arizona Press, Tucson, 1995.
11. Kliore, A. J., Moroz, V. I., and Keating, G. M. (eds.), "The Venus International Reference Atmosphere", *Advances in Space Research*, vol. 5, no. 11, 1985, Pergamon Press, Oxford, 1986, pp. 1-304.
12. Hourdin, F., Talagrand, O., Sadourny, R., Courtin, R., Gautier, D., and McKay, C.P., "Numerical simulation of the general circulation of the atmosphere of Titan", *Icarus*, vol. 117, no. 2, Oct. 1995, pp. 358-74.
13. Mueller-Wodarg, I. C. F., "The Application of General Circulation Models to the Atmospheres of Terrestrial-Type Moons of the Giant Planets", in *Comparative Atmospheres in the Solar System*, American Geophysical Union, 2002.
14. Keating, G.M., Bougher, S. W., Zurek, R. W., Tolson, R. H., Cancro, G. J., Noll, S. N., Parker, J .S., Schellenberg, T. J., Shane, R. W., Wilkerson, B. L., Murphy, J. R., Hollingsworth, J. L., Haberle, R. M., Joshi, M., Pearl, J. C., Conrath, B J., Smith, M. D., Clancy, R .T., Blanchard, R. C., Wilmoth, R. G., Rault, D. F., Martin, T. Z., Lyons, D. T., Esposito, P. B., Johnston, M. D., Whetzel, C. W., Justus, C. G., Babicke, J. M., "The structure of the upper atmosphere of Mars: in situ accelerometer measurements from Mars Global Surveyor", *Science*, **279**(5357), 1672-6, 1998.
15. López-Valverde, M. A. and López-Puertas, M., "A non-local thermodynamic equilibrium radiative transfer model for infrared emissions in the atmosphere of Mars. 1. Theoretical basis and nighttime populations of

vibrational states", *Journal of Geophysical Research,* **99**, 13, 093-13,115, 1994.

16. Ye, Z. J., Segal, M., and Pielke, R. A., "A comparative study of daytime thermally induced upslope flow on Mars and Earth", *Journal of the Atmospheric Sciences*, **47**(5), 612-628, 1990.

17. Rafkin, S. C. R., Haberle, R. M., and Michaels, T. I., "The Mars Regional Atmospheric Modeling System: model description and selected simulations", *Icarus*, **151**, 228-256, 2001.

18. Toigo, A. D., and Richardson, M. I., "A mesoscale model for the Martian atmosphere", *Journal of Geophysical Research*, **107**(E7), 3-1-21, 2002.

19. Rafkin, S. C. R. and Michaels, T. I., "Meteorological predictions for 2003 Mars Exploration Rover high-priority landing sites", *Journal of Geophysical Research*, **108**(E12), 32-1-22, doi:10.1029/2002JE002027, 2003.

20. Kass, D. M., Schofield, J. T., Michaels, T. I., Rafkin, S. C. R., Richardson, M. I., and Toigo, A. D., "Analysis of atmospheric mesoscale models for entry, descent, and landing", *Journal of Geophysical Research*, **108**(E12), 31-1-10, doi:10.1029/2003JE002065, 2003.

An Investigation of Aerogravity Assist at Titan and Triton for Capture into Orbit About Saturn and Neptune
2nd International Planetary Probe Workshop
August 2004, USA

Philip Ramsey[1] and James Evans Lyne[2]

[1]Department of Mechanical, Aerospace and Biomedical Engineering
The University of Tennessee
Knoxville, TN 37996 USA
pramsey@utk.edu

[2]Department of Mechanical, Aerospace and Biomedical Engineering
The University of Tennessee
Knoxville, TN 37996 USA
jelyne@utk.edu

ABSTRACT

Previous work by our group has shown that an aerogravity assist maneuver at the moon Titan could be used to capture a spacecraft into a closed orbit about Saturn if a nominal atmospheric profile at Titan is assumed. The present study extends that work and examines the impact of atmospheric dispersions, variations in the final target orbit and low density aerodynamics on the aerocapture maneuver. Accounting for atmospheric dispersions substantially reduces the entry corridor width for a blunt configuration with a lift-to-drag ratio of 0.25. Moreover, the choice of the outbound hyperbolic excess speed (with respect to Titan) strongly influences the corridor width. Given the influence of these two parameters, certain mission scenarios may be feasible using a blunt aeroshell, while other mission designs would likely require a biconic vehicle with a higher lift-to-drag ratio. Preliminary simulations indicate that the same technique may be feasible for capture into orbit about Neptune using the tenuous atmosphere of Triton.

1. INTRODUCTION

Aerocapture and aerogravity assist maneuvers have long been recognized as methods by which otherwise impractical interplanetary missions could be accomplished. Our group recently proposed that an aerogravity assist maneuver at the moon Titan could be used to capture a probe into orbit about Saturn, using an aeroshell with a with a low to moderate lift-to-drag ratio (0.25 to 1.0).[1] This approach provides for capture into orbit about the gas giant, while avoiding the very high entry speeds and aerothermal heating environment inherent to a trajectory thru the atmosphere of Saturn itself.

Titan is unique among moons in the solar system in that it has an atmosphere considerably thicker than Earth's, with a ground level density of about 5.44 kg/m^3. Moreover, its atmosphere extends much higher above the surface than Earth's atmosphere, with the density at 800 km above the surface being approximately the same as that on Earth at 132 km.

Titan has a near-circular, equatorial orbit about Saturn at a radius from the planetary center of 1.22 (10^6) km and an orbital velocity of approximately 5.57 km/s. This orbit is well outside the ring system, which extends in the equatorial plane to a radius of approximately 480,000 km. A wide range of target orbits about Saturn can be achieved by means of a Titan aerogravity assist maneuver. The final orbit will depend on both the orientation and the magnitude of the outbound hyperbolic excess speed with respect to Titan after the AGA maneuver.

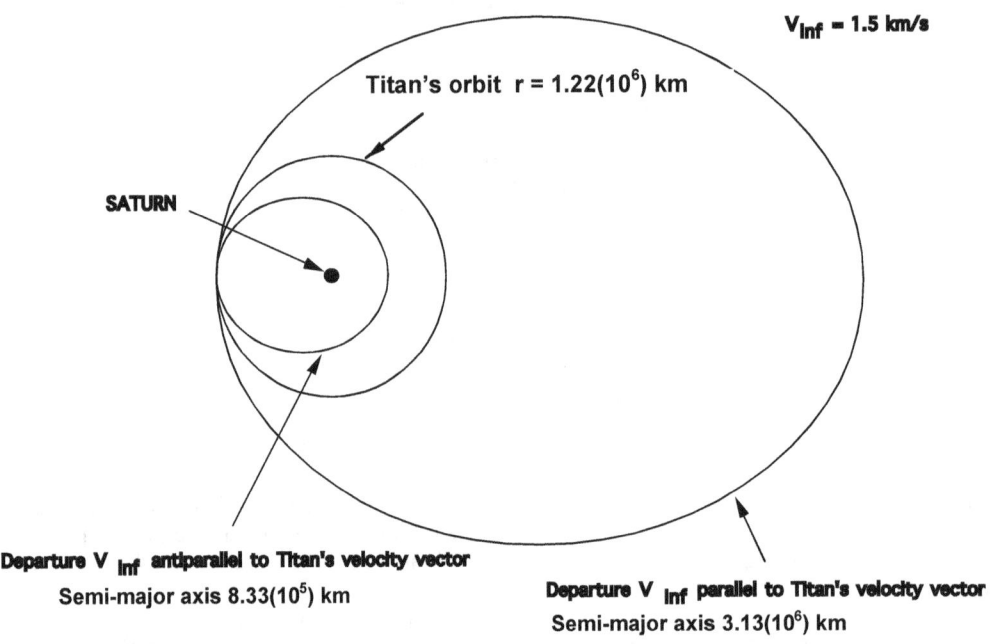

Figure 1. Potential variation in Final Saturn Orbit for an Outbound V_{inf} of 1.5 km/s

By lining up the outbound V_{inf} with Titan's orbital velocity vector, an orbit can be reached with a periapsis radius equal to that of Titan's orbit and a higher apoapsis. If the outbound V_{inf} is in the opposite direction, the final orbit will have an apoapsis radius equal to that of Titan's orbit and a lower periapsis radius (Fig. 1). An orbital periapsis of approximately 160,000 km would allow the probe to pass through the gap between rings F and G, a maneuver which the Cassini spacecraft recently accomplished without any apparent damage. Achieving this periapsis radius would require an outbound V_{inf} of approximately 2.89 km/s, opposite in direction to Titan's orbital velocity vector. However, since Titan's orbit lies virtually in the same plane as Saturn's rings, any mission using this strategy will be complicated by the need not to fly through the debris field. If Titan's orbital velocity vector and the outbound V_{inf} are co-linear, the final spacecraft trajectory about Saturn would lie very near or in the ring plane. Directing an outbound V_{inf} of 3 km/s 15 degrees above the ring plane would result in a final orbit about Saturn with its apoapsis coincident with its ascending node at Titan's orbital distance, its periapsis coincident with the descending node and in the Cassini division (between rings F and G), and an orbital inclination of approximately 16 degrees (Fig. 2 and 3).

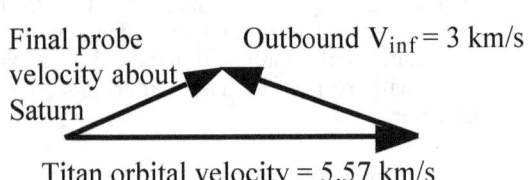

Figure 2. Velocity vector diagram for insertion into inclined orbit about Saturn

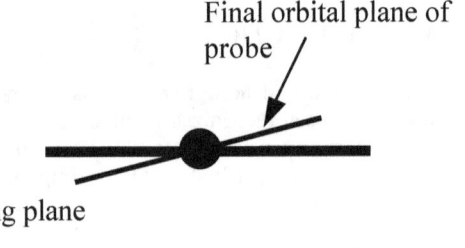

Figure 3. Edge view of possible final orbital geometry

2. METHODOLOGY

The atmospheric entry corridor of an aerocapture maneuver is bounded by the overshoot and undershoot limits. These represent the shallowest and steepest angles at which the spacecraft can enter the atmosphere and successfully complete a capture into the desired orbit. The width of the corridor depends on many factors, including the entry velocity, the atmospheric density profile, the vehicle's lift-to-drag ratio and limits which may be imposed on the vehicle's deceleration or aerodynamic heating. For this study, trajectory simulations were run using the three-dimensional version of POST.[2] No constraints were placed on the vehicle's deceleration or aerothermal heating. All simulations were begun at an altitude of 905 km at zero degrees latitude and with a due east azimuth. Entry angles are measured at the 905 km altitude.

Overshoot boundaries were found by directing the vehicle's lift vector toward Titan's surface (a vehicle roll angle of 180 degrees) and adjusting the entry angle until the desired outbound V_{inf} was achieved. Various target values of the outbound V_{inf} were evaluated, corresponding to different final orbits about Saturn. The undershoot boundaries were found by flying a full lift up trajectory and adjusting the entry angle until the target outbound V_{inf} was achieved.

Nominal, minimum and maximum atmospheric density profiles seen in Fig. 4 were based on the work by Yelle.[3] Atmospheric winds were not considered, nor were horizontal density dispersions.

For Titan, four vehicle configurations were evaluated. The first was a blunt aeroshell with an L/D of 0.25 that has been considered by other investigators for use on a conventional Titan aerocapture maneuver.[4] An ellipsled with an L/D of 0.39, an Apollo-derived capsule flying at a high angle of attack (L/D of 0.482), and a biconic with an L/D of 1.0 were also evaluated. (It is recognized that flank heating may prevent an Apollo configuration from flying at such a high alpha, but the configuration was chosen simply as a "place holder" to determine the corridor available to a vehicle with an L/D in this range.) The vehicle mass was assumed to be 600 kg, and a reference area of 12.56 m^2 (corresponding to a base radius of 2 meters) was used for all three blunt configurations. For the biconic, a base radius of 1 m was used, giving a reference area of 3.14 m^2.

3. RESULTS

The corridor width for the capsule with an L/D of 0.25 is shown in Fig. 5 as a function of entry velocity for both the nominal and extreme atmospheric models and for three target values of the outbound V_{inf}. In general, the high density atmosphere produced a shallower undershoot bound than the nominal atmosphere did, and the low density atmosphere produced a steeper overshoot bound than was allowed with the nominal atmospheric profile. Since the density profile which will be encountered is not known prior to entry, it is necessary to consider the corridor width with these narrower limits. The corridor defined in this manner must be wide enough to allow for off-nominal atmospheric entry angles, knowledge errors and uncertainties in the expected aerodynamic performance of the vehicle. Current estimates indicate that insertion angle errors of +/- 0.9 degrees are to be expected.[5] Therefore, a corridor width of 2.0 degrees is the minimum considered acceptable for this study. Improvements in interplanetary navigation capabilities that result in a more precise insertion angle would allow this corridor requirement to be relaxed.

Our original study considered only an outbound V_{inf} of 1.5 km/s in determining the corridor bounds (Ref. 1). From Fig. 5 it is apparent that an increase in the targeted outbound V_{inf} from 1.5 to 3.0 km/s substantially reduces the aerodynamic corridor width. This finding follows the trend seen in conventional aerocapture studies, where corridor widths often decrease as the target apoapse (and orbital energy) increase. This reflects the decreased control the vehicle's aerodynamics can exert over the trajectory as the duration of the atmospheric pass and the required energy loss are reduced.

Thus, if we assume that two degrees of corridor width are required to allow for off-nominal entry angles and aerodynamic dispersions, the choice of entry velocity and outbound V_{inf} will determine whether a given L/D provides a sufficient corridor. If a target orbit about Saturn with a periapsis in the Cassini division is desired, the outbound V_{inf} will need to be near 3 km/s, and a lift-to-drag higher than 0.25 will be required.

Fig. 6 shows corridor widths for all four vehicles, assuming an outbound V_{inf} of 3.0 km/s (this is considered the most likely target value of V_{inf}, since it provides a close approach to Saturn, opportunities for repeated Titan flybys and a close view of the ring system). From this Fig., it is apparent that the Ellipsled (with an L/D of 0.39) can provide a satisfactory corridor only at entry speeds of 9.5 km/s or more. The modified Apollo configuration at an angle of attack of 30 degrees (L/D of 0.482) provides a sufficient corridor width at an entry speed of 8 km/s or higher. As the entry speed decreases, vehicle configurations with higher lift-to-drag ratios (such as a biconic) become necessary unless better targeting of the atmospheric entry angle can be achieved. The arrows in Fig 6 indicate the reduction in corridor width caused by atmospheric uncertainty, with the solid lines representing the nominal atmosphere and the dashed lines showing the results for Yelle's low and high density profiles.

While all the results presented up to this point have assumed continuum aerodynamics, we conducted an evaluation of the impact of free molecular and transitional aerodynamics on vehicle trajectories and corridor bounds. For the vehicles considered here, there was no appreciable difference between results obtained neglecting and accounting for low-density influences on aerodynamic characteristics. We assume this results from the fact that the vehicles' deceleration almost exclusively occurs in the continuum flow regime (Fig. 7)

4. POTENTIAL APPLICATION AT NEPTUNE AND TRITON

NASA has conducted fairly detailed studies in recent years of aerocapture at Neptune;[6,7] an important conclusion of this work has been that the required aeroshell would have a much higher mass fraction than those considered for use at Mars, Earth or Titan, with 50% or more of the total probe mass being devoted to the aeroshell. This situation greatly decreases the appeal of aerocapture for Neptune missions.

Preliminary calculations indicate that the approach discussed in this paper for use at Titan/Saturn may also be feasible for capturing a probe into orbit about Neptune using an aerogravity assist maneuver at the moon Triton. This maneuver would be substantially different from the Titan scenario, in that the atmosphere of Triton is extremely tenuous, with a surface pressure of approximately 16 microbars and a pressure at 48 km altitude of 2 microbars (Fig. 8).[8] Despite the very thin atmosphere, trajectory simulations show that either a blunt aeroshell (L/D = 0.25) or a ballute could be used to accomplish a capture into orbit about Neptune, assuming the vehicle encounters atmospheric conditions similar to those shown in Fig. 8. Figures 9 shows the corridor width for a 600 kg aeroshell (L/D of 0.25, reference area of 12.56 m^2), with the vehicle targeted to an atmospheric exit velocity of 4.8 km/s. As can be seen, the corridor width for this vehicle is unlikely to prove adequate for the maneuver in light of the high degree of temporal variability in Triton's atmospheric density profile, In addition, the vehicle often flies to altitudes less than 10 km above Triton's surface during its atmospheric trajectory. Therefore, we also considered the use of a non-releasing ballute with a reference area of 100 m^2. Without modulation of the release time, variation in the atmospheric entry angle produces alterations in the outbound energy as shown in Fig 10. The ballute has a significant advantage in that it flies at very high altitudes, thereby providing a greater "margin of error" with regard to surface impacts.

4. CONCLUSIONS AND FUTURE WORK

The use of an aerogravity assist maneuver at Titan to capture a vehicle into orbit about Saturn appears to offer the potential for a less severe aerothermal enviroment than a direct capture using Saturn's atmosphere, while allowing for a variety of final orbital geometries. Depending on the choice of final Saturn orbit and the atmospheric entry speed at Titan, blunt aeroshells or vehicles with mid-range lift-to-drag ratios (such as biconics) offer adequate entry corridor width, even when accounting for atmospheric dispersions. Low density aerodynamics seem to have minimal impact on typical trajectories, but the choice of outbound V_{inf} strongly influences entry corridors.

Future studies of the topic must address optimal approach geometries, aerodynaic heating during the Titan atmospheric trajectory, thermal protection system requirements and TPS mass fractions. Guidance algorithms must be developed that can target both the magnitude and direction of the outbound hyperbolic excess speed in order to achieve the desired orbit about Saturn. In addition, the use of ballutes for capture at both the Titan/Saturn and

Neptune/Triton systems should be further examined.

5. REFERENCES

1) Ramsey, P. and Lyne, J.E., "An Investigation of Titan Aerogravity Assist or Capture into Orbit About Saturn," AAS Paper AAS-03-644, presented at the AAS/AIAA Astrodynamics Scecialists Conference, Big Sky, Montana, Aug. 2003.

2) Brauer, G.L. at al, "Program to Optimize Simulated Trajectories (POST)," NASA contract report NAS1-18147, Sept. 1989.

3) Yelle, R.V., Strobel, D.F., Lellouch, E. and Gautier, D., "Engineering Models for Titan's Atmosphere." In Huygens: Science, Payload and Mission," European Space Agency, ESA SP 1177, pp.243-256, 1997.

4) Way, D.W., Powell, R.W., Edquist, K.T., Masciarelli, J.P. and Starr, B.R., "Aerocapture Simulation and Performance for the Titan Explorer Mission," AIAA paper 2003-4951, presented at the 39th AIAA/ASME/SAE/ASEE Joint Propulsion Conference, July 2003.

5) Private communication, Jeff Hall, Jet Propulsion Laboratory, August 2004.

6) Lemmerman, L. and Wercinski, P., "Small Neptune Orbiter Using Aerocapture," Space Technology and Applications International Forumn (STAIF-97), Albuquerque, NM, Jan. 1997.

7) Wercinski, Paul, Chen, Y.K., Loomis, M., Tauber, M., McDaniel, R., Wright, M., Papadopoulos, P., Allen, G., and Yang, L., "Neptune Aerocapture Vehicle Preliminary Design," AIAA Atmospheric Flight Mechanics Conference, Monterey, CA, Aug. 2002.

8) Elliot, J.L. et al, "The Prediction and Observation of the 1987 July 18 Stellar Occultation by Triton: More Evidence for Distortion and Increasing Pressure in Triton's Atmosphere," *Icarus*, 148, pp. 347-369, 2000.

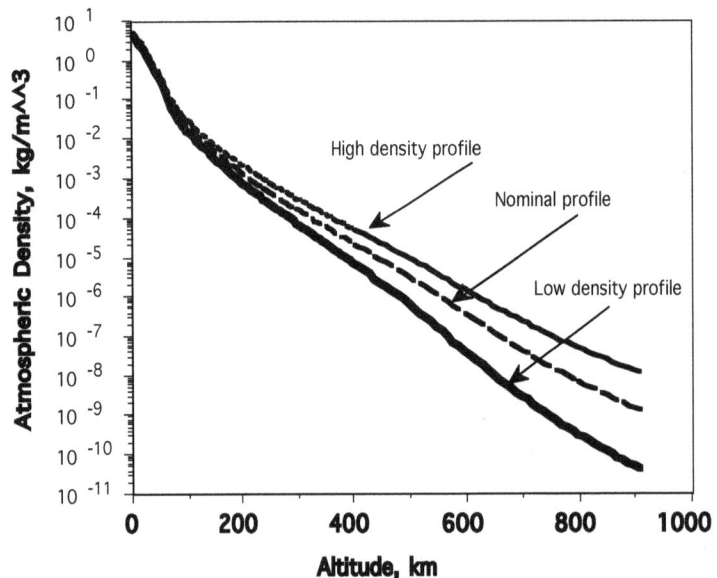

Figure 4. Titan atmospheric density profiles from Yelle (Ref. 4)

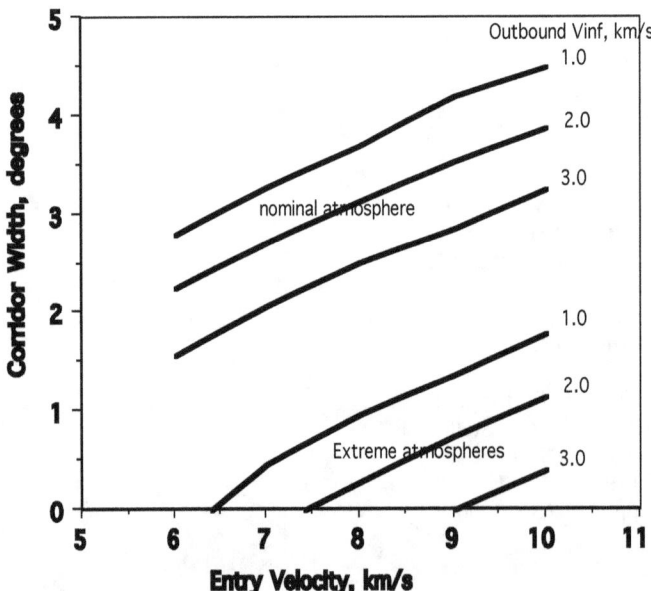

Figure 5. Corridor width at Titan vs entry velocity for a probe with an L/D of 0.25

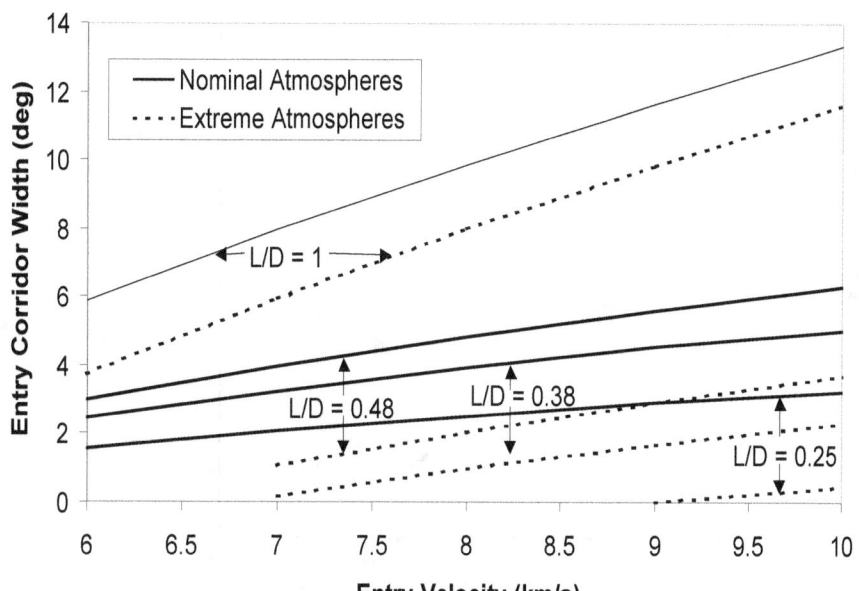

Figure 6 Titan/Saturn AGA maneuver corridor width vs. entry angle for various vehicles

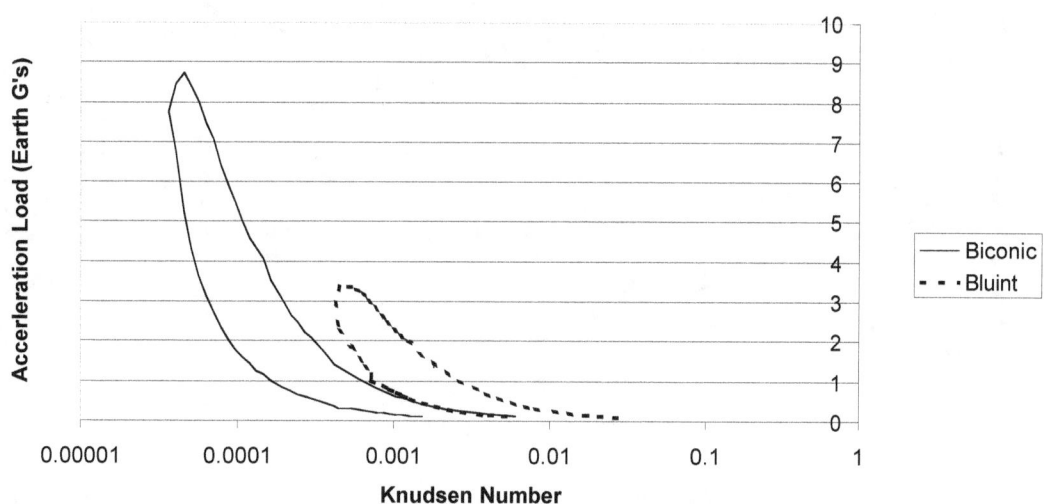

Figure 7. Acceleration Load vs. Knudsen Number

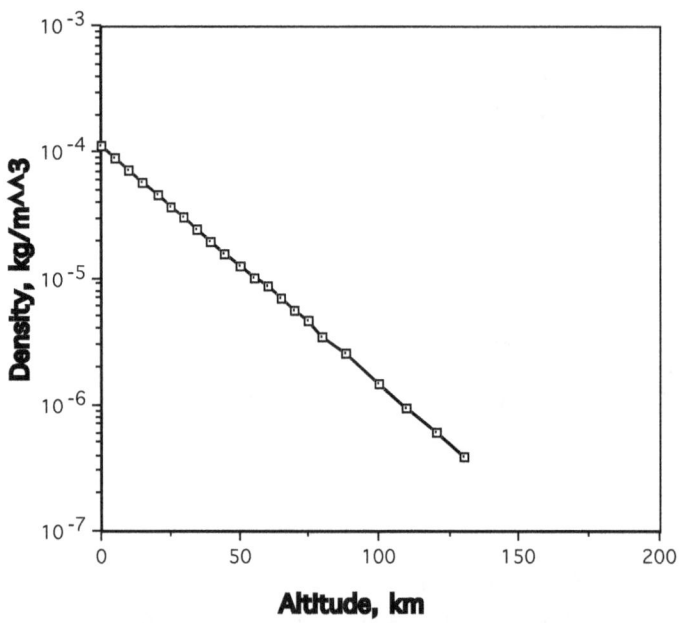

Figure 8. Triton atmospheric density profile

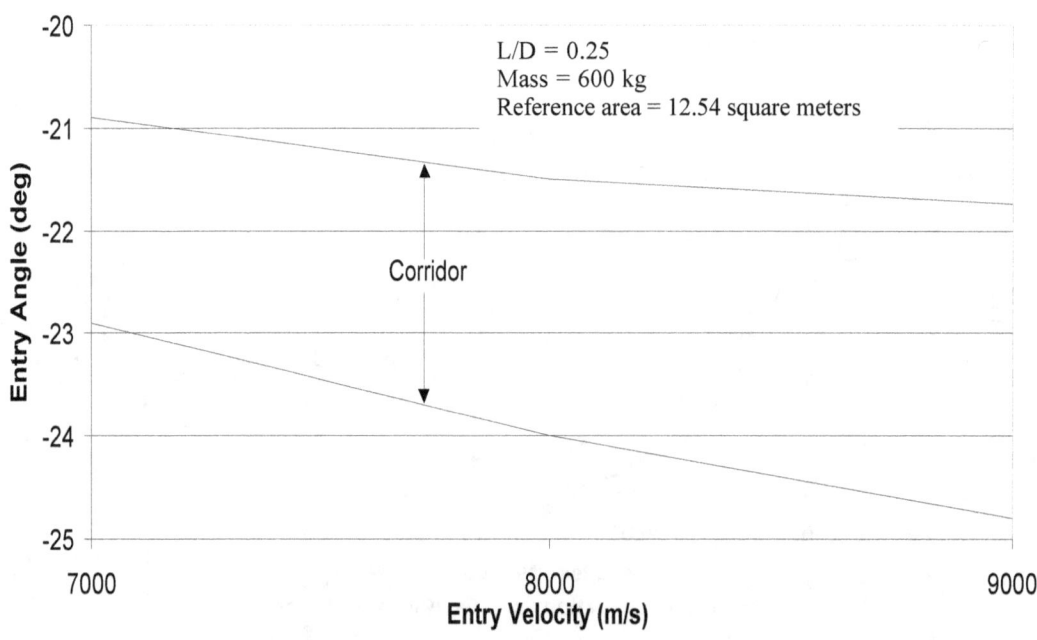

Figure 9. Corridor Bounds for Triton/Neptune AGA Maneuver

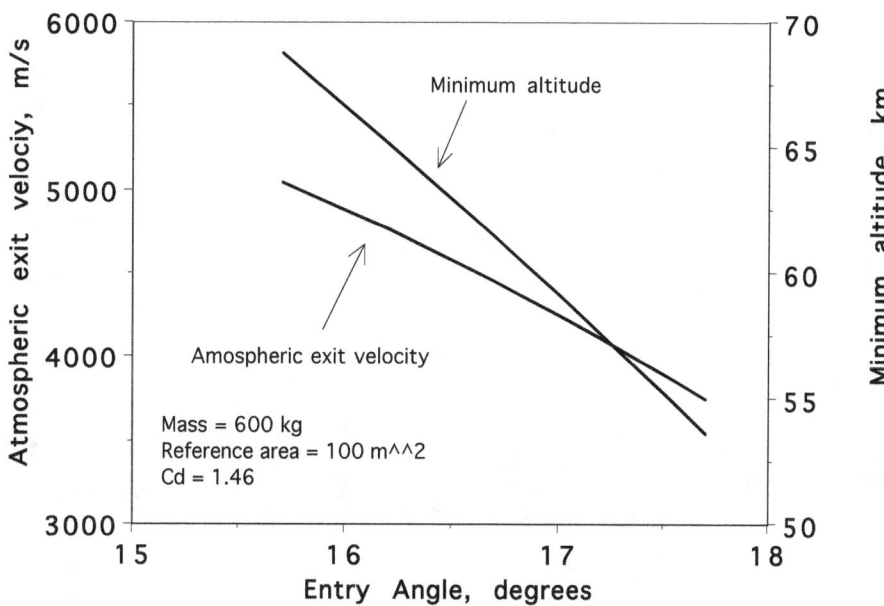

Figure 10. Aerocapture of a ballute at Triton as a function of entry speed

Software Risk Identification for Interplanetary Probes

Robert J. Dougherty [1], Periklis E. Papadopoulos. Ph.D.[2]

[1] *Lockheed Martin Space Systems, 1111Lockheed Way, Sunnyvale, CA 94089, USA,*
Email:robert.j.dougherty@lmco.com

[2] *Professor, Mechanical and Aerospace Engineering, San José State University, One Washington Square, San Jose, CA 95192, USA, Email: ppapado1@email.sjsu.edu*

ABSTRACT

The need for a systematic and effective software risk identification methodology is critical for interplanetary probes that are using increasingly complex and critical software. Several probe failures are examined that suggest more attention and resources need to be dedicated to identifying software risks. The direct causes of these failures can often be traced to systemic problems in all phases of the software engineering process. These failures have lead to the development of a practical methodology to identify risks for interplanetary probes. The proposed methodology is based upon the tailoring of the Software Engineering Institute's (SEI) method of taxonomy-based risk identification. The use of this methodology will ensure a more consistent and complete identification of software risks in these probes.

1. INTRODUCTION

Software is a critical component for command, control, and communications in interplanetary probes. Many probe failures and other problems have been traced to software errors. Future probe software will become more complex as it is used to perform novel tasks such as the optimization of landing and exploration sites. This software will become increasing difficult to effectively debug and test and will be prone to new risks.

The first step of the Software Engineering Institute's (SEI) risk management process is risk identification [1]. The purpose of this step is to anticipate risks before they materialize as actual problems. For example, an unclear requirement is a risk, but the incorrect software implementation due to this lack of clarity is a problem. Uncertainties and issues can be transformed into a list of risks that specify the cause of the concern and its potential impact. For a risk to be useful once it is identified, it must be clear, concise and informational within a particular context. This context describes the circumstances, contributing factors and related background information of the identified risk.

Risks that are not identified are not available during the risk management process to analyze and mitigate. The identification of risks is therefore the most important step in this process. Risks identified late in the software development process may require additional, expensive software rework and testing. Risks that are not identified can have catastrophic consequences.

Certain software risks can be identified by analyzing and applying lessons learned from previous missions and by proactively anticipating new risks from evolving technologies. Historically, risks have often been identified in a reactive mode - in response to a failure, change, or problem. These reactive risk identification methods may detect historical problems, but will not adequately detect latent risks from new technologies. The detection of these types of risks in interplanetary probes is essential to a successful mission.

2. UNIQUE ASPECTS OF SPACE PROBE SOFTWARE

Many software risks found in interplanetary probes are common to all software applications. These risks can be avoided through a careful and thorough application of best software engineering practices. However, certain factors make space, and more specifically probe, software different from other software.

Early space software had few commands placed in small memories, and the development process was poorly documented and idiosyncratic. Software size and complexity has greatly increased since then.

Interplanetary probes now contain and are a part of complex systems that have embedded computers that must interact with sensors, ground systems, and other on-board computers. The interactions and coupling of the software systems have critical time-dependent processes. Probe software is also becoming more autonomous and can compensate for mission impacting problems and limitations through self-diagnosis error detection and correction routines.

These new capabilities are increasing the probe's software complexity, making it impossible for a single person to understand all the software components and their interactions. Many of the interactions among the software systems and onboard hardware are essential to mission success. An error in any one of a number of flight parameter values, lines of code, or software interfaces could translate into a total mission loss.

Interplanetary probe software is only used once operationally. Since no two missions are exactly the same, field data from longer-term usage of the software is unavailable to help identify risks. Therefore, complete and accurate models, simulations and testing are important for a successful mission.

The software factors previously noted are common to all space software. The software unique to probes is the entry, descent and landing software that controls the most perilous phases of the mission. This software, for example, may need to make important autonomous realtime decisions to select a landing site that could include altitude ranges, surface slopes, number and size of rocks, temperature, and winds. The software will need to balance scientific objectives with safety objectives to optimize the site location.

3. MISSED OPPORTUNITIES

Risk identification processes are often believed to be adequate until a problem occurs. Accident investigations have shown that the reasons for many software problems are systemic issues that fall outside of the traditional risk identification processes [2]. The root causes of many space accidents are surprisingly simple and seemingly avoidable.

The first U.S. interplanetary mission was Mariner 1 in 1962. This was intended to be the first planetary flyby and was destined for Venus. However, the spacecraft was destroyed shortly after liftoff after it had gone off-course. The Post Flight Review Board later attributed the error to an omission of a hyphen in the code, which allowed incorrect guidance signals. It is interesting to note that the Soviet Union attempted four planetary missions prior to Mariner 1, which were all failures [3]. Both of these early space programs had more failures than successes.

The following three missions illustrate missed opportunities to find critical problems prior to launch. Although not discussed here, secondary and less probable causes of accidents also need to be understood and their lessons applied to future missions. These lessons should become part of the feedback loop to the risk management process to inform the space community of new risks. Of course, populating a risk database following an accident is not the preferable method of identifying risks.

3.1 Mars Climate Orbiter (MCO)

The Mars Climate Orbiter (MCO) was lost in 1998 when it entered the thin Martian atmosphere at a lower elevation than expected during an aerobraking sequence. According to the investigation board, the root cause of this accident was the failure to use metric units for the thruster performance data as described in the software interface specification. This resulted in a navigation error that placed the spacecraft at about a 57 km periapsis, instead of the intended 140-150 km. This altitude was too low for the spacecraft to survive [4].

3.2 The Mars Polar Lander (MPL)

This probe was launched in 1999 and failed in the entry, deployment, and landing (EDL) sequence on Mars. The legs of the probe had Hall Effect sensors that were programmed to shut down the descent engines within 50 milliseconds after touchdown on the planet. It was determined that the most probable cause of the accident was that the software interpreted the leg deployments as a touchdown, shutting off the engines. The vehicle then likely dropped about 40 meters and was destroyed. Although the Hall effect is well understood, the software requirements were not flowed down properly to the software engineers.[5].

3.3 Solar Heliospheric Observatory (SOHO)

SOHO was launched in 1998 to study the solar atmosphere, corona and wind. Although not an interplanetary probe, it is relevant as another example of the risks common to all space software. Two months after completing a successful two-year primary mission in 1998, contact with SOHO was lost. The NASA/ESA Investigation Board concluded that there were no anomalies on-board the vehicle, but instead the problems had been caused by a number of ground errors that led to a significant loss of spacecraft attitude. The loss was caused by operational errors, the failure to adequately monitor the vehicle's status, and a decision to disable part of the on-board autonomous failure detection system. These multiple, avoidable factors created the circumstances that directly caused the errors [6].

4 ROOT AND AUXILIARY CAUSES

Many software failures are the result of the systemic factors listed below. Systemic factors look beyond the particular technical causes for an accident and describe or identify the underlying reason or root cause for the problem. Although the risks associated with some of these factors are difficult to quantify, they can still be incorporated into a risk management plan to help ensure that a similar problem is not repeated. Various systemic software risk factors have been listed in the past [2, 7]. These factors have been divided into technical and non-technical categories. The non-technical category includes communication, management and organizational factors. The technical category includes causes related to software engineering.

4.1 Non-Technical Factors

4.1.1 Complacency

One factor that results in increased risk is the complacency due to past successes and the feeling that risk decreases over time because the software has become more "mature." This complacency can cause safety requirements to be relaxed and increased risks to be inadvertently accepted. One example is the risk resulting from a reduction in the monitoring of the software development and testing process. Typically reductions in safety, quality assurance, operations and training can lead to such failures. Warning signs often occur before accidents, but are often unreported or not acted upon. The failed MCO, MPL and SOHO missions were all affected by this culture of complacency.

4.1.2 The Diffusion of Responsibility

A diffusion of responsibility may result in an incomplete coverage of issues by an organization. A program is best served by leadership that demonstrates a broad understanding of all systems and interfaces, and by qualified individuals that take responsibility for each system, subsystem and interface. Reviews and technical discussions should have representatives from all affected disciplines. Software engineers need to interact with other disciplines to understand the source of their software requirements and to identify gaps that may not have been properly communicated to them. However, under principles of good management, only one individual can ultimately be responsible for each and every decision and that individual needs to feel accountable for his or her decisions.

4.1.3 Information Transfer

A clear communication path is needed to transfer important information among project team members. However, a filter is also needed to avoid information overload. Excessive information can confuse or bury more important information. This problem has been exacerbated by email and inexpensive data storage. The transfer of critical information should be handled in a formalized manner to ensure that the message is clear, that the impact and significance is understood by the recipient, and that the issue is properly managed. The use of voice mail and email may provide an illusion that a problem is being actively managed, but the casual nature and volatility of both these communication methods may interfere with an effective communication process.

4.1.4 Employee Turnover

The loss of key technical and management personnel contributes to risk as project continuity is lost and the experience base shrinks. This problem has become worse as new programs replace legacy programs, the aging aerospace workforce moves toward retirement, cost cutting measures reduce the number of employees on a program, and aerospace employees are lost to higher-paying non-industry companies and projects. It is an enormous challenge in the industry to find key individuals with the right experience and training for the new positions created by these trends.

4.1.5 Legacy Software, Reuse and COTS

The use of Commercial off-the-shelf (COTS), reused, or legacy software is generally believed to reduce risk and the cost of software development because of its previous usage and testing. However, a hidden risk can be that the internal architecture and code in the software is not an exact match for the new probe application. For example, the software may have been developed for a different environment and worked correctly with the original, but not current, set of requirements. The behavior of such software may not be linear or stable in the new environment. COTS or third party software is especially a concern because the source code may be proprietary. The developer may not have access or understand the internals of the software except from limited vendor information. Software engineers must understand the assumptions and limitations of any COTS or legacy code before adoption. An independent verification and validation (IV&V) program is especially important to reduce risk for this type of software.

4.1.6 Requirements Creep and Unneeded Code

Requirements creep occurs when features are added to the software that are not true requirements. Instead of converting these features to actual requirements, or removing them, the inadequately documented features are left in the software. Often requirements from a previous mission are retained under the assumption that removing the unnecessary software code would create a higher risk than leaving the code unchanged. However, this unneeded code increases complexity during development and increases the possibility of untested states.

4.2 Technical Factors

4.2.1 Inadequate Engineering

Software activities are sometimes inadequate due to a lack of resources or critical processes that are flawed. One such problematic process is the incorrect flow-down of system requirements. A variety of other inadequacies and engineering problems include: interfaces that are defined after the software coding has already begun, programmers who do not have the systems experience or science background to understand the problem they are solving, inconsistent or incomplete error handling, and testing that does not cover all of the possible exceptions. The example of the MPL failure discussed earlier illustrates inadequate engineering; i.e. there was no reasonableness check for the altitude when the descent engines were turned off.

4.2.2 Inadequate Reviews

Every review needs to have a clear goal and outcome. The responsibilities of all review participants should be clearly understood. Quality Assurance (QA), for example, should not approve documents or processes by merely checking signatures without understanding the real quality of the documents. QA also needs to ensure that peer reviews take place and that a "second set of eyes" reviews processes, coding, and parameters at appropriate points in the development process. It is often the case that a second review by a qualified person will be sufficient to greatly increase quality. However, by itself, QA cannot ensure that all critical software values, analyses and processes are correct. An independent verification and validation (IV&V) organization is necessary to double-check all software engineering work performed. A contributory factor to the failure of the SOHO mission was that important reviews were bypassed because of tight schedules and compressed timelines.

4.2.3 Inadequate Specifications

Many software-related problems are caused by flawed requirements and their flow-down, rather than by actual coding problems. Requirements are sometimes incomplete or unclear, assumptions may be incorrect or unstated, and system-states may not be understood and controlled. Good specifications that follow an effective process and have traceable requirements are important to minimize risks. The MPL accident report suggested that the software engineers on the project would have prevented the accident if they had understood the rationale behind the design.

4.2.4 Inadequate Education

Software engineers are rarely taught safe design principles, such as eliminating unused functionality, designing appropriate error detection and correction, and conducting reasonability checks. In a complex system such as interplanetary space probes, small errors can have catastrophic consequences. Education should be on-going so that software engineers continue to remain current on important advances in their discipline. A typical undergraduate education does not typically provide the level of skills needed for architecting software for complex space systems.

4.2.5 Inadequate Testing

Although the space industry attempts to "test like you fly, fly like you test," in reality, simulations and test environments have limitations and may not be able to meet that goal. The understanding of these limitations and testing assumptions are important to reduce risks. Testing is only as good as the test plan, test equipment, and the test cases chosen. In addition, testing may not catch every problem, especially when COTS or legacy software is being used. The internals of the software need to be well understood so that assumptions are known and appropriate test cases can be chosen. Software is too complex for test cases to be comprehensive, and instead need to include representative non-nominal and stressed conditions. Inadequate testing could be considered a contributing factor in almost every probe accident.

4.2.6 Inability to Understand Software Risks

Some engineers do not realize the complexity of the software and believe testing will reveal any latent problems. The software may be assumed to be correct unless testing shows otherwise. Practical testing, however, is limited to a subset of possible states. This inability to understand software risks also allows the inclusion of unnecessary software functionality. The over-reliance on testing to uncover problems and unused functionality increases risk. The removal or modification of software fault protection for reasons like performance also increases risk and have led to mission failures.

4.2.7 Use of Hardware Techniques for Software

Software is pure design, and its failure modes are different then physical devices. Software doesn't have safety margins like hardware - a single bit flip may cause a mission loss in a poorly designed system. Most software problems are flaws in the requirements specification and not in coding errors. Failure modes are often not clearly defined for software. A statement like, "Flight software doesn't execute properly" is too vague to be useful in identifying risks. Redundancy in software is no help in mitigating these risks if the problem is in the hardware. In fact, such software redundancy may give a false sense of security if the solution is merely duplicating a design error, and the additional complexity may actually increase the risk.

5.0 ANTICIPATING FUTURE RISKS

Lessons learned and accident reports are useful in understanding and avoiding past errors, but latent software risks are always a concern. Increased software complexity and new technologies increase these risks. A compilation of potential risks of new software technologies used in a probe can be used to address and mitigate these risks. New modeling and analysis tools also will be needed as these technologies are introduced in the future.

6.0 THE MODIFIED TAXONOMY-BASED RISK IDENTIFICATION METHODOLOGY

There are several commercial and government software tools that are advertised to help identify risks - none of them do so effectively. They instead allow risks already identified to be classified according to areas like severity and probability as part of an overall risk management process.

6.1 Risk Identification Methods

There are many ways to identify risks - all of them can be useful, but each is limited in the number and types of risks that they can effectively identify. Here are six approaches to identify risks:

1. **Identification Based on Experience** – The engineers, scientists and others who have the most experience and understand the details of the probe are the best suited to identify its risks. A process can be developed to allow these experts to report risks.

2. **Identification Based on Historical Data** – After each mission, the data is analyzed by the various subsystems and risks are identified either directly or by analogy. Applicable risks are identified and tracked for each mission. NASA, for example, maintains a Lessons Learned/Best Practices database that may be useful in identifying risks. Data from as many sources as possible should be compiled for a comprehensive list of possible risks.

3. **Brainstorming** – All stakeholders may add value to the risk identification process. The

synergy of a diverse group may help expose a new risk.

4. **Voluntary Risk Reporting** – A method of reporting risks, anonymously if desired, should be in place. Potential risks that are submitted are analyzed for applicable missions. This provides an avenue for risks to be identified that may be politically unpopular or without another outlet to be raised.

5. **Reviews** – Requirements, code and other reviews and working groups are common methods of identifying risks.

6. **Testing** – Testing can expose flaws in both processes and assumptions and is a primary method of risk identification, although it is preferable for risks to be exposed earlier in the software development cycle.

6.2 The SEI Taxonomy-Based Risk Identification Method

A systematic method to ask the right risk-related questions is needed that can be tailored for a particular interplanetary probe. A process needs to be in place that goes beyond assigning blame, and instead concentrates on helping understand the sources of risks. One method is based on the Software Engineering Institute's (SEI) Taxonomy-Based Risk Identification method (SEI/CMU Technical Report CMU/SEI-93-TR-6 ESC-TR-93-183).

The SEI Taxonomy-Based Risk Identification method is a process to identify risks associated with the development of a software-dependent project in a systematic and repeatable way. It has been tested in both government and civilian projects and shown to be useful, usable, and efficient. In general, the method consists of a series of questions that helps surface risks by using an interviewing process with subsets of the project team.

Some engineers and scientists may express resistance to this method because many of the questions are open-ended and some risks cannot be easily quantified. Technical people prefer to have a method that can assign a numerical value to the severity and probability of risks, but the quantification of some of the identified risks may be misleading. For example, bringing a new technical lead into a project certainly introduces risk, but it is difficult to place a numerical value on a person's education and experience as compared to the previous technical lead - to do so would be meaningless. However systems have been proposed that quantify every risk in such a way. A list of questions similar to the SEI risk identification taxonomy process, for example, has been proposed by Karolak [8]. He quantifies each risk identification question with a number between 0 (none) and 1 (all).

The Taxonomy-Based Risk Identification method consists of 194 questions divided into three main divisions (Product Engineering, Program Engineering, and Program Constraints) and 13 further sub-divisions. The sub-divisions are listed below with a sample question from each one to illustrate the type and level of sample questions. The numbers in brackets before the sample questions are the numbers assigned by SEI.

A. Product Engineering

1. Requirements - [6] Are the external interfaces completely defined?

2. Design - [16] How do you determine the feasibility of algorithms and designs?

3. Code and Unit Test - [42] Is the development computer the same as the target computer?

4. Integration and Test - [51] Are the external interfaces defined, documented, and baselined?

5. Engineering Specialties - [66] Are safety requirements allocated to the software?

B. Development Environment

1. Development Process - [85] Is there a requirements traceability mechanism that tracks requirements from the source specification through test cases?

2. Development System - [94] Does the development system support all aspects of the program?

3. Management Process - [106] Are there contingency plans for known risks?

4. Management Methods - [126] Does program management involve appropriate program members in meetings with the customer? (including Technical Leaders, Developers, Analysts)

5. Work Environment - [139.b] Are members of the program able to raise risks without having a solution in hand?

C. Program Constraints

1. Resources - [147] Are there any areas in which the

required technical skills are lacking?

2. Contract - [165] Are there dependencies on external products or services that may affect the product?

3. Program Interfaces - [175] Are the external interfaces changing without adequate notification, coordination, or formal change procedures?

6.3 Questionnaire Creation Process

Specific software risks are often known by project personnel but are not always communicated. Effective risk identification requires a process and culture of open communication to encourage all stakeholders to use their knowledge of the project to help with its success. The basic risk identification methodology for a particular probe should be tailored from the generic SEI Risk Taxonomy Method questionnaire to include risks peculiar to that probe.

The tailoring of the generic SEI software questionnaire should follow a systematic process to create a probe-specific questionnaire:

1. Review each generic risk taxonomy question and tailor it to the particular space probe domain.

2. Review the literature for other potential risks from new technologies used in the probe. Add or tailor the generic risk taxonomy questions for these additional risks.

3. Review the literature on lessons learned from other similar missions. Add or tailor the generic risk taxonomy questions for these additional risks. Note that these lessons learned do not need to be from similar probes or even from aerospace risks, but may include any software lesson that can be applied to the particular probe.

4. Have the Probe's Software Team and other experts review the list for further modifications.

These steps will result in a baseline risk questionnaire, which can be modified on an on-going basis as risks are identified from reviews and other sources. This dynamic baseline can also be modified and used as a starting point for other similar missions.

7.0 PROCESS FOR APPLYING THE METHODOLOGY

The reviewers must prepare before the interviews are conducted. If possible, they should understand the processes that are being used by the organization and initially assess whether the processes are complete, accurate, and being followed. This will help facilitate the interview process.

The following steps describe a process to apply the proposed methodology after the modified baseline questions have been completed:

1. Set up Interview Groups

Interview groups should be determined and sorted by functional area or subsystem. For example, the interview groups might be divided into command and control software, mission data software, sequencing software, guidance and navigation software, flight parameters, scientific experiment, testing, and IV&V. There is no single approach or perfect grouping, but it may prove more efficient to keep personnel together that work on similar tasks.

Ideally, there would be no reporting relationships in a particular interview group. Management commitment is needed for the interview process, but management should not be present so the interviewees feel they can speak freely. If possible, groups would be limited to 5 participants [1], although this may be difficult to ensure in practice.

2. Divide the Questions

Certain questions are more appropriate for some groups than others. The questions for each interview group should be a subset of the total risk identification questions. A list of questions tailored to each group should be provided to the group members before the interviews.

3. Conduct the Interviews

The risk identification session begins with briefing the participants with the methodology and its purpose. During the interview, the tailored questions can lead to other issues, concerns and risks. Items of risk noted during the interviews are used to update the risk database for the mission and the questionnaire template for the mission. For practical reasons, the interviews should not normally last more than 2 hours.

4. Provide Feedback

A summary of the potential issues is provided back to the participants at the end of the interview, and a hard copy is subsequently provided to the project manager.

5. Repeat Process

Depending on the size and complexity of the project, this process may need to be repeated several times to different groups during the software development life cycle. If the methodology is only performed once near the beginning of the project, many risks will still be unknown, but if it is only performed near the end of the project, then the risks may be expensive and difficult to correct.

8.0 TESTING THE METHODOLOGY

The best test of the usefulness of a method is whether it actually works. The lessons learned from the failed missions in Section 3 were used to tailor some of the baseline questions to test the methodology. Obviously these examples are somewhat contrived since the problems have already occurred, but they are still useful to illustrate how the process works and ensures that these type of systemic problems are addressed and are less likely to re-occur on future missions.

Risk 1: Complacency because of past mission successes

Question: Do you feel that your past successes have reduced your risk of failure?

Risk 2: The constants definition process not well defined

Question: Are there formal, controlled plans for all development activities? Are developers familiar with the plans?

Risk 3: There was a lack of anyone in charge of the entire process

Question: Is there a qualified person responsible for every process and the overall processes?

Risk 4: No communication channel for relaying the problem when discovered

Question: Are the realtime interfaces among the different organizations sufficient for status and anomaly resolution?

Risk 5: No formal anomaly reporting and tracking system

Is there a formal anomaly reporting and tracking system for all phases of the development and deployment process?

9.0 CONCLUSION

Early software risk identification is important for the success of interplanetary probes. Potential risks must be considered not only from other probe missions, but also from other sources within the aerospace industry as well as from other industries. There are currently no commercial or government products that are completely adequate for probe software risk identification. Most available risk management applications only provide a shell for previously identified risks.

The SEI risk identification method can be used as a starting point to create a tailored true risk identification methodology for interplanetary space probes. The general software risk questions are tailored through a series of steps that ultimately provides a mission with a unique questionnaire for risk identification. A process to apply the methodology transforms it from a theoretical suggestion to a practical process for identifying risks for a particular probe. The software lessons learned from other missions are incorporated into the questions to ensure the past problems are not repeated.

This proposed methodology could replace ad-hoc, undocumented, incomplete or reactive methods that are typically used. This methodology will enable a more consistent identification of risks in interplanetary probe software systems with an ultimate goal of more successful missions.

10. REFERENCES

[1] Carr, Marvin J., et al. *Taxonomy-Based Risk Identification*, Carnegie Mellon University, Pittsburgh, Pennsylvania, 1993.

[2] Leveson, Nancy G., *Systemic Factors in Software-Related Spacecraft Accidents*, American Institute of Aeronautics and Astronautics, 2001.

[3] Williams, David, *Planetary Sciences at the National Space Science Data Center,* 19 May 2004, NASA Goddard Space Flight Center, 17 August 2004, <http://nssdc.gsfc.nasa.gov/planetary/chronology.html>.

[4] Stephenson, Arthur G., *Mars Climate Orbiter: Mishap Investigation Board Phase I Report,* NASA, November 10, 1999.

[5] JPL Special Review Report, *Report on the Loss of the Mars Polar Lander and Deep Space 2 Missions,* Jet Propulsion Laboratory, March 22, 2000.

[6] Joint NASA/ESA Investigation Board, *SOHO Mission Interruption,* NASA/ESA, 31 August 1998.

[7] Leveson, Nancy G., *The Role of Software in Spacecraft Accidents*, American Institute of Aeronautics and Astronautics, 2001.

[8] Karolak, Dale W, *Software Engineering Risk Management,* IEEE Computer Society Press, Los Alamitos, California, 1996.

STAGNATION POINT HEAT TRANSFER WITH GAS INJECTION COOLING

B. Vancrayenest[*,†], M. D. Tran[*], and D. G. Fletcher[*]

[*]von Karman Institute for Fluids Dynamics, Chaussée de Waterloo 72, 1640 Rhode-Saint-Genèse, Belgium,
Email: vancraye@vki.ac.be
[†]Centre National d'Études Spatiales, 18 avenue Edouard Belin, 31401 Toulouse, France

ABSTRACT

The present paper deals with an experimental study of the stagnation-point heat transfer to a cooled copper surface with gas injection under subsonic conditions. Test were made with a probe that combined a steady-state water-cooled calorimeter that allows the capability to study convective blockage and to perform heat transfer measurements in presence of gas injection in the stagnation region. The copper probe was pierced by 52 holes, representing 2.4% of the total probe surface. The 1.2 MW high enthalpy plasma wind tunnel was operated at anode powers between 130 and 230 kW and a static pressures from 35 hPa up to 200 hPa. Air, carbon dioxide and argon were injected in the mass flow range 0-0.4 g/s in the boundary layer developed around the 50 mm diameter probe. The measured stagnation-point heat transfer rates are reported and discussed.

1. INTRODUCTION

One type of thermal protection system, which is used by many capsules and probes employs ablative composite material. At high temperatures the organic resin decomposes and vaporizes, absorbing some of the thermal energy; this process is known as pyrolysis. The pyrolysis gas is then injected into the flow creating a thin layer of cooled gas over the vehicle which blocks additional thermal load. The effect of blowing mass through porous walls, holes or slits has been studied in the literature but shows a large scattering of data. Related to the ablation re-entry flow regime, the convective blockage is the topic of this study. The release of pyrolysis gas will be simulated by injecting gas into the flow through multiple ports; this is known as transpiration cooling. The goal of this project is to investigate the reduction in heat flux caused by transpiration cooling for a probe in a subsonic plasma flow. Attempts will be made to correlate the change in heat flux to the mass flow rate and composition of the gas injected.

Extensive plasma tests were performed for different transpiration and plasma conditions. Different gases such as air, argon, and carbon dioxide were injected through the surface of the probe into the air plasma flow. The probes were also instrumented to measure temperatures, pressures, and heat flux within the probe. A significant part of this effort was focused on the development and verification of the gas injection system.

2. FACILITIES AND TESTING

The Plasmatron is a high enthalpy facility in which a jet of plasma is generated in a test chamber kept at sub-atmospheric pressure (typically between 7 and 200 mbar). The plasma is generated by heating a gas (in the present study, only air plasma was considered) to temperatures up to about 10.000 K, using electrical current loops induced inside a 160 mm diameter plasma torch. The inductively-coupled plasma wind tunnel uses a high frequency, high power, high voltage (400 kHz, 1.2 MW, 2 kV) solid state (MOS technology) generator.

The accurate quantitative measurement of heat transfer rates in high enthalpy plasma facilities has always been a challenging task. Coupling gas injection systems with this type of measurement is even more difficult because of the need for the same space. The multi-point gas injection probe (Fig. 1) was based off the 14 mm diameter water-cooled calorimeter, which has been used successfully in the past at VKI to measure heat fluxes. In addition to measuring heat flux, this probe must inject gas uniformly into the plasma stream at the stagnation point for known conditions. The settling chamber for the gas had to be immediately after the front face; therefore, the water chamber could only be placed after the settling chamber. Thus, heat would is transferred by conduction through the copper side walls to the calorimeter. Since this heat conduction path from the front face to the water calorimeter has to be short and to allow for sufficient circulation in the water calorimeter, the volume of the gas settling chamber was significantly reduced (this raised concern over whether the injected gas would still be uniform across the front face. In order to assess the validity of this assumption, velocity profiles using hot wire velocimetry were performed latter and showed reasonable uniformity within 1 mm of the surface).

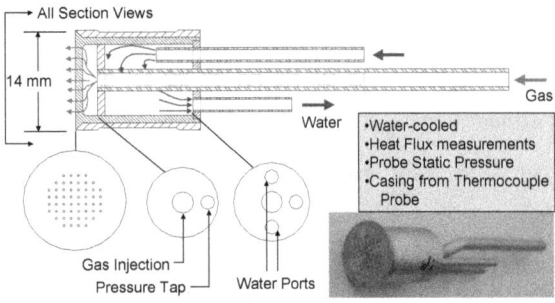

Figure 1. Steady-state water-cooled calorimeter with gas injection

In [4], different amounts of holes for the gas injection on the front face were tested: 21, 37, 57, and 77. The probe with 21 holes showed significantly less heat transfer efficiency compared to the others. There was little difference, however, between the 37 and 77 holes. Therefore, The 52 holes configuration was chosen, with a hole diameter of 0.3 mm (for a 2 mm gas injection pipe diameter) and a 1 mm distance between the holes, giving approximately a 2 hole diameter spacing between the holes.

A teflon piece is used to separate the probe from the holder. It limits heat flux loss due to conduction through the sidewalls, which are then considered negligible. For the water calorimeter, conduction losses due to the proximity of the water inlet and outlet pipes must also be considered. The radiated heat flux of the cold-wall probe is also considered small, since the probe wall is kept at a low surface temperature. This effect was measured using a Gardon gage for air and CO_2 plasma at various static pressures. For the most extreme case, CO_2 at 70 mbar, maximum radiative loss was 5.4% of the total heat flux and represented, on average 4.6%.

The probe was inserted into the ESA sample holder, 50 mm diameter cylindrical blunt body. The gas was injected through a port in the back of the probe. The mass flow rate of the transpiration gas was measured using a G0-100 rotameter with a range of 0 to 1 g/s. The pressure transducer was located inside the arm of the model, so the gage pressure would be measured relative to the Plasmatron test chamber pressure. The mass flow of the cooling water is measured using a L16-630 rotameter. The pressure transducer for the pressure tap in the probe was an SM5415 with a 15 psi (1030 mbar) range. The rotameter for the transpiration gas was also switched after the initial tests. The G0-100 was used for the initial tests from 0.1 to 0.4 g/s. For measurements at a lower mass flow, the rotameter was switched to the L16-630 which has a range from 0 to 0.4 g/s.

3. EXPERIMENTAL RESULTS

Fig. 2 shows the heat flux measurements for air as the transpiration gas. The initial tests with the G0-100 rotameter were performed for the flow rates between 0.1 and 0.4 g/s. The results from these tests were as expected, as the transpiration mass flow rate was decreased the heat fluxed increased. Injected mass flow was found to be too high to study the region where the rise of heat flux is exponential as in [4], so the smaller rotameter was used to measure lower mass flow rates between 0 to 0.1 g/s. In Fig. 2, one can note differences in the two heat flux measurements at 0.1 g/s which can mostly be explained by this switch in rotameters. This difference will be discussed later in the uncertainties of the rotameter calibration.

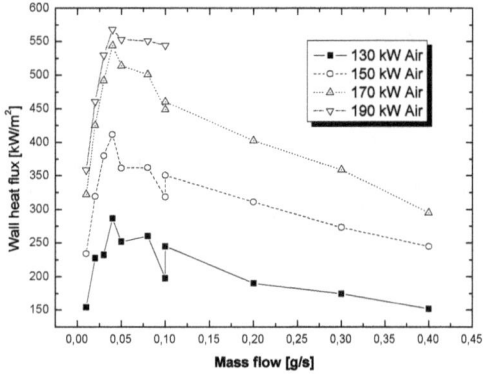

Figure 2. Heat flux vs. transpirational $\dot{m}_{air}$

The heat flux was found to unexpectedly decrease for low mass flow rates. The monotonic behavior observed in [4] was not reproduced. For all the power settings, the heat flux rises to a maximum heat flux at 0.04 g/s. For flow rates below 0.4 g/s, the heat flux decreases with decreasing transpiration flow rates, but then rises again somewhere between 0 and 0.01 g/s. Fig. 3 and 4 are the heat flux plots for carbon dioxide and argon (zero mass flow values in Fig.4 were obtained turning off the gas injection). They exhibit the same trend as air. The heat flux always reaches a maximum at 0.04 g/s.

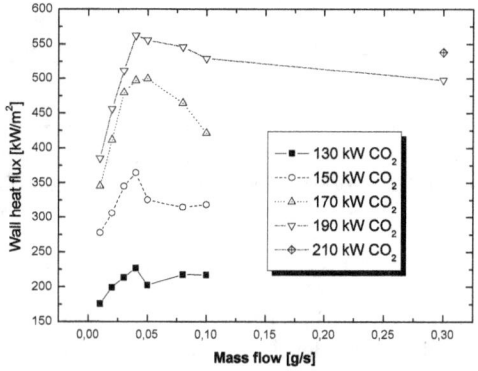

Figure 3. Heat flux vs. transpirational $\dot{m}_{CO_2}$

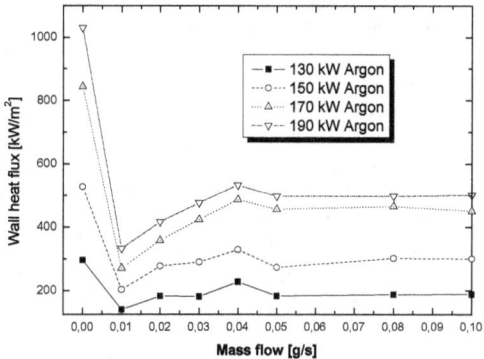

Figure 4. Heat flux vs. transpirational $\dot{m}_{argon}$

4. DISCUSSION OF RESULTS

Sensitivity studies for the injected gas pressure were performed in the transpiration gas rotameter and also for the static pressure in the test chamber. For the last test case with nitrogen, a thermocouple was added to the gas injection pipe of the probe to determine whether the gas was getting preheated before reaching the settling chamber. The temperature measurements varied between 24 to 34 °C, so its effect was determined to be negligible.

Pressure measurements inside the settling chamber of the probe were also made. These measurements were taken to determine the conditions of the gas before being injected. Unfortunately, the temperature of the gas could not be measured so that the gas in the settling chamber could not be completely characterized.

The pressure measurements could also be used to determine whether the velocity in the injection holes reached sonic flow using the following equation:

$$\frac{p_0}{p} = \left(1 + \frac{\gamma-1}{2}M^2\right)^{\gamma(\gamma-1)} \quad (1)$$

where γ equals 1.4 for air, 1.29 for carbon dioxide, 1.4 for nitrogen, and 1.67 for argon. To check for the onset of sonic conditions, Mach number M, is set to 1. Even though the holes are choked, mass flow through the holes can still be increased because according to mass flow rate at a choked throat can be found by:

$$\dot{m}^* = \left(\frac{2}{\gamma+1}\right)^{(\gamma+1)/[2(\gamma-1)]} \sqrt{\gamma/R}\frac{A^*P_0}{\sqrt{T_0}} \quad (2)$$

This equation shows that for a choked flow the mass flow rate is directly proportional to the throat area and the stagnation pressure and inversely proportional to the square root of the stagnation temperature. Therefore, if the stagnation pressure is increased the mass flow rate is increased. Choking seemed to be a possible cause for the unexpected heat flux measurements so the pressure ratios were increased by raising the test chamber static pressure to 200 mbar. Still the heat flux plot had the same trend. Therefore, possible sonic flow in the holes appears to not affect the trend in heat flux. Fig. 5 compares test chamber result for the static pressures of 200 mbar with nitrogen as the transpiration gas and 35 mbar for the other transpiration gases. The case for nitrogen at 35 mbar is not shown since a leak is suspected for this run. It should be similar to the other gases though, especially air which is 79% nitrogen.

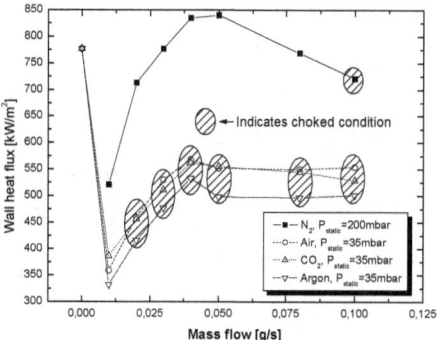

Figure 5. Heat flux with and without chocking

The plots of the heat flux show three distinct regions. Although data points were not measured in the first region due to the limitations of the rotameter, heat flux was measured for mass flow rates of 0 and 0.01 g/s. From these two points, we can conclude that in general the heat flux decreases with increased transpiration mass flow within this range. For mass flows between 0.01 to 0.04 g/s, the heat flux rises with increased transpiration mass flow. For mass flows above 0.04 g/s, the heat flux becomes more constant, only slightly decreasing with increased transpiration mass flow. A reasonable explanation was established after a survey of film cooling [1] and a review of the videos from the tests.

Film cooling, typically for turbine blades, is similar to transpiration cooling in that gas is injected into the flow through many holes or slots. The difference is that film cooling is specifically for crossflows and generally intended to protect regions downstream of the flow. Still certain analogies can be made for transpiration and film cooling, and much more literature is available for film cooling.

For film cooling, there are two flow regimes: low and high injection rates. Injection rates are characterized by a blowing ratio M, defined as:

$$M = \frac{\rho_{gas} U_{gas}}{\rho_\infty U_\infty} \quad (3)$$

where, the $(\cdot)_{gas}$ subscript is for injected gas properties and the $(\cdot)_\infty$ subscript for freestream conditions. At low injection rates, the momentum of the impinging jet causes the injected gas to immediately bend along the surface of the probe. This creates a thin film over the surface which is very effective at cooling the surface. At high injection rates the jet penetrates into the mainstream and eventually separates from the wall. This is not as effective at cooling the surface.

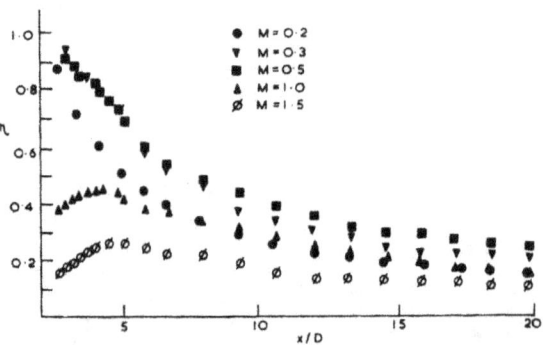

Figure 6. Effect of blowing ratio for film cooling

For the case of film cooling, there exists an optimum blowing ratio, where the injected mass flux is high yet the jet remains attached to the wall. For example in Fig. 6 from [1], the optimum blowing ratio for this particular case was around 0.5. In this graph, η is defined as the wall film cooling effectiveness and x/D is a streamwise coordinate. We are not concerned with x/D for our study, but concentrating on one coordinate, you can see that effectiveness at first increases for higher blowing ratios then decreases after about 0.5.

A similar situation could be argued for the current results. The dynamic pressure of the plasma jet at the location of the probe has been measured using a pitot probe and values are low (62.44 Pa for the 130 kW and 75.47 Pa for 150 kW). The velocities measured for the injected gas, on the other hand, were fairly large even in atmospheric conditions. Also, since the plasma jet is in a low pressure and high temperature condition the density of the gas is much lower. Ultimately, this leads to a high blowing ratio which means the transpiration flow could well be penetrating into the plasma jet. The carbon dioxide injection was shown to be an excellent tool for visualizing the transpiration flow and the carbon dioxide run videos support this hypothesis as seen in Fig. 7.

By dividing the mass flow rate of the transpiration gas by the area of the injected holes, the numerator of the blowing ratio can be determined. The denominator of the blowing ratio can be found using the dynamic pressure and plasma mass flow rate information. The blowing ratio M, for transpiration mass flow rate of 0.01 g/s and 0.10 g/s is approximately 3 and 30, respectively. Blowing ratios for the Mars Pathfinder were estimated according to results from a numerical study [2]. Based on a given trajectory and heat shield made of silicone elastomeric charring ablator, known as SLA-561V, the maximum blowing ratio was approximately 0.01.

Although heat flux measurements were not taken at the low flow rates of the first region, there must be a steep decrease in heat flux with increasing mass flow, because the heat flux for zero mass flow is significantly larger than for 0.01 g/s. In this region, the transpiration gas encounters the impinging plasma jet causing a thin film of cool, transpiration gas to envelope the surface. This convective blockage shields the probe from the heat load, decreasing the heat flux the probe experiences. As the transpiration mass flow increases in this regime, the film becomes thicker and there is more mass to transport the heat load, so the heat flux continues to decrease.

For mass flows in the second regime, the jets from the transpiration probe are penetrating into the plasma jet. This is shown in Fig. 7, where the jets from the transpiration probe are creating a cone shape. This is a less effective heat shield because the transpiration gas is being used to cool a larger volume rather than a thin layer over the most critical region where heat transfer is the highest. The transpiration gas is not as effective at cooling this larger volume, and it is the hotter gases in this recirculation region that are now in contact with the surface of the probe. As the injection rate is increased in this regime, the heat flux increases because the transpiration jets are penetrating further into the plasma jet increasing the volumes and becoming less effective.

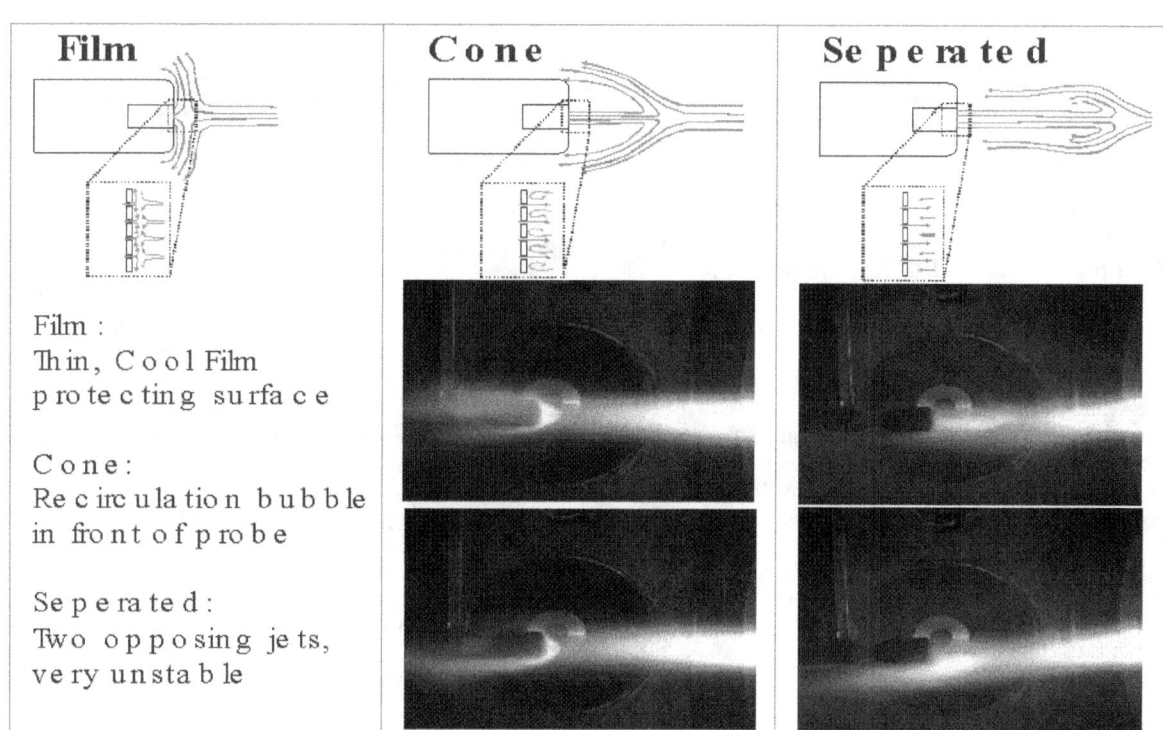

Figure 7. Three flow regimes

In the third regime, the jets of transpiration gas have penetrated so far into the plasma jet that the cone has become detached. A region covering the probe surface is no longer defined. Instead, the transpiration gas is a jet in front of the probe. Heat from the plasma flow can come in direct contact with the surface due to instabilities in the opposing jets. Increases in injection rates decrease heat flux only slightly since the additional mass is less effective so far away from the probe surface.

The results from [4] showed an exponential decrease in heat flux with increasing mass flow rates. This would correspond to the results expected for the first regime. Unfortunately, we were not able to achieve blowing ratios that low.

5. HEAT FLUX MEASUREMENT UNCERTAINTIES

To calculate the heat flux the probe experiences the following energy balance, as discussed previously, is applied:

$$q_w = \frac{\dot{m} \cdot Cp \cdot (T_{out} - T_{in})}{A} \quad (4)$$

Therefore, the uncertainties in the heat flux measurements arise for a combination of the uncertainties in the measurement chain with:

- $\dot{m}$ is the mass flow rate of the water through the calorimeter, measured with a ROTA L16/630-6404. Uncertainty of mass flow rate will typically be $\delta(\dot{m}) = 1/50 = \pm 0.02$ g.

- Cp is the specific heat of water. For the range of temperatures the water experiences, the change in Cp was estimated to be $\delta(Cp) = 0.01$ J/kg-K.

- T_{out} and T_{in}, are the temperatures at the outlet and inlet of the calorimeter. The temperatures are measured using thermocouples.

- A, is the area of the sensing element. The diameter of the face is 14 mm and uncertainty in the diameter is estimated to be ± 0.1 mm. $\delta(T_{read,in}) = \pm 0.3$ °C, uncertainty in temperature going into the calorimeter. $\delta(T_{read,out}) = \pm 0.7$ °C, uncertainty in temperature leaving the calorimeter.

The most probable error in a measurement can be calculated using the following equation:

$$\left(\frac{\delta q}{q}\right)^2 = \left(\frac{\delta \dot{m}}{\dot{m}}\right)^2 + \left(\frac{\delta Cp}{Cp}\right)^2 + \left(\frac{\delta A}{A}\right)^2 + \left(\frac{\delta T_{in}}{\Delta T}\right)^2 + \left(\frac{\delta T_{out}}{\Delta T}\right)^2 \quad (5)$$

where ΔT is the temperature difference between $T_{out} - T_{in}$.

Table 1. Breakdown of heat flux uncertainties.

Breakdown of Uncertainties Percentage of Uncertainty/Total Heat flux						
Run	$\dot{m}_w$	Cp_w	A	T_{out}	T_{in}	total
Air	1,8	0,2	1,4	6,9	2,9	13,3
CO2	1,5	0,2	1,4	7,5	3,2	13,8
Argon	2	0,2	1,4	6,3	2,7	12,6
Combined	1,7	0,2	1,4	6,9	2,9	13,2

Table 1 shows a breakdown of the uncertainties. For each test condition, the uncertainties were calculated then divided by the total heat flux measured to obtain a percentage. These percentages were then averaged for the different gases tested. The last row shows the average of the three runs. The temperature measurements are the largest source of error, because the fluctuations in the temperature make it difficult to determine the steady state value. The last column shows the total uncertainty for the measured heat flux.

6. IMPROVEMENTS AND FURTHER STUDIES

In order to be able to perform measurements in the first transpiration film cooling regime which is more representative of ablating re-entry flight conditions, the easiest way to proceed is to decrease the transpiration mass flow. Adaptations to the existing setup are currently in progress. Another way to measure this first regime accurately without having to go to lower transpiration mass flow rates would be to decrease the blowing ratio but working conditions affecting this ratio are more uncertain.

As mentioned previously in the probe design section, the heat flux measured is only relative to the front face, because there is a cool layer of gas in between the majority of the area between the front face and the calorimeter. The heat is primarily transferred to the calorimeter through the cooper walls. Heat pipes could be implemented but one can imagine that the best option for the next generation of combined transpiration/heat flux probe is to eliminate the settling chamber behind the front face and make it coaxial with the heat flux sensor. The front face and the water/slug calorimeter could be made of the same block. The front face could be made thicker so that ducts can be machined into it to feed the injection holes. In this way, the calorimeter would be covered by a thicker that will create the settling chamber for the injected gas.

7. CONCLUSIONS

The goal of this study was to gain experience in transpiration cooling in a subsonic plasma stream. This was achieved by a multistage design and test approach. The probes were able to inject gas uniformly into the plasma flow. This was verified by measuring the velocity profiles of the injected gas. The probes successfully measured conditions of the gas (temperature, pressure, and heat flux) before being injected into the flow. Relationships were then developed and verified for various plasma conditions, transpiration gases and mass flow rates in the Plasmatron facility.

Some of these results were unexpected. However they do appear to be consistent with other research on film cooling. Through this investigation, it was determined that the tests were performed at higher blowing ratios than expected, and the transpiration gas was actually penetrating the plasma jet. Recommendations were made for improvements and verification of the three flow regimes assumption.

New probe designs for the Plasmatron facility are never straightforward. By taking a multistage approach, though, we have successfully gained experience in transpiration cooling in a subsonic plasma flow. Eventually, this probe and knowledge will be applied to ablation research for a better understanding of pyrolysis gas injection.

8. ACKNOWLEDGEMENTS

This work was supported by funding from the Belgian Federal Office for Scientific, Technical and Cultural Affairs through a grant from ESA, administered by CNES (Dr J.M. Charbonnier technical monitor). The support of M. Tran by USAF fellowship is gratefully acknowledged.

REFERENCES

G. G. Bergeles, A. D. Gosman, and B. E. Launder. Near-field character of a jet discharge through a wall at 30 degrees to a maintream. *AIAA journal*, 15(4):499–504, 1977.

Y. K. Chen, W. D. Henline, and M. E. Tauber. Mars Pathfinder trajectory based heating and ablation calculations. *Journal of Spacecraft and Rockets*, 32(2):225–230, March-April 1995.

W. C. Davy, G. P. Menees, J. H. Lundell, and R. R. Dickey. *Hydrogen-helium ablation of carbonaceous materials: numerical simulation and experiment*, volume 64 of *Progress in Astronautics and Aeronautics*. Raymond Viskanta, 1978.

M. I. Yakushin, I. S. Pershin, and A. F. Kolesnikov. An experimental study of stagnation point heat transfer from high-enthalpy reacting gas flow to surface with catalysis and gas injection. In *Proceedings of the fourth European Symposium on Aerothermodynamics for Space Vehicles*, pages 473–479, Capua, Italy, 15 - 18 October 2001.

CHEMISTRY MODELING FOR AEROTHERMODYNAMICS AND TPS

Dunyou Wang[1], James R. Stallcop[2], Christopher E. Dateo[1], David W. Schwenke[2], Timur Halicioglu[1], Winifred M. Huo[2]

NASA Ames Research Center, Mail Stop T27B-1, Moffett Field, CA 94035-1000, USA
[1]*Eloret Corporation, Email:* dwang@mail.arc.nasa.gov, cdateo@mail.arc.nasa.gov, haliciog@nas.nasa.gov
[2]*NASA Advanced Supercomputing Division, Ames Research Center, Email:* James.R.Stallcop@nasa.gov, David.W.Schwenke@nasa.gov, Winifred.M.Huo@nasa.gov

ABSTRACT

Recent advances in supercomputers and highly scalable quantum chemistry software render computational chemistry methods a viable means of providing chemistry data for aerothermal analysis at a specific level of confidence. Four examples of first principles quantum chemistry calculations will be presented. Study of the highly nonequilibrium rotational distribution of a nitrogen molecule from the exchange reaction $N + N_2$ illustrates how chemical reactions can influence rotational distribution. The reaction $C_2H + H_2$ is one example of a radical reaction that occurs during hypersonic entry into an atmosphere containing methane. A study of the etching of a Si surface illustrates our approach to surface reactions. A recently developed web accessible database and software tool (DDD) that provides the radiation profile of diatomic molecules is also described.

1. INTRODUCTION

In an aerocapture mission, there is a tradeoff between the mass of the propellant needed for orbital insertion and the mass of the Thermal Protection System (TPS) needed to shield against aerodynamical heating. Thus the design of such a mission must include careful aerothermal analysis carried out at a specific level of confidence. This in turn calls for a corresponding level of confidence in the chemistry data used in the analysis.

The chemistry data needed in aerothermal analysis include chemical reaction rates, radiation emission and absorption probabilities, and probabilities of surface catalytic reactions. There are two sources for these data — laboratory measurements, and first principles quantum chemistry calculations. Traditionally, experimental measurements were the major source of data. However, measurements of high temperature radical reactions and photoemission probabilities for high-lying rotational and vibrational states of a molecule pose experimental problems due to the difficulties in the unambiguous measurement of a specific reaction rate or molecular transition probability at a high temperature. Under such circumstances, first principles calculations frequently are the only source of reliable data. While first principles calculations have been used in the past to provide selected data, the approach has been plagued by the lengthy turn-around time. Recently, the availability of powerful new supercomputers such as Project Columbia at Ames Research Center and the development of highly scalable chemistry software make quantum chemistry calculations more efficient than those of even just a few years ago. Thus first principles calculations can now be used as a reliable source of chemistry data for aerothermal analysis.

To illustrate recent advances in modern quantum chemistry methods and how they can be used to serve the need of mission design for planetary probes, this paper presents examples of four recent chemistry calculations. The rotational temperature of N_2 is a determining factor of its radiation profile and has to be included in the calculation of the radiative heat load to a vehicle or probe during high-velocity entry into a nitrogen-containing atmosphere. Our calculation of the $N + N_2$ exchange reaction is carried out to reconcile the rotational temperature of N_2 observed in a recent shock tube experiment [1]. The ethynyl radical C_2H is an important intermediate product in the entry shock of a methane containing atmosphere and the $H_2 + C_2H$ reaction presented in this paper illustrates current quantum chemical capability in calculating high-temperature radical reactions. The etching of a silicon surface by halogens serves as an example of a computational study of surface catalytic reactions. A newly developed web-based tool, Dynamic Database for Diatomics (DDD) provides the user with the unprecedented capability of determining the complete radiation profile of a diatomic molecule.

Even with recent software and hardware advances, the amount of chemistry data required in aerothermal analysis is still far too numerous for a reasonable turn-around time.

In this respect, the recent sensitivity study of thermochemical modeling by Bose et al. [2] for Titan atmospheric entry points to sensitivity analysis as an important tool in selecting a subset of chemical data particularly important in reducing the uncertainties in the modeling. Sensitivity studies serve as an important guide to selecting the needed data before quantum chemistry calculations are carried out.

2. N + N_2 EXCHANGE REACTION

In a recent shock tube experiment using nitrogen gas, Fujita et al. [1] measured the rotational and vibrational temperatures using the N_2 second positive and N_2^+ first negative bands and found significant nonequilibrium between the measured rotational and translational temperatures. So far, proposed explanations [3,4] of this nonequilibrium rotational distribution using N_2–N_2 collisions require physically unrealistic features in the N_2–N_2 interaction potential and are likely to be unreliable [5]. Thus it is likely that the large Δj transitions indicated in the shock tube data of Reference [1] do not come from N_2–N_2 collisions. Instead, our study [5] shows that the N + N_2 exchange reaction can lead to large Δj transitions at the high-temperature regime of the experiment.

In the atom exchange reaction

$$N + N_2(v,j) \rightarrow N_2(v',j') + N, \quad (1)$$

the N-N bond in the original N_2 molecule is broken and a new N-N bond is formed. Due to the nature of the N + N_2 interaction potential [6], the new N_2 molecule may result in a highly excited rotational state in comparison with the rotational state of the original molecule.

A theoretical determination of exchange scattering must be based on a quantum mechanical description. For example, the N–N^+ resonance charge exchange [7] would be underestimated by almost an order of magnitude using classical mechanics. The first *ab initio* potential energy surface for the N–N_2 system (denoted as WSHDSP PES) that includes variation in the interatomic separation distances of the N_2 molecules and the first quantum dynamics study of the cumulative reaction probabilities and chemical reaction rates of the N + N_2 exchange reaction have recently been reported by our group [6].

The WSHDSP PES [6] for N–N_2 is constructed from the results of high-level quantum electronic structure calculations. These calculations are primarily based on the same methods [8] that were used earlier to determine the *ab initio* data for the construction of the very accurate N–N_2 rigid-rotor PES of Stallcop et al. [9]; long-range dispersion force data and nearly 4,000 *ab initio* data points are applied to construct the N–N_2 WSHDSP PES. This potential has a number of special features. In particular, strong enhancements in the collision cross section due to the formation of a short-lived metastable N_3 state have been reported [10,11].

2.1 Rotational Excitation

State-to-state cross sections for scattering energies at 2.5 eV are presented in Fig. 1 to illustrate the distribution of rotational state excitation. Note that we plot two sets of cross sections — one with nuclear spin of the N_3 system properly accounted for, and the other without spin. N_3 obeys the Bose-Einstein statistics. Since the incorporation of nuclear spin introduces rapid oscillations in the cross sections, the underlying features in the cross sections are more easily recognizable using the data without nuclear spin. Note, however, the rapidly varying cross sections of even versus odd j states in N_2 will be reflected in its radiation profile.

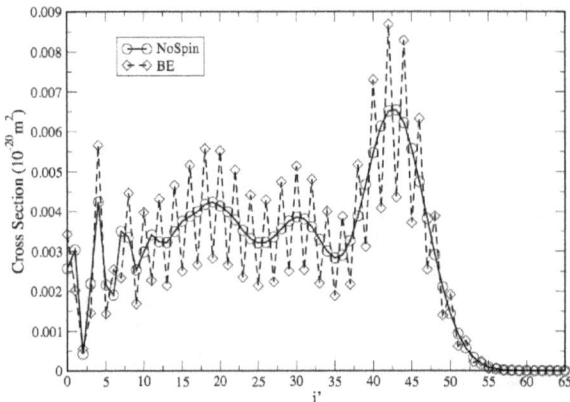

FIG. 1. State-to-state cross sections for exchange scattering with rotational excitation from initial ground rotational and vibrational state of N_2, N + $N_2(v=0, j=0) \rightarrow N_2(v'=0, j')$ + N, vs. j' for total scattering energy at 2.5 eV. The symbols o and ◊ represent the results obtained by neglecting nuclear spin and using Bose-Einstein statistics for the nuclear spin, respectively. The discrete data points are connected by a straight line for clarity.

Fig. 2 shows the cross section $\sigma_{v'j' \leftarrow vj}$ for scattering from an $(v = 0, j = 0)$ initial ro-vibrational state to various final ro-vibrational states. The curves of $\sigma_{0j' \leftarrow 00}$ for low-lying final ro-vibrational states, such as $(v' = 0, j' = 0)$ and $(v' = 0, j' = 2)$ states, have weak oscillating behavior arising from the formation of short-lived metastable N_3 states [10,11]. Note that the cross section for $j = 0 \rightarrow j' = 48$ is larger than for $j = 0 \rightarrow j' = 2$ at most energies. The variation of the large cross sections for producing high j' with kinetic energy verifies that the N + N_2 exchange reaction is quite efficient for producing high rotational states of N_2 molecules.

The kinetic energies and average energies of the rotational state distributions of our studies are comparable to the translational and rotational temperatures of the experiment of [1]. For example, in our study the average N_2 rotational energy for a total scattering energy at 3.0 eV corresponds to rotational temperature (T_r) at 5,200 K; the kinetic energy in Fig. 2 corresponds to translational temperatures in the range 33,000 K to 50,000 K. Considering that the vibrational and rotational motions are not coupled for low vibrational excitations, we conclude that T_r for the lower vibrational states is close to the value 4,500 K found from the rotational distribution in [1]. Furthermore, the translational temperature of the experiment, estimated [1] to be above 40,000 K, falls within our estimate of 33,000 K to 50,000 K.

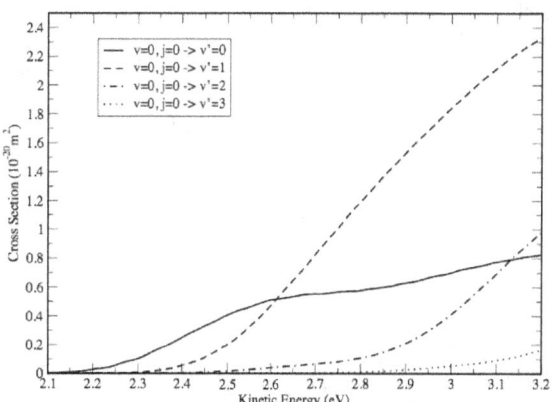

FIG. 3. Cross sections for exchange scattering with vibrational excitation from initial ground rotational and vibrational state of N_2, $N + N_2(v = 0, j = 0) \rightarrow N_2(v') + N$ as a function of kinetic energy; $v' = 0$-3.

3. $C_2H + H_2$ REACTION

Hydrogen is a major component in the atmosphere of all giant planets and methane is a minor component. The Titan atmosphere also contains 2–8 percent methane. The ethynyl radical C_2H is an intermediate product of methane chemistry upon high temperature shock. The $H_2 + C_2H$ reaction is an important step leading to acetylene, C_2H_2.

$$H_2 + C_2H \rightarrow C_2H_2 + H. \quad (3)$$

Because this reaction also plays an important role in combustion chemistry, extensive experimental studies have been reported in the literature [12-21], making it a good candidate for benchmarking the capability of first principles calculations of radical reactions of interest to giant planet missions. For this study, the first ever seven-dimensional (7D) quantum dynamics calculation is carried out. This is done by fixing two of the nine dimensional space based on quantum chemistry transition state calculation, thus reducing the nine-degree-of-freedom calculation to a seven-degree-of-freedom calculation. Then, using the time-dependent wave packet method [22,23], we calculate the initial state selected reaction probabilities for angular momentum J=0; these results are used with J-shifting and energy averaging to determine the rate constant. Fig. 4 compares the thermal rate constants from the present calculation with those from various experimental measurements. It is seen that in most cases the calculated rate constants not only agrees with the measured data to within experimental error, but also reproduce the temperature dependence over a wide range of temperatures. There are two exceptions. The single measurement of Renlund et al. [13] disagrees with both the calculated rate constant and all other measurements in the temperature range by almost an order of magnitude. Also, the high-temperature data of Kruse and Roth [19] not only are higher than the

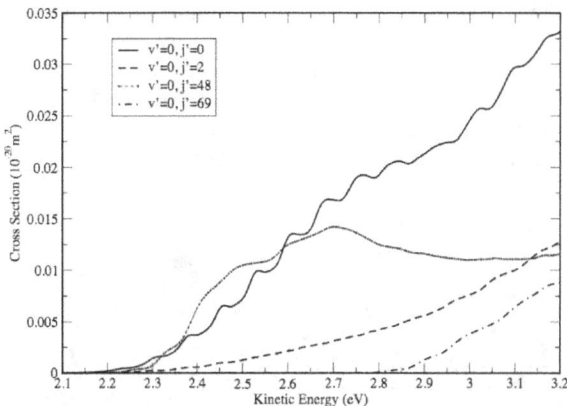

FIG. 2. State-to-state cross sections for exchange scattering with rotational state excitation from initial ground rotational and vibrational state of N_2, $N + N_2(v = 0, j = 0) \rightarrow N_2(v' = 0, j') + N$, vs. j' for total scattering energy at 2.5 eV.

The above results for rotational excitation indicate that the exchange reaction of $N + N_2$ collisions provides a mechanism to pump N_2 molecules at low rotational states into N_2 molecules at higher rotational states. Because this exchange reaction has a propensity for producing highly excited rotational states, it provides a route that may lead to a nonequilibrium rotational temperature behind a strong shock.

2.2 Vibrational Excitation

The cross section for v' excitation from a $(v = 0, j = 0)$ initial ro-vibrational state is obtained by summing over all the final rotational states; i.e.,

$$\sigma^*_{v'\leftarrow 00} = \sum_{j'} \sigma_{v'j'\leftarrow 00}. \quad (2)$$

The values of $\sigma^*_{v'\leftarrow 00}$ are shown in Fig. 3 for $v' = 0$-3. This figure shows that significant vibrational excitation of N_2 is produced by exchange reactions at higher kinetic energies.

calculated curve but also are higher than the general temperature dependence extrapolated from the lower temperature data. The present result clearly demonstrates our 7D reduced-dimensional calculations can yield accurate rate constants for radical reactions.

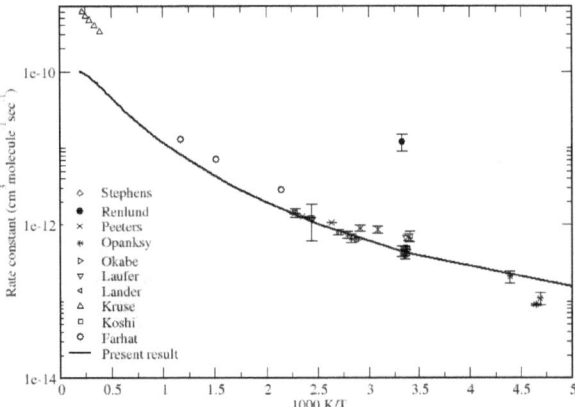

FIG. 4. Calculated rate constant of $H_2 + C_2H \rightarrow C_2H_2 + H$ vs. inverse temperature. Also presented are various experimental results from ◊ [12], ♦ [13], × [14], * [15], ▷ [16], ▽ [17], ◁ [18], △ [19], □ [20], and ○ [21].

4. ETCHING OF A Si SURFACE

The etching of a Si surface by halogens is used to illustrate our approach to the study of surface catalytic reactions. In these studies molecular dynamics (MD) simulations were used to investigate etching of Si surfaces by halogen species [24,25]. For complete descriptions of surface reactions taking place during the etching (and epitaxial growth) process, it is necessary to understand the bonding structures and stabilities of chemisorbed halogen species on the Si surface. MD studies today can reproduce short-time behavior of an etching process quite successfully. Generally speaking, simulation calculations are consistent with experiments.

MD calculations provide detailed understanding of the local physics and chemistry taking place during the adsorption and subsequent desorption of halogen species from the exposed Si surfaces. First and second chlorination (and fluorination) steps of surface Si atoms were analyzed separately, and the role played by surface vacancies was investigated. Steric effects coming from repulsive forces among the halogen atoms were found to be quite significant. Results indicate that the second step that leads to the formation of doubly chlorinated species is very important in the overall etching process. In MD simulations, energies and forces are calculated employing existing empirical potential functions for systems containing Si and halogen species. These potentials are based on two- and three-body interactions.

In general, calculations were carried out in two steps. First, we used a simulated annealing procedure based on a MD method. For complete equilibration, the temperature of the system was first increased and then linearly cooled down to a low temperature limit. Typical runs in these cases were about a picosecond long. The final configuration of the low-temperature result was then used as an input to a minimization routine based on a quasi-Newton algorithm to find the local minimum. This method is very efficient in finding low-lying minima for systems containing large numbers of moving particles.

Fig. 5 illustrates the mechanisms of surface defect etching of Si(001)-2×1 surface using MD simulations. Our results show that MD calculations provide an atomic level understanding of the etching process, which is the removal of halogenated Si particles from the surface region. Furthermore, the outcome of these studies strongly indicates that MD simulations are able to predict the thermodynamics and energetics of surface etching.

FIG. 5. Etching of SiF_3 from A type step edge. (a) The side and front views of the sample at the beginning of a room temperature MD trajectory calculation. (b) Same as in (a) at the end of the MD trajectory. The arrows indicate the dimerization of the lower step edge Si atoms as the SiF_3 on the top is etched away.

5. DYNAMIC DATABASE FOR DIATOMICS

A proper account of re-entry heating requires an accurate description of radiation effects. The radiation field is due to emission from atoms and molecules that are either heated by the shock layer or formed by reactions. Both neutrals and ions will participate. The transitions of importance are electronic, and modeling requires a database of opacity data for each species expected to be present in the flow field.

Our present focus is on the opacity data of diatomics. The diatomics involved depend on the entry atmosphere, and the number of species can be quite large. For example, entry into the earth's atmosphere produces excited states of CN, NO, N_2, as well as their ions.

The conventional way to obtain the opacity data is from experimental analysis. This, however, is a slow process requiring many man-years for a single diatomic. The measurements produce line positions and line strengths. While the line positions can be obtained with a very high degree of accuracy, line strengths are much more difficult to measure accurately. In fact, many spectroscopic studies do not even attempt to measure line strengths.

Another source of opacity data is from calculations carried out from first principles. In contrast to experimental measurements, theory can yield accurate line strengths without too much difficulty, but accurate line positions are much more difficult. However, progress has been made on that front; see, for example, Polyansky et al. [26]. In the past, a calculation for a single diatomic also required a number of man-years of work. Now we can exploit the recent increases in computer power to introduce a new paradigm whereby automatic computation of accurate opacities for diatomics becomes possible. We call our software package for carrying this out DDD.

DDD will be accessible as a world-wide web interface, and will work in one of two ways. The first mode, the retrieval mode, occurs when a user requests data for a diatomic for which calculations had been previously carried out. In this case, DDD will return the data essentially instantly and in one of several forms indicated by the user. The second mode, the data creation mode, occurs when a diatomic with no stored data is requested. In this case DDD will start the necessary calculations to determine the opacity required by the user. Depending on the diatomic and the load on the parallel computer used, this calculation may require a few minutes to several days. If the calculations are not completed in a short period of time, DDD will inform the user that calculations are being carried out, and will e-mail the user when the calculations are completed.

We have been testing our preliminary version of DDD, and the results are quite promising. In Table I we present some preliminary results for CN, T_e, the electronic excitation energy, ω_e, the vibrational frequency in the harmonic approximation, and B_e, the rotational constant for a rigid rotor. These are low-order spectroscopic constants describing the line positions, but our calculations also yield line strengths. Numbers in parenthesis are experimental values from Huber and Herzberg [27]. The agreement is quite satisfactory for this level of calculation. Of interest is the large number of states that are predicted for which experiments have not been carried out. Note that these calculations were carried out without any user intervention after the initial specification of the atoms C and N.

The algorithm for DDD is quite complicated. Given a database for atoms consisting of orbital occupations of the ground and valence excited states, DDD constructs a preliminary list of binding configurations of the diatomic. These orbital occupations are then used in trial calculations. Based on these results, more sophisticated calculations are carried out on the strongly bound levels — potential energy curves, transition moment curves, and coupling matrix elements are computed. For these calculations, the highly reliable multi-reference configuration interaction method is used. These are then used in a coupled rotation-vibrational-electronic code [28] to generate a line list.

TABLE I: Summary of Electronic States of CN. Numbers in parenthesis are from [27].

	T_e(cm^{-1})	ω_e	B_e
$^4\Sigma^+$	67917	1073	1.316
$^2\Pi_r$	67620	1351	1.174
$^4\Pi_r$	66585	2393	1.714
$^2\Delta_r$	60704 (60096)	1549 (1239)	1.375(1.383)
$^2\Sigma^-$	59638	1242	1.347
$^2\Sigma^+$	59439 (59151)	1721 (1681)	1.456(1.487)
$^2\Phi_r$	59291	931	1.095
$^2\Pi_r$	58747	2187	1.429
$^4\Sigma^-$	52655	1260	1.348
$^2\Pi_r$	51761 (54486)	1133 (1004)	1.045(1.162)
$^4\Delta_r$	46864	1309	1.367
$^4\Pi_r$	44243	923	1.104
$^4\Sigma^+$	36466	1326	1.367
$^2\Sigma^+$	27128 (25752)	2059 (2163)	1.868(1.973)
$^2\Pi_i$	8401 (9245)	1766 (1812)	1.642(1.715)
$^2\Sigma^+$	0 (0)	2028 (2068)	1.818(1.900)

6. CONCLUDING REMARKS

This paper illustrates the capability of present day first principles quantum chemistry calculations. By making use of the recent advances in hardware and software, first principles calculations can now provide reliable chemistry data needed for aerothermal analysis of aerocapture missions and other entry problems. The N + N_2 calculation supports the experimental observation by Fujita et al. [1] of strong nonequilibrium between the kinetic and rotational temperatures of a high-energy shock. Furthermore, our theoretical calculations provide insight into the origin of this nonequilibrium and indicate that future aerothermal modeling of high-velocity entry should incorporate these nonequilibrium conditions. The scatter in the measured $C_2H + H_2$ reaction data at high temperatures illustrates the risk of using limited experimental data. First principles calculations can be validated by comparisons with available low-temperature experimental data and, moreover, provide otherwise unavailable high-temperature data. Our capability is not limited to gas-phase studies, as illustrated by the silicon etching study. These techniques can be used to study surface catalytic reactions. The web-based DDD is a pioneering tool that provides a powerful means for the aerothermodynamicists to do their modeling of radiation heat load. We also point out that sensitivity analysis should be used as a guide for selecting the critical subset of required data. This approach will significantly shorten the turn around time, minimize the computational effort, and maximize the benefit of the calculations.

7. REFERENCES

1. Fujita K., Sato S., Abe T., and Ebinuma Y., Experimental Investigation of Air Radiation from behind a Strong Shock Wave, *JOURNAL OF THERMOPHYSICS AND HEAT TRANSFER*, Vol. 16, 77-82, 2002; also AIAA paper 98-2467, June 1998.

2. Bose D., Wright M., and Gökçen T., Uncertainty and Sensitivity Analysis of Thermchemical Modeling for Titan Atmosphere Entry, AIAA paper 2004-2453, June 2004.

3. Park C., Review of Chemical-Kinetic Problems of Future NASA Missions I: Earth Entries, *JOURNAL OF THERMO-PHYSICS AND HEAT TRANSFER*, Vol. 7, 385-398, 1993.

4. Fujita K., and Abe T., State-to-State Nonequilibrium Rotational Kinetics of Nitrogen Behind a Strong Shock Wave, AIAA paper 2002-3218, 8th AIAA/ASME Joint Thermophysics and Heat Transfer Conference, June 2002.

5. Wang D. Y., Stallcop J. R., Dateo C. E., Schwenke D. W., and Huo W. M., Quantum Scattering Study of Ro-Vibrational Excitations in $N+N_2$ Collisions Under Re-Entry Conditions, AIAA paper 2004-337, Jan. 2004.

6. Wang D. Y., Stallcop J. R., Huo W. M., Dateo C. E., Schwenke D. W., and Partridge H., Quantal Study of the Exchange Reaction for $N+N_2$ Using an *Ab Initio* Potential Energy Surface, *JOURNAL OF CHEMICAL PHYSICS*, Vol. 118, 2186-2189, 2003.

7. Stallcop J. R., Partridge H., and Levin E., Resonance Charge Transfer, Transport Cross Sections, and Collision Integrals for $N^+(^3P)-N(^4S^0)$ and $O^+(^4S^0)-O(^3P)$ Interactions, *JOURNAL OF CHEMICAL PHYSICS*, Vol. 95, 6429-6439, 1991.

8. Stallcop J. R., Partridge H., and Levin E., Effective Potential Energies and Transport Cross Sections for Interactions of Hydrogen and Nitrogen, *PHSICAL REVIEW A*, Vol. 62, 062709(1-15), 2000.

9. Stallcop J. R., Partridge H., and Levin E., Effective Potential Energies and Transport Cross Sections for Atom-Molecule Interactions of Nitrogen and Oxygen, *PHYSICAL REVIEW A*, Vol. 64, 042722(1-12), 2001.

10. Wang D. Y., Huo W. M., Dateo C. E., Schwenke D. W., and Stallcop J. R., Reactive Resonances in the $N+N_2$ Exchange Reaction, *CHEMICAL PHYSICS LETTERS*, Vol. 379, 132-138, 2003.

11. Wang D. Y., Huo W. M., Dateo C. E., Schwenke D. W., and Stallcop J. R., Quantum Study of the $N+N_2$ Exchange Reaction: State-to-State Reaction Probabilities, Initial-State Selected Probabilities, Feshbach Resonance, and Product Distributions, *JOURNAL OF CHEMICAL PHYSICS*, Vol. 120, 6041-6050, 2004.

12. Stephens J. W., Hall J. L., Solka H.,Yan W.-B., Curl R. F., and Glass G. P., Rate Constant Measurements of Reactions of C_2H with H_2, O_2, C_2H_2 and NO Using Color Center Laser Kinetic Spectroscopy, *JOURNAL OF PHYSICAL CHEMISTRY*, Vol. 91, 5740-5743, 1987.

13. Renlund A. M., Shokoohi F., Reisler H., and Wittig C., Gas-Phase Reactions of C_2H ($X^2\Sigma^+$) with O_2, H_2, and CH_4 Studied via Time-Resolved Product Emissions, *CHEMICAL PHYSICS LETTERS*, Vol. 84, 293-299, 1981.

14. Peeters J., Van Look H., and Ceursters B., Absolute Rate Coefficients of the Reactions of C_2 with NO and H_2 between 295 and 440 K, *JOURNAL OF PHYSICAL CHEMISTRY*, Vol. 100, 15124-15129, 1996.

15. Opansky B. J., and Leone S. R., Rate Coefficients of C_2H with C_2H_4, *JOURNAL OF PHYSICAL CHEMISTRY*, Vol. 100, 19904-19910, 1996.

16. Okabe H., Photochemistry of acetylene at 1470Å, *JOURNAL OF CHEMICAL PHYSICS*, Vol. 75, 2772-2778, 1981.

17. Laufer A. H., and Bass A. M., Photochemistry of Acetylene. Bimolecular Rate Constant for the Formation of Butadiyne and Reactions of Ethynyl Radicals, *JOURNAL OF PHYSICAL CHEMISTRY*, Vol. 83, 310-313, 1979.

18. Lander D. R., Unfried K. G., Glass G. P., and Curl R. F., Rate Constant Measurements of C_2H with CH_4, C_2H_6, C_2H_4, D_2, and CO, *JOURNAL OF PHYSICAL CHEMISTRY*, Vol. 94, 7759-7763, 1990.

19. Kruse T. and Roth P., Kinetics of C_2 Reactions during High-Temperature Pyrolysis of Acetylene, *JOURNAL OF PHYSICAL CHEMISTRY*, Vol. 101, 2138-2146, 1997.

20. Koshi M., Fukuda M., Kamiya K., and Matsui H., Temperature Dependence of the Rate Constants for the Reactions C_2H with C_2H_2, *JOURNAL OF PHYSICAL CHEMISTRY*, Vol. 96, 9839-9843, 1992; Koshi M., Nishida N., and Matsui H., Kinetics of the Reactions of C_2H with C_2H_2, H_2, and D_2, *ibid*, Vol. 96, 5875-5880, 1992.

21. Farhat S. K., Morter C. L., and Glass G. P., Temperature Dependence of the Rate of Reaction of C_2H with H_2, *JOURNAL OF PHYSICAL CHEMISTRY*, Vol. 97, 12789-12792, 1993.

22. Zhang D. H., and Zhang J. Z. H., Full-Dimensional Time-Dependent Treatment for Diatom-Diatom Reactions: the H_2+OH Reaction, *JOURNAL OF CHEMICAL PHYSICS*, Vol. 101, 1146-1156, 1994.

23. Zhang D. H., and Zhang J. Z. H., Quantum Reactive Scattering with a Deep Well: Time-Dependent Calculation for $H+O_2$ Reaction and Characterization for HO_2, *JOURNAL OF CHEMICAL PHYSICS*, Vol. 101, 3671-3678, 1994.

24. Srivastava D., Halicioglu T., and Schoolcraft T. A., Fluorination of Si(001)-2×1 Surface Near Step Edges: A Mechanism for Surface Defect Induced Etching, *JOURNAL OF VACUUMN SCIENCE AND TECHNOLOGY A*, Vol. 17, 657-661, 1999.

25. Halicioglu T., Calculated Energetics for Adsorption and Desorption Steps During Etching of Si(110) Surface by Cl, *JOURNAL OF VACUUMN SCIENCE AND TECHNOLOGY A*, Vol. 19, 372-375, 2001.

26. Polyansky O. L., Császár A. G., Shirin S. V., Zobov N. P., Barletta P., Tennyson J., Schwenke D. W., and Knowles P. J., High-Accuracy *Ab Initio* Rotation-Vibration Transitions of Water, *SCIENCE*, Vol. 299, 539-542, 2003.

27 Huber K. P. and Herzberg G., *Constants of Diatomic Molecules*, Van Nostrand Reinhold Co., New York, 1979.

28. Schwenke D. W., Opacity of TiO from a Coupled Electronic State Calculation, *FARADAY DISCUSSIONS*, Vol. 109, 321-334, 1998.

BENEFITS OF APPLICATION OF ADVANCED TECHNOLOGIES FOR A NEPTUNE ORBITER, ATMOSPHERIC PROBES, AND TRITON LANDER

Alan Somers[1], Luigi Celano[2], Jeffrey Kauffman[3], Laura Rogers[4], and Craig Peterson[5]

[1]California Institute of Technology, MSC 891 Pasadena, California 91126, somers@caltech.edu
[2]California Institute of Technology, MSC 1010 Pasadena, California 91126, luigi@caltech.edu
[3]California Institute of Technology, MSC 1071 Pasadena, California 91126, jeffreyk@caltech.edu
[4]California Institute of Technology, MSC 831 Pasadena, California 91126, lrogers@caltech.edu
[5]Jet Propulsion Laboratory, 4800 Oak Grove Drive M/S: 301-180 Pasadena, CA 91109-8099, craig.peterson@jpl.nasa.gov

ABSTRACT

Missions with planned launch dates several years from today pose significant design challenges in properly accounting for technology advances that may occur in the time leading up to actual spacecraft design, build, test and launch. Conceptual mission and spacecraft designs that rely solely on off the shelf technology will result in conservative estimates that may not be attractive or truly representative of the mission as it actually will be designed and built. This past summer, as part of one of NASA's Vision Mission Studies, a group of students at the Laboratory for Spacecraft and Mission Design (LSMD) have developed and analyzed different Neptune mission baselines, and determined the benefits of various assumed technology improvements. The baseline mission uses either a chemical propulsion system or a solar-electric system. Insertion into orbit around Neptune is achieved by means of aerocapture. Neptune's large moon Triton is used as a tour engine. With these technologies a comprehensive Cassini-class investigation of the Neptune system is possible. Technologies under investigation include the aerocapture heat shield and thermal protection system, both chemical and solar electric propulsion systems, spacecraft power, and energy storage systems.

Key words: Neptune; Triton; Aerocapture; Technology.

1. INTRODUCTION

The aim of this study is to quantify the benefits of advanced technology for a Cassini-class mission to Neptune. This data would be useful in guiding a technology development program leading up to such a mission. Given a price tag of almost $3B and a launch date beginning in 2017, this mission would likely have its own technology development program.

The requirements for this mission are that it cost no more than $5B, that a nuclear reactor not be used, that it must be launchable by a Delta IV Heavy or lesser rocket, and that the trip time not exceed 12 years.

Our methodology was to construct a baseline mission satisfying these requirements. For each technology area under study, the baseline was reevaluated assuming a reasonable range of improvement in the technology. The benefits were quantified in terms of launch mass, and where possible, cost. The cost estimates do not, however, include the cost of developing the technology to the specified level. This was beyond the scope of our study. Cost estimates were made using the 2003 JPL (Jet Propulsion Laboratory) Cost Model.

1.1. Science Goals

Current models suggest that Uranus and Neptune have similar compositions and histories, and that exploring either one will yield useful information about the other, and about the primordial solar system. Triton is believed to be a Kuiper belt object captured by Neptune. Thus a mission to Neptune would also gather information about the Kuiper belt. For this reason Neptune is considered a more desirable target than Uranus, despite the distance.

The Neptune system has 4 major targets of investigation: the planet, the rings, the magnetosphere, and Triton. The

I. Neptune	1.	Measure the composition of Neptune's deep atmosphere
	2.	Measure the thermal structure of Neptune's deep atmosphere
	3.	Measure the winds of Neptune's deep atmosphere
	4.	Image the entire planet at various spatial locations and times
	5.	Spectrally image the planet in UV to far-IR at various locations and times
	6.	Measure the three-dimensional structure of the magnetic field
	7.	Measure the three-dimensional structure of the gravitational field
	8.	Measure atmospheric properties of upper atmosphere
II. Triton	1.	Image Triton globally at high resolution (100 m)
	2.	Image areas of Triton surface at very high resolution (10 m)
	3.	Spectrally image surface in UV to far-IR for surface composition (100 m)
	4.	Measure magnetic field of Triton
	5.	Measure gravitational field of Triton
	6.	Measure atmospheric properties of Triton
	7.	Examine relevant geologic properties, including plumes and surface features
	8.	Map surface temperatures of Triton
III. Rings	1.	Image rings at high resolution (100 m) and determine orbital characteristics of rings
	2.	Image minor satellites and determine orbital characteristics of satellites
	3.	Image ring arcs in UV to far-IR at high resolution (100 m)
	4.	Image Proteus, Larissa, and Nereid in UV to far-IR at high resolution (100 m)
	5.	Determine composition of large ring bodies and minor satellites
	6.	Determine ring particle size and composition
	7.	Determine composition and mass of Proteus, Larissa, and Nereid
	8.	Measure magnetic fields produced by ring bodies or minor satellites, if any
IV. Magnetosphere	1.	Observe magnetosphere at various spatial locations and times
	2.	Determine composition, energy, temperature, and distribution of particles trapped in magnetosphere

Table 1. Measurement objectives

measurement objectives of the mission are presented in Table 1.

1.2. Model Overview

The mission is modeled using ICEmaker (Integrated Concurrent Engineering), a software tool developed at the LSMD. It is a medium fidelity model. The spacecraft is modeled at the component level, with components inherited or extrapolated for predicted technology advances. Components are sized according to first principles subject to reasonable approximations. For example, the structural bus is modeled with rules of thumb based on continuous mechanics, not finite element analysis. The thermal balance is based only on radiative calculations, with margins to accommodate conduction through the bus. Orbital mechanics are modeled as a series of two-body problems, but the SEP trajectory is selected from a set of trajectories developed at the Jet Propulsion Laboratory (JPL) for NASA's In Space Propulsion (ISP) program. Contingency is applied at the system level, based on standard AIAA mission classes. Aerocapture is not modeled computationally. Aerocapture parameters were estimated based mainly on [3], [4], and [15].

2. BASELINE OVERVIEW

The baseline mission consists of four modules: a Neptune orbiter, an atmospheric probe, a Triton lander, and a SEP (Solar Electric Propulsion) carrier. The total wet mass without contingency is 4224 kg. This is launched into a 10.26 year trajectory with a 4 year science tour at Neptune and Triton. Insertion is accomplished by aerocapture. Further details of the baseline are covered by subsystem below.

2.1. Mission Design

A Boeing Delta IV launch vehicle lifts the spacecraft to a C3 of 18436000 m^2/s^2. The mission then uses solar electric propulsion with a VJGA (Venus Jupiter Gravity Assist) to reach Neptune in 10.26 years. The SEP engines are shut off at 3 AU (Astronomical Units), but the SEP module is retained until just prior to insertion, to carry the probe and a downlink antenna. 5 months prior to insertion, the probe is released from the carrier. The probe enters Neptune's atmosphere and relays its data to Earth through the carrier just before aerocapture.

Aerocapture takes place with an entry velocity of 22 km/s

and a Δv of 6667 m/s . Peak deceleration is 22 g . The design uses a slender body ellipsled aeroshell. The mass fraction of the aeroshell was assumed to be 28% of the entry mass [3]. In light of more recent studies such as [4], a mass fraction of 44% would be more realistic for current TPS (Thermal Protective System) technology. However, materials advancements could reasonably lower this to 36% and current estimates of trailing ballute aeroshells are much lower.

The science phase of the mission lasts for 4 years, during which the orbiter will make a flyby of Triton once every 12 Earth days. It will use Triton as an engine to increase the inclination of its orbit from near 0° to ~75°.

The orbiter releases the Triton lander prior to one of the flybys. The lander uses chemical rockets to guide its descent with a Δv of 1125m/s. No aeroshell is used for the lander. The lander will survive on the surface for 8 hours while relaying its data to the orbiter.

2.2. Thermal

Because of the wide range of thermal environments, from Venus to Neptune, the spacecraft was designed for a slight cold bias at Neptune where the thermal environment is the most stable, and the orbiter is operating at its highest power levels. The craft is designed for a target operating temperature between 285 K and 308 K.

The resulting configuration uses a moderate heater array with a total of 145 RHUs at 1 W each, supplemented with 75-100 W of cartridge heaters for colder areas of the spacecraft.

The orbiter uses a thermal coating with emissivity in the range of 0.1 - 0.07 (anodized titanium, some vapor deposited metals). The solar absorptivity is not a driving factor in Neptune orbit. For transit, the aeroshell and SEP carrier stage use a Ag-AlO overcoat ($A = 0.08, E = 0.19$) due to its low dependence on solar and IR radiation to maintain temperature while still keeping the spacecraft warm enough during eclipses and ballistic cruise beyond Jupiter.

The orbiter and SEP stage are also equipped with deployable heat pipe radiators totaling 9 m^2 coated with MgO/AlO white paint ($A = 0.09, E = 0.92$). Additionally, the power processors are mounted on the outer surface SEP stage with 0.4 m^2 of fixed radiator area per unit.

A Freon-12 pumped fluid loop is used to transport heat from the RTGs to either the interior of the spacecraft or to the radiators (modeled after the system used on MER). Internal orbiter and SEP components and tanks are wrapped with up to 7 kg of multi-layer mylar insulation (MLI).

The atmospheric probe uses MLI on the body, as well as a blunt conical heat shield with backshell for Neptune entry, with 28 passive RHUs for internal heating.

The Triton lander, because of the extreme cold environment of Triton (34 K) uses a 1 cm layer of aerogel where possible (weight < 0.001 kg).

2.3. Propulsion

The orbiter propulsion system serves mainly to provide trajectory corrections and manuevers throughout the mission. Upon reaching Neptune, the orbiter propulsion system puts the spacecraft into the proper entry trajectory and performs the periapsis raise maneuver. Within the Neptune system, it changes the orbit's plane from equatorial to polar by using Triton as a cranking engine. Combined, these manuevers require a Δv of 1770 m/s. This is provided with a dual mode, N_2O_4/Hydrazine propsulsion system. The orbiter has a single 5 kg thruster capable of 445 N thrust, an analog to a TRW DMLAE (Dual Mode Liquid Apogee Engine). The propulsion requirements are met with 277 kg of hydrazine and 363 kg of N_2O_4. Two tanks are used to store the main and ADACS (Attitude Determination and Control) propellant assuming a PV/W figure of 10,000 m. To maintain proper pressure levels in these tanks, roughly 3.2 kg of pressurant and an 18.7 kg pressurant tank are also present. In addition, the propulsion system uses another 32 kg of support components (plumbing, pressure transducers, etc.).

Because the Neptune probe uses a passive attitude control system, it has no need for a propulsion system.

The Triton lander is ejected from the spacecraft on a Triton approach while within the Neptune system. Its propulsion system slows the lander to a point several meters above the Triton surface, at which point the lander will drop and soft-land on its compressible landing pads. This sequence of manuevers requires a Δv of 1125 m/s. The lander uses a monopropellant system with a single 5 kg thruster capable of 44.5 N of thrust. Assuming an ISP of 285 s, 56.5 kg of propellant (hydrazine) is needed to meet this requirement. Also, roughly 0.3 kg of pressurant are used to maintain proper storage of the hydrazine in its tank. Using the same PV/W as before of 10,000 m, two 1.7 kg tanks are used to store the hydrazine and pressurant. Propellant lines, propulsion system management, and other support components add 2.8 kg of mass to the system.

2.4. Telecom

The science instruments included in the baseline require an average transmission rate of 164 kbps from Neptune, assuming 8 hours per day of downlink time is available. To meet this goal the orbiter carries a 3.6 m Ka-band dish antenna. It broadcasts with 98 W RF (Radio Frequency) power and an antenna efficiency of 65%. The data is encoded with a rate 1/2 turbo code requiring $E_b/N_o = 0.8$ for a BER (Bit Error Rate) of 10^{-5}. The ground station was assumed to be a 70 m DSN (Deep Space Network) antenna with 70% efficiency, but an additional 3 dB increase in gain was assumed to account for planned upgrades scheduled to be complete well before Neptune insertion [12]. An omnidirectional emergency antenna is not included because even at Ka band with 98 W RF power (a dubiously possible power level), the maximum achievable data rate is approximately 1 bps.

The orbiter includes a smaller 1.2 m S band antenna for communication with the Triton lander. The lander uses a wide-angle antenna with 1.4 W RF power for uplink and no downlink. It has no active pointing system, and the design assumption is that it can passively point the antenna to within 45° of the orbiter during its short life.

The atmospheric probe uses a similar S band wide angle antenna with 45° pointing accuracy, but it broadcasts at 3.5 W RF power. Since the orbiter is still within its aeroshell at this point, the carrier stage includes a 2 m S band antenna to relay the probe data. S band is used as opposed to a higher frequency to reduce atmospheric losses.

The link between the carrier stage and earth is accomplished by means of a 1.3 m X band antenna with 2.5 W RF power for downlink.

2.5. C&DH

The C&DH (Control and Data Handling) system was modeled in low fidelity. No improvements in this area were considered, since it is believed that the private sector will substantially develop C&DH technologies without NASA's help. The orbiter used 2 Harris RH-3000 computers for redundancy and 24GB of flash memory from SEAKR. The carrier module shared the orbiter's C&DH subsystem. The probe used 1 Harris RH-3000 computer and needed no external storage. The lander used 1 Harris RH-3000 computer with 768 Mb of external flash memory.

2.6. Power

The spacecraft has been designated both an average and peak power during each of 8 mission phases. The primary driver for the power system is the last phase, Science, both due to larger peak power requirements, and power decay associated with radioisotope power sources.

The science phase is tabulated with a peak power of 895 W and an average power of 685 W (including battery charge), both including a 40% contingency factor. Secondary batteries reduce the maximum power load by up to 82 W with contingency, bringing the power supplied by the RTGs to 813 W end-of-life.

The beginning-of-life (BOL) power requirement, given ∼14 years of mission time, is met by 8 advanced stirling RTGs, producing a total power of 992 W, with a total weight of 128 kg (124 W and 16 kg each). The orbiter carries a 15 kg secondary lithium-ion battery for load distribution in Neptune orbit.

The SEP stage is equipped with 77.5 m^2 of quad-junction solar arrays to meet the specified trajectory maximum power of 31 kW BOL at 1 AU, with a weight of 240 kg. Additionally, the SEP stage carries 21 kg of primary lithium thionyl-chloride batteries to power both the orbiter and SEP stage during launch, until the RTGs are brought online.

The atmospheric probe is powered during its descent by 14 kg of lithium thionyl chloride batteries. During cruise, the probe is connected to the orbiter power system via an umbilical connection. The Triton lander carries 17 kg of batteries, and is also powered by the orbiter during transit.

2.7. ADACS

The pointing control requirement during cruise is driven by the pointing requirements of the SEP stage antenna. Near Neptune, the pointing accuracy needed is ∼.2°. During this phase of the mission, attitude control is provided by a set of twelve .22 N hydrazine thrusters on the SEP stage. The SEP stage also carries a full complement of attitude control sensors, including 3 sun sensors and 3 star trackers. Inertial measurements are provided by gyroscopes and accelerometers (in an IMU) within the orbiter. Major trajectory control maneuvers can be accomplished by altering the thrust direction of the gimbaled NEXT ion engines.

After the SEP stage disengages, the orbiter performs aerocapture. Altitude control is necessary during aerocapture to compensate for uncertainties in atmospheric density and to maintain an acceptable aerocapture corridor (i.e., to not go so low into the atmosphere that the spacecraft burns up, or so high in the atmosphere that the spacecraft does not successfully capture into an appropriate Neptune orbit). Control is provided by six 70 N thrusters piercing the cooler side of the elipsled aeroshell. Venting hydrazine from these thrusters should reduce the heat conducted to the spacecraft through the metal plumbing. The expected heating rates were not calculated, however.

After aerocapture, the aeroshell is shed, exposing a set of sixteen .22 N thrusters that provide full 3-axis control in a perfect couples configuration (the thrusters fire in pairs, such that there is no net translatory motion of the spacecraft, only a net torque). The imager becomes the driver for the pointing control of the spacecraft during this phase of the mission. To satisfy this finer requirement as well as to improve pointing stability, the orbiter also carries a set of 4 reaction wheels. To maximize the duration of both science and telecom operations, the science payload is divided between two scan platforms. The power budget was sized to allow simultaneous data collection and telecom transmission, thus greatly increasing the total quantity of data taken in the mission.

2.8. Science & Instruments

The nominal science tour of Neptune is 4 years long. No science observations are made before arrival at Neptune because all instruments are enclosed within the aeroshell. All instruments are heritage or extrapolated from other missions. The total science return is 21 Tb.

The orbiter baseline includes the following instruments: Radar altimeter (Cassini), USO (Ultra Stable Oscillator) (Cassini), wide and narrow angle imager (Mars Observer Camera (MOC)), IR (InfraRed) spectrometer (Cassini Composite InfraRed Spectrometer (CIRS)), visible/near IR mapping spectrometer (Cassini Visible and Infrared Mapping Spectrometer (VIMS)), UV (UltraViolet) spectrometer (Galileo UltraViolet Spectrometer (UVS) and Cassini UltraViolet Imaging Spectrograph (UVIS)), magnetometer (Galileo and Cassini), dust instrument (Galileo Dust Detector System (DDS) and Cassini Cosmic Dust Analyzer (CDA)), plasma subsystem (Galileo and Cassini Plasma Spectrometer (CAPS)), ion detector (Galileo Energetic Particle Detector (EPD)), cosmic ray detector (Voyager Cosmic Ray System (CRS)), ion & neutral mass spectrometer (Cassini Ion and Neutral Mass Spectrometer (INMS)), plasma wave instrument (Cassini Radio and Plasma Wave Science instrument (RPWS)), energetic neutral atom instrument (Cassini Ion and Neutral Camera (INCA), gamma ray spectrometer (Near Earth Asteroid Rendezvous (NEAR) Gamma Ray Spectrometer (GRS)), and microwave radiometer (NPOESS (National Polar-orbiting Operational Environmental Satellite System) Preparatory Project (NPP) Advanced Technology Microwave Sounder (ATMS)). The average data rate for the orbiter is 167 kbps and the maximum rate is 342 kbps.

The atmosphere probe carries: Doppler wind instrument (Cassini USO), atmospheric structure package (Huygens Atmospheric Structure Instrument (HASI) and Galileo Atmospheric Structure Instrument (ASI)), net-flux radiometer (Galileo Net Flux Radiometer (NFR) and Cassini Descent Imager Spectral Radiometer (DISR)), neutral mass spectrometer (Galileo Probe Mass Spectrometer (GPMS)), nephelometer (Galileo), and radio emission detectors (Galileo Lightning and Radio emission Detectors (LRD)). The maximum data rate is 132 bps, and the total return is 570kb.

The lander carries: atmospheric structure package (Huygens HASI and Galileo ASI), mass spectrometer (Galileo GPMS), imagers (DISR), APXS (Mars Pathfinder Alpha Proton X-ray Spectrometer). The maximum data rate is 2492 bps, and the total return is 68.5Mb.

3. RESULTS

Aerocapture is an enabling technology for this mission. Using SEP injection and chemical insertion, it was necessary to eliminate the probe, orbiter, radar altimeter, dust instrument, cosmic ray detector, and energetic neutral atom instrument. The launch margin was just 17 kg with contingency. Using chemical injection and chemical insertion, we made the aforementioned sacrifices and also lengthened the cruise time to 15.84 years. We concluded that without aerocapture, we could not meet the science objectives.

Fig. 1 shows the trade space between aeroshell mass fraction and payload. Here the payload is defined as the mass of the orbiter's instruments and the entire lander, since the lander is carried until after aerocapture but the probe is not. The baseline has an aeroshell mass fraction of 28% and a payload of 472 kg. We now believe that contemporary technology is capable of no better then 40–44%. An improvement to 36%, just an 8% improvement in technology, would allow for 26kg additional payload.

Of all technology areas, the instruments have the greatest marginal mass yield. That is, a 1 kg change in instrument mass on the orbiter yields a 5.7 kg change on the mission mass, assuming that the instruments' power consumption scales with their mass. Most important are the instruments carried on the lander. There is a small-scale delta of 2.8 kg lander wet mass for every 1 kg of instruments added to the lander. This yields a $16\times$ rollup from the lander's instruments to the mission's total wet mass.

Fig. 2 shows the effects of increased RTG energy density on the wet mass of the spacecraft. The baseline used Stirling 2.0 generators, with 7.75 W/kg. For comparison, Cassini used solid state SiGe thermoelectric generators

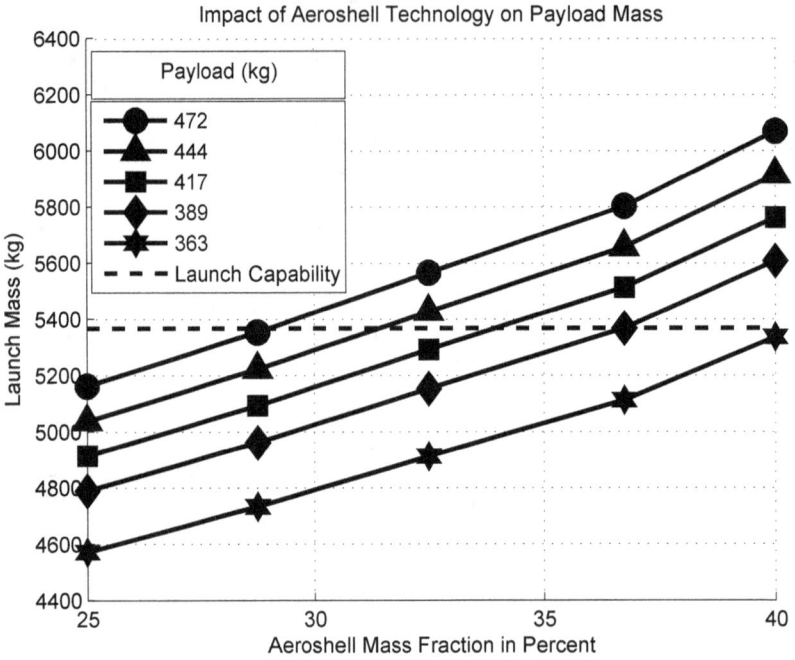

Figure 1. Allowable payload by aeroshell mass fraction.

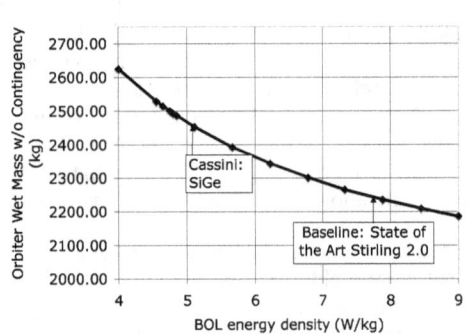

Figure 2. Effects of RTG technology on orbiter mass.

that achieved 5.07 W/kg. RTGs show a large range of potential improvement, corresponding to a mission wet mass change of ∼200kg.

The structural material was also a high yield area. 11% of the orbiter's wet mass is composed of the structural material, so this should not be surprising. The baseline uses an aluminum bus, but a graphite/epoxy composite, if manufactured properly, has the potential to reduce spacecraft structure mass by up to 66% due to its high tensile strength, high modulus of elasticity (stiffness) and low density - as demonstrated by state-of-the-art composite propellant tanks. See Fig. 3.

However, replacing the baseline's propellant tanks with advanced composites has only moderate yield. The baseline assumed a slightly conservative PV/W of 10,000 m for the tanks. State of the art composite overwrap tanks can have PV/W as high as 21,600 m, and corporations claim that they can develop PV/W to as high as 100,000 m. However, as Fig. 4 shows, there are diminishing returns in developing the tanks past 30,000 m.

Solar cell efficiency is a moderate yield area. Since the solar cells are not inserted into Neptune orbit, there is a smaller roll-up to the mission mass. Also, deficiencies in the power of the SEP stage can be accomodated by tradeoffs in mission design. For example, a VEEJSGA (Venus-Earth-Earth-Jupiter-Saturn) trajectory was found

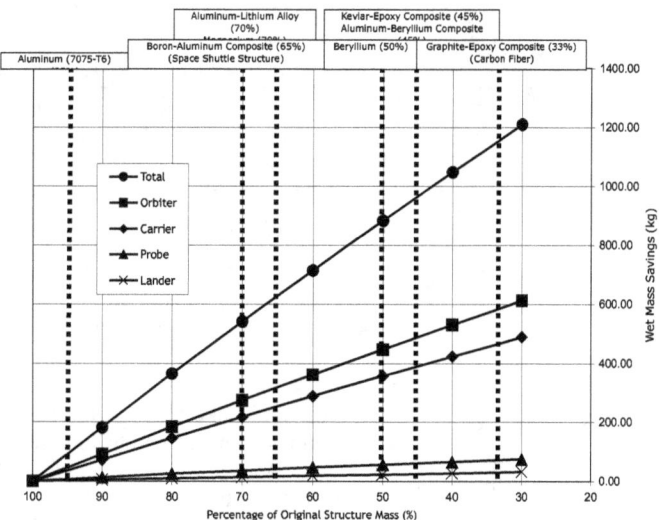

Figure 3. Advanced structural materials.

that reaches Neptune in the same time as the baseline. Using this trajectory, the same payload could be sent to Neptune with no SEP at all.

Advanced telecom technologies are also modeled. Inflatable and mesh antennas are unable to directly reduce mission mass. However, increasing antenna size allows for marked reduction in transmission power, allowing the orbiter to use 1 fewer RTG. See Table 2 for the summary. Antenna mass estimates are based on [9]. Inflatable antennas have low aperture efficiencies, complex and heavy deployment mechanisms, and little mass savings over fixed antennas for small and moderate sized antennas. Mesh antennas simply cannot achieve low areal densities at high frequencies. These limitations, coupled with the modest bandwidth requirements of the mission, lead us to believe that advanced antenna design will have minimal payoff.

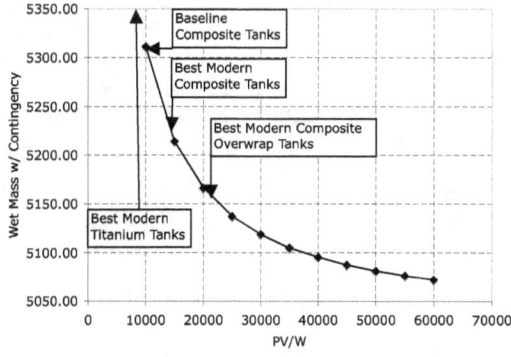

Figure 4. Effects of tank material on launch mass.

However, improvements to the DSN are extremely beneficial. An increase in link time from 8–20 hours per day saves 225 kg total on the mission. This is at a nominal cost of $20M for the added time [22].

Antenna Type	Fixed	Inflatable	Mesh	Fixed
Diameter (m)	3.60	5.05	4.50	3.62
Antenna System Mass (kg)	45.4	52.3	59.3	46.7
Aperture Efficiency (%)	65	40	65	65
RF Power (W)	97.5	68.0	69.0	32.0
DSN time (h/day)	8	8	8	20
Launch Mass (kg)	5312	5234	5265	5087
Savings (kg)	0	78	47	225

Table 2. Advanced antenna technologies

4. CONCLUSIONS

We conclude that a decade from now, for less than Cassini's cost, a deep-space mission could answer key questions about Neptune, Triton, and by extension the Kuiper belt. Aerocapture and RTGs are key enabling technologies for this mission. Aerocapture in particular is in need of significant development to support this mission. TPS material, aeroshell design, and aerocapture guidance algorithms all require work. Batteries, structural composites, DSN upgrades, and RTG energy densitites are also enhancing technologies. We conclude that development in these technology areas will yield the greatest benefits to a Neptune mission in the next 13 years.

5. FURTHER STUDY

The obvious and necessary extension of this study is to estimate the R&D cost required for these advancements. This was beyond the scope of the present study, but necessary to make the results truly useful. Only in the case of DSN time was even an estimate possible. Though we did estimate mission cost with the 2003 JPL Cost Model, the model is not calibrated for nonexistent technology.

Additionally, we would like to consider RTG-powered probes and landers. An RTG-powered lander would be able to collect useful data for weeks or months. From a science standpoint, this would enable seismic studies, shedding light on Triton's geysers and internal composition. An RTG-powered atmospheric probe could float around the planet on a balloon, measuring temporal changes in the atmosphere.

Inflatable ballutes are also a tempting option for study. They have the potential to greatly reduce the mass of the aerocapture system.

ACKNOWLEDGMENTS

We would like to thank the following people for contributions to this research: Joel Sercel and the AE/CDS 125 class, Tom Spilker, Muriel Noca, Andy Ingersoll, Robert Bailey, Robert Haw, Richard Cowley, and Craig Peterson. Credit is also given to JPL's Team X, upon whose unpublished work this research was substantially based.

The research described in this paper was carried out by the California Institute of Technology, under a contract with the National Aeronautics and Space Administration.

REFERENCES

1. Advanced Radioisotope Power Systems Report, March 2001.

2. Aerojet Successfully Completes Manufacturing and System Integration Milestones For NASA's NEXT Ion Engine Development Program, 9 September 2003. http://www.corporate-ir.net/ireye/ir_site.zhtml?ticker=GY&script=460&layout=1&item_id=447134 Available 03 June 2004.

3. Alagheband A., Corazzini T., Duchemin O., Henny D., Mason R., Noca M., Neptune Explorer: An All Solar Powered Neptune Orbiter Mission, AIAA 96-2980.

4. Bailey R. W., Hall J. L., Spilker T. R., Okong'o N. O., "Neptune Aerocapture Mission and Spacecraft Design Overview," AIAA-2004-3842, *Conference Proceedings of the 40th AIAA/ASME/SAE/ASEE Joint Propulsion Conference and Exhibit*, Huntsville, AL. July 2004

5. Bird R. B., Stewart W. E., Lightfoot E. N., *Transport Phenomena*, 2nd ed. New York: John Wiley and Sons, 2001. 912 p.

6. Brophy, et al. Ion Propulsion System (NSTAR) DS1 Technology Validation Report. 2000.

7. Cassini-Huygens Home. `http://saturn.jpl.nasa.gov` Available 03 June 2004.

8. Cassini Mission to Saturn. NASA Facts. JPL. `http://www.jpl.nasa.gov/news/fact_sheets/cassini.pdf` Available 14 March 2004.

9. Chandler C. W., 2004, The Case for Deep Space Telecommunications Relay Stations, NASA/CR–2004-213053.

10. Consultative Committee for Space Data Systems report concerning space data systems standards: "Lossless Data Compression" May 1997.

11. Delta IV - Payload Planners Guide. October 2000. Boeing. `http://www.boeing.com/defnese-space/space/delta/docs/DELTA_IV_PPG_2000.pdf` Available 03 June 2004.

12. Geldzahler B. and Deutsch L., Deep Space Mission System Roadmap.

13. Gilmore D. G., Spacecraft Thermal Control Handbook Volume I: Fundamental Technologies, AIAA, 2002.

14. Griffin M. D. and French J. R., Space Vehicle Design, AIAA, 2004.

15. Hall J. L., Noca M. A., Bailey R. W., Cost-Benefit Analysis of the Aerocapture Mission Set (AIAA 2003-4658), 39th AIAA/ASME/SAE/ASEE Joint Propulsion Conference & Exhibit, 2003.

16. Hammel H. B., Baines K. H., Cuzzi J. D., de Pater I., Grundy W. M., Lockwood G. W., Perry J., Rages K. A., Spilker T., Stansberry J. A., *Exploration of the Neptune System*. ASP Conference Series, Vol. 272, pp. 299-319, 2002.

17. Humble R. W., Henry G. N., Larson W. J., *Space Propulsion Analysis and Design*, McGraw-Hill, Inc. 1995.

18. Ingersoll, et al., Study of a Neptune Orbiter with Probes Mission, Proposal for NRA 03OSS01VM, 2003.

19. James B. Technology Assessment Group. Aerocapture. March 2004.

20. Lide D. R., ed., *CRC Handbook of Chemistry and Physics*, 76th ed., Boca Raton: CRC Press, 1995.

21. Pollara F., JPL Turbo Codes Web Page. 04 February 2004. `http://www331.jpl.nasa.gov/public/` Available 03 June 2004.

22. Rosenberg L., et al. Parametric Mission Cost Model v3 (costmodel6d10.xls). JPL. 17 June 2003.

23. Russell C. T., ed., *The Galileo Mission*, Space Science Reviews, Volume 60, Nos. 1-4, Boston, MA: Kluwer Academic Publishers, 1992.

24. Sarafin T. P. *Spacecraft Structures and Mechanisms - From Concept to Launch*. Microcosm Press, Kluwer Academic Publishers. 1995.

25. Sovey J. S., Rawlin V. K., Patterson M. J. Ion Propulsion Development Projects in U.S.: Space Electric Rocket Test 1 to Deep Space 1. *Journal of Propulsion and Power*, Vol. 17, No. 3, May-June 2001, pp. 517-526.

26. Structure and Evolution of the Universe Subcommittee of the Space Science Advisory Committee Meeting Report. `http://spacescience.nasa.gov/adv/minutes/SEUS0402.pdf` Available 03 June 2004.

27. Wertz J. R. and Larson W. J., ed., *Space Mission Analysis and Design*, 3rd ed. Torrance, CA: Microcosm Press, 1999

ON NONEQUILIBRIUM RADIATION IN HYDROGEN SHOCK LAYERS

Chul Park

Department of Aerospace Engineering, Korea Advanced institute of Science and Technology, Guseong-dong, Yuseong-gu, Daejeon, 305-701 Korea, cpark216@kaist.ac.kr

ABSTRACT

The influence of thermochemical nonequilibrium in the shock layer over a vehicle entering the atmosphere of an outer planet is examined qualitatively. The state of understanding of the heating environment for the Galileo Probe vehicle is first reviewed. Next, the possible reasons for the high recession in the frustum region and the low recession in the stagnation region are examined. The state of understanding of the nonequilibrium in the hydrogen flow is then examined. For the entry flight in Neptune, the possible influence of nonequilibrium is predicted.

1. INTRODUCTION

Entry flight into Jupiter has been accomplished in Project Galileo. In the future, exploration into other outer planets such as Saturn, Uranus, or Neptune is possible. The atmospheres of these outer planets consist of a mixture of hydrogen and helium. In the shock layer over the entry vehicle, hydrogen tends to be ionized. Ionization of hydrogen produces radiation which, depending on the condition, may be strong. In order to design the heatshields for these outer planet probe vehicles efficiently, one needs to examine how the heatshield for the Galileo Probe vehicle performed.

In the case of Galileo Probe, heating was expected to be mostly by radiation.. Prior to the Galileo Probe mission, theoretical prediction of the surface recession of the heatshield for the Probe vehicle has been made by, among others, Moss and Simmonds [1]. The surface recession of the Galileo Probe heatshield in the flight data was found to be substantially different from the predictions, as shown in Fig. 1.

In the figure, the abscissa is the ratio of the radial chord length along the surface s to the nose radius R. As seen in the figure, the recession in the flight was greater than the prediction by Moss and Simonds by a ratio of 1.74 to 1 in the frustum and smaller than the prediction by a ratio of 0.77 to 1 at the stagnation point.

Shown also in Fig.1 is the latest calculation by Matsuyama et al [2]. Their calculation reproduced the recession values in the frustum region closely. The difference between their calculation and the calculation by Moss and Simmonds [1] is in the assumed intensity of turbulence; turbulence is more intense in the latest calculation in the region adjacent to the ablating wall. In the calculation by Moss and Simmonds, turbulence intensity was assumed to be zero at wall, following the existing turbulence model for flows over a smooth wall. In the calculation by Matsuyama et al, it is assumed to be finite at wall. There are several possible reasons why turbulence intensity could be finite at wall. Among these possible reasons, Matsuyama et al have chosen the injection-induced turbulence model of Park [3] to explain it.

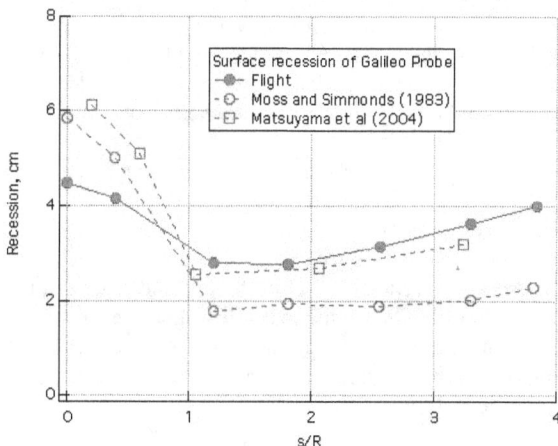

Fig. 1. Comparison of the predicted and measured recessions of the Galileo Probe heatshield.

According to the injection-induced turbulence model, the ablation product gas is inherently turbulent when it emerges from the surface of an ablating heatshield. The finite turbulence intensity at wall disperses the ablation product species, C_3, faster into the flow, and thus reduces its concentration in the region near the wall, as shown in Fig. 2. This leads to smaller radiation absorption by C_3, and hence larger radiative heat flux reaching the wall.

It is to be noted here that Matsuyama et al did not account for the increase in convective heating rate due to surface roughness. If it is accounted for, the calculated heating rate in the frustum region will become even larger.

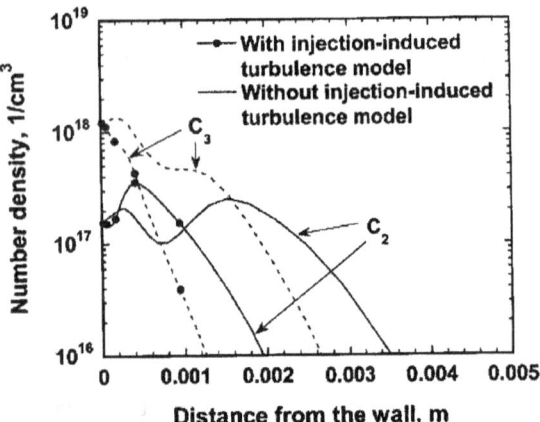

Fig. 2. Distribution of C_3 and C_2 in the boundary layer at the location of ARAD 7-8 in the frustum region calculated by Matsuyama et al [2].

The agreement between the theory and calculation in the frustum region leaves only the stagnation region behavior in the recession data to be explained. There are three possible explanations as to why the recession in the stagnation region is over-predicted. They are:

a) *Spallation*: Carbonaceous heatshield materials produce solid particles at their surface and inject them into the shock layer flow with a finite speed [4-6]. This phenomenon is referred to commonly as spallation. These solid particles travel to the inviscid region of the shock layer. There, the particles vaporize and the resulting carbon atoms partially ionize. The process of vaporization and ionization absorb energy, and thereby cool the flow and reduce radiation emission.

c) *Radiation Absorption*: The ablation product gas contains species that are not accounted for in the calculation but absorb radiation. The phenolic resin in the heatshield contains a substantial concentration of hydrogen and a small concentration of oxygen. These will form CH and CO. CO will absorb in the wavelength region shorter than 2000 A. CH will absorb from about 3500 to about 5000 A. CO is long-lived because of its strong bonding. But CH will decompose rapidly because of its weak bonding. Exactly how much of these two species exist in the boundary layer and how much radiation they will absorb are unknown at this time.

c) *Nonequilibrium*: The region immediately downstream of the bow shock wave is undergoing dissociation and ionization. A finite time is required for the ionization equilibrium is reached. In the nonequilibrium region prior to reaching equilibrium, electron density will be low or nonexistent. This nonequilibrium region will emit radiation smaller than the equilibrium region or none at all. Radiative heat flux reaching the ablating wall will be reduced correspondingly. This possibility was first proposed by Howe [7].

None of these three possible explanations has been explored in detail. It is the purpose of the present work to explore the nonequilibrium explanation, c).

2. EQUILIBRATION TIME

Howe based his arguments on the shock tube experiment conducted by Leibowitz [8]. Leibowitz studied the evolution of electron density behind a shock wave in a 21%H_2-79%He mixture. The shock tube was driven by an electric arc-heated driver gas shown schematically in Fig. 3. The driver section was in the shape of a cone. In the experiment, the intensity of the radiation emitted behind the moving shock wave at 5145 A, which is known to be proportional to the square of electron density, was measured as a function of time as shown in Fig. 4. The ionization equilibrium time, or relaxation time for ionization, t_{lab}, was defined from the oscillogram trace as shown. The true relaxation time, ∞ is t_{lab} multiplied by the density ratio across the shock wave, which was typically 5 in the experiment.

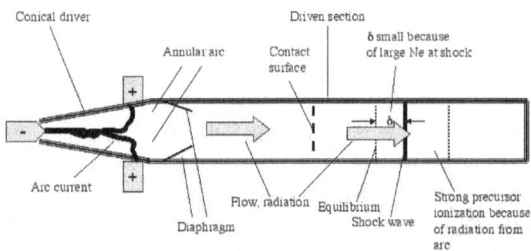

Fig. 3. Schematic of a conical electrically-driven shock tube used by Leibowitz [8] and Bogdanoff and Park [11].

The true relaxation time ∞ determined in Leibowitz's experiment, multiplied by the number density of H_2 molecules behind the shock wave, is shown as a function of the reciprocal of the post-shock translational temperature prior to vibrational excitation or chemical reaction (dissociation and ionization) T_s in Fig. 5.

Subsequently, Livingston and Poon [9] repeated the experiment in a shock tube driven by an electric arc configured in an annular geometry, as shown schematically in Fig. 6. Their data are compared with Leibowitz's data in Fig. 5.

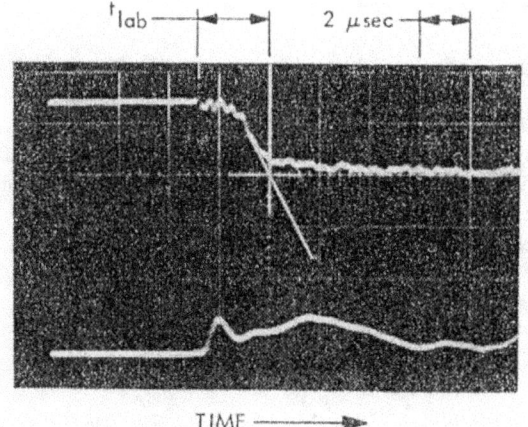

Fig. 4. Oscilloscope trace of continuum radiation at 5145 A in the shock experiment of Leibowitz [7] showing the definition of the ionization equilibrium (relaxation) time in the laboratory frame t_{lab}.

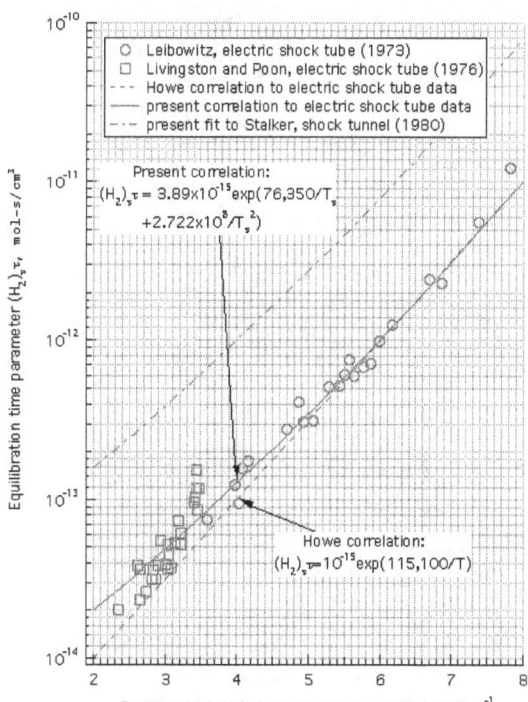

Fig. 5. The product of the number density of H2 behind shock and the equilibration time t plotted as a function of $1/T_2$.

Still later, Stalker [10] measured the relaxation time in a shock tunnel experiment. In that experiment, the test flow was produced in the test section of a shock tunnel. An inclined flat plate was placed in the test flow, and density variation behind the oblique shock wave was determined by an interferometer, as shown schematically in Fig. 7. The equilibration distance so obtained is shown by a dash curve in Fig. 5.

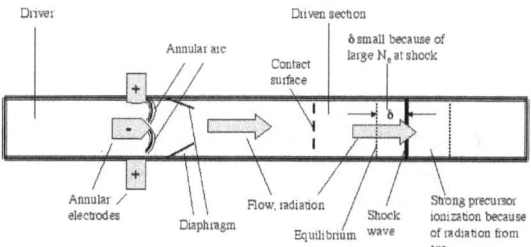

Fig. 6. Schematic of the annular-arc driven shock tube used by Livingston and Poon [9].

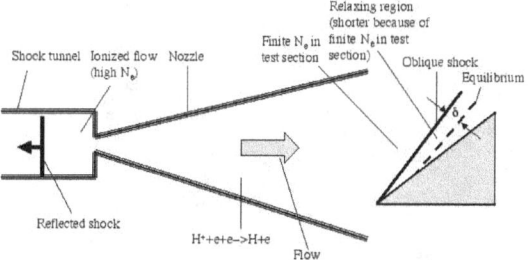

Fig. 7. Schematic of the shock tunnel experiment of Stalker [9].

As seen in Fig. 5, the data by Leibowitz and that by Livingston and Poon lie roughly on a straight line. Howe fitted Leibowitz's data by a straight line in this semi-log plot producing the correlation formula shown in the figure. A slightly more accurate correlation is derived here as

$$(H_2)_s \infty = 3.89 \times 10-15 \exp(76,350/T_s + 2.722 \times 108/T_s^2) \text{ mol-sec/cm}^3 \quad (1)$$

The relaxation time values obtained by Stalker [10] is 7.9 times longer than the values given by this expression, as shown in Fig. 5.

The substantial disagreement between the two sets of data on relaxation time casts doubt on both sets. A clue to the inadequacy of the two sets of data obtained by an arc-driven shock tube is in the measured values of electron density in the experiment by Livingston and Poon shown in Fig. 8. In the figure, the solid curve is the theoretical equilibrium value determined from the Rankine-Hugoniot relations assuming no radiative cooling. The dash curve shows the equilibrium values accounting for radiative cooling by the emission of Balmer lines in the wavelength range from 4300 to

6600 A that occur during the travel of 3 meters. As seen in the figure, measured electron density is substantially larger than the theoretical values.

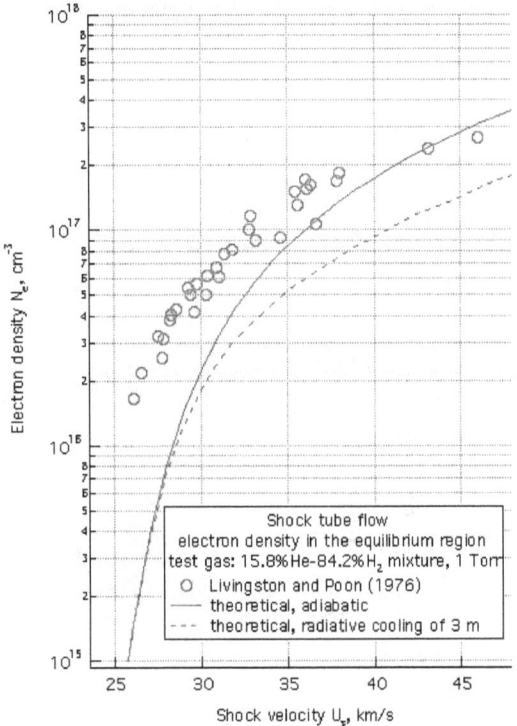

Fig. 8. Comparison between the measured and calculated electron density in the equilibrium region behind the shock wave in the experiment by Livingston and Poon [9].

Bogdanoff and Park [11] attempted to recreate the works of Leibowitz and Livingston and Poon in a shock tube of a design very similar to that used by Leibowitz. They found that the measured electron densities are much larger than the theoretical equilibrium values. They investigated the cause of this discrepancy, and found that the radiation emitted in the arc in the driver ionized the test gas flow behind the shock wave. Thus, the data by Leibowitz and Livingston and Poon are rendered inaccurate.

The data by Stalker is not totally trustworthy either. In his experiment, the flow in the reflected region is highly ionized. In the expanding nozzle, electron recombination occurs. But the recombination is not expected to be completed: finite concentration of electrons is bound to exist in the test section. The rate of the ionization process

$$H + e \rightarrow H^+ + e + e$$

Depends very strongly on the concentration of electrons in front of the shock wave. Presence of electrons in the test section will tend to shorten the relaxation time behind the shock wave.

Thus, none of the existing data on the ionization relaxation time can be trusted. The true value of ∞ should be longer than the Stalker value shown in Fig. 5. A new experiment is being carried out at NASA Ames Research Center which is free from the impediments encountered in those past experiments is presently being carried out. One must wait for the outcome of the experiment to correctly assess the nonequilibrium problem.

3. NONEQUILIBRIUM PROCESSES

The nonequilibrium processes occurring in the shock layer are shown schematically in Fig. 9. These processes are described in one-dimensional flow in Fig. 10.

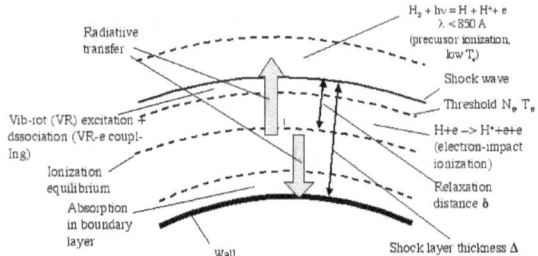

Fig. 9. Schematic of the nonequilibrium processes in shock layer.

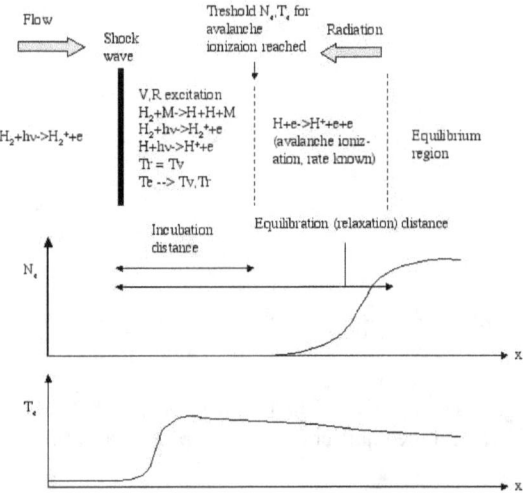

Fig.10. Schematic of nonequilibrium process behind a normal shock wave.

As shown in these figures, in front of the bow shock wave, H$_2$ starts to be dissociated and ionized by

H$_2$ + h∞ → H + H (800 < ∞ < 850 A)
H$_2$ + h∞ → H$_2^+$ + e (∞ < 800 A)

The second process, photo-ionization, leads to the so-called precursor ionization. The cross section for these two processes have been measured by Cook and Metzger [12], and are shown in Fig. 11. According to the data, the strongest absorption occurs at 700 A with a cross section of about 10^{-17} cm^2.

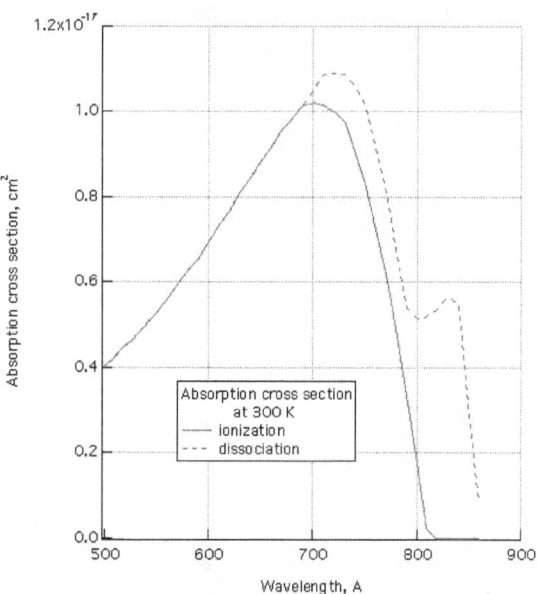

Fig. 11. Absorption cross section of H$_2$ for photo-dissociation and photo-ionization, from Cook and Metzger [12].

The radiation emitted by the ionized flow behind the shock wave propagates upstream as shown in Figs. 9 and 10. In the freestream flow in front of the shock wave, the radiation is absorbed by the cross section sown in Fig. 11. The radiation intensity decays there exponentially with distance according to the Beer's law. The e-folding distance of the radiation is given by

e-folding distance = 1/(∞n)

where ∞ is the cross section and n is the number density of H$_2$. The e-folding distance is shown for the Galileo Probe entry in Fig. 12. As shown, the e-folding distance is nearly 1 cm at the peak heating point. In the experiment by Bogdanoff and Park [11], the precursor ionization was detected to a distance of several centimeters. Electrons produced by the precursor ionization process is at a low temperature, but its exact value is unknown.

Behind the shock wave, vibrational and rotational excitation and dissociation of H$_2$ occur. According to Furudate et al [13], vibrational and rotational temperatures are locked together here. Electron temperature will tend to couple with the vib-rotational mode here. However, the exact extent of the coupling is unknown at this time. As vib-rotational temperature rises, electron temperature will rise also. At a certain point, the threshold value of electron density will be reached to trigger the avalanche ionization process.

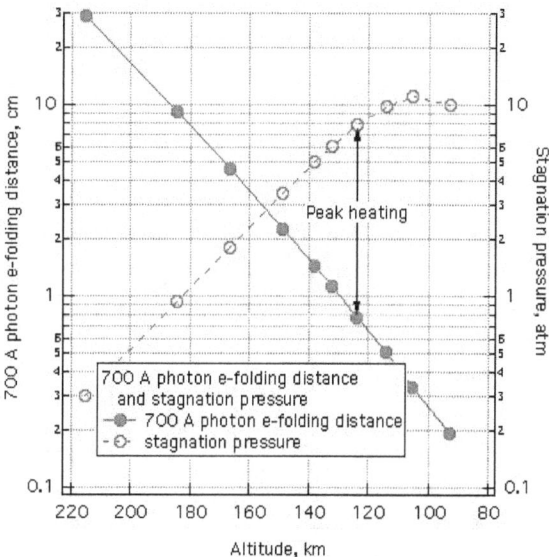

Fig. 12. The e-folding distance for decay of 700 A radiation in front of the shock wave.

The distance from the shock wave to the threshold point is the incubation distance. It is this distance which is unknown at this time. All three existing sets of experiment [8-10] suffered from uncertainties as described above. Currently, theoretical work is being carried out on this topic by Furudate and Chang [15]. Once the avalanche ionization is started, the process can be predicted relatively accurately, because the ionization process of H is well known from both theory and experiment. It is to be noted here that the rate of the avalanche ionization will be influenced by the radiative transfer phenomenon, as described by Park [14].

In the incubation region, density of the flow will not change rapidly. Density will change rapidly as the avalanche ionization occurs. The interferogram of Stalker [10] shows this trend. The difference between the two sets of data obtained by an arc-driven shock tube [8,9] and the shock tunnel data [10] is the difference in the length of the incubation distance.

4. INFLUENCE ON RADIATIVE FLUX

For the Galileo Probe at its peak heating point, the distance to the ionization equilibrium point ∞ is calculated using the relaxation time values of Stalker [10], and are shown as a function of the normalized chord length s/R in Fig. 13. The thickness of the radiating layer, i.e. the layer with equilibrium ionization, is shown as a ratio to the shock layer thickness. As seen here, the thickness of radiating layer is substantially smaller than the shock layer thickness.

In Fig. 15, the estimated surface recessions for the Galileo Probe at the two points are shown. These estimated values are obtained by multiplying the recession values of Matsuyama et al in Fig. 1 by the ratio of the emitting thickness to the shock layer thickness at the peak heating point. The exact value can be known, of course, through a detailed calculation.

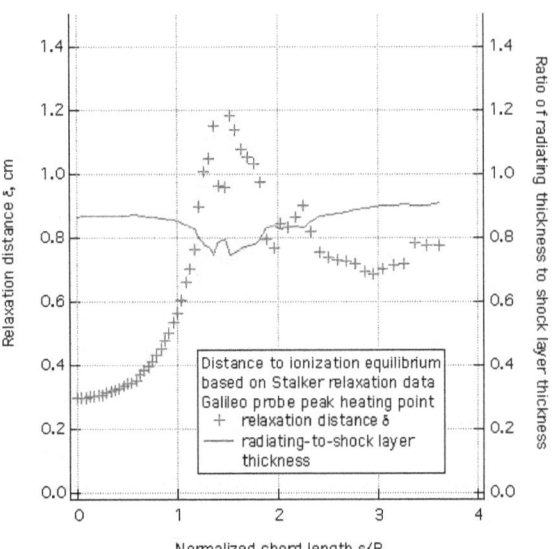

Fig. 13. Relaxation distance behind the bow shock for Galileo at its peak heating point calculated by Stalker's [10] relaxation data and the resulting radiating thickness expressed by its ratio to the shock layer thickness.

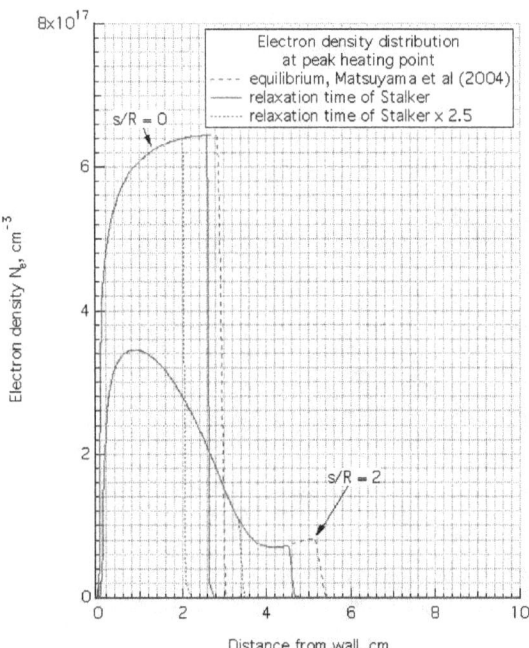

Fig. 14. Electron density profile across the shock layer at stagnation point and a frustum point accounting for the nonequilibrium phenomenon.

In Fig. 14, the electron density distribution across the shock layer is shown at two points, the stagnation point and a frustum point s/R = 2, resulting from the ionization nonequilibrium phenomenon. Two sets of relaxation distance were used here: the Stalker value and 2.5 times the Stalker value. For both cases, the electron density distribution becomes truncated because of the nonequilibrium phenomenon. For the stagnation point, the total number density of electrons has been significantly reduced by the truncation, especially for the 2.5 times Stalker value case. One can imagine that the radiative flux reaching the ablating wall will be reduced accordingly, because intensity of radiation emitted by a unit volume is approximately proportional to the square of electron density. However, for the frustum point, the truncation occurs in the region of low electron density. Therefore, one expects little decrease in radiative flux reaching the wall at the frustum point.

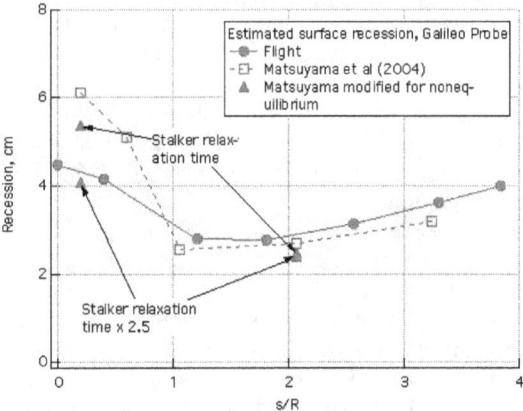

Fig. 15. Estimated surface recessions accounting for nonequilibrium, determined by Stalker's relaxation time and 2.5 times Stalker's relaxation times.

As seen in Fig. 15, if the relaxation time is 2.5 times Stalker's value, then the calculated recession should approximately agree with the flight data.

In Fig. 16, the ionization equilibration distance behind the normal shock wave is calculated for a typical aerobraking flight through the atmosphere of Neptune. The atmosphere is considered to consist of 20%He-80%H_2 mixture. The relaxation time data of Stalker [10] is used here.

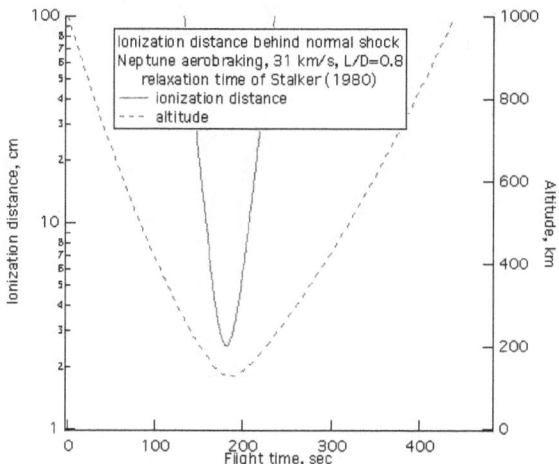

Fig. 16. Ionization relaxation distance behind a normal shock wave during the aerobraking flight through Neptune calculated using the relaxation time data of Stalker [10]; ballistic coefficient = 400 kg/m^2, L/D = 0.8, entry velocity at 1000 km altitude = 31.3 km/s.

As seen in the figure, the relaxation distance is about 2.5 cm at the perigee. This means that the nonequilibrium phenomenon will be very significant in the Neptune entry flight.

5. DISCUSSION

As seen in Fig. 15, the nonequilibrium phenomenon will successfully explain the low recession of Galileo Probe heatshield in the stagnation region without raising the recession in the frustum region, if the true relaxation time is longer than that given by Stalker [10]. It is highly desirable that the true relaxation time be determined experimentally. Theoretical works, such as that by Furudate and Chang [15] is desirable also.

As mentioned in Introduction, ionization nonequilibrium is one of the three possible causes of the observed low recession of the Galileo Probe heatshield in the stagnation region. The other two possible explanations should be pursued also.

6. CONCLUSIONS

The existing experimental data on ionization equilibration time taken in arc-driven shock tubes greatly underestimate because of the radiation from the driver. The experimental data obtained in a shock tunnel, which is 8 times that taken in arc-driven shock tubes, is likely to be underestimating still because of the nonequilibrium in nozzle flow. If the true equilibration time is 2.5 times that determined in a shock tunnel, the low surface recession at the stagnation point of Galileo Probe is explained. For Neptune entry, nonequilibrium effect will be even more significant. Uncertainty regarding the equilibration time concerns the processes prior to reaching the threshold of avalanche ionization.

7. REFERENCES

1. Moss, J. N., and Simmonds, A. L., "Galileo Probe Forebody Flowfield Predictions," *Entry Vehicle Heating and Thermal Protection Systems: Space Shuttle, Solar Starprobe, Jupiter Galileo Probe,* Progress in Astronautics and Aeronautics, Vol. 85, edited by P. E. Bauer and H. E. Collicott, AIAA new York, 1983, pp. 419-445.

2. Matsuyama, S., Ohnishi, N., Sasoh, A., and Sawada, K., "Numerical Simulation of Galileo Probe Entry Flowfield with Radiation and Ablation," AIAA Paper 2002-2994; to be published in Journal of Thermophysics and Heat Transfer.

3. Park, C., "Injection-Induced Turbulence in Stagnation Point Boundary Layers," *AIAA Journal*, Vol. 22, No. 2, February 1984, pp. 219-225.

4. Lundell, J. H., and Dickey, R. R., "Response of Heat-Shield materials to Intense Laseer Radiation," AIAA Paper 78-0138, January 1978.

5. Davies, C. B., and Park, C., "Trajectories of Solid Particles Spalled from a Carbonaceous Heat Shield," *Entry Vehicle Heating and Thermal Protection Systems; Space Shuttle, Solar Starprobe, Jupiter Galileo Probe,* Progress in Astronautics and Aeronautics, Vol. 85, edited by P. E. Bauer and H. E. Collicott, AIAA, 1983, pp. 472-495.

6. Park, C., Raiche, G. A., and Driver, D. M., "Radiation of Spalled Particles in Shock Layers," to b published in Journal of Thermophysics and Heat Transfer.

7. Howe, J. T., "Hydrogen Ionization in the Shock Layer for Entry into the Outer Planets," AIAA Journal, Vol. 12, No. 6, June 1974, pp. 875-876.

8. Leibowitz, L. P., "Measurements of the Structure of an Ionizing Shock Wave in a Hydrogen-Helium Mixture," The Physics of Fluids, Vol. 16, No. 1, January 1973, pp. 59-68.

9. Livingston, F. R.., and Poon, P. T. Y., "Relaxation Distance and Equilibrium Electron Density Measurements in Hydrogen-Heliium Plasmas," AIAA Journal, Vol. 14, No. 9, September 1976, pp. 1335-1337.

10. Stalker, R. J., "Shock Tunnel Measurement of Ionization Rates in Hydrogen," AIAA Journal, Vol. 18, No. 4, April 1980, pp. 478-480.

11. Bogdanoff, D. A., and Park, C., "Radiative Interaction Between Driver and Driven Gases in an Arc-Driven Shock Tube," Shock Wave, Vol. 12, No. 3, November 2002, pp. 205-214.

12. Cook, and Metzger, Journal of Optical Society of America, Vol. 54, No. 8, August 1964, pp. 968-972.

13. Furudate, M., Fujita, K., and Abe, T., "Coupled Rotational-Vibrational Relaxation of Molecular Hydrogen at High Temperatures," AIAA Paper 2003-3780, 2003.

14. Park, C., "Effect of Lymann Radiation on Nonequilibrium Ionization of Atomic Hydrogen,"
AIAA Paper 2004-2277, 2004.

15. Furudate, M., and Chang, K. S., "Calculation of H_2-He Flow with Nonequilibrium Ionization and Radiation: Interim Report," paper presented at the 2nd Planetary Prlbe Workshop.

Calculation of H_2-He Flow with Nonequilibrium Ionization and Radiation : an Interim Report

Michiko Furudate[1] and Keun-Shik Chang[2]

[1]Postdoctoral Fellow, [2]Professor
Korean Advanced Institute of Science and Technology
373-1 Guseong-dong, Yuseong-gu, Daejeon, 305-701, Korea
[1]furu@kaist.ac.kr, [2]kschang-ks@kaist.ac.kr

Abstract

The nonequilibrium ionization process in hydrogen-helium mixture behind a strong shock wave is studied numerically using the detailed ionization rate model developed recently by Park which accounts for emission and absorption of Lyman lines. The study finds that, once the avalanche ionization is started, the Lyman line is self-absorbed. The intensity variation of the radiation at 5145 Å found by Leibowitz in a shock tube experiment can be numerically reproduced by assuming that ionization behind the shock wave prior to the onset of avalanche ionization is 1.3%. Because 1.3% initial ionization is highly unlikely, Leibowitz's experimental data is deemed questionable. By varying the initial electron density value in the calculation, the calculated ionization equilibration time is shown to increase approximately as inverse square-root of the initial electron density value. The true ionization equilibration time is most likely much longer than the value found by Leibowitz.

1. Introduction

In the past, on-going, and future entry flight missions to the outer planets, heating rates to the vehicle's surface tend to be large, and therefore the extent of ablation becomes also large. Therefore, accurate prediction of the heating rates becomes imperative. Computational Fluid Dynamics (CFD) can be a helpful tool in predicting the heat transfer rate if its reliability can be validated against a flight data.

The atmospheres of outer planets consist of hydrogen and helium. An entry flight into the planet Jupiter has already been accomplished in the Galileo Probe mission. Surface recession data have successfully been obtained in that mission [1]. This data is the only flight data available to validate the CFD methodology in designing the heatshield.

The Galileo Probe data on surface recession has shown a surprisingly low surface recession in the stagnation region, and a surprisingly large recession in the downstream frustum region compared with the pre-flight predictions [2]. Very recently, Matsuyama et al [3] were able to explain the high surface recession in the downstream region by injection-induced turbulence. However, the low surface recession in the stagnation region has not yet been explained.

Park [4] has earlier speculated that the low heating in the stagnation region might be due to thermochemical nonequilibrium. The pre-flight predictions were made assuming thermochemical equilibrium. The nonequilibrium idea was first introduced by Howe [5], who derived his concept from the shock tube data of Leibowitz [6]. Leibowitz's data showed that the region immediately behind the shock wave was not ionized and did not radiate. Howe derived an empirical expression for the thickness of the non-radiating nonequilibrium region, i.e. ionization equilibration time, in a hydrogen-helium mixture.

In a companion paper to the present paper, Park [7] examined the nonequilibrium issue still further. He points out that the experiment by Leibowitz and subsequent experiment by his colleagues Livingston and Poon [8], which extended Leibowitz's data to higher flight speeds, are mostly likely erroneous because of the interaction of radiation emanating from the arc-heated driver gas. Park [7] points out that the experiment by Stalker [9] in a shock tunnel produced an equilibration time 8 times longer than that by the Leibowitz-Livingston-Poon group. Park speculates that the true equilibration time may be even longer than that determined by Stalker.

According to the experiment conducted by Bogdanoff

and Park [10], the freestream flow in front of the normal shock wave in a hydrogen-helium mixture is ionized by photo-ionization of H_2. The temperature of the electrons so-produced is speculated to be determined indirectly by the temperature of the vibrational-rotational mode of H_2 through the electron-vibration-rotation coupling [7]. The relaxation of vibrational-rotational mode of H_2 has been studied by Furudate et al [11]. The study shows that the two modes are strongly coupled, and the temperature of the combined vib-rotational mode approaches the translational temperature relatively slowly, i.e., with a collision number of the order of hundred. The slowly rising electron temperature leads to slow initiation of the so-called avalanche ionization which sets off the rapid electron-impact ionization process that brings about the ionization equilibrium. The rate of avalanche ionization is dictated partly by the absorption of Lyman radiation [12]. Thus, the time for equilibration is dictated by the photo-ionization rate, vib-rotational relaxation rate, the coupling rate between electron temperature and vib-rotation temperature, and the rate of Lyman line absorption.

It goes without saying that further experiments should be carried out to determine the true value of the time for ionization equilibrium in hydrogen. Along with such an experiment, a CFD modeling of the phenomenon would be required. The purpose of the present paper is to carry out such a CFD modeling, and investigate how the photo-ionization, vib-rotation relaxation, electron-vib-rotational coupling, and nonequilibrium ionization proceed in a hypersonic H_2-He flow. Because of the enormous complexity and because most of the relevant parameters are unknown, a very first simple calculation is carried out in the present work. Here, a new set of the ionization rate coefficients by Park [12] is employed. The ionization rate values used by Leibowitz in explaining his experimental data were arbitary, and contained little physical ground. In comparison, Park's new ionization rate values are firmly based on the state-of-the-art knowledge of such a process. The electron density and electron density were first chosen arbitrarily in order to numerically reproduce Leibowitz's shock tube data [6]. The parameters controlling the initial electron density and temperature were varied to show that the ionization equilibration times can be longer than that determined by Leibowitz.

2. Methods of Calculation

2.1 Governing equations

The governing equations are the one dimensional Euler equations,

$$\frac{\partial Q}{\partial t} + \frac{\partial F}{\partial x} = W . \quad (1)$$

In the equations, the global mass, the momentum, the total energy, the species mass, and the electron energy conservation equations are included. Hence, the conservative variables Q, the convective flux vector F, and the source term W are given respectively as follows;

$$Q = \begin{pmatrix} \rho \\ \rho u \\ E \\ \rho_s \\ E_{el} \end{pmatrix}, \quad F = \begin{pmatrix} \rho u \\ \rho u^2 + p \\ (E+p)u \\ \rho_s u \\ E_{el} u \end{pmatrix}, \quad W = \begin{pmatrix} 0 \\ 0 \\ 0 \\ W_s \\ W_{el} \end{pmatrix} \quad (2)$$

where s stands for the chemical species, H_2, H, H^+, He, He^+, e. Electron-electronic energy E_{el} in the present study is defined by

$$E_{el} = 1.5 N_e k T_e + \sum_i E_H(i) N_H(i) \quad (3)$$

where N_e is electron number density, k is Boltzmann constant, T_e is electron temperature, $E(i)$ is energy level of state i. Number density of atomic hydrogen in the state i, $N_H(i)$, is given by

$$N_H(i) = \frac{g_i \exp(-E_i/kT_e)}{\sum_i g_i \exp(-E_i/kT_e)} N_H \quad (4)$$

In the present study, first three electronic states are considered. The electron-electronic energy source term, W_{el}, can be written by

$$W_{el} = 2N_e \sum_k \nu_{ek} \frac{m_e}{m_k} \frac{3}{2} k(T - T_e) - D W_H \\ + \sum_i E_H(i) N_H(i) \quad (5)$$

where D is ionization energy and W_H is chemical source term for atomic hydrogen. Elastic collision frequency, ν_{ek}, is defined by $\nu_{ek} = N_k Q_{ek} \sqrt{8kT_e/\pi m_e}$. In the present study, the elastic collision cross section, Q_{ek}, are given by the same formula as in [6].

2.2 Chemical reaction rates coefficients

The dissociation of molecular hydrogen, the ionization of atomic hydrogen, and the ionization of helium are considered. They are summarized in Table 1.

2.2.1 Rates coefficients used by Leibowitz

In [6], Leibowitz employs the idea of the two-step excitation-ionization process in his calculations; The atomic hydrogen is firstly excited by collisions (Reaction 3 and 5), and then ionizes rapidly [6]. The Electron-electronic energy E_{el} is defined by $E_{el} = 1.5kN_e T_e$ in [6]. The reaction rate coefficients for these reactions are summarized in Table 2. The backward reaction rates are given by the fraction of the forward reaction rates and the equilibrium constants. The equilibrium constants are summarized in Table 3. Hereafter, this approach is called the Leibowitz's method.

Table 1. Reaction

	Reaction		
1	$H + e$	$\rightleftarrows$	$H^+ + e + e$
2	$He + e$	$\rightleftarrows$	$He^+ + e + e$
3	$H + e$	$\rightleftarrows$	$H^* + e$
4	$He + e$	$\rightleftarrows$	$He^* + e$
5	$H + H$	$\rightleftarrows$	$H^* + H$
6	$H + He$	$\rightleftarrows$	$H^* + He$
7	$H_2 + He$	$\rightleftarrows$	$H + H + He$
8	$H_2 + H_2$	$\rightleftarrows$	$H + H + H_2$
9	$H_2 + H$	$\rightleftarrows$	$H + H + H$
10	$H_2 + H^+$	$\rightleftarrows$	$H + H + H^+$
11	$H_2 + e$	$\rightleftarrows$	$H + H + e$

Table 2. Reaction rate coefficients [6].

Reaction	Reaction Rates [m³/mole-s]	Ref.
1	$6.09 \times 10^{-23} \sqrt{8kT_e/\pi m_e} \exp(-15782/T_e)$	13
2	$3.56 \times 10^{-23} \sqrt{8kT_e/\pi m_e} \exp(-285248/T_e)$	14
3	$7.5 \times 10^{-22} \sqrt{8kT_e/\pi m_e} \exp(-11605/T_e)$	15, 16
4	$6.00 \times 10^{-23} \sqrt{8kT_e/\pi m_e} \exp(-23210/T_e)$	17
5	$4.0 \times 10^{-23} \sqrt{8kT/\pi m_e} \exp(-11605/T)$	
6	$4.0 \times 10^{-23} \sqrt{8kT/\pi m_e} \exp(-11605/T)$	
7	$6.93 \times 10^{-12}/T \exp(-52340/T)$	18
8	$2.5 k_7$	18
9	$20.0 k_7$	18
10	$20.0 k_7$	
11	$20.0 k_7$	

Table 3 Equilibrium constants [6].

i	Equilibrium constants, K_i [mole/m³]
H^+	$4.05 \times 10^{-3} T^{3/2} \exp(-15872/T)$
He^+	$1.62 \times 10^{-2} T^{3/2} \exp(-285248/T)$
H	$3.7 \times 10^6 [1.0 - \exp(1.50 \times 10^8 T^{-2})] \exp(-52340/T)$

2.2.3 Park's ionization rate coefficients of atomic hydrogen accounted for radiation absorption

In order to account for the effect of the radiation absorption, the rate coefficient for the electron-impact ionization (Reaction 1) is taken from [12], and Reaction 3 is discarded. The change of electron number density due to the electron impact ionization of atomic hydrogen is expressed by

$$\frac{dN_e}{dt} = k_f N_H N_e - N_e^2 \alpha_1 - N_e^2 \alpha_c \alpha(1), \quad (6)$$

$$\alpha_1 = \alpha - \alpha(1) = N_e k_r + \sum_{i=2}^{m} \alpha(i), \quad (4)$$

where k_f is the collisional ionization rate coefficient, k_r is the collisional three-body recombination rate coefficient, N_H is the atomic hydrogen number density, α is the collisional-radiative recombination rate coefficient, $\alpha(i)$ is the rate coefficient for radiative recombination into state i, α_c is Lyman continuum radiation fraction, and m is the highest bound state quantum number. Using the source code provided in [12], the parameters, k_f, k_r, α and $\alpha(i)$ can be obtained from the number density N_H and N_e, the electronic temperature T_e, and the Lyman line absorption factor α_L. The Lyman ∞ line absorption factor α_L is defined by

$$\alpha_L = \frac{B(1,2)}{B(1,2)_E} = \frac{P/1.634 \times 10^{-11} N_H(1)}{2^2 \exp[-E_H(2)/kT_e] A(2,1)}. \quad (5)$$

where $B(i, j)$ is rate coefficient for radiative transition from lower state i to upper state j and subscript E stands for a equilibrium state. Power absorbed by the Lyman ∞ line, P, can be written by

$$P = 1.634 \times 10^{-11} A(2,1) N(2) - \nabla q_{rad}, \quad (6)$$

where ∇q_{rad} is divergence of radiative heat flux determined by solving the radiative heat transfer equation.

2.2.4. Initial electron density and temperature

Initial electron density and temperature at the onset of avalanche ionization were varied arbitrarily because they are totally unknown at this time. The relative initial electron concentration $(N_e / N_{H2})_0$ was varied between 5×10^{-4} to 0.05, and the initial electron temperature was varied from 300 to 6000 K. The true initial electron density value is most likely lower than the lowest value considered here. However, lower initial electron density values led to numerical instability. This problem needs to be solved in the future.

2.3 Numerical method

The convective fluxes are given by AUSM-DV upwind scheme. The dependent variables are interpolated by the MUSCL approach An explicit Runge-Kutta method is used for time integration.

3. Test condition

A shock tube problem is numerically solved for the experimental condition of Leibowitz.[3] The test gas is a mixture of 20.8% H_2-79.2% He. The driver gas is He. The measured shock velocity was 15.5 km. Using this value of shock velocity, it was impossible to numerically reproduce the measured electron density: the measured electron density values were higher. The measured electron density values were numerically reproduced assuming the shock velocity to be 16.6 km/s. The reason for such a high effective shock speed is believed to be the radiative heating of the driven gas by the radiation emitted by the driver gas [10]. The static pressure at the test section is 133 Pa. The initial translational and the electron temperature in the test section are set to 300K. The pressure ratio between the test section and the driver section is determined from the Rankine-Hugoniot relation. The computational space is 2.0 m in length and divided by 1000 cells with a constant space. The first 120 cells are considered to be the driver section.

4. Results and discussions

4.1. Leibowitz data

In order to verify the present numerical code, the reproduction of the calculated results by Leibowitz is attempted using the same reaction rates as in Leibowitz's [6]. The initial electron density is assumed to be 0%: avalanche ionization is assumed to be initiated by electrons produced mostly by Reaction 5, as in Leibowitz's work. In Fig.1, the calculated temperature profiles behind the shock wave are compared with the Leibowitz's numerical result. In the present calculation, the position of shock wave is defined by the position where translational temperature begins to increase. Differences in the temperature variation between these two calculations are initiated immediately behind the shock wave. They may be caused by differences in the numerical methods. Note that, in Leibowitz's calculation, the equations of fluid mechanics are solved in a shock fixed coordinate system, and convections of fluids are ignored. A good agreement in the temperature is obtained in the downstream where the transnational and the electron temperature are in equilibrium. Fig. 2 shows the corresponding number density profiles for atomic hydrogen and electrons. Deviation in the atomic hydrogen number density is also believed to come from the differences in the numerical methods.

In order to qualitatively examine the effect of radiation absorption on ionization of atomic hydrogen, calculations with Park's ionization rate at three fixed values of the Lyman line absorption factor α_L are implemented. The values of $\alpha_L = 0$, 1 and 100 are chosen, representing an optically thin case, a

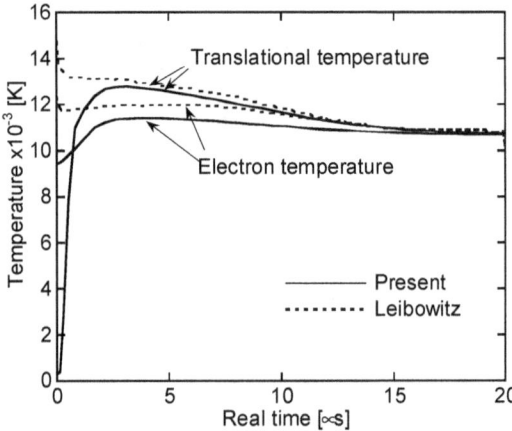

Fig. 1 Temperatures profile obtained in the present calculation.

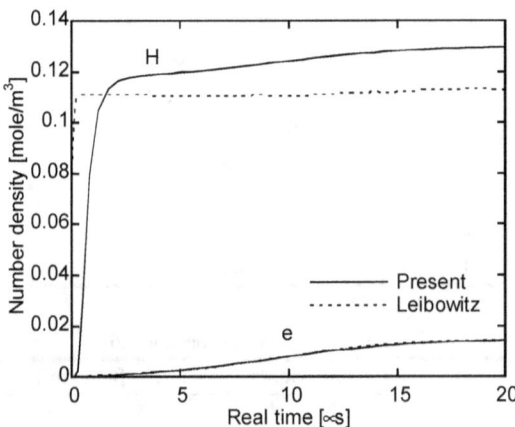

Fig. 2 Number density profile obtained in the present calculation.

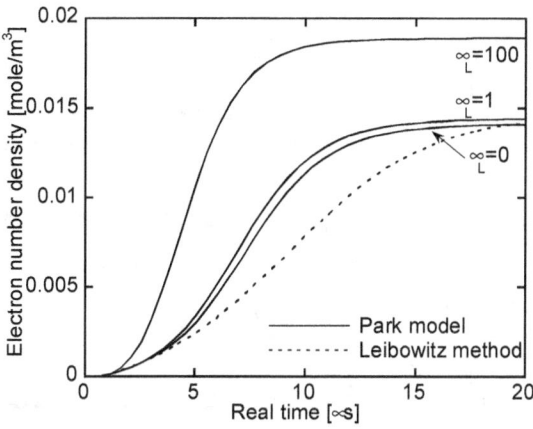

Fig. 3 Effect of Lyman radiation on electron number density profile.

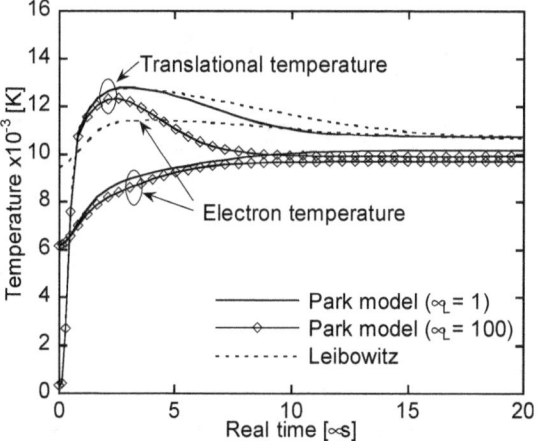

Fig. 4 Effect of Lyman radiation on temperatures profile.

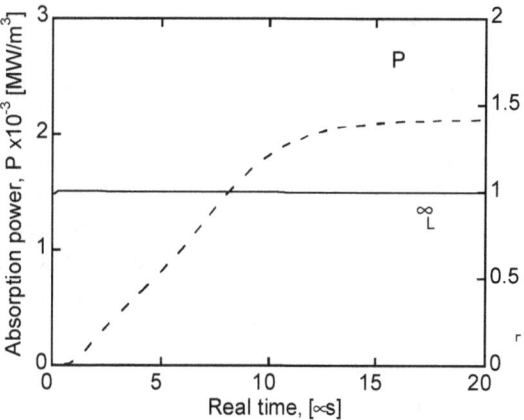

Fig. 5 Radiative power absorbed by the Lyman ∞ line and Lyman line absorption factor behind shock wave.

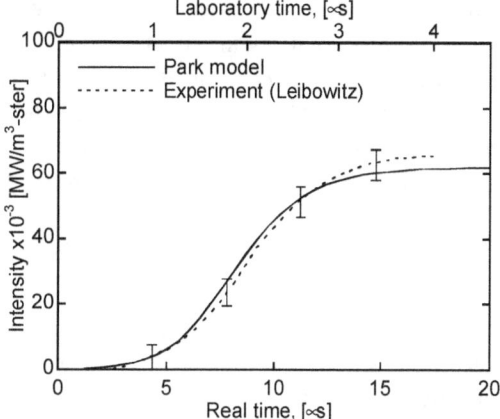

Fig. 6 Calculated radiative intensity and the experimental data by Leibowitz (∞=5145 Å).

self-absorbed case, and a strongly irradiated case, respectively. For all the cases here, the Lyman continuum radiation fraction $\circ_c$ is set to 1. Fig. 3 shows the number density profiles obtained in the present calculations. The electron number density stays low in the first few microseconds behind shock wave, where the collisional ionization (Reaction 5) is dominant. Then, an avalanche ionization zone appears where the electron-impact ionization is dominant. In the cases for higher $\circ_L$, incubation time to initiate the avalanche ionization is shorter, and the rate of electron production during the avalanche ionization is faster. The electron number density at the quasi-steady state (QSS), which appears after the avalanche ionization ends, is also affected by $\circ_L$. In the strongly irradiated case of $\circ_L = 100$, the electron number density at the QSS is 30 % larger than the case of $\circ_L = 0$. The temperature profiles are also drastically changed according to the value of $\circ_L$, as shown in Fig 4. The high degree of electron impact ionization with the large value of $\circ_L$ lowers the electron temperature in the ionization zone. Accordingly, equilibrations between the transnational and the electron temperature proceed faster.

The variation of the value of $\circ_c$ gives a miner effect on the flowfield. Although the figure is not shown here, the electron number density profiles with $\circ_c = 0.5$ are almost identical to the results with $\circ_c = 1$ when the value of $\circ_L$ is fixed.

A calculation using local values of $\circ_L$, which is determined by solving the heat transfer equation, is performed next. The heat transfer equation is solved using a multi-band model. Absorption coefficients for the Lyman ∞ line are evaluated at 23 wavelength points in the range from 1200 to 1230 Å. The calculated radiation power of absorption by the Lyman ∞ line and the Lyman line absorption factor α_L are shown in Fig. 5. The radiation power of absorption varies from 0 in the non-ionization zone behind the shock wave to about 2000 MW/m^3 in the QSS zone in the down stream. The Lyman ∞ line absorption factor α_L keeps constant at the value nearly equal to 1 after slight increase immediately behind the shock wave; the emission from Lyman ∞ line is totally absorbed. Therefore, the electron number density and the temperatures profiles are almost identical to those in the case of $\alpha_L = 1$ in Fig 3 and 4. The calculated radiative intensity profile for the wavelength of 5145 Å is compared with the experimental data in Fig. 6. The same approach as in [3] is employed to calculate the intensity. The obtained intensity profile agrees well with the experimental data measured by Leibowitz.

4.2. Effect of initial electron density and temperature

In this subsection, effects of initial electron number density and electron temperature are examined by a parametric study. For the calculations in this study, Reaction 5 and 6 in Table 1 is discarded, and the initial number density at the point of initiation of avalanche ionization is given at a certain value. From the observation in the previous subsection, $\alpha_L = 1$ and $\alpha_q = 1$ are employed for all the cases in this subsection.

First, the ratio of the initial electron number density to the initial molecular hydrogen is varied, while the initial electron temperature is remained to be 300K. As shown in Fig. 7, the intensity profile for 5145 Å is changed according to variation of $(N_e/N_{H2})_0$. Leibowitz's experimental data can be reproduced with $(N_e/N_{H2})_0 = 0.013$. The initiation and the degree of avalanche ionization of atomic hydrogen are greatly affected by the initial electron number density, as shown in the electron number density profile in Fig. 8. Larger non-ionization region is obtained with lower initial electron number density.

Next, the initial electron temperature, T_{e0}, is varied from 300 to 6000 K, while the initial $(N_e/N_{H2})_0$ is

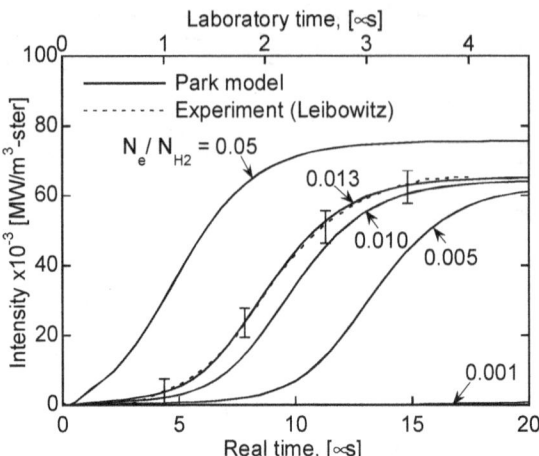

Fig. 7 Effect of initial electron number density on intensity profile.

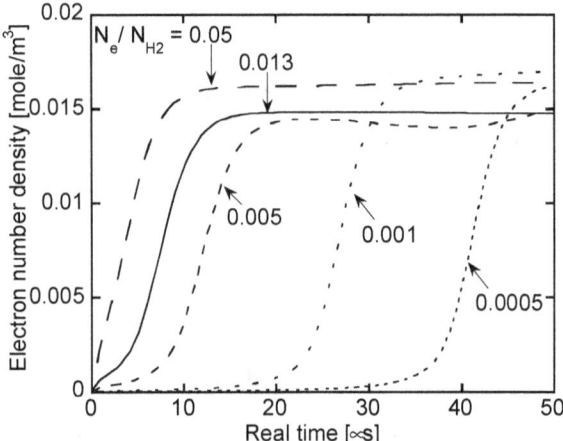

Fig. 8 Effect of initial electron number density on electron number density profile. ($T_{e0} = 300$ K).

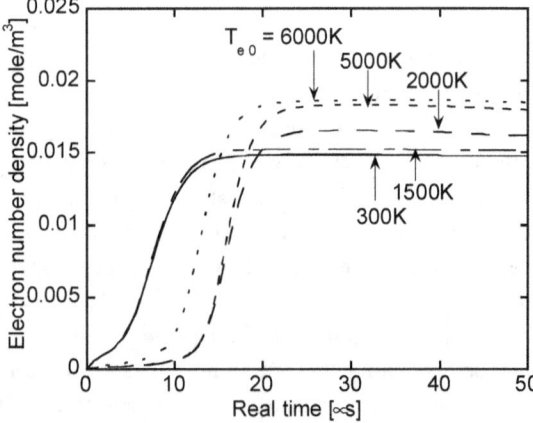

Fig. 9 Effect of initial electron temperature on electron number density profile ($(N_e/N_{H2})_0 = 0.013$).

fixed at the value of 0.013. The initial electron temperature at the point of initiation of avalanche ionization is at presently totally unknown. The electron number density profiles obtained for the various value of T_{e0} are shown in Fig. 9. For 2000 < T_{e0} < 5000K, times to initiate the avalanche ionization are almost identical, although higher temperature gives more production of electrons. The time to reach QSS with T_{e0} = 2000 is about two times large than that of 300 K. For the T_{e0} higher than 2000K, the time to QSS becomes shorter as the temperature increases. It is interesting to see that the effect of high initial electron temperature is not monotonic: the longest delay of ionization occurs at electron temperature of 2000 K.

Fig. 10 summarizes the effect of the initial electron number density on the e-folding ionization equilibrium time. For all the initial electron temperature cases, the equilibration time increases according to the $(N_e/N_{H2})_0$ decrease. When the initial electron number density is lower, the effect of initial electron temperature is miner. The equilibration time for $(N_e/N_{H2})_0$ = 0.0005 is about 5 times longer than that for $(N_e/N_{H2})_0$ = 0.013 at T_{e0} =300K. For T_{e0}=300K, the ionization equilibrium time varies approximately as inverse square-root of the initial electron density.

5. Discussion

In this study, only the avalanche ionization process is modeled. We find first that, once avalanche ionization process is started, the absorption of Lyman lines is of little consequence, because it is nearly totally self-absorbed. However, we do not yet know how Lyman radiation will affect the region prior to the avalanche ionization.

Another important finding here is that the time to reach ionization equilibrium is affected strongly by the initial electron density and temperature at the point of avalanche ionization. In the range of the initial electron density values considered in the present work, a factor of 5 variation is seen. As mentioned in 2.2.4, the lowest value of initial electron density was set in the present work by the numerical instability. The true initial electron density is most likely lower than the value considered in the present work. In the region ahead of avalanche ionization, photo-ionization process is the only mechanism to produce the initial electrons. Because electron temperature will be low in

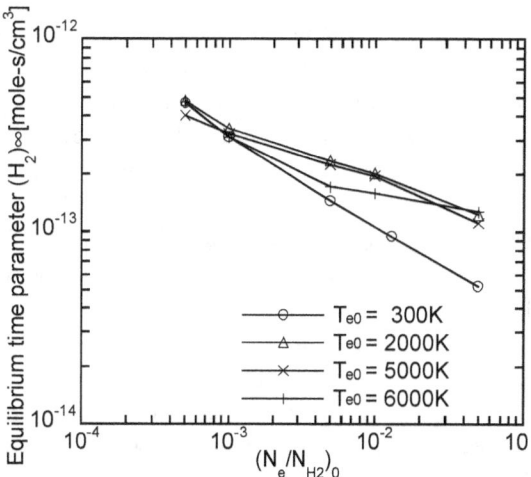

Fig. 10 Effect of initial electron number density and initial electron temperature on the ionization equilibrium time.

the region, the free electrons produced by photo-ionization will be rapidly recombining. The present work finds that Leibowitz's shock tube data can be numerically reproduced with an initial $(N_e/N_{H2})_0$ value of 0.013, i.e. 1.3% ionization. This value of 1.3% is much higher than the value found experimentally by Bogdanoff and Park [10], and should be considered impossible. The only explanation is the photo-ionization by the strong radiation transmitted from the arc-heated driver gas [10].

According to Fig. 10, equilibration time increases roughly as an inverse square-root of the initial electron density, as mentioned. If the true value of $(N_e/N_{H2})_0$ is 10^{-6}, the equilibration time will be two orders of magnitude longer than Leibowitz's value. The equilibration time value of Stalker [9], which is 8 times larger than the value by Leibowitz, is well within the explainable range. This means that, at this time, we do not know how to predict the thickness of the nonequilibrium, un-ionized, non-radiating region behind the shock wave in outer planet entry flights. This points to the need to know more about the processes occurring in the region prior to the initiation of avalanche ionization.

6. Conclusions

Using the detailed ionization rate model developed recently by Park, it is found that the absorption of Lyman alpha has a great influence on the initiation and the rate of avalanche ionization behind a shock wave. For a strong absorption, avalanche ionization initiates earlier and faster. Once the avalanche ionization is started, the Lyman line absorption behind shock wave is almost constant at the value of 1, which represents the emission from Lyman line is self-absorbed by the atomic hydrogen gas. The calculated intensity profile for 5145 Å agrees well with the experimental data by Leibowitz, when the starting degree of ionization is assumed to be 1.3%, which is unrealistically high. The ionization equilibration time is roughly inversely proportional to the square-root of the initial electron density, which are presently unknown. The true ionization equilibration time could be up to two orders of magnitude longer than that determined by Leibowitz.

7. Acknowledgments

The authors would like to express their appreciation to Prof. Chul Park, Korean Advanced Institute of Science and Technology, Korea, for helpful comments and suggestions.

8. References

1. Milos, F. S., "Galileo Probe Heat Shield Ablation Experiment," *Journal of Spacecraft and Rockets*, Vol. 31, No. 6, November-December 1997, pp. 705-713.

2. Moss, J. N., and Simmonds, A. L., "Galileo Probe Forebody Flowfield Predictions," Entry Vehicle Heating and Thermal Protection Systems: Space Shuttle, Solar Starprobe, Jupiter Galileo Probe, *Progress in Astronautics and Aeronautics*, Vol. 85, edited by P. E. Bauer and H. E. Collicott, AIAA new York, 1983, pp. 419-445.

3. Matsuyama, S., Ohnishi, N., Sasoh, A., and Sawada, K., "Numerical Simulation of Galileo Probe Entry Flowfield with Radiation and Ablation," *AIAA Paper 2002-2994*; to be published in Journal of Thermophysics and Heat Transfer.

4. Park, C., "Heatshielding Problems of Planetary Entry, A Review," *AIAA Paper 99-3415*, June 1999.

5. Howe, J. T., "Hydrogen Ionization in the Shock Layer for Entry into the Outer Planets," *AIAA Journal*, Vol. 12, No. 6, June 1974, pp. 875-876.

6. Leibowitz., L. P., "Measurements of the Structure of an ionizing shock wave in a hydrogen-helium mixture," *The Physics of Fluids*, Vol. 16, No. 1, January 1973, pp. 59-68.

7. Park, C., "On Nonequilibrium Radiation in Hydrogen Shock Layers," paper presented at the 2nd Probe Workshop, NASA Ames Research Center, August 2004.

8. Livingston, F. R., and Poon, P. T. Y., "Relaxation Distance and Equilibrium Electron Density Measurements in Hydrogen-Heliium Plasmas," *AIAA Journal*, Vol. 14, No. 9, September 1976, pp. 1335-1337.

9. Stalker, R. J., "Shock Tunnel Measurement of Ionization Rates in Hydrogen," *AIAA Journal*, Vol. 18, No. 4, April 1980, pp. 478-480.

10. Bogdanoff, D. A., and Park, C., "Radiative Interaction Between Driver and Driven Gases in an Arc-Driven Shock Tube," *Shock Wave*, Vol. 12, No. 3, November 2002, pp. 205-214.

11. Furudate, M., Fujita, K, and Abe, T. "Coupled Rotational-Vibrational Relaxation of Molelcular Hydrogen at High Temperatures," *AIAA paper 2003-3780*.

12. Park, C., "Effect of Lyman Radiation on Nonequilibrium Ionization of Atomic Hydrogen," *AIAA Paper 2004-2277*, June 2004.

13. Wade L. Fite and R. T. Brackmann, "Collisions of Electrons with Hydrogen Atoms. I. Ionization," *Physical Review*, vol. 112, 1958, pp. 1141-1151.

14. H. Massey and E. Burhop, *Electronic and Ionic Impact Phenomena*, Oxford, 1952, p.38.

15. J. W. McGowan, J. F. Williams[*], and E. K. Curley, "e-H Resonances in the $2p$ Excitation Channel," *Physical Review*, vol. 180, 1969, pp. 132-138.

16. W. E. Kauppila, W. R. Ott, and W. L. Fite, "Excitation of Atomic Hydrogen to the Metastable $2^2 S_{1/2}$ State by Electron Impact", *Physical Review A*, vol.1, 1970, pp. 1099-1108.

17. H. Holt and R. Krotkov, "Excitation of $n=2$ States in Helium by Electron Bombardment," *Physical Review*, vol. 144, 1966, pp. 82-93.

18, T. Jacobs, R. Giedt, and N. J. Cohen, "Kinetics of Hydrogen Halides in Shock Waves. II. A New Measurement of the Hydrogen Dissociation Rate", *The Journal of Chemical Physics*, vol.47, 1967, pp. 54-57.

CLOUDS OF NEPTUNE AND URANUS

Sushil K. Atreya and Ah-San Wong

Department of Atmospheric, Oceanic, and Space Sciences, University of Michigan, Ann Arbor, MI 48109-2143, USA, Email: atreya@umich.edu, aswong@umich.edu

ABSTRACT

We present results on the bases and concentrations of methane ice, ammonia ice, ammonium hydrosulfide-solid, water ice, and aqueous-ammonia solution ("droplet") clouds of Neptune and Uranus, based on an equilibrium cloud condensation model. Due to their similar p-T structures, the model results for Neptune and Uranus are similar. Assuming 30–50× solar enhancement for the condensibles species, as expected from formation models, we find that the base of the droplet cloud is at the 370 bars for 30× solar, and at 500 bars for 50× solar cases. Despite this, entry probes need to be deployed to only 50–100 bars to obtain all the critical information needed to constrain models of the formation of these planets and their atmospheres.

1. INTRODUCTION

Comparative planetology of deep well-mixed atmospheres of the outer planets is the key to the origin and evolution of the solar system, and by extension, extrasolar systems. Critical factors to constrain the formation models are abundances of heavy elements (heavier than helium) below cloud levels of the giant planets. Much has been written previously about the two gas giants, Jupiter and Saturn (e.g., Atreya et al. [1, 2]). In this paper, we focus on the two icy giants, Neptune and Uranus. Methane ice is the only other condensible species on these two planets, in addition to the clouds of ammonia ice, ammonium hydrosulfide (NH_4SH) solid, water ice, and aqueous-ammonia solution ("droplet") that form also on the gas giants. To the first order, cloud structure can be calculated using an equilibrium cloud condensation model (ECCM) that employs basic principles of thermodynamics. Based on the measured methane (CH_4) mixing ratio, the C/H is 30–50× solar at Neptune, and 20–30× solar at Uranus. Assuming similar enhancement for the other condensibles, as expected from formation models, we find that the base of the droplet cloud is at 370 bars for 30× solar, and at 500 bars for 50× solar cases. Not only such high pressure levels pose immense technological challenges to entry probe missions, the N/H and O/H ratios deduced at these pressures are not even representative of their well-mixed values. On the other hand, noble gases, methane (CH_4), hydrogen sulfide (H_2S), as well as D/H and $^{15}N/^{14}N$ can be accessed and measured at much shallower levels, and would still permit the retrieval of information critical to the formation of Neptune and Uranus and their atmospheres, especially when combined with the elemental abundance information for the gas giants.

2. THERMOCHEMICAL CLOUD MODEL

ECCM was first developed by Weidenschilling and Lewis [3], and improved by Atreya and Romani [4]. The lifting condensation level (LCL), i.e., the base of the cloud, is calculated by comparing the partial pressure and the saturation vapor pressure of the condensible volatile. The LCL is reached at the altitude where 100% relative humidity is attained. The amount of condensate in the ECCM is determined by the temperature structure at the LCL and vicinity. The release of latent heat of condensation modifies the lapse rate, hence the temperature structure, of the atmosphere. The composition and structure of the clouds depend on the composition of the atmosphere, and in particular the distribution of condensible volatiles. For details of the current model, see Atreya and Wong [5].

Thermochemical equilibrium considerations suggest that CH_4, NH_3, and H_2O are the only species likely to condense in the atmospheres of Neptune and Uranus, if the composition were solar. H_2S does not condense even if it were enriched substantially. In the gas phase, H_2S can combine with NH_3 to form NH_4SH, i.e., $NH_3(g) + H_2S(g) \rightarrow NH_4SH$, or ammonium sulfide $((NH_4)_2S)$ which is less likely. NH_4SH would condense as a solid in the environmental conditions of Neptune and Uranus. NH_3 could also dissolve in H_2O, resulting in an aqueous solution (droplet) cloud in the atmosphere. The extent of such a cloud depends on the mole fractions of NH_3, and H_2O.

2.1 Model Inputs

The presently known elemental abundance information for Neptune and Uranus along with that for Jupiter is given in Table 1. The heavy element ratios for Uranus and Neptune are taken to be the same as C/H from CH_4 measurements on these planets, i.e., N (from NH_3), S (from H_2S), and O (from H_2O) are enriched 30–50 times relative to solar at Neptune, and 20–30 times at Uranus. The progressively larger enrichment in the heavy elements from Jupiter to Neptune is consistent with predictions of the core accretion model. For purposes of cloud structure modeling, it is reasonable to assume factors of 30 and 50 enrichment over solar for all of Neptune's condensible species, CH_4, NH_3, H_2S, and H_2O. A 20–30 times solar enrichment is expected at Uranus.

Table 1a. Elemental Abundances

	Sun	Jupiter/Sun	Uranus/Sun	Neptune/Sun
He/H	0.0975	0.807±0.02	0.92–1.0	0.92–1.0
Ne/H	1.23×10^{-4}	0.10±0.01	20–30 (?)	30–50 (?)
Ar/H	3.62×10^{-6}	2.5±0.5	20–30 (?)	30–50 (?)
Kr/H	1.61×10^{-9}	2.7±0.5	20–30 (?)	30–50 (?)
Xe/H	1.68×10^{-10}	2.6±0.5	20–30 (?)	30–50 (?)
C/H	3.62×10^{-4}	2.9±0.5	20–30	30–50
N/H	1.12×10^{-4}	3.0±1.1	20–30 (?)	30–50 (?)
O/H	8.51×10^{-4}	0.29±0.1 (hotspot)	20–30 (?)	30–50 (?)
S/H	1.62×10^{-5}	2.75±0.66	20–30 (?)	30–50 (?)
P/H	3.73×10^{-7}	0.82	20–30 (?)	30–50 (?)

Table 2b. Relevant Isotopic Abundances

Isotopes	$^{15}N/^{14}N$	D/H
Sun	$< 2.8 \times 10^{-3}$	$2.1 \pm 0.5 \times 10^{-5}$
Jupiter	$2.3 \pm 0.3 \times 10^{-3}$	$2.6 \pm 0.7 \times 10^{-5}$
Saturn		$2.25 \pm 0.35 \times 10^{-5}$
Uranus		$5.5 (+3.5, -1.5) \times 10^{-5}$
Neptune		$6.5 (+2.5, -1.5) \times 10^{-5}$

See Atreya and Wong [5] for reference.

The initial temperature profile of Neptune below 1 bar pressure level is calculated with the model using a solar composition for heavy elements but without accounting for heat of condensation or chemical reaction. The temperature at 1 bar is 72 K, consistent with the temperature profile from Voyager [6]. The temperature profile is shown in Fig. 1.

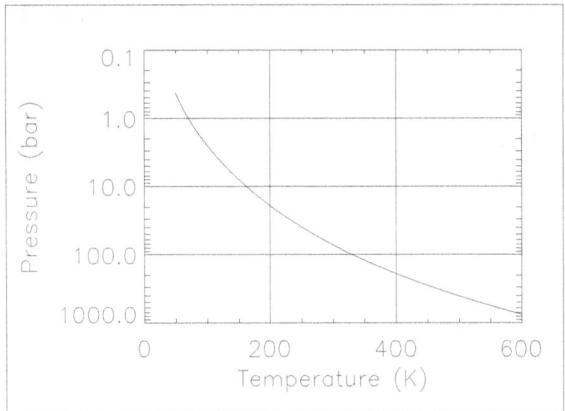

Fig. 1. Calculated p-T profile of Neptune.

2.2 Van der Waals corrections

The behavior of gas at high pressures departs from that given by the Ideal Gas Law. Under high pressure, hydrogen atoms repel each other and the real pressure is greater than pressure predicted by Ideal Gas Law

$$p = nRT/V \quad (1)$$

where p is the pressure, n the number of moles, R the gas constant, T the temperature, and V the volume. After the quantities of n, T and V are determined from Eq. 1, the modified pressure is calculated using Van der Waals equation

$$p = [nRT/(V-nb)] - a(n/V)^2 \quad (2)$$

where for hydrogen, $a = 0.2453$ bar L^2 mol^{-2}, and $b = 0.02661$ L mol^{-1}. Due to the Van der Waals effects, in the case of 30× solar enrichment of elements, the "ideal gas pressure" of 600 bars increases to 860 bars, 400 bars to 515 bars, and 200 bars to 226 bars.

3. MODEL RESULTS

According to the ECCM, the topmost cloud layer at ~1 bar level is made up of CH_4 ice. Voyager radio occultation observations did in fact infer a cloud layer at ~1 bar level. The base of the water-ice cloud for solar O/H is expected to be at ~40 bar level, whereas for the NH_3-H_2O solution clouds it is at approximately twice this pressure. We present cases with 1×, 30×, and 50× solar enrichment of the condensible volatiles (CH_4, NH_3, H_2S, H_2O) in Fig. 2 for Neptune. The NH_3-H_2O aqueous solution cloud base is calculated to be at 370 bars and 500 bars, respectively for 30× and 50× solar cases. The 30× solar case of Neptune represents very closely the cloud structure at Uranus where the heavy element enrichment is predicted to be 20–30× solar.

Some models (e.g. [7]) predict the presence of an ionic ammonia ocean in the 0.1 megabar region, much deeper than even the solution cloud. Such an ocean is most likely also responsible for the depletion of ammonia in the upper troposphere, which is significantly more severe than can be explained by the loss of this species in the formation of an NH_4SH cloud. Therefore NH_3 (as well as H_2O) will have been depleted well below their predicted LCLs.

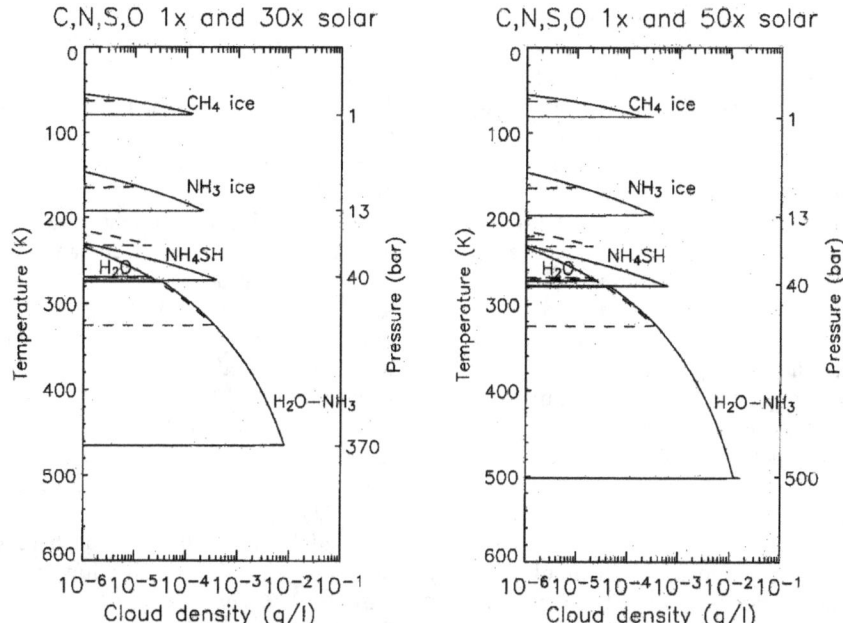

Fig. 2. ECCM results for Neptune, assuming 1× (dashed lines), and 30× (left panel) or 50× solar enrichment (right panel), of condensible volatiles (CH_4, NH_3, H_2S, H_2O ratioed to H) relative to solar. Cloud bases for 30× and 50× solar cases are marked on the right ordinates. The cloud densities represent upper limits, as cloud microphysical processes (precipitation) would almost certainly reduce the density by factors of 100–1000 or more. The cloud bases will not be affected, however. The structure and locations of the clouds at Uranus are very similar to the 1× and 30× solar (left panel) cases for Neptune due to similar thermal structure (p-T) and 20–30× solar enrichment of condensible volatiles, noble gases and the other heavy elements.

4. ENTRY PROBES

Much still remains mysterious about the clouds of the giant planets. It is only by having access to the region well below the main cloud layers that the abundances of key heavy elements can be determined.

Comparative study of the gas giants, when combined with a similar study for the icy giants, can provide the most comprehensive constraints for the models of formation of our solar system. Determination of the water abundance on Uranus and Neptune is much more challenging than that on Jupiter and Saturn. The colder atmospheres of the icy giants result in their cloud water

Uranus and Neptune seem insurmountable also in the near future. Survival of the probe structure and scientific payload to kilobar levels (as in Marianas Trench) where temperatures reach 500 K or greater, combined with the difficulty of data transmission from such great depths are only two of a multitude of obstacles. However, even if the entry probes could be designed to survive to only a hundred bar level, critical composition and dynamics information can still be collected. All heavy elements, except O, can be measured. As explained earlier, the O/H and N/H even at the kilobar level are not representative of their well-mixed abundance on Neptune and Uranus. On the other hand, noble gases, He, Ne, Ar, Kr, Xe, as well as C/H, S/H, $^{15}N/^{14}N$, and D/H, all of which can be accessed and measured at shallower depths with pressures of 50–100 bars, are fully adequate for constraining models of the formation of the icy giants and their atmospheres, especially when combined with the elemental and isotope abundance measurements, including O/H, at Jupiter and Saturn. Complementary information on disequilibrium species, PH_3, GeH_4, and AsH_3, as well as cloud, wind, and lightning characteristics would greatly enhance the value of the compositional data.

Multiple probes to the giant planets are critical for collecting the data required for understanding the formation of our solar system. Either in a single grand tour or on individual spacecraft missions, 2–3 probes deployed to 50–100 bars at all giant planets is recommended. The deployment of entry probes and proper operation of scientific payloads even to these depths must overcome enormous technological challenges. The transmission of probe radio signal from 100 bars at Neptune is also much more challenging than from 100 bar level at Jupiter. This is due to the 10–20 times greater abundance of the highly microwave absorbing molecules, ammonia and water (and perhaps also phosphine), at Neptune than at Jupiter at corresponding pressure levels (30–50× solar on Neptune, while only approximately 3× solar on Jupiter). Microwave remote sensing from spacecraft in the shorter term can provide a valuable guide to the development of probe missions.

5. REFERENCES

1. Atreya, S. K., Wong, M. H., Owen, T. C., Mahaffy, P. R., Niemann, H. B., de Pater, I., Drossart, P. and Encrenaz, T., A comparison of the atmospheres of Jupiter and Saturn: deep atmospheric composition, cloud structure, vertical mixing, and origin, *Planet. Space Sci.*, Vol. 47, 1243–1262, 1999.
2. Atreya, S. K., Mahaffy, P. R., Niemann, H. B., Wong, M. H. and Owen, T. C., Composition and origin of the atmosphere—an update, and implications for the extrasolar giant planets, *Planet. Space Sci.*, Vol. 51, 105–112, 2003.
3. Weidenschilling, S. J. and Lewis, J. S., Atmospheric and cloud structure of the Jovian planets, *Icarus*, Vol. 20, 465-476, 1973.
4. Atreya, S. K. and Romani, P. N., Photochemistry and clouds of Jupiter, Saturn and Uranus, in *Planetary Meteorology* (ed. G. E. Hunt), pp. 17-68, Cambridge University Press, 1985.
5. Atreya, S. K. and Wong, A. S., Coupled chemistry and clouds, in *Outer Planets* (eds. R. Kallenbach, Th. Encrenaz, T. Owen), Kluwer Academic Publisher, in press, 2004.
6. Lindal, G. F., The atmosphere of Neptune: an analysis of radio occultation data acquired with Voyager 2, *Astron. J.*, Vol. 103, 967-982, 1992.
7. Podolak, M., Hubbard, W. B. and Stevenson, D. J, Models of Uranus interior and magnetic field, in *Uranus* (ed. J. Bergstralh et al.), pp 48-49, University of Arizona Press, 1991.

Titan

THERMAL PROTECTION OF THE HUYGENS PROBE DURING TITAN ENTRY: LAST QUESTIONS

Jean-Marc BOUILLY [1]

[1] EADS SPACE Transportation - BP 11, 33 165 SAINT-MÉDARD-EN-JALLES Cedex, France
jean-marc.bouilly@space.eads.net

ABSTRACT

CASSINI-HUYGENS mission is a cooperation between NASA and ESA, dedicated to the exploration of the Saturnian system. In the framework of this mission, the entry of the HUYGENS probe in the atmosphere of TITAN will be of major scientific interest.

One of the essential points of the HUYGENS mission is therefore the good behavior of the thermal shield designed to maintain the aerodynamic shape and to protect the probe from excessive heating during the atmospheric entry on TITAN.

The design and the qualification of this thermal shield were carried out between 1992 and 1995 (development phase).

Currently, the final definition of mission parameters is being completed. As the performance of the thermal shield is one of all the parameters considered at system level, it is therefore necessary to reassess the thermal response of the TPS, taking into account some updated information that was not yet available during the development phase.

After some recall of the results of 1992 to 1995, the paper will present a status of the current work on TPS.

1. GENERALITIES ABOUT HUYGENS TPS

1.1 CASSINI - HUYGENS mission

Cassini-Huygens is a joint NASA/ESA planetary exploration mission to the Saturnian system. The mission has two components, an orbiter, which will explore the entire system, and a descent probe, which will investigate Saturn's largest moon, Titan.

Cassini, the Saturn orbiter, was built by NASA's Jet Propulsion Laboratory. Huygens, the descent probe, and the associated communications equipment on the orbiter were supplied by ESA.

The Cassini-Huygens spacecraft was launched on October 15th 1997. After a 7 years interplanetary journey, it has been inserted into orbit around Saturn on July 1st 2004. The Huygens probe will be separated from Cassini on December 25th 2004, and will finally enter the atmosphere of Titan on January 14th 2005.

Several experiments will be activated during the descent phase, which will be between 2 and 3 hours long.

1.2 Industrial organization (HUYGENS TPS)

The prime contractor of the ESA program HUYGENS is ALCATEL SPACE.

EADS SPACE Transportation is responsible for aerodynamic and aerothermodynamic analyses of entry and descent phases, as well as for design, justification and manufacturing of the thermal protections of two subsystems: the Back-Cover and the Frontshield. This is depicted on fig.1 below.

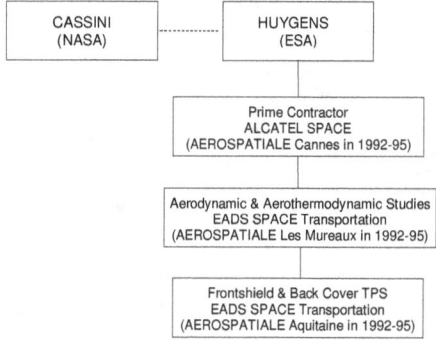

Fig. 1. Industrial Organization (TPS only)

1.3 TPS Architecture

The frontshield is made of a sandwich structure (aluminium honeycomb + CFRP skins) -(CFRP = carbon fibers reinforced plastic) and of two ablative thermal protection materials developed and produced by EADS-ST:
- AQ60/I on the front face is made of tiles bonded on the structure and jointed by a silicone glue.
- PROSIAL on the rear face (moderate heat flux level) is implemented using a spraying process

The back cover is made of an aluminum shell covered with PROSIAL.

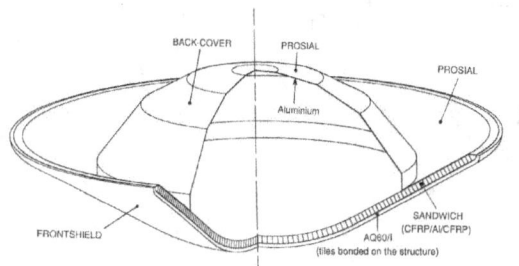

Fig. 2. TPS architecture

1.4 Entry conditions: (atmosphere, heat flux, shear stress, pressure)

The main constituent of TITAN atmosphere is nitrogen (N_2). Two other constituents are identified: argon (Ar) and methane (CH_4).

During entry, the methane dissociates in the shock layer, leading to the formation of CN. This molecule generates a high radiation in the narrow UV band. Though the convective heat flux is not very sensitive to atmosphere composition, the radiative heat flux can on the contrary reach very high values, especially for trajectories with highest FPA (Flight path angle).

The table 1 below summarizes the main characteristics of the probe, TPS, and environment during entry.

Table 1 - Main characteristics of HUYGENS TPS

HUYGENS Mission		Entry on Titan (Saturn's moon) after a 7 years travel with CASSINI
Entry characteristics (development phase values)	duration	300 sec.
	max. heat flux (front face)	1400 kW/m^2 (20 sec.)
	max. heat flux (rear face)	30 to 120 kW/m^2
	max. shear stress	135 Pa (area close to edge of decelerator)
	max. pressure	0.1 atm. (stagn. point)
	worst atmosphere	77% N_2, 20% Ar, 3% CH_4
Frontshield	T.P. material	AQ60/I
	Density	d = 0.28
	Thickness	17.4 to 18.2 mm
	T.P. mass	30 kg + 9 kg glue & joints
	Structure	CFRP honeycomb
	Structure mass	32 kg
	total mass	76 kg (including 5kg PROSIAL on back face)
Rear part and back-cover	T.P. material	PROSIAL
	Density	d = 0.54 to 0.60
	Thickness	0.3 to 3.1 mm
	T.P. mass	5.2 kg
	Structure	stiffened aluminium (0.8 mm)
	total mass	17 kg
Whole Entry Module	Total height	0.97 m
	Max. diameter	2.70 m
	Total mass of the vehicle	320 kg (actual mass) 335 kg (1992 hypothesis)

1.5 Thermal protection materials

AQ60, an EADS-ST trademark, is a felt made of short fibers. It is obtained (fig.3) by vacuum processing of an aqueous suspension of silica fibers. The material is then reinforced by an impregnation of Phenolic resin (representing 30% of the total mass).

AQ 60/I is the reinforced felt. The final density is 0.28, with a total porosity around 84 %. The volumic ratio of Silica is 10% and the one of resin is 6 %.

Remark: it must be noticed that the material used for HUYGENS is **AQ60/I** even though it is more often called AQ60, using an abusive contraction.

Due to its non-mineral bonding agent, this material undergoes pyrolysis for temperatures between 200° C and 1000° C. However, following pyrolysis the material is still self-supporting and becomes a very efficient thermal insulator with a quite high ablation temperature. It is also worth noting the quite good mechanical properties. This is of interest for withstanding thermomechanical entry loads.

For its industrial applications, this material has been applied following two main ways:
- thermal protection covers directly molded with the shape of equipment (military programs).
- tiles or panels, after a precise machining (HUYGENS)

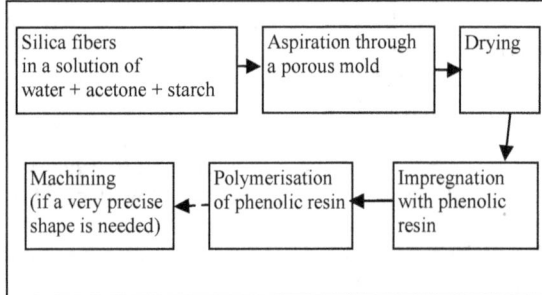

Fig. 3. Short description of AQ60/I manufacturing

PROSIAL 1000 is also an EADS-ST trademark: It is made up of a silicone elastomer with excellent thermal properties and of silica hollow spheres, making it possible to decrease density down to values of 0.6 to 0.54. Prosial 1000 is directly sprayed onto the surface to be protected.

2. THERMAL QUALIFICATION TESTS of HUYGENS TPS

2.1 General logic of the tests

Before being used for HUYGENS, these two materials had been developed for a quite different application. During the development phase, from 1992 to 1995, it was thus necessary to update and

complete their characterizations, particularly for AQ60, for which solicitations are indeed much higher than on PROSIAL.
The following objectives were reached successfully during the study, in order to demonstrate the satisfactory behavior of AQ60 in conditions representative of the HUYGENS entry:
- validation of the choice of this material
- update of the material data set thanks to thermal and thermomechanical characterization tests
- qualification of the tile arrangement (joints, steps, possible defects, micrometeoroid impact)
- thermomechanical qualification of the whole heatshield

The most specific physical aspects to consider were the following:
- high heat fluxes in a non oxidizing atmosphere as representative as possible of the Titan's one (gas mixture N_2, Ar, CH_4, or pure N_2)
- combination of high heat flux and aerodynamic shear
- thermomechanical effects

The corresponding tests are recalled in next sections.

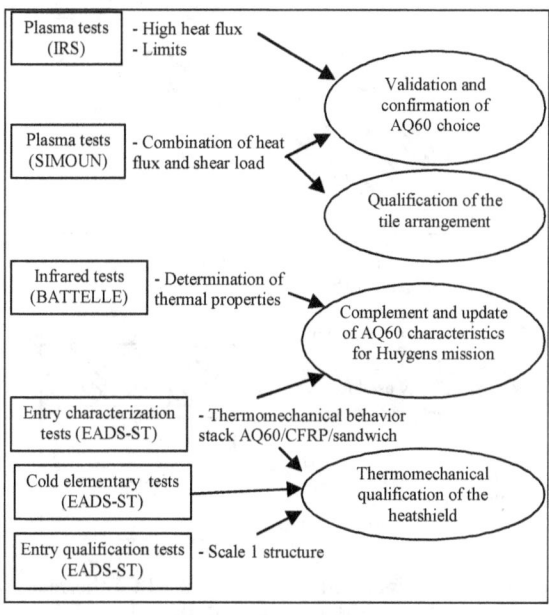

Figure 4. General logic of the tests

2.2 IRS stagnation point plasma tests

Plasma tests in stagnation point configuration [1,2,3] were performed in the plasma wind tunnel PWK2 of IRS (Institut für Raumfahrtsysteme) of the University of Stuttgart.
A series of 17 tests was performed under various aerothermodynamic conditions
- Titan atmosphere (77% N_2, 20% Ar, 3% CH_4)
- Stagnation pressure: from 0.015 to 0.020 atm.
- Heat flux: from 600 to 2500 kW/m²

Several other parameters were also analyzed:
- Influence of atmosphere (Titan or pure N_2)
- Influence of joints
- Influence of coating
- Heat flux: constant value, or flight evolution

The sample is positioned on a support fixed on a platform that can be moved with a high precision and speed, which allows realizing heat flux evolutions versus time, with various possible maximum values.

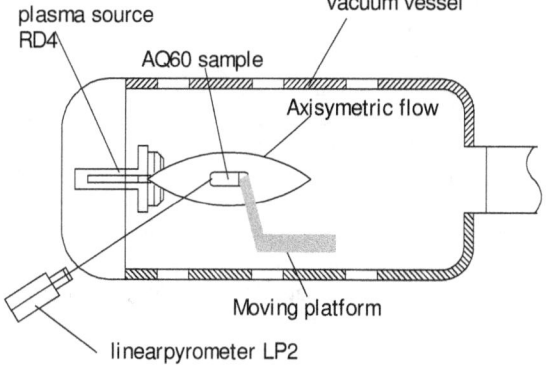

Fig. 5. Experimental setup for plasma tests at IRS-PWK2

Fig. 6. AQ60 sample during IRS plasma test

These tests permitted to demonstrate the good behaviour of AQ60/I submitted to heat fluxes up to 2500 kW/m² in an atmosphere representative of Titan's one (77% N_2, 20% Ar, 3% CH_4). They also demonstrated a good margin (without reaching an upper limit) with regard to heat flux level, which was necessary because of uncertainties on heat flux computations. Furthermore, they evidenced a good ablative behavior, with surface temperature and surface recession increasing regularly versus heat flux value. The exploitation of these tests permitted to determine the ablation law associated to this behavior. Finally, they proved the major influence of the atmosphere, with a better behavior than in air, by comparison with previous results not detailed in this paper, obtained on SIMOUN with similar samples, configuration, and heat flux values.

2.3 SIMOUN PLASMA TESTS

Two series of plasma tests in tangential flow configuration [1,2,3] were performed on the SIMOUN facility of EADS-ST. The test conditions were the following:
- Pure N_2 atmosphere
- Pressure: from 0.100 to 0.140 atm.
- Shear load = 500 Pa (estimation)
- Heat flux: 740 to 973 kW/m²
- Samples : 300 x 300 mm
- Board inclination : 16.5 degrees

They allowed demonstrating the good behavior of the material itself, when submitted both to heat flux and aerodynamic shear. They also permitted to qualify some particular points, such as MLI fixations, micrometeoroids holes, and manufacturing defects like wide joints, steps or local repairs. The main conclusion was thus that AQ 60 could withstand the Huygens entry conditions without any critical damage.

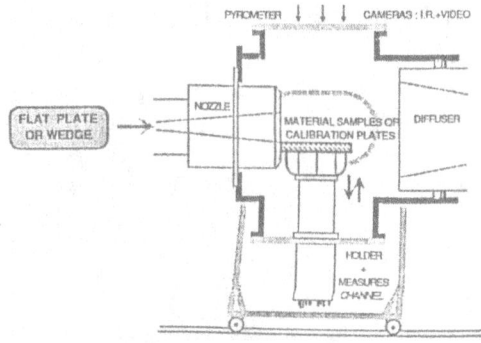

Figure 7. SIMOUN test facility in flat plate configuration

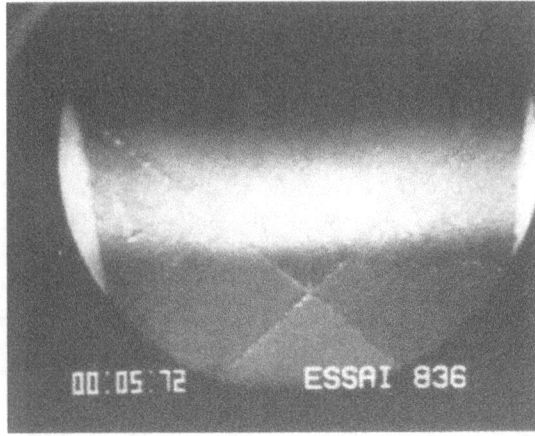

Figure 8. AQ60 sample during SIMOUN test

2.4 THERMAL CHARACTERIZATION TESTS at BATTELLE

This series of characterization tests was performed at Geneva center of the Battelle Institute.
Test samples were introduced in a crucible heated at a constant temperature of about 1580°C, generating a heat flux of 650 kW/m² in a controlled N_2 atmosphere. The AQ60 samples were cylinders of Ø 20 mm and 20 mm thick with 8 thermocouples in the depth. These samples were designed to obtain a 1D heat transfer.

The exploitation of these tests permitted to determine a global but simplified thermal data set. Indeed, it takes into account the pyrolysis reactions occurring during the heating only in an equivalent way. In addition, these tests provided some indications on the dispersions of the material properties.
Similar tests were performed on PROSIAL 1000, which allowed adjusting the thermal properties and associated uncertainties for this material.

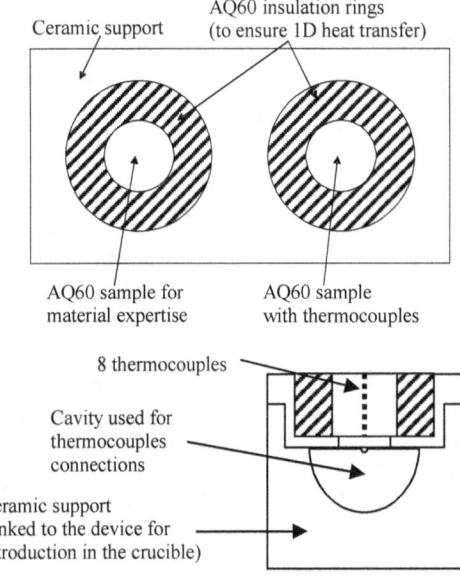

fig.9. description of the samples used for thermal characterization tests at Battelle.

2.5 Entry Characterization Tests

The Entry Characterization Test [1,2,3], or ECT was a series of thermomechanical tests performed in the test department of EADS-ST.
After some elementary tests at the beginning of the test campaign, i-e a simple concave or convex deflection, then a heating without mechanical solicitation, the ECT tests themselves were performed. The main characteristics were the following:

- Application by infrared lamps of a heat flux representative of the real mission
- Simultaneous application of a mechanical bending leading to tensile or compressive stressing according to the applied deflection.
- non oxidizing controlled atmosphere of pure N_2
- 12 samples representative of the complete stack, with actual design thickness: CFRP Honeycomb (long piece 800 x 40 mm) - glue - AQ60/I

These tests permitted to overcome the impossibility to determine mechanical properties at high temperature because of chemical transformations in the material. They also validated the thermo-mechanical behavior of AQ60/I tiles during entry, and permitted to deduce equivalent material data for theoretical analyses.

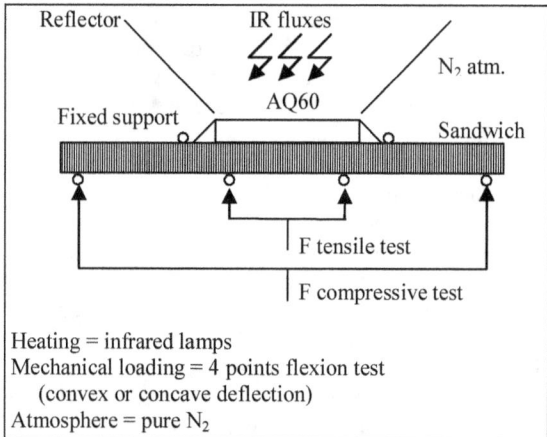

fig.10: schematic description of Entry Characterization Tests

2.6 Cold Elementary Test

The Cold Elementary Test [2,3], or CET showed that there was no failure mode under cold environment (-120°C), and permitted to determine the actual influence of glue and PROSIAL thermal expansions on the stack deflections. These tests were performed on samples representative of the stack of the frontshield.

2.7 Entry Qualification Test

The Entry Qualification Test [2,3], or EQT, permitted to demonstrate the good global behavior of the Frontshield during the entry phase. This was done by a simulation of the following parameters: initial cold temperatures (-60°C), neutral N2 atmosphere, entry heat fluxes simulated by infrared lamps, mechanical loads (i.e. external pressure and deceleration) simulated by a pressure difference between the back and front faces.

The test specimen was made up of a flight representative structure partly covered on its front face by AQ60/I tiles. This permitted to carry out two tests with the same thermal solicitation: the first one at the qualification level for mechanical loading, the second one at a higher value of mechanical loading to evidence some margins.

This specific test facility (fig.11) was developed and operated by the test department of EADS-ST. It was composed of a large chamber that included the cooling, depressurization and heating devices.

fig.11: picture of EQT device

2.8 Conclusions at the end of development phase

After analyses and tests performed during the development phase (1992-1995), the qualification of the Huygens TPS was considered as successfully completed and both frontshield and back-cover were accepted for flight.

3. MISSION PREPARATION PHASE

In order to prepare the impending Huygens entry, the final definition of mission parameters is being completed. The performance of the thermal shield is one of all the parameters considered at system level and it is thus necessary to reassess the thermal response of the TPS. These last analyses must obviously take into account some updated information that was not yet available during the development phase 10 years ago. More particularly, two points have to be considered:
- A possible transparency of the AQ60 material in the UV wavelengths
- Updated heat fluxes, with expected values significantly higher than during the development phase.

Several actions are still on-going on these two topics, and a status of this current work is presented in the following sections.

4. AQ60 POSSIBLE TRANSPARENCY

4.1 Overview of the problem

As mentioned previously, due to the atmosphere composition, the entry velocity, and the shape of the probe, the heat shield undergoes both convective and radiative heat fluxes. More precisely, the radiative emission of the shock layer occurs in the narrow UV band. In the framework of studies about aerocapture mission at Titan [4,5], NASA experts identified possible uncertainties on performance of lightweight materials. Indeed, a general trend was suggested from Laser tests performed in the 80's on several dozens of TP materials. The shorter the wavelength was, the larger became the absorption length. There is no available test result in UV wavelength for lightweight materials. The potential for in-depth absorption could thus be of concern for AQ60, since it could lead to char spallation that would significantly reduce its efficiency and lead to eventual additional heating of the underlying substructure. In order to evaluate the performance of candidate Titan TP materials exposed to UV radiation, NASA has decided to develop a specific facility based on a high-power Mercury-Xenon lamp that has a strong emission in the UV range.

4.2 Action plan

Based on above mentioned information, this topic was analyzed during the Delta-FAR (Flight acceptance review) held at the beginning of 2004. It was decided to initiate several actions in order to evaluate the influence of this phenomenon on the performance of the Huygens Frontshield.
- Status on representativeness of development phase tests wrt radiative emission in UV band
- Status on representativeness of IRS test wrt radiative emission of the flow in UV wavelength
- Low intensity radiation exposure tests at ESTEC
- High intensity radiation exposure tests at NASA

The corresponding results are presented hereafter.

4.3 Representativeness of development phase tests

Two families can be identified among the tests of the development phase: radiative and plasma tests.
The radiative tests (BATTELLE, ECT, and EQT) were performed with an Infrared radiant source. No information about UV radiation can therefore be deduced from these tests.
The SIMOUN plasma tests were carried out in a pure N_2 tangential flow. There was therefore no radiation effect during these tests.
On the other hand, most of the IRS tests were performed in an atmosphere comprised of N_2, Ar and CH_4. In addition, it must be highlighted that the introduction of methane was very spectacular, inducing a high brightness of the flow [6]. Thus only IRS tests can be relevant with regard to UV radiation.

4.4 Representativeness of IRS tests

During the development phase, this problem of UV radiation had not been considered, and only the total heat flux had been measured for this test campaign.
In order to evaluate if the test performed in 1992 could provide information about AQ60 possible transparency, some actions have been proposed but have not been selected for several reasons (delay, cost and uncertainties on the results).
- reexploitation of the 1992 tests in order to identify if such a phenomenon occurred. This would require a very precise thermal model while current data only lead to a satisfactory global resetting, remaining on purpose slightly conservative. Furthermore, very accurate measurements would be necessary whereas only one thermocouple was installed on each IRS sample.
- theoretical analysis and evaluation of the radiative emission of the flow
- complementary tests with same conditions as in 1992 and specific measurements (knowing that it is considered as difficult to perform radiative flux measurement in UV)

However, a synthetic analysis can be established, relying on experimental works conducted by IRS after the end of the Huygens development [6,7,8]. Indeed, an extensive characterization of Nitrogen/Methane plasma flows was undertaken from 1992 to 1998.
A specific radiometer was developed and used to measure the radiation emitted by the flow [6]. In addition, a set of emission spectroscopy measurements was done for various combinations of N_2/CH_4 mixtures [7]. This allowed to evidence that some radiative heat flux occurred during these experiments, and that some emission could be observed around 380 nm, which corresponds to CN violet. A direct quantitative interpretation of these tests is not easy, because these are mainly local measurements in reduced solid angles. An estimation of the integrated value is provided in [8]: the radiative heat flux is 377 kW/m², which represents ≈20% of the corresponding total heat flux equal to 1800 kW/m².
Even though some uncertainty must obviously be associated to this result, it shows that the radiative component of the flux can be considered as significant for the tests that were performed in 1992.
However, no evident influence on material behavior was identified. This point is thus quite positive, even though the worst expected value of the radiative heat flux could be much higher than the experienced one of 377 kW/m².

4.5 Low intensity radiation exposure tests at ESTEC

As recommended by the board of the Delta-FAR, elementary characterization tests on AQ60 were performed by ESTEC in March and April 2004 [9].
AQ60 samples of 40 x 40 mm x 1 to 5 mm thick (fig.12) were illuminated by a spectral Xenon lamp radiating at a wavelength of 377 nm, and the intensity of the light transmitted through the samples was recorded. The transmission was then calculated by comparison with transmission obtained with calibrated neutral density filters.

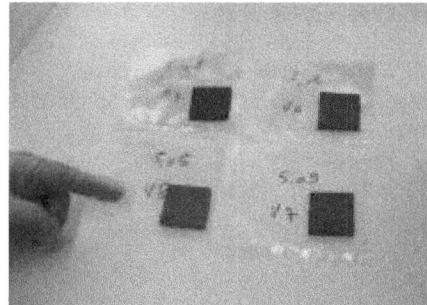

Fig. 12: AQ60 samples

The test device (fig.13) is operated at room temperature.

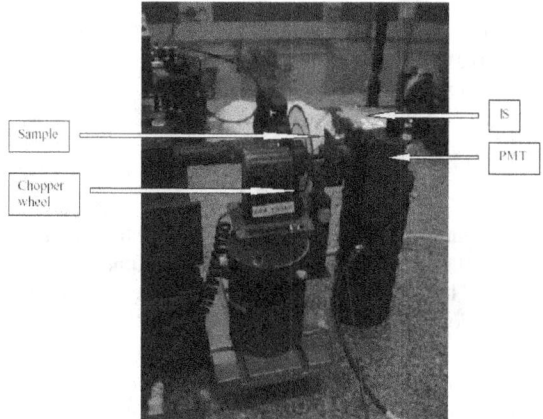

Fig. 13: Picture of test device
(IS = Integrating Sphere; PMT=Photomultiplier)

Tests were carried out on 10 samples provided by EADS-ST. 8 of these samples were made of virgin material without coating, with different thickness between 2 and 5 mm (two 1 mm thick samples were machined by ESTEC from already tested specimen in order to complete a first set of results). The 2 other samples were made of char material issued from tested samples remaining from the 1993 ECT tests.

The results presented in the table 2 show a very low transmission in the UV through AQ60. This transmission is even lower for the pyrolised samples.

Table 2: samples thickness and transmission calculated from the measured signal

Sample	Thickness (mm)	Transmission
V1(bis)	1.00	1.79E-04
V7(bis)	1.10	1.35E-04
V2	2.12	7.77E-06
V3	3.10	1.20E-06
V4	3.10	1.27E-06
V5	4.08	1.82E-07
V6	4.10	1.57E-07
V8	5.05	5.20E-08
P1	4.06	8.00E-09
P2	4.08	6.00E-09

(V= virgin material. P= pyrolised material).

These results are very positive and they constitute the first step of the demonstration that the transparency of AQ60 in UV can be considered as a negligible phenomenon for the Huygens mission.

4.6 High intensity radiation exposure tests at NASA

It is considered as very relevant to complete the previous results by tests at high temperature. With this aim, NASA proposed to include AQ60 samples in a test campaign prepared at NASA Ames for analyzing the performance of lightweight TP materials when exposed to high intensity UV radiation (cf. §4.1).
EADS-ST provided 8 AQ60 samples (75 x 75 x 20 mm) for these tests. These samples include a central plug insert (diameter 30mm) in which several thermocouples must be installed by NASA.

Fig. 14: AQ60 samples for UV tests at NASA

Tests will be performed at different heat flux levels: 500, 1000 (tbc) and 1500 kW/m², generated by a high-power Mercury-Xenon lamp. The detailed missions and their durations are still to be precised.
The performance of these tests is planned within a few weeks, as soon as the test device is available.

4.7 Conclusion

Available experimental results do not show any significant transparency of AQ60 in the UV:

- No influence was noticed during IRS tests despite a significant radiative component of the flux
- The transmission at room temperature measured during low intensity radiation exposure tests at ESTEC is negligible

This trend is expected to be confirmed by the high intensity radiation exposure tests at NASA. Afterwards, this phenomenon of transparency in UV would no longer be a problem for the AQ60 frontshield of the Huygens mission.

5. INFLUENCE OF HEAT FLUX UPDATE ON TPS

Several elements contribute to update the mission and refine the entry corridor:
- Communications between orbiter and probe
- Selection of a new atmosphere model (Yelle) associated to the Strobel Gravity Wave model

The associated aerothermal environment is therefore rather different from the one used during C/D phase.

The corresponding reassessment work performed in 2003 by the industrial team has been reviewed in the frame of the Delta-FAR in February 2004. In the course of this review, different heating levels have been observed between various contributions, namely EADS-ST, ESTEC-MPA and NASA ARC. Following one recommendation of the review board, it was thus created an Aeroheating Convergence Working Group (ACWG) with the following objective:
- After a correct understanding of the differences, to reconcile the various aerothermal inputs and consolidate a single aerothermal environment.

The worst cases have been determined in a first step of this activity [10], but work is still on-going in order to achieve this task.

The influence of this heat flux update on TPS thermal response will be analyzed as soon as the heat flux reassessment is completed.

In addition, it will perhaps be necessary to analyze a last update of atmosphere models, if some new information is brought by Titan flyby (scheduled on July 3rd & October 26th)

6. CONCLUSION

Work is still in progress on the two following topics:
- possible transparency of the AQ60 material in the UV wavelengths
- Last update of heat fluxes and influence on TPS

They will be completed within the following weeks and hopefully will show that the Huygens TPS is ready to fulfill its role in the success of the last and critical phase of the mission in January 2005.

7. ACKNOWLEDGEMENTS

This paper has been prepared in the framework of the Huygens program of the European Space Agency.

The author wishes to thank people from ESA/ESTEC, NASA, Alcatel Space, IRS and EADS-ST who helped for the preparation and the verification of this article.

8. REFERENCES

1. Bouilly J-M, Entry testing of AQ60 for Huygens, 1st ESA/ESTEC Workshop on Thermal Protection Systems, Noordwijk, The Netherlands, 5-7 May 1993.

2. Test reports and exploitations issued during the Huygens development phase (1992-1995)

3. Bouilly J-M., Guerrier D., Lautissier P., Saguet J-P., Design, development and qualification of thermal protection systems for Reentry vehicles (Huygens, ARD, future vehicles), IAF-96-I.6.09, 47th International Astronautical Congress, Beijing, China, October 7-11, 1996.

4. Laub B., Venkatapathy E., Thermal protection system technology and facility needs for demanding future planetary missions, International Workshop on planetary probe atmospheric entry and descent trajectory analysis and science, Lisbon, Portugal, 6-9 October 2003

5. Laub B., Venkatapathy E., Thermal protection concepts and issues for aerocapture at Titan, 39th AIAA/ASME/SAE/ASEE Joint Propulsion Conference and Exhibit, AIAA 2003-4954, July 20-23 2003, Huntsville, Alabama.

6. Röck W., Auweter-Kurtz M. - Experimental Investigation of the Huygens Entry into the Titan Atmosphere within a Plasma Wind Tunnel - 30th AIAA Thermophysics conference - June 19-22, 1995 - San Diego, California

7. Röck W, Auweter-Kurtz M.-Spectral measurements in the Boundary Layer of Probes in Nitrogen/Methane Plasma Flows - 32nd AIAA Thermophysics conference - June 23-25, 1997 - Atlanta, GA

8. Röck W. - Simulation des Eintritts einer Sonde in die Atmosphäre des Saturnmondes Titan in einem Plasmawindkanal – Dissertation– Universität Stuttgart – 1998 *(partial oral information about this document)*

9. Witasse O., Daddato R., Blancquaert T., Transparency test of the AQ60 material, ESTEC report *HUY-RSSD-RP-001-1_0-AQ60-Transparency-Test 2004, May21*

10. Walpot L., Molina R., Huygens Aeroheating Status Report - ESTEC report HUY-RP-191-MSM/MXA (2/1) - June 4th, 2004

REVALIDATION OF THE HUYGENS DESCENT CONTROL SUB-SYSTEM

J.C. Underwood, J.S. Lingard and M.G. Darley

Vorticity Ltd., Chalgrove, Oxfordshire OX44 7RW, United Kingdom

ABSTRACT

The Huygens probe, part of the Cassini mission to Saturn, is designed to investigate the atmosphere of Titan, Saturn's largest moon. The passage of the probe through the atmosphere is controlled by the Descent Control Sub-System (DCSS), which consists of three parachutes and associated mechanisms.

The Cassini / Huygens mission was launched in October 1997 and was designed during the early 1990's. During the time since the design and launch, analysis capabilities have improved significantly, knowledge of the Titan environment has improved and the baseline mission has been modified. Consequently, a study was performed to revalidate the DCSS design against the current predictions.

Nomenclature

a_{11}	Added mass (kg)
C_D	Drag coefficient
C_{mq}	Pitch damping coefficient
D_p	Projected diameter
g	Acceleration due to gravity (m/s^2)
m_s	System mass (kg)
S_p	Projected Area (m^2)
V	Velocity (m/s)
ρ	Density (kg/m^3)

1 INTRODUCTION

In the 10 years since the DCSS was designed much has changed: the knowledge of the Titan atmosphere has improved, parachute inflation analysis tools have been refined and the entry conditions of the probe into the Titan atmosphere have been revised.

As the release of Huygens from Cassini approaches, this is the last opportunity to revalidate the design of the Huygens probe using the latest information gained from Cassini and predict the performance we expect on 14th January 2005.

2 SEQUENCE

The Huygens DCSS sequence starts at a nominal Mach number of 1.5, 157 km above the surface of Titan (Fig 1a). At this point, approximately 260 seconds after first encountering the atmosphere, the probe is still encased in its protective aeroshell. The first function of the DCSS is to remove the rear portion of the aeroshell using a mortar deployed, 2.59 m Disk-Gap-Band (DGB) parachute of a similar design to that used on Viking (Fig 1b).

As the pilot chute separates the rear aeroshell from the probe (Fig 1c), a second, 8.3 m parachute is deployed by a lanyard. This parachute, a DGB of a slightly different design, provides stability as the probe decelerates through Mach 1 (Fig 1d) and sufficient drag to allow the front aeroshell to fall away from the probe when it is released 32.5 seconds into the sequence (Fig 1e).

Once the aeroshell has separated from the probe, the science instruments start to take data as the probe descends through the upper atmosphere (Fig 1f).

If the probe were to remain in this configuration the probe descent to the surface would take over 5 hours. Since the Cassini orbiter is only visible for 2.5 hours, the main parachute must be released 15 minutes after the start of the descent sequence by means of three pyrotechnic cutters and a 3.03 m stabilising drogue deployed (Fig 1g).

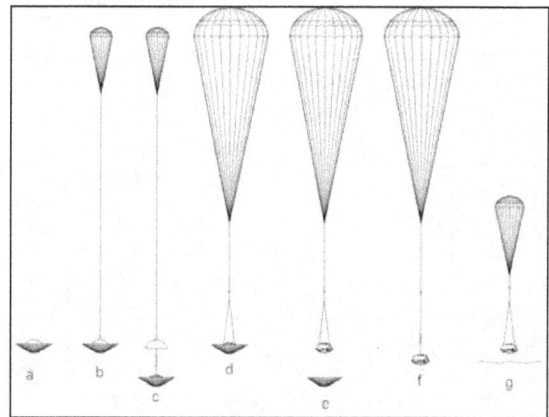

Fig. 1 Probe Sequence

The sequence has been documented in detail elsewhere [1].

3 CHANGES
3.1 Atmospheres

In order to design the Huygens probe, assumptions had to be made about the environment it would experience on arrival at Titan. The most crucial of these related to the atmosphere. Its temperature and density profiles determine the overall trajectory and the conditions at initiation of the descent sequence whilst its composition is critical for the design of the heat shield used to protect the probe during the initial entry into the atmosphere.

The original 1987 Lellouche-Hunten atmosphere model [2] defined three profiles (minimal, nominal and maximal) and was used throughout the development of the Huygens DCSS. This was superseded by a new model [3] derived by Yelle in 1994. However, since the DCSS was already designed and tested it was not fully revalidated against the new atmosphere.

Subsequently, the atmosphere was modified by the addition of coherent gravity waves [4] which could be adjusted to give a worst case atmosphere and an alternative atmosphere, TitanGRAM, was produced independently [5] which generates random perturbations which are more suited to Monte-Carlo analysis. Both use the Yelle values as their nominal profiles.

Over the next few months it is expected that these models will be further refined using data from Cassini's targeted flybys of Titan in October 2004 and December 2004

3.2 Initial Conditions

During the design of Huygens the responsibility for trajectory analysis was spilt between the prime contractor (Entry phase) and the DCSS contractor (Descent phase). This split made it impossible to carry out large numbers of simulations from the atmosphere interface to the surface. In order to design the DCSS, seven design cases were defined which gave extreme conditions at initiation. These were derived using the extreme cases tabulated below.

Table 1. Design Initiation Cases

Condition	Entry Angle	Atmos	Mass	T0
Nominal	Nom	Nom	Nom	Nom
Min Mach	Steep	Min	Max	Late
Max Mach	Shallow	Max	Min	Early
Max q	Steep	Min	Max	Early
Min q	Shallow	Max	Min	Late
Failure Max q	Steep	Min	Max	FoA
Failure Max Mach	Shallow	Max	Min	FoA

These seven cases were derived from three trajectories: nominal, steep entry and shallow entry. Three points were generated on each trajectory representing the earliest nominal initiation (based on sensing uncertainty), latest nominal initiation and fire on arm (FoA) – a single failure case.

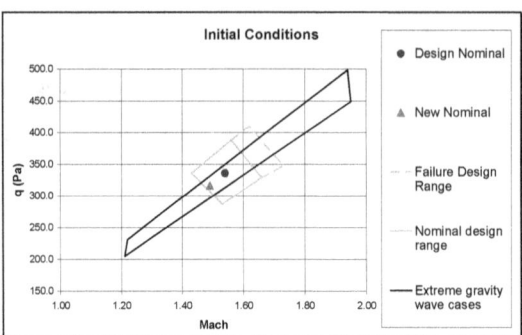

Fig. 2. Revised initial conditions

As a result of the change of atmosphere model and a change in entry angle brought about by a change in the Cassini / Huygens geometry during the mission, the entry trajectory and thus DCSS initiation points changed. New initiation cases were derived using trajectories modelled with the new atmosphere and entry conditions. Since the effects of gravity waves on the sequence sensing logic are not straightforward, a number of cases were examined in order to determine the worst case conditions. The new envelope is shown graphically in Fig. 2.

It should be noted that the new analysis takes account of failure cases which were not considered during the design of the DCSS. This explains in part the large increase in the range of initial conditions.

It can be seen that the ranges of Mach number and dynamic pressure at DCSS initiation have increased significantly from those it was designed for. The new analysis used only five initial conditions (one nominal and four worst cases), the high Mach and high q non-failure cases were discarded since they were enveloped by the other four cases.

3.3 Masses

During development all modelling was performed using the extreme design range of masses, including margins. During the final integration of the probe all items were weighed, so the "as flown" masses are now available to refine the predicted performance. The only remaining mass uncertainty is the loss from the aeroshell during entry.

3.4 Aerodynamic Databases

Since the completion of the development programme, data obtained during the programme and on other subsequent programmes have allowed refinement of the aerodynamic databases for the probe and parachutes. The current study provided an opportunity to incorporate these data into the analyses.

4 THE TOOLS

4.1 Timeline and deployments

Originally, the overall descent timeline, deployments and parachute inflation loads were each modelled using separate, dedicated software. For the early stages of the re-validation new models were written and validated against output from the original software. However, although this approach was adequate for checking a small number of design cases, it was inadequate for running Monte-Carlo entry and descent simulations.

4.2 Inflations

Parachute inflations were modelled using an engineering model derived from wind tunnel and full scale test data. The design of this software is described later in this paper.

4.3 Stability

One important function of the DCSS is to stabilise the probe as it decelerates through Mach 1. Since the probe incidence at parachute deployment, the deployment Mach number and the aerodynamics of the probe had all changed since the original analysis, the stability of the probe had to be reassessed.

The stability model used during development was created using a software package which is no longer available, so a new multi-body, 6 degree of freedom model was constructed to perform this analysis and was validated against the original model.

4.4 Full Entry and Descent Model

The analysis during development and the early stages of revalidation involved modelling of different stages of the Huygens sequence in isolation. This was time consuming and made it very difficult to investigate the effects of parameter changes to the overall performance of the system.

In order to investigate the effects of the new atmosphere models on the performance of the Huygens probe it was necessary to model a large number of sequences through differing atmospheres from entry interface to deployment of the main parachute to identify the most extreme conditions. To accomplish this, an existing 3 degree of freedom simulation code was extended for the purpose. The new code includes:

- Four planetary atmosphere models in addition to arbitrary look-up tables;
- Flexible sequencing based on Mach, accelerations, timing and height;
- Accurate parachute inflation models;
- The ability to model the decision logic and voting of the Huygens CDMUs; and
- The option to vary any input parameter according to statistical distributions.

The code may be used to explore the envelope of extreme values of each input parameter or to run Monte-Carlo analyses randomising the whole system.

5 INITIAL ASSESSMENT

5.1 Deployment Conditions

The five new initiation cases were used to predict descent trajectories, the deployment conditions for the parachutes and the release time for the front shield. The simulations were performed using the extreme worst case masses and aerodynamics in order to obtain the most extreme conditions at each stage. These were then used to calculate deployment times and loads for the parachutes and thence inflation loads.

The increase in range of conditions seen at mortar firing was also evident at the time of main parachute release (only 2.5 seconds later). The lowest Mach number at this time was now 1.12, which compares with 1.30 for the design sequence. The reduced Mach number gives less time for the main parachute to deploy and stabilise the probe before Mach 1, where the probe is dynamically unstable.

At the time of front shield release, 32.5 seconds after the pilot chute deployment, the range of conditions (Mach 0.34 to 0.55) still exceeds the values predicted during the development programme (Mach 0.39 to 0.51); however, they are well within the design limits (Mach<0.6).

By the time of main parachute release, 15 minutes from pilot chute deployment, the conditions were indistinguishable from those predicted during the development programme.

The extreme limits of descent time were calculated to be 2 hours, 0 minutes and 2 hours, 31 minutes. This compares favourably with the design aim of 2 hours, 15 minutes ± 15 minutes.

5.2 Deployment dynamics

The pilot chute is deployed by a pyrotechnic mortar through a region of recirculating flow behind the probe. Following ejection by the mortar it first decelerates as it passes through the near wake and then accelerates towards lines taut. Too slow a deployment could result in the chute becoming caught in the wake and not deploying; too fast a deployment will increase the snatch loads at lines taut and potentially damage the parachute.

The pilot chute deployment predictions indicated the velocity at lines taut was slightly increased from the design value, giving maximum snatch forces of just under 600 N. Since the inflation load exceeds this by a factor of three, these loads are not considered to be an issue.

The main parachute is deployed by the pilot chute, which pulls the back cover away from the probe, thus deploying the parachute from its bag. The back cover / pilot chute combination and probe were designed to have a ballistic coefficient ratio of no worse than 0.7. In fact, when the new conditions and known masses are taken into account, the worst case ballistic coefficient ratio is 0.45. This indicates a very positive separation and suggests there is a possibility that the main parachute may deploy too quickly, thus causing high snatch loads and potential searing damage to the canopy.

The deployment model was used to predict the deployment velocities and snatch loads during parachute deployment. The maximum velocity at bag strip was predicted to be 54 m/s using the new conditions, the highest velocity predicted during development was 51 m/s. Although the new velocity is higher, the difference is small enough that there is little increase in the likelihood of damage.

The peak main parachute snatch load was predicted to be a maximum of 2.7 kN, up from 2.1 kN during the DCSS design. This is much lower than the parachute design inflation load of 14.7 kN so is not a concern.

The front shield release was designed such that the worst case ballistic coefficient ratio at release was 0.7. The main source of uncertainty in this ratio during the development programme was the mass uncertainty of the front shield and probe. Since this has now been eliminated, the uncertainty in ballistic coefficient ratio has been reduced such that the predicted ratio now lies in the range 0.50 to 0.54.

It was concluded that all the component deployments and separations are robust with respect to the new conditions.

5.3 Stability

The reduced Mach number and increased dynamic instability at main parachute deployment leads to a possibility that the oscillations may increase during the transonic deceleration to the point that parachute bridles could become slack and wrap around other items on the probe back cover (for example the communications antennae). In order to assess the possibility, a number of simulations were carried out from the lowest expected Mach numbers with the maximum expected probe initial incidence.

Fig 3 shows the probe (heavy line) and parachute incidence when started from an initial probe incidence of 10° and parachute incidence of 2°. The parachute motion quickly damps out, while the probe oscillation is controlled quickly, reducing to less than 1° by Mach 0.7. This oscillation is well within the capabilities of the system.

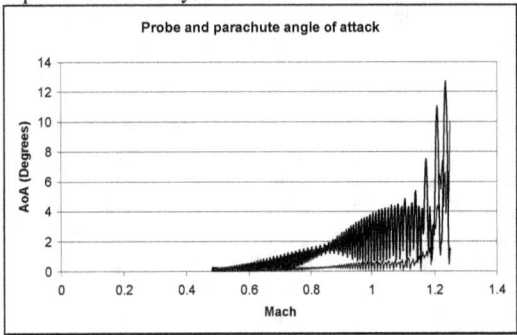

Fig. 3 Probe and Parachute incidence

The Huygens main parachute has been designed to be stable at zero incidence. However, some variants of the DGB design (including the Huygens pilot chute) tend to glide, having a non-zero stable incidence. A second simulation (fig 4) was carried out assuming the parachute had a stable incidence of 10°. This shows that the probe and parachute both start to glide with a stable incidence but the oscillation about this stable angle rapidly damps out.

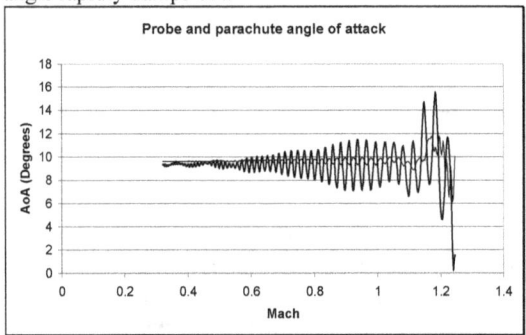

Fig 4. Unstable parachute

The stability analyses indicate that the DCSS is capable of controlling the probe as it decelerates through Mach 1 even with the new, low parachute inflation mach numbers.

6 INFLATION LOADS

During the DCSS development, the parachutes were tested in wind tunnels [6], in low level drop tests [7] and finally in a high altitude system drop test [8, 9]. These tests provided valuable information to assess the aerodynamic coefficients and inflation characteristics of the parachutes. During the development of the parachute for MER (a DGB of a slightly different design) a test anomaly occurred, where the opening load was significantly higher than predicted by existing models. A re-analysis of Huygens data in combination with the new MER data allowed a refinement of the opening load model.

The parachute inflation loads have been remodelled using a code that explicitly includes added mass terms. The code is based on work published in [10] which in turn is similar to the work of Cruz [11].

The fundamental equation of motion is written:

$$m_s \frac{dV}{dt} = m_S g - \frac{1}{2}\rho V^2 C_D S_p - a_{11}\frac{dV}{dt} - V\frac{da_{11}}{dt}$$

Canopy drag area evolution ($C_D S_P$) was extracted from Huygens and MER test data. The added mass was defined as:

$$a_{11} = 2.136\rho \frac{\pi D_P^3}{12}$$

Inflation time estimates were based on the Huygens test data extrapolated to the supersonic regime using Greene [12].

A typical force profile is shown in fig. 5 for the main parachute.

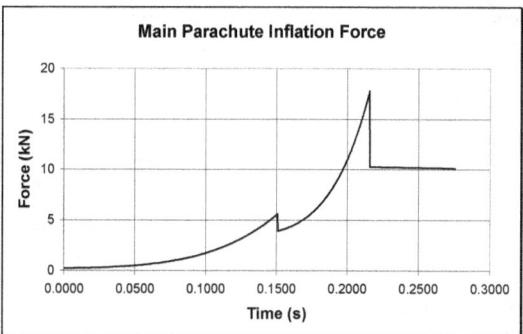

Fig. 5 – Typical parachute inflation force profile

The maximum inflation loads for the pilot and main parachutes in the worst case conditions were predicted to be 2,350 N and 17.6 kN respectively. These compare with design loads of 1,800 N and 14.7 kN.

Following the increase in predicted loads, the strength of the parachutes and probe structure were reassessed. The new loads were found to be within the structural capabilities of all components.

7 DETAILED ENTRY ASSESSMENT

The analyses performed during the first part of the reassessment involved analysing the performance of the descent phase of the sequence with discrete atmospheres from a very limited number of initiation conditions. This approach has two drawbacks:
- The value of each parameter which produces the worst case outcome must be correctly chosen (this is not always obvious, for instance in the case of gravity waves, and the "worst case" for one subsystem may not be the worst for another);
- Worst cases can be identified without any idea of their probability of occurrence – it is not worthwhile designing for a 1 in a million occurrence.

In order to explore the envelope of potential outcomes and determine the probability of an extreme event occurring a Monte-Carlo simulation was carried out.

7.1 Model

Advances in computer power over the last decade and improved software using the latest object-oriented capabilities of languages such as C++ have allowed the fidelity of entry simulations to be improved and made large Monte-Carlo runs, which would previously have required the use of supercomputers, accessible to ordinary PCs. The simulation used for the re-validation has the capability of modelling the internal logic of the Huygens sequencer controllers as well as the inaccuracies of the sensors and uncertainty in probe parameters. The DCSS was designed using seven discrete trajectories but over 55 independent variables were available for the re-validation exercise.

7.2 Analysis

In order to produce a valid set of simulations it was necessary to start the simulation outside the atmosphere at an altitude of 1,270 km above the surface. This allowed specification of an initial state and 6x6 covariance matrix which produced the expected range of initial flight path angles. The software then modelled the entry and descent phases using the probe mass and aerodynamic databases.

For this analysis TitanGRAM was used in preference to the discrete models used earlier in the analysis. This produces random density and temperature profiles based on the same nominal profile used in the previous models.

Simulations were run from atmosphere interface until main parachute inflation to determine the probability of exceeding the pilot chute or main parachute inflation loads and to investigate a potential failure mode where the mortar is fired before the system is armed. A smaller number of simulations were run from atmosphere interface to landing in order to assess the potential variability in mission length.

7.3 Initial Conditions

The DCSS initiation points generated from the analysis are shown in fig. 6. It can be seen that the gravity wave cases used in the initial analysis (solid black line) produce initial conditions which are well outside the design range (grey lines). Two sets of Monte-Carlo results are shown: nominal sequence (crosses) and fire on arm (a single failure case – circles). These predict less extreme conditions than the worst case gravity waves but even these lie outside the design limits for the mission.

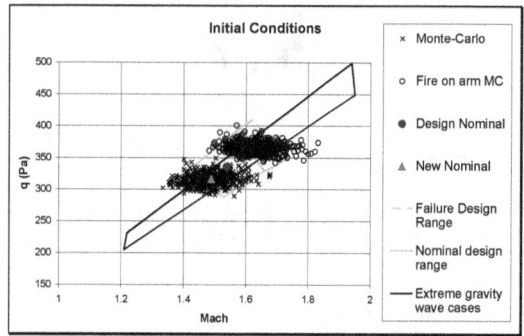

Fig. 6. DCSS Initiation Conditions

7.4 Pilot Chute Inflation Loads

The pilot chute inflation force was predicted using the high fidelity inflation model within the trajectory software. Two families of inflation forces can be seen, corresponding to the fire on arm and nominal firing cases.

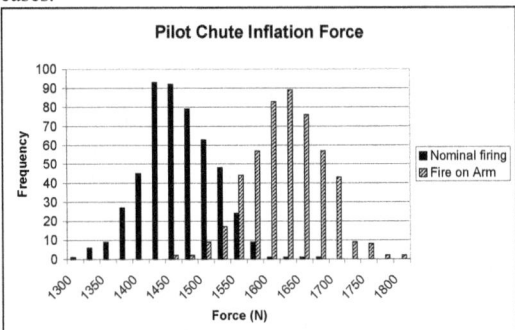

Fig. 7. Pilot Chute Inflation Force

The maximum inflation force seen during this analysis was 1800 N. This is significantly less than the value of 2334 N predicted by using all the combined worst cases.

7.5 Main Parachute Inflation Loads

The main parachute is deployed 2.5 seconds after pilot chute deployment in the nominal sequence. In the fire-on-arm case this delay is increased, since the main parachute deployment time is related to the time when the pilot chute should have been deployed. The result is that the main parachute is deployed at a slightly lower Mach number and dynamic pressure than the nominal and the worst case inflation loads for the main parachute occur for the nominal sequence.

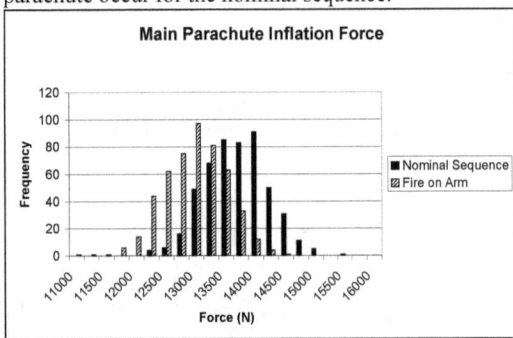

Fig 8. Main Parachute Inflation Force

Once again, the peak inflation force of 15.5 kN is much less than the value of 17.6 kN predicted using the extreme worst cases.

7.6 Descent times

The overall descent time can easily be derived from the analysis. The results indicate a nominal descent time of 2 hours and 20 minutes with the 3σ extremes lying 10 minutes either side.

8 DYNAMIC STABILITY

During the original development, limited testing of the dynamic stability of the probe during entry was performed in order to determine the dynamic aerodynamic coefficients (C_{mq}). This type of testing is very expensive and it is very difficult to determine the coefficients precisely from the data. Furthermore, the flow in the base region, which largely determines the damping coefficients, is influenced by the model support. As a result the uncertainties assigned to these coefficients were large. The latest computational tools incorporating fluid/structure interaction offer the potential to determine these coefficients analytically.

Recent studies using the Arbitrary Lagrangian-Eulerian capabilities of LS-DYNA suggested that it might be possible to simulate the dynamics of the probe in a supersonic flow in order to derive C_{mq}. An Eulerian fluid mesh was created and configured with an equation of state to represent the atmospheric properties. A rigid body shell structure, free to rotate about the centre of gravity in all axes, was then created based on the probe geometry and mass properties. By prescribing a flow velocity to the fluid domain, it was then possible to simulate the flow development about the probe and any induced oscillatory motion. Extracting the angular position, rate and acceleration data then enables derivation of dynamic aerodynamic coefficients.

During development of the simulation, published test data from the Viking ballistic range tests [13] were used to verify the dynamic oscillation of the probe. The motion of this model was found to be in good agreement with the test data.

For Huygens, stability characteristics of the probe are being investigated at a range of eight Mach numbers from 0.7 to 3.0. Titan atmospheric properties with a static pressure of 200 Pa and density of 0.0042 kg/m³ are defined for the fluid domain. The simulations are initiated with the probe pitched up by 2° to avoid solving for a completely symmetrical case. An oscillation sequence with flow velocity vectors for the probe at Mach 1.5 is shown in Fig 9. Pressure contours at an instant in time in the same sequence is shown in Fig 10.

The initial results are encouraging and analysis is ongoing to verify performance.

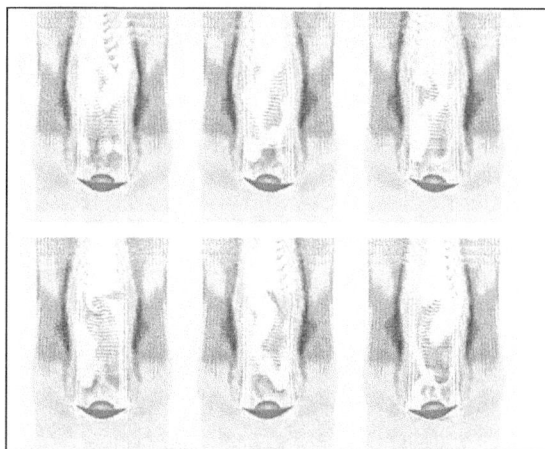

Fig 9. Mach 1.5 Oscillation Sequence

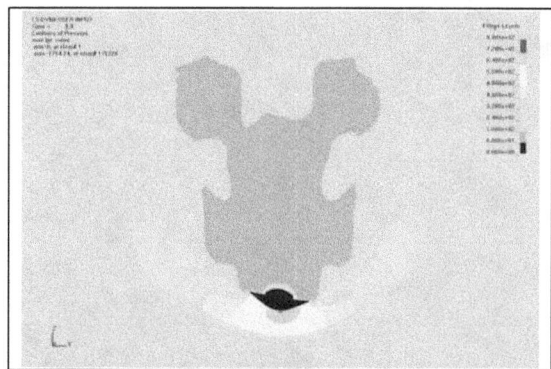

Fig 10. Mach 1.5 Pressure Contours

9 CONCLUSIONS

Updated analysis of the Huygens DCSS shows that it is robust with respect to the environments predicted using the latest data.

Ongoing analysis of the aerodynamic damping offers the potential to improve the analysis of entry stability.

A model has been derived which will allow the performance to be reassessed rapidly if atmosphere models are updated prior to the release of Huygens from Cassini.

ACKNOWLEDGEMENTS

This work was funded by the European Space Agency through Alcatel Space, the Huygens prime contractor.

REFERENCES

1. Neal, M.F. and Wellings, P.J., *Descent Control Subsystem for the Huygens Probe*, AIAA-95-1533.

2. Lellouch, E. and Hunten, D.M., *The Lellouch-Hunten Models for Titan's Atmosphere*, ESA SP-1177.

3. Yelle, R. et al., *Engineering Models for Titan's Atmosphere*, ESA SP-1177.

4. Strobel, D.F. and Sicardy, B., *Gravity Wave and Wind Shear Models*, ESA SP-1177.

5. Justus, C.G. and Duvall, A.L., *Atmospheric Models for Aerocapture*, AIAA 2004-3844.

6. Underwood, J.C. and Sinclair, R.J. *Wind Tunnel Testing of Parachutes for the Huygens Probe*, The Aeronautical Journal, October 1997.

7. Underwood, J.C, *Development Testing of Disk-Gap-Band Parachutes for the Huygens Probe*, AIAA 95-1549

8. Jäkel, E. et al., *Drop test of the Huygens Probe from a stratospheric balloon*, Advances in Space Research 21 (1998).

9. Underwood, J.C., *A System Drop Test of the Huygens Probe*, AIAA 97-1429.

10. Lingard, J.S., *The Effects of Added Mass on Parachute Inflation Force Coefficients*, AIAA 95-1561 (1995)

11 Cruz J.R.et al., *Opening Loads Analyses for Various Disk-Gap-Band Parachutes*, AIAA 2003-2131 (2003)

12. Greene, G.C., *Opening Distance of a Parachute*. J. Spacecraft V7 No 1 (Jan 1970)

13. Sammonds, R.I., *Transonic Static and Dynamic Stability Characteristics of Two Large-Angle Spherically Blunted High-Drag Cones*, AIAA70-564.

THE HUYGENS ATMOSPHERIC STRUCTURE INSTRUMENT (HASI): EXPECTED RESULTS AT TITAN AND PERFORMANCE VERIFICATION IN TERRESTRIAL ATMOSPHERE

F. Ferri[1], M. Fulchignoni[2], G. Colombatti[1], P.F. Lion Stoppato[1], , J.C. Zarnecki[3], A.M. Harri[4], K. Schwingenschuh[5], M. Hamelin[6], E. Flamini[7], G. Bianchini[1], F. Angrilli[1] and the HASI team

[1]CISAS "G.Colombo" - Università di Padova, Via Venezia, 1 35131 Padova – Italy
francesca.ferri@unipd.it
hasi@dim.unipd.it

[2] Université Paris VII - LESIA, Observatoire de Paris-Meudon, 5 place Jules Janssen, 92190 Meudon, France
[2] PSSRI – Open University, Walton Hall, Milton Keynes, MK7 6AA, United Kingdom
[4] FMI (Finnish Meteorological Institute), Helsinki, Finland
[5] IWF (Space Research Institute, Austrian Academy of Sciences), Graz, Austria
[6] CETP (Centre d'étude des Environnements Terrestre et Planétaires), Saint Maur des Fossés, France
[5] Agenzia Spaziale Italiana, Viale Liegi 16, 00198 Roma – Italy

ABSTRACT

The Huygens ASI is a multi-sensor package resulting from an international cooperation, it has been designed to measure the physical quantities characterizing Titan's atmosphere during the Huygens probe mission.

On 14th January, 2005, HASI will measure acceleration, pressure, temperature and electrical properties all along the Huygens probe descent on Titan in order to study Titan's atmospheric structure, dynamics and electric properties. Monitoring axial and normal accelerations and providing direct pressure and temperature measurements during the descent, HASI will mainly contribute to the Huygens probe entry and trajectory reconstruction.

In order to simulate the Huygens probe descent and verify HASI sensors performance in terrestrial environment, stratospheric balloon flight experiment campaigns have been performed, in collaboration with the Italian Space Agency (ASI). The results of flight experiments have allowed to determine the atmospheric vertical profiles and to obtain a set of data for the analysis of probe trajectory and attitude reconstruction

1. INTRODUCTION

The Cassini/Huygens mission, currently touring the Saturnian system, will devote particular attention to Titan, Saturn's largest moon. On the 14th January 2005 the Huygens probe [1] will descend through the thick atmosphere of this satellite down to its surface. Measurements will be performed during the entry, descent and landing phases in order to characterize the atmosphere and surface of Titan. In particular, vertical profiles of atmospheric density, pressure and temperature will be derived using accelerometric and direct pressure and temperature measurements.

The Huygens Atmospheric Structure Instrument (HASI) [2] is one of the six instruments on board the Huygens probe that actually will provide these key measurements to investigate Titan's environment.

The HASI sensors are devoted to study Titan's atmospheric structure and dynamics measuring of acceleration, pressure and temperature and electrical properties [3].

An outline description of the HASI instrument, its subsystems and sensors, operations and expected results at Titan are reported. The actual performance of HASI is discussed especially on the basis of the results of stratospheric balloon flight experiments.

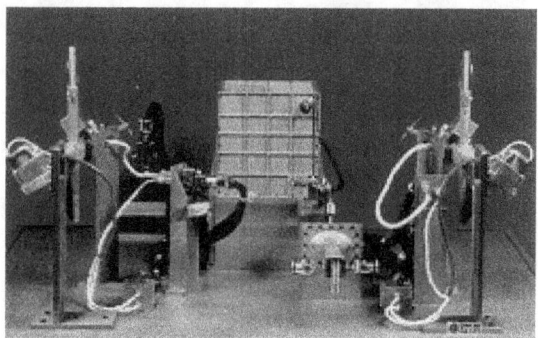

Fig.1   HASI experiment

2. HASI OVERVIEW

HASI is a multitask experiment package that has been designed to measure physical quantities characterizing Titan's atmosphere, specifically to determine the density, temperature and pressure profiles, and to study winds, turbulence and electric properties. HASI data will contribute also to the analysis of atmospheric composition and to provide information on surface, whatever its phase: liquid or solid.

The main scientific objectives are to:

- Determine the atmospheric density, pressure and temperature profiles.
- Determine the atmospheric electric conductivity and charge carrier profiles.
- Investigate ionization processes.
- Survey wave electric fields and atmospheric lightning; analyse quasi-static electric fields leading to storm formation.
- Detect acoustic noise due to turbulence or storm.
- Characterize the roughness, mechanical and electric properties of the surface material whatever its phase, solid or liquid.

HASI will monitor the acceleration experienced by the Huygens probe during the whole descent phase and will provide the only direct measurements of pressure and temperature through sensors having access to the atmospheric flow. HASI accelerometers are the only sensors operating during the entry phase, they will allow to determine the main physical properties of the upper atmosphere of Titan. Thus HASI data will also represent the unique contribution to the Huygens probe entry trajectory reconstruction.

Electrical measurements will be performed in order to characterise the electric environment on Titan and to detect effects connected to electrical processes, such as lightning and thunders.

In situ measurements are essential for the investigation of the atmospheric structure and dynamics. The estimation of the temperature lapse rate can be used to identify the presence of condensation and eventually clouds, to distinguish between saturated and unsaturated, stable and conditionally stable regions.

The variations in the density, pressure and temperature profiles provide information on the atmospheric stability and stratification, on the presence of winds, thermal tides, waves and turbulence in the atmosphere. Moreover, the descent profile can be derived from temperature and pressure data as a function of pressure and altitude. The return signal of the Huygens altimeter radar is processed by the HASI electronics, providing an independent estimation of altitude and spectral analysis of the signal yields information on satellite's surface.

HASI experiment (fig.1) is divided in four subsystems: the accelerometers (ACC); the deployable booms system (DBS); the stem (STUB) carrying the temperature sensors, a Kiel probe for pressure measurement and an acoustic sensor and the data processing unit (DPU).

The scientific measurements are performed by four sensor packages: the accelerometers (ACC), the temperature sensors (TEM), the Pressure Profile Instrument (PPI) and the Permittivity, Wave and Altimetry package (PWA). PWA perform also the spectral analysis of the return signal of the Huygens radar altimeter to derive information on altitude and surface properties.

HASI has been proposed by an international collaboration including 17 institutions from 11 countries. HASI has been funded by the Italian Space Agency (ASI) and by other European institutions who provided hardware elements. HASI subsystems and elements have been designed, developed and built in the different institutes and by Galileo Avionica (GA, Firenze, Italy), the industrial contractor, which has had the responsibility for the Assembly, Integration and Verification activities.

3. HASI-ACC: ACCELEROMETER PACKAGE

The HASI accelerometer subsystem (ACC) is a three axial accelerometer consisting of a one-axis highly sensitive accelerometer (Servo) and three piezoresistive accelerometers aligned to the principal axes of the Huygens probe. ACC is mounted as close as possible to the Huygens probe's center of mass, in order to sense acceleration along the probe's descent axis (X axis).

ACC sensors could sense a wide range of accelerations from the limiting resolution (around 0.3 mg) during the high speed entry phase to a maximum up to 20 g at landing [4]. The accuracy on the acceleration measurement is of the 1% and the resolution, for the X-servo accelerometer, is of the order of $0.3\mu g$ in high range and 0.3 mg in low range.

The main objective is monitoring the axial and normal accelerations experienced by the Huygens probe and thus to derive Titan's atmospheric density profile, but also record the impact signature.

Moreover being the only sensors operating during the entry phase, ACC measurements will provide the unique contribution to the Huygens probe entry trajectory reconstruction.

4. HASI TEM: TEMPERATURE SENSORS

The HASI temperature sensors (TEM) are two redundant dual element platinum resistance thermometers (TEM) mounted on the STUB in order to be appropriately located and oriented with respect to the gas flow during the measurements.

TEM design [5] derives its heritage from the *Pioneer Venus* temperature sensors [6].

Each TEM unit has two sensing elements (Pt wire): the primary sensor (FINE) is directly exposed to the air flow, while the secondary sensor (COARSE) is designed as spare unit in case of damage of the primary sensor. Temperature measurements are performed by monitoring resistance variation of TEM sensors.

Sensors can resolve 0.02 K with an accuracy of 0.2 K at best [3].

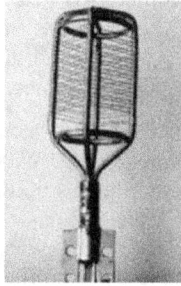

Figure 2 HASI STUB and TEM sensor

5. HASI PRESSURE PROFILE INSTRUMENT (PPI)

The Pressure Profile Instrument (PPI) includes sensors for measuring the atmospheric pressure during descent and surface phase. The atmospheric flow is conveyed through a Kiel probe, mounted on the STUB tip, inside the DPU where the transducers and related electronics are located. The PPI transducers are silicon capacitive absolute pressure sensors (Barocap) and temperature capacitive sensors (Thermocap) for thermal compensation [7]. PPI has three different pressure sensitivity ranges for low, medium, high pressure from 0 to 1600 hPa, with a resolution of 0.01 hPa.

6. HASI PERMITTIVITY, WAVE AND ALTITUDE (PWA) ANALYSER

The Permittivity, Wave and Altimeter package (PWA) [8] consists of six electrodes and an acoustic transducer.

Fig.3 PWA sensors on the Deployable Boom

The electrodes placed on the deployable booms (DBS) form a quadrupolar probe consisting of a pair of mutual impedance transmitter (TX) and receiver (RX) to measure atmospheric electric conductivity due to free electrons and detect wave emission. The two electrodes placed at the tip of the booms are relaxation probes (RP) for measuring ion electric conductivity and quasi-static electric field. The acoustic sensor (ACU) is mounted on the STUB and should detect sound waves to correlate with acoustic noise, turbulence and meteorological events.

PWA processes also the radar echo signals of the Huygens Proximity Sensor (RAU) [9] in order to derive information on surface properties and altitude.

7. HASI MISSION AT TITAN: OPERATIONS AND EXPECTED RESULTS

HASI will be the first instrument to perform measurements during Huygens entry phase in Titan's atmosphere. During the high speed entry phase, from an altitude 1300 km down to 160 km (4 min duration) acceleration data are sampled. Following the maximum acceleration near 270 km, the entry phase ends and the Probe device deployment sequence begins. In a period of three minutes pilot parachute is fired to lift off the Probe after cover and inflate the main parachute; the frontal thermal shield is released falling away. From this moment, starting from 160 km altitude, Huygens scientific instruments are exposed to Titan's atmosphere. The Probe descent continues under chute dragging. At 120 km altitude the main parachute is cut away and replaced by a smaller one. The complete descent will last 2 hours and 15 min; at least other 15 minutes are foreseen before the batteries run out and the loss of Probe relay link to Orbiter, to perform surface measurements after landing [10].

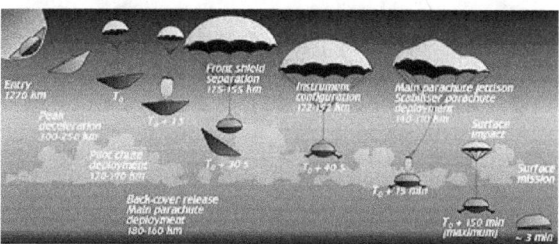

Fig. 4 Huygens mission scenario

About one minute after the frontal shield release (at about 160 km), HASI booms are deployed and direct measurements of pressure, temperature and electrical properties will be performed.

The sampling of HASI sensors is driven by timeline and triggered by environment conditions during descent. Huygens proximity sensor sampling will start from 60 km altitude level, but the Probe system will continue to use the altitude table until both the two radars will lock (expected around 30-km level). In that part of the atmosphere PPI will switch from low to medium and then to high pressure range measurements. In the last km (HASI IMPACT state) the ACC will be triggered to impact detection, no ACC data will be transmitted until the impact. In this last part of atmosphere only TEM fine sensor will be sampled every 1.25 s to achieve a better vertical resolution; the normal sampling (4 measurements alternating fine and coarse sensors sample of TEM1 and 2 on a period of 5 s) will be selected again at surface.

8. ENTRY PHASE: THE UPPER ATMOSPHERE

HASI will be the first instrument to be operative, measuring the deceleration of the Huygens probe as function of time. ACC measurements will be used for determining the main physical properties of Titan's upper atmosphere.

Knowing the atmospheric mean molecular weight and the Probe aerodynamics, vertical profiles of density, pressure and temperature will be derived from the accelerometer data with similar techniques already used for atmospheric entry probe on Venus [11], Mars [12] and Jupiter [13]

The density profile will be derived from

$$\rho = -\frac{2ma}{C_D A V^2}, \quad (1)$$

where m and A are the mass and frontal area of the probe, a the measured deceleration, C_D the drag coefficient and V the velocity of the probe relative to the atmosphere in the direction of the descent trajectory.

Given the accelerometer sensitivity (0.3 µg in the most sensitive range) [4] and the engineering Titan's atmospheric model [14], HASI ACC should start detect the atmospheric drag at an altitude of ~1300 km (corresponding to a threshold atmospheric density of ~7x 10^{-12} kg/m^3).

HASI-ACC Servo accelerometer is one of the most sensitive and stable launched to date [4], promising to deliver excellent performance during entry into Titan's atmosphere.

HASI data represent also the unique contribute to the Huygens probe trajectory reconstruction.

9. DESCENT PHASE: THE LOWER ATMOSPHERE

Starting from ~162 km direct measurements of temperature, pressure, electrical properties and acoustic recording will be performed by sensors having access to the unperturbed field outside the probe boundary layer.

The definition of temperature and pressure profiles, also combining HASI measurements with data from other experiments, will help to define the atmospheric structure, layer by layer composition (in particular to evaluate the CH_4 mixing ratio in saturation region), the vertical concentration profile of organic and inorganic compounds, and the partial pressure of saturated gas in order to detect presence of tropospheric clouds. Quantities depending on temperature (e.g. viscosity coefficient, scale height) and the temperature lapse rate dT/dz will be derived from local temperature and pressure measurements in order to detect transition between stable and conditionally stable regions and investigate the existence and the extent of a convective zone in the lower atmosphere.

Wind gusts in the atmosphere can be observed by monitoring with the accelerometers, the period oscillations on the Probe-parachute system, and thus detecting any perturbations on these oscillations caused by wind [15]. Variations on density and pressure profile, as well as on the temperature gradient, will provide information regarding atmospheric stability, wave propagation and saturation and atmosphere layering.

The electrical properties of Titan's environment will be investigated through PWA measurements in order to study ionisation processes related to galactic cosmic radiation, Saturn magnetospheric electrons, and micrometerites and to determine electrical charges density profiles. Electric fields and electrons and ions conductivities will be measured; acoustic noise and lightning effects connected to thunder storm, if any, will be detected.

10. LANDING AND SURFACE PHASE: THE SURFACE

HASI will contribute also to the characterization of Titan's surface partly redundant with the Huygens Surface Science Package (SSP) [16]. ACC will detect the impact and record the trace of the impact signature, yielding information on surface hardness and allowing distinguishing between liquid and solid surface. Electrical properties, pressure and temperature will be sampled as well at ground level. The spectral analysis of the Huygens radar return signals (blanking and echo signals) will allow to derive surface topography along ground track, small scale structures, dielectric properties and to contribute to discriminate between liquid and solid surface.

Moreover HASI data will be used as ground truth for calibration of remote sensing measurements (e.g. radio occultation, IR spectra).

Fig. 5   Huygens/HASI balloon experiment on the launch pad

11. HUYGENS-HASI BALLOON FLIGHT EXPERIMENTS: HASI PERFORMANCE VERIFICATION

A stratospheric balloon flight experiment campaign has been organized in collaboration with the Italian Space Agency (ASI) to verify and test HASI performance in dynamic conditions similar to the descent phase.

In order to simulate the last part Huygens probe parachuted descent on Titan and test HASI, a mock up of the Huygens probe carrying onboard HASI instrument and other instrumentation is launched with a stratospheric balloon from the ASI launch base of Trapani, Sicily. The probe is lifted up high altitude (>30 km) by means of a stratospheric balloon (see fig. 6). Once the balloon is cut away, the probe starts to descend dragged by the parachute and spinning, till impact to the ground (fig.7). The gondola and the payload are recovered after landing and, if possible, are refurbished to be flown in the next flight experiment.

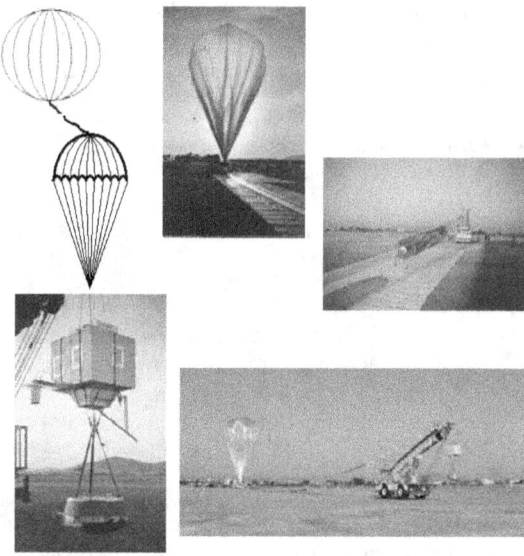

Fig. 6   The flight train: balloon, parachute, ASI telemetry gondola and Huygens mock-up with onboard HASI.

The measurements carried out during the ascending and descending phases are transmitted to ground by telemetry and recorded and stored on board.

Fig. 7   Descent under the parachute and landing

The flight experiments are aimed to a better understanding of the different HASI subsystems actual performance and in preparation of the mission data interpretation, but also to the study of the physical properties of the terrestrial atmosphere.

Several balloon flight tests with onboard some subsystems composing HASI and the reference payload have been successfully performed from the ASI launch base "Luigi Broglio" in Sicily (June 2001, May 2002 [17] and again on June 7th, 2003 [18]) and from Léon, Spain in December 1995 the COMAS SOLA experiment [19].

Together with HASI instrument, other equipment of the Huygens probe, and some sensors for Mars exploration (namely UV Beagle2/MarsExpress sensor and Mars TEM) housekeeping and add-on package have been flown providing data, in some case redundant, to improve our ability in interpreting the Huygens data that will be recorded at Titan.

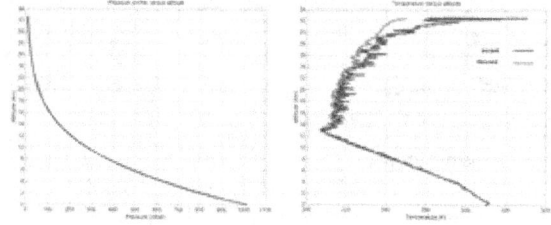

Fig. 8   Atmospheric vertical profiles as measured by HASI PPI and TEM sensors.

The results of flight experiments allow determining the atmospheric vertical profiles (fig.8) [20], to obtain a set of data for the analysis of probe trajectory [21] and attitude reconstruction [22] and to test the algorithm developed for the Huygens trajectory reconstruction [23].

While waiting for Huygens probe arriving at Titan, we are running in parallel a balloon flight experiment from Antarctica. The scope of this experiment is to enlarge the scientific investigation and test the instrument performance in an environment more similar to that one expected at Titan. The environmental conditions in a polar region are more similar to Titan's ones, the atmosphere is drier (no of water vapour is present in Titan's atmosphere), there the thermal structure of terrestrial atmosphere is more similar to that one of Titan (temperature are lower than anywhere else on Earth), the terrestrial electromagnetic background, that could affect electrical measurements, is lower than in other more populated terrestrial zones, the fly-byed region is covered by ice/snow; the permafrost is more similar to the expected Titan's surface (e.g. for checking the radar altimeter performance).

The balloon launch is scheduled for November – December 2004 from the Italian Antarctic base of Baia Terra Nova in cooperation with ASI and the PNRA (the National Program of Researches in Antarctica).

12. CONCLUSIONS

Taking the heritage of prior atmospheric structure experiments (e.g. Pioneer Venus and Galileo ASI), HASI will provide an in-sight in Titan's atmosphere determining atmospheric profiles of density, temperature, and pressure. In addition to similar ASI experiment, HASI will measure atmospheric electrical conductivity, wave and electric fields and investigate ionisation processes. HASI data are a unique contribution to the Huygens probe entry and trajectory reconstruction. Moreover HASI will contribute also to the characterization of Titan's surface, whatever its phase, solid or liquid and to monitor the environmental conditions at ground level.

In flight data and HASI/Huygens balloon flight experiments demonstrate and assess HASI performance, promising to deliver excellent results during the descent into Titan's atmosphere.

Looking forward for the Huygens mission at Titan on 14th January 2005, we are running in parallel a new balloon experiment from Antarctica that will help for the analysis and interpretation of the data that HASI will record at Titan.

13. ACKNOWLEDGEMENTS

HASI and HASI/Huygens balloon flight experiment campaign is funded by the Italian Space Agency (ASI), whose contribution is acknowledged. The authors also wish to acknowledge HASI co-investigators, the personnel of their institutes and Huygens scientists who provide hardware and support. The technical staff of the ASI balloon launch base "Luigi Broglio", Trapani-Milo in Sicily deserves a special mention for the successful launch and execution of the balloon flights, for their technical support and fruitful cooperation.

14. REFERENCES

1. Lebreton, J.-L., D.L. Matson The Huygens probe: Science, Payload and mission overview *ESA-SP-1177*, 5-24, 1997
   Lebreton, J.-L., D.L. Matson The Huygens mission to Titan: overview and status *ESA-SP-544*, 21-30, 2004

2. Fulchignoni, M., F. Angrilli, G. Bianchini, A. Bar-Nun, M.A. Barucci, W. Borucki, M. Coradini, A. Coustenis, F. Ferri, R.J. Grard, M. Hamelin, A.M. Harri, G.W. Leppelmeier, J.J. Lopez-Moreno, J.A.M. McDonnell, C. McKay, F.H. Neubauer, A. Pedersen, G. Picardi, V. Pironello, R. Pirjola, R. Rodrigo, C. Schwingenschuh, A. Seiff, H. Svedhem, E, Thrane, V. Vanzani, G. Visconti, J. Zarnecki. The Huygens Atmospheric Structure Instrument (HASI), *ESA SP 1177*, 163-176, 1997.
   Ferri, F., F. Angrilli, G.Bianchini, M. Fulchignoni & HASI team The Huygens Atmospheric Structure Instrument of the Huygens probe on the Cassini mission, *Acta Astronautica Journal*, **50**, No. 4, 249-255, 2002

3. Fulchignoni, M., et al. The Characterisation of Titan's Atmospheric Physical Properties by the Huygens Atmospheric Structure Instrument (HASI). *Space Sci. Rev.* **104**(1), 395-431, 2002.

4. Zarnecki, J.C., F. Ferri, B. Hathi, M.R. Leese, A.J. Ball, G. Colombatti, M. Fulchignoni,. In-Flight Performance of the HASI Accelerometer and Implications For Results at Titan, *ESA-SP-544*, February 2004

5. Ruffino, G., A. Castelli, P. Coppa, C. Cornaro, S. Foglietta, M. Fulchignoni, F. Gori and P. Salvini, The temperature sensor on the Huygens probe for the Cassini mission: design, manufacture, calibration and tests of the laboratory prototype, *Planet. Space Sci.,*. **44**, Issue 10,1149-1162, 1996

6. Seiff, A., Jurgen, D. W. and Lepetich, J. E., Atmosphere structure instruments on the four Pioneer Venus entry probes. *IEEE Trans. Geosci. Remote Sensing* **GE-18**(1), 105-111, 1980.

7. Harri, A. M., B. Fagerström, A. Lehto, G. W. Leppelmeier, T. Mäkinen, R. Pirjola, T. Siikonen and T. Siili, Scientific objectives and implementation of the Pressure Profile Instrument

(PPI\HASI) for the Huygens spacecraft, *Planetary and Space Science,* **46**, Issues 9-10, 1383-1392,1998.

8. Grard, R., H. Svedhem, V. Brown, P. Falkner, and M. Hamelin, An experimental investigation of atmospheric electricity and lightning activity to be performed during the descent of the Huygens probe onto Titan, *J. Atmos. Terr. Phys.,* **57**, 575, 1995

9. Jones, J.C. and F. Giovagnoli The Huygens Probe System Design, *ESA-SP*-**1177**, 25- 45, 1997

10. McCoy, D., M. Brisson, M. Verdant, H. Hassan, Huygens probe Cassini programme Experiment Interface Document part A, Issue 1 Rev. 6, 1997, *internal document*

11. Seiff, A., D. B. Kirk, R.E. Young, R.C. Blanchard, J.T. Findlay, G.M. Kelly, and C. Sommer, Measurements of thermal structure and thermal contrast in the atmosphere of Venus and related dynamical observations: Results from the four Pioneer Venus probes, *J. Geophys. Res.*, **85**, 7903-7933, 1980.

12. Seiff, A., Kirk, D.B., Structure of the atmosphere of Mars in summer at mid-latitudes. *J. Geophys. Res.* **82**, 4364–4378, 1977.
Magalh˜aes, J.A., Schofield, J.T., Seiff, A., Results of the Mars Pathfinder atmospheric structure investigation. J. Geophys. Res. 104, 8943–8956, 1999.

13. Seiff, A., D. B. Kirk, T.C.D. Knight, R.E. Young, J.D. Mihalov, L. Young, F.S. Milos, G. Schubert, R.C. Blanchard, and D. Atkinson, Thermal structure of Jupiter's atmosphere near the edge of a 5 μm hot spot in the north equatorial belt, *J. Geophys. Res.*, **103**, 22,857-22,889, 1998
Magalhaes, J.A., A. Seiff and R. E. Young, The Stratification of Jupiter's Troposphere at the Galileo Probe Entry Site, *Icarus,* **158**, 410-433, 2002.

14. Yelle, R.V., D.F. Strobell, E. Lellouch, D. Gautier, Engineering Models for Titan's Atmosphere, *ESA SP 1177*, 243-256, 1997

15. Seiff, A., Mars atmospheric winds indicated by motion of the Viking Landers during the parachute descent, *J. Geophys. Res.*, **98**, 7461-7474, 1993

16. Zarnecki, J.C., M.R. Leese, J.R.C. Garry, N. Ghafoor, B. Hathi, Huygens' Surface Science Package *Space Sci. Rev.* **104**(1), 593-611, 2002

17. Fulchignoni, M., A. Aboudan, F. Angrilli, M. Antonello, S. Bastianello, C. Bettanini, G. Bianchini, G. Colombatti, F. Ferri, E. Flamini, V. Gaborit, N. Ghafoor, B. Hathi, A-M. Harri, A. Lehto, P.F. Lion Stoppato, M.R. Patel, J.C: Zarnecki, A stratospheric balloon experiment to test the Huygens Atmospheric Structure Instrument (HASI) *Planet.Space Scie.***52**, 867-880, 2004
Bettanini, C., M.Fulchignoni, F.Angrilli, P.F.Lion Stoppato, M.Antonello, S.Bastianello, G. Bianchini, G.Colombatti, F. Ferri, E. Flamini, V.Gaborit, A. Aboudan, Sicily 2002 balloon campaign: a test of the HASI instrument *Advanc. Space Scie.*, **33**, 1806–1811, 2004.

18. Lion Stoppato, P.F., Ferri, C. Bettanini, G. Colombatti, M. Antonello, S. Bastianello, A. Aboudan, E. Flamini, V. Gaborit, J.C. Zarnecki, B. Hathi, A.M. Harri, A. Lehto, G. Bianchini, F. Angrilli, M. Fulchignoni, Stratospheric Balloon Flight Experiment Campaign for the Simulation of the Huygens Probe Mission: Verification of HASI Performance in Terrestrial Atmosphere, *ESA-SP-544*, 2004

19. López-Moreno, J.J., G.J. Molina-Cuberos, M. Hamelin, V.J.G. Brown, F. Ferri, R. Grard, I. Jernej, J.M. Jerónimo, G. W. Leppelmeier, T. Mäkinen, R. Rodrigo, L. Sabau, K. Schwingenschuh, H. Svedhem, and M. Fulchignoni, The COMAS SOLA mission to test the Huygens/HASI instrument on board a stratospheric balloon, *Adv. in Space Scie,* **30** 5, 1359, 2002

20. Colombatti, G., F. Ferri, F. Angrilli, M. Fulchignoni and the HASI balloon team Atmospheric stability & turbulence from temperature profiles over Sicily during Summer 2002 & 2003 HASI balloon campaigns, *ESA SP* (this issue)

21. Gaborit, V., M. Fulchignoni, G. Colombatti, F. Ferri, C. Bettanini, Huygens/HASI 2002 balloon test campaign: probe trajectory and atmospheric vertical profiles reconstruction *Planet.Space Scie*, **52**, 887-895, 2004

22. Bettanini, C., M. Antonello, F. Ferri, F. Angrilli, G. Bianchini, Attitude determination of planetari probes: the Huygens mock up role in determinino reconstruction strategies and perturbations durino descent in lower atmosphere, *ESA SP* (this issue)
Gaborit, V., M. Antonello, G. Colombatti, M. Fulchignoni HASI 2002 balloon campaign: Probe-Parachute system attitude, in press *J.Aircraft*. 2004.

23. D. Atkinson, B. Kazeminejad, V. Gaborit, F. Ferri, J-P. Lebreton, Huygens probe entry and descent analysis and reconstruction techniques, in press *Planet.Space Scie.*, 2004

A SURFACE SCIENCE PARADIGM FOR A POST-HUYGENS TITAN MISSION

Wayne Zimmerman, Jet Propulsion Laboratory
wayne.f.zimmerman@jpl.nasa.gov
Dr. Jonathan Lunine, University of Arizona
jlunine@lpl.arizona.edu
Dr. Ralph Lorenz, University of Arizona
rlorenz@lpl.arizona.edu

ABSTRACT

With the Cassini-Huygens atmospheric probe drop-off mission fast approaching, it is essential that scientists and engineers start scoping potential follow-on surface science missions. This paper provides a summary of the first year of a two year design study [1] which examines in detail the desired surface science measurements and resolution, potential instrument suite, and complete payload delivery system. Also provided are design concepts for both an aerial inflatable mobility platform and deployable instrument sonde. The tethered deployable sonde provides the capability to sample near-surface atmosphere, sub-surface liquid (if it exists), and surface solid material. Actual laboratory tests of the amphibious sonde prototype are also presented.

1. TITAN SCIENCE

Since the discovery of a methane atmosphere around Titan by Gerard Kuiper in 1944, Titan has attracted much exobiological interest. In the 1970s, before spacecraft reconnaissance established surface conditions with certainty, much of this interest centered around the possibility that the greenhouse effect of a thick atmosphere could permit hospitable surface conditions. The Voyager 1 encounter in 1980 showed that while the atmosphere was a familiar one in some respects (1.5 bar, mostly molecular nitrogen), the surface temperature of 90K was far too low for liquid water [2]. Many organic compounds have been detected in Titan's atmosphere. These form as a result of the recombination of molecular fragments produced by methane (and nitrogen) photolysis. Ultraviolet solar radiation is primarily responsible, although irradiation by electrons in Saturn's magnetosphere is an additional energy source. The ultimate fate of these compounds is to condense at or near the base of Titan's stratosphere and be deposited as liquids or solids on the surface [3]. Titan's photochemistry is of keen interest for modelers of atmospheric chemistry. However, by itself it goes but a little ways towards addressing the origins of life, a key goal of astrobiology and of Titan exploration. If there is to be any chemistry relevant to pre-biotic synthesis on Titan it must be occurring on or near the surface, acting on the products of stratospheric chemistry and powered by sources other than direct solar ultraviolet radiation. *Thus, we must go to the surface and analyze the organics there in order to explore organic systems that might be direct precursors to life.* Titan's atmosphere is essentially bereft of oxygen or oxygen-containing compounds like water. Even the most abundant (CO) is present only at a few tens of ppm. Although the early Earth similarly lacked molecular oxygen, carbon dioxide was an important or even dominant constituent of its atmosphere. The lack of oxygen-bearing compounds in Titan's atmosphere is a crucial point for two reasons. First is that the chemistry that sustains life on Earth is mediated in liquid water - it requires liquid water as a solvent. Second, virtually every organic molecule of biochemical interest contains some oxygen (that means every amino acid - and therefore every protein and enzyme - every sugar, every fatty acid, and DNA itself). Were there no way to incorporate oxygen into the chemical chain in the atmosphere then the organics that made in Titan's atmosphere would be sterile nitriles and hydrocarbons. On the other hand, experiments have shown that tholins and nitriles are readily hydrolyzed by liquid water into amino acids; e.g., Titan tholin yields about 1% amino acids by mass on hydrolysis [4]. Hence, if the accumulating organics on Titan's surface are exposed to liquid water, an entirely new step in chemical synthesis is introduced. *It is of keen astro-biological interest to find locations on the surface that bear evidence of past episodes of liquid water.*

Although we do not yet understand the nature of Titan's surface very well, it is clear that organic products of methane photochemistry must have been deposited on the surface over geologic time. By inference, Titan is thought to be half rock and half water-ice based on the Voyager-derived density and cosmic abundance considerations. Rock-ice moons surveyed in the Jupiter system similar in size to Titan show partial or full differentiation that migrates water-ice to the surface. Hence, two key ingredients for life (i.e., organic molecules and water in the form of ice) almost certainly exist today at the surface of Titan---*scientists need to identify ways to investigate the chemical/ice nature of the Titan surface.* Although the surface temperature of Titan at 90K is too cold to support liquid water in steady state, sources of transient liquid water at the surface of Titan might include cryo-volcanism (extrusion of liquid water or--more plausibly--water ammonia solutions on the surface) or medium- to large-sized impacts. Such impacts would gouge out craters in the water ice crust of Titan and leave behind approximately 1% by volume melt water [5]. If ammonia is present, complete refreezing of the liquid water would take up to 10^4 years [6]. We have no experience with the products of aqueous organic chemistry on such long timescales, or large spatial scales, yet such materials may lie preserved by deep freezing in the near-surface crust of Titan.

2. STRUCTURE OF THE TITAN DESIGN STUDY

The overall design study architecture started with the definition of desired science measurements, which were then networked into various implementation options, starting with launch/cruise, payload insertion, payload delivery to the surface, culminating in a surface/sub-surface sample acquisition and delivery system design. The overall system approach taken included the following steps:

1. Team based science requirements definition;
2. Science measurement-to-instrument mapping/ instrument payload definition;
3. Use of a cadre of functional design models that assess thrust, power/thermal, communication, and EDL requirements for different flight architectures needed to deliver the science payload, and;
4. Development of surface sampling robotic/ mobility kinematic/dynamic design models to implement the actual mission;

Once the overall science mission goals were defined and a potential instrument suite scoped, the first model developed was the optimal manner in which to launch and inject the desired payload (e.g., we examined the array of launch vehicles that might be available, examined possible fuel savings from Earth/Venus orbital assist, and then determined the cruise/EDL orbital mechanics needed to get us there). This design effort determined exactly how the science payload would be placed on the Titan surface. The results of this modeling work also defined disjoints between the desired science-driven mass and volume of the science payload vs. what the launch/orbital mechanics models say we could inject with the array of launch/cruise options available. Depending on desired launch windows, these two potentially conflicting parameters usually required an iterative approach to establish the "likely" subset of phased launch options and instrument suites. The surface component of the flight architecture was derived from examining an array of surface options for delivering the science payload, followed by developing functional models of the surface system The surface system dynamic models also encompassed communications, thermal control, power, structural analysis, materials, avionic/ mechanical component selection, navigation/ buoyancy and cooperative control (i.e., lander or aerial platform, science instrument/mobility/ sampling platform). The results of this analysis were used to size the surface power system as well as define the power conversion mechanism (e.g., one large RTG vs. many mini-radioisotope power sources or thin film boost batteries). Ultimately, the separate subsystem modeling results were iteratively worked to establish a complete trade-space science/surface system configuration envelope which was bounded by mass, power, and volume. Once fed back into the space transport and orbital mechanics models, a "likely" subset of system design(s) was developed which fell within the constraint envelope for "near-term" and "far-term" missions. The technology "tall poles" were filtered out of this solution subset.

3. RESULTS OF THE DESIGN/TRADE-SPACE ANALYSIS

Science Measurements and Supporting Instrument Suite

The science measurements discussed in the previous section are expanded here. We wish to understand the make-up of the atmospheric column close to the surface as a function of organic compounds present, densities, temperature, pressure. It is essential to perform complete gas/liquid/solid analysis of constituents, particularly examining make-up of tholins and possible presence of chirality. We also wish to perform a complete physical study of the subsurface liquid column in a shallow crater lake. This includes analyzing the liquid column as well as bottom material since it may be ancient deep material brought to the surface by an impact event. Again, temperature, pressure, opacity/particle suspension, and chemical properties will be studied. Last, we wish to perform shallow subsurface (<5cm depth) analysis of crater rim solid icy conglomerate material bordering shallow lakes. Measurements include physical characteristics of the rim area (microscopy/far field images, hardness), chemical make-up, presence of H20, and trace mineralogy.

The minimum science instrument payload deemed necessary to make the above measurements includes:
- Engineering sensors like temperature/pressure sensors, acoustic ranging (provides liquid depth), pH/conductivity/ dielectric, and turbidity measurements using off-angle LED's;
- Micro-scopic/near field imagers mounted in the shell of the sonde;
- Gas Chromatography coupled with a mass spectrometer (GCMS) to measure organic and chiral signatures (considered top priority);
- Either a Raman mass spectrometer or ion-micro electrodes for analyzing minerals/bulk material (if we can meet volume constraints);

Launch/Cruise/Entry Results

The tradeoff analysis looked at different launch and cruise options for getting the total surface payload to Titan. For the near term, it appeared that the most viable launch option was an Atlas 551, coupled with solar electric propulsion (SEP) and passive aerocapture at Titan. Use of a single or double Venus gravity assist greatly added mass margin to the total system. While an orbiter was considered useful as both a communication relay and global positioning system (GPS), it was found that the mass of the orbiter taxed the ability to deliver the desired science payload (~100kg, 3 sondes + instruments @ 33kg ea.). Therefore, the option of finding a trajectory and entry latitude which would enable a direct-to-earth (DTE) communication link was also examined. In this case, the aerial platform would be used as a large antenna for the DTE connection. The net results of the Atlas 5 vs. Delta 4 tradeoffs are summarized in Table 1.

One critical variable considered in the tradeoff analysis was the trip time. The science team all concurred that it was essential to keep the trip time in the 6-7 year range in order to maximize the science return as a function of the scale of the mission relative to cost. It was felt that this time period would frame the mission as a more cost effective alternative given the grand scope of the proposed in-situ science element. The last critical variable was the entry velocity. Since a passive aerocapture (AC) system saved considerable mass, it was necessary to pick a trajectory and entry vector which kept the entry velocity in the 5-7km/sec range—well within existing aeroshell material strength limits. Fig 1 shows the actual mass breakdown for the Atlas 5/SEP option considered optimal.

The total system mass was slightly over 5000kg which allowed injection of the aerial platform and 100kg science payload to the surface, with approximately 30% mass margin which is typically desired in the early phase of mission design.

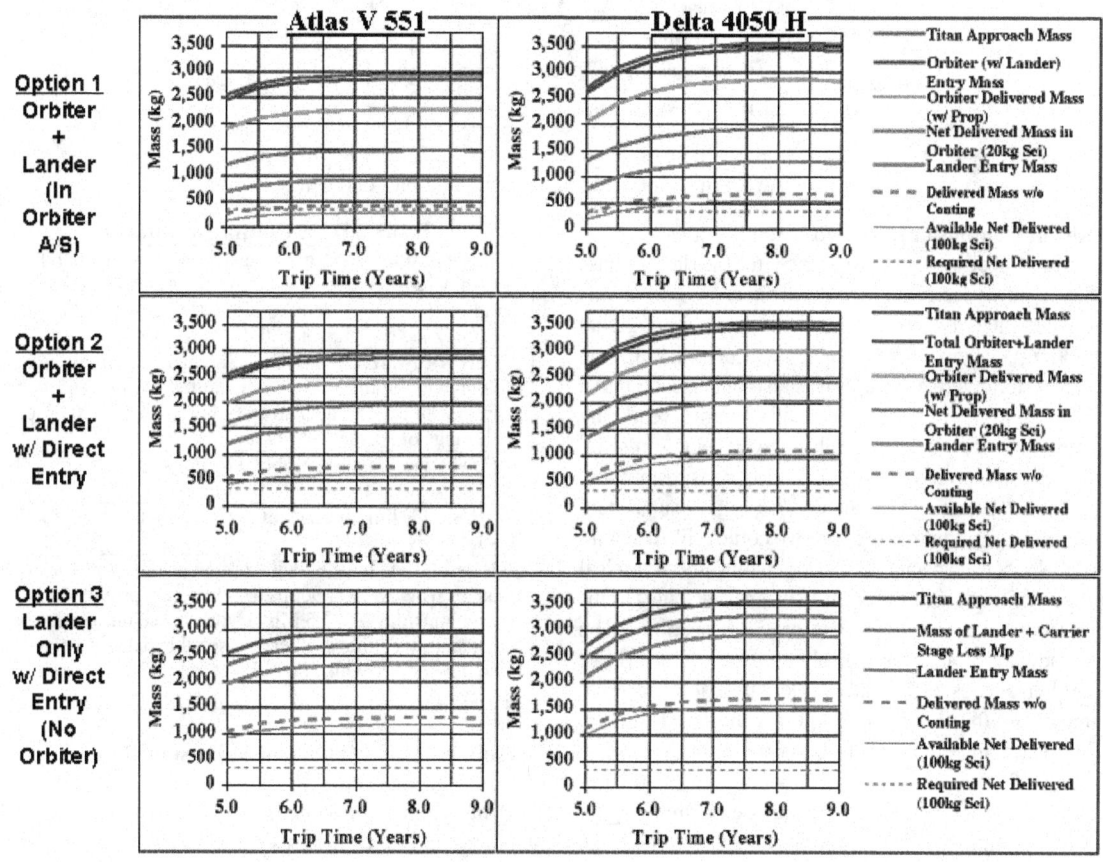

Table 1. Comparison of Injected Payload Capacity as f(launch vehicle, trip time)

SEP Mission Analysis

Summary SEP-A/C system mass breakdown for Option 1 (Orbiter w/ Lander inside Orbiter) for 6 year flight time and Atlas V 551 launch vehicle

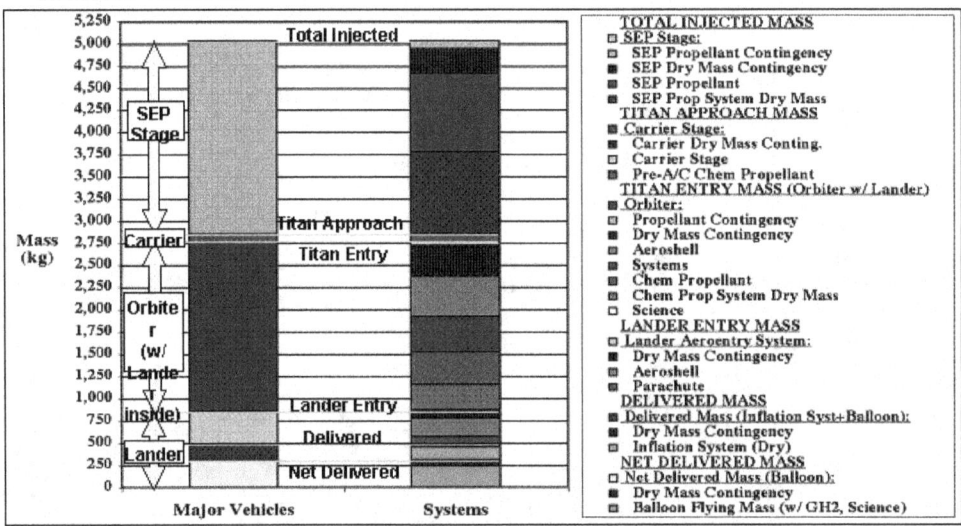

Fig 1. Final Atlas 5/SEP/Aerocapture Configuration Offering Best Trip Time and Highest Injected Payload Mass

Results of Communication Subsystem Trades

The above section made reference to the large mass penalty paid by carrying a relay orbiter. While it was understood that the orbiter relay capability and global positioning system (GPS) capability for tracking the surface system were critical, it was also understood that a surface mission of this scale would benefit greatly by reducing mass as much as possible. By reducing mission complexity and mass, cost could also be substantially cut. The trade-space analysis for this subsystem concentrated first on making sure we could still fit within a total launch capacity envelope using the classical approach of both an orbiter and surface package. This was successful as shown in the previous section. The final trade-space analysis then focused on getting rid of the orbiter and using a quad-dipole patch antenna array mounted on the aerial platform for a direct-to-earth (DTE) link. Since the aerial platform would experience natural drift due to surface winds (not large at 1km altitude), the final patch array design used a "max signal strength" polling/switching control system to home in on Earth. The design effort paid off as shown in Table 2. By setting the entry trajectory for touchdown at 85degrees N. latitude, 22W of transmit power enables 1kbps to be transmitted for 8hrs while Earth is above the horizon. This was a significant finding.

Table 2 DTE Comm Architecture

1- **Mission time-** 2016, no orbiter (mass savings of 600kg), quad-dipole patch array mounted on the gondola of an aerial platform, transmit altitude at 1km above the surface, transmit zone on Titan at >80 degrees latitude;
2- **Direct-to-Earth** (70 m DSN antenna) link from the balloon will require ~ 22 W transmitted power (~ 65 W DC power) with 35 cm patch arrays – based on a worst case DTE range of ~ 1.65×10^9 km;
3- **Operation** at latitudes above 85 degs allows Earth to be in direct view of the aerial pltform;
4- **Max DTE link availability** is 8 hours/day when Earth is rotated toward Titan
5- **Result-** The above 85degN option provides an 8hr/day transmit window at 1000bps data rate, and provides a max data volume/day of ~3.6 Mbytes/day based on worst case range analysis—this is a very reasonable data return at fairly low power with a significant reduction in launch mass;

Results of Power/Thermal Subsystem Trades

The design and trade-space analysis for the power and thermal control subsystems were tightly coupled. The power subsystem design considered both a dual string and single string architecture. The single string architecture was considered the likely implementation due to mass/volume constraints. However, it was understood that for a mission of this scale, we needed to consider redundancy as a viable alternative. It was for this reason that the science payload was designed to include 3 sondes with internal instrument packages. This redundancy would allow 2 sondes to fail and still meet the mission goals. The single string architecture considered internal redundancy from the standpoint of employing not only the tether supplied power from the aerial platform 100We radioisotope power source (RPS), but also employing rechargeable secondary batteries trickle charged off a single RPS general purpose heat source (GPHS) with thermo-electric converters mounted in the nose of the sonde. Fig 2 provides the general power architecture for the sondes.

Fig 2 shows the use of staged voltage regulators vs. DC-DC converters. This selection was made because extreme low temperature DC power converter technology is maturing at a very slow rate. However, low temperature voltage regulator technology is well developed and offers the advantage of less mass and volume with better low temperature performance. The most significant design element of this architecture was the realization that the many of the instruments in the surface/subsurface payload could not survive the extreme 90K Titan environment without staying warm. Further, although tether loses over the tether length of ~100m at 90K were minimal, the 100We aerial platform power source could not deliver sufficient

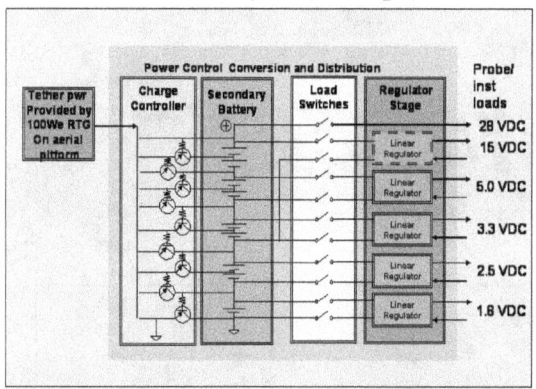

Fig 2. Sonde/Instrument Power Architecture

power to operate the sonde/ instruments and power the aerial platform as well. An elegant low mass solution was required. After considering vacuum bottle approaches, use of a GPHS/thermo-electric conversion system in the sonde nose, we finally closed on a hybrid thermal solution. The problem with the vacuum bottle approach was that the sonde wall had to have at least a 2cm vacuum gap between the outer wall and inner wall. This reduced the internal volume of the sonde down to a point where the instrument payload would not fit. While the GPHS solution provided sufficient heat to keep the sonde warm while allowing the probe internal volume to be adequate for the science payload, it was determined that the sonde could overheat when

suspended in the Titan atmosphere. Not only did the design complexity increase for radiating the heat to the atmosphere, but additional heat pipes were needed to remove the heat during the long cruise phase. The final solution was a combination of a small vacuum gap (5mm) coupled with use of a phase change material (e.g., water) which would convert to super-heated vapor when heated by the 100We (1250Wt) aerial platform RPS. This thermal jacket, augmented by only 20We of aerial platform power, allowed the sonde to maintain an internal temperature of 0-10degC (ideal operating temperature for the electronics) for up to 7hrs of surface mission time---more than adequate to obtain surface/sub-surface samples. Spot heat could be used for select instrument components.

Surface Mobility/Instrument Delivery and Sample Acquisition Results

Section 1 described the desired science measurements. Given the science requirement to sample atmospheric+liquids+solids over a region, we expanded our design/trade-off envelope to look at multiple sampling/in-situ analysis options—in all cases the vehicle would be deployed off an aerial platform. The reason an aerial platform was picked was two fold:

1. Titan has a significant atmosphere which makes use of an aerial platform attractive;

2. Being an icy body, the topography and potential orgainic sludge of Titan is expected to be challenging for a long range rover type vehicle and, perhaps, too constraining for a stationary lander (i.e., the landing site only offers a single location which may not allow access to other more interesting sites like hovering over a methane lake);

The overall surface mission concept is shown in Fig 3. The surface system design effort looked at small harpoon sampling devices fired from a tether when in proximity to the liquid or solid environment, followed by retrieval/delivery of a small sample canister to a sample transfer/distribution facility on the gondola. We also examined a passive drop sonde lowered via tether from the aerial platform to the surface, in which the sonde center of gravity (c.g.) is offset from the long axis of the probe. By off-setting the c.g., the probe automatically orients itself so the sampling mechanism always faces the surface. The last option we examined was a limited mobility amphibious vehicle capable of operating in all three environments.

Fig 3 illustrates the aerial platform deployment, survey, and sampling phases of the in-situ surface mission. Two primary surface instrument delivery options were considered:

1. The aerial platform does an aerial survey of a large area first before selecting a prime science target (i.e., a cratered organic lake with rich rim deposits of organic precipitates and mineralogy)---three sondes are separately deployed (one tethered for analyzing the atmospheric column, one free swimming for subsurface liquid column analysis, and one crawler for crater rim solids analysis)—multiple harpoons could also be deployed which retrieve/transfer sample to the aerial platform gondola instruments;

2. The aerial platform carries one tethered deployable sonde or harpoon suite which samples the atmospheric column, samples the liquid lakes, and samples the crater rim material as the aerial platform moves from site to site.

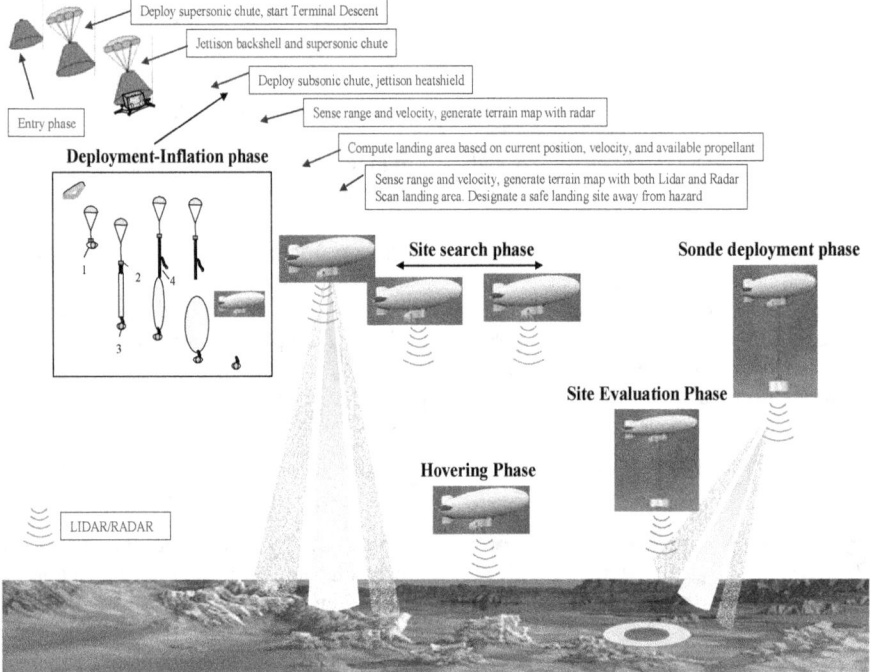

Fig 3. Aerial Platform Deployment/Survey Scenario

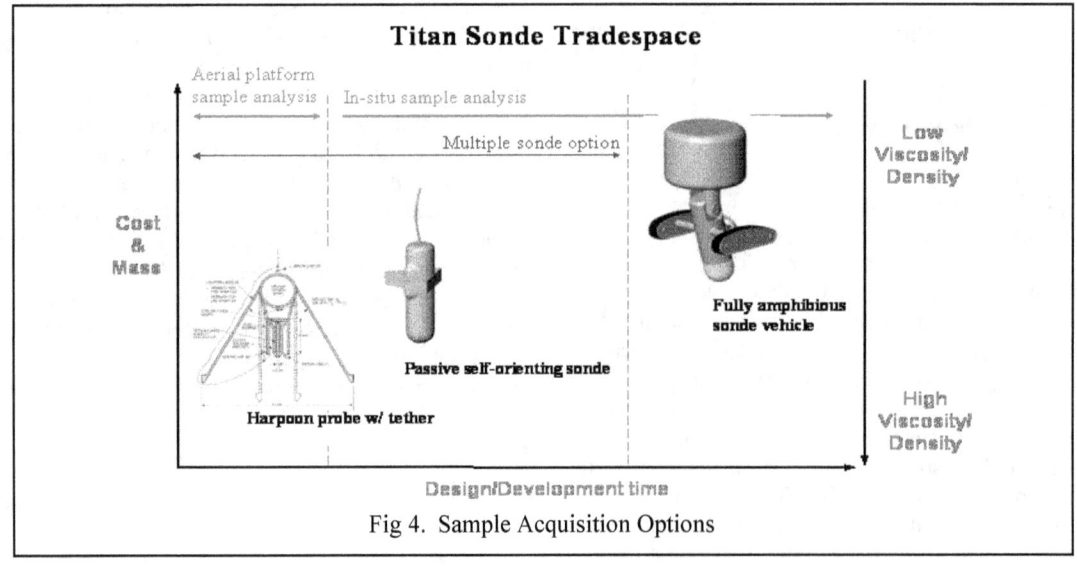

Fig 4. Sample Acquisition Options

Harpoon Sample Acquisition System

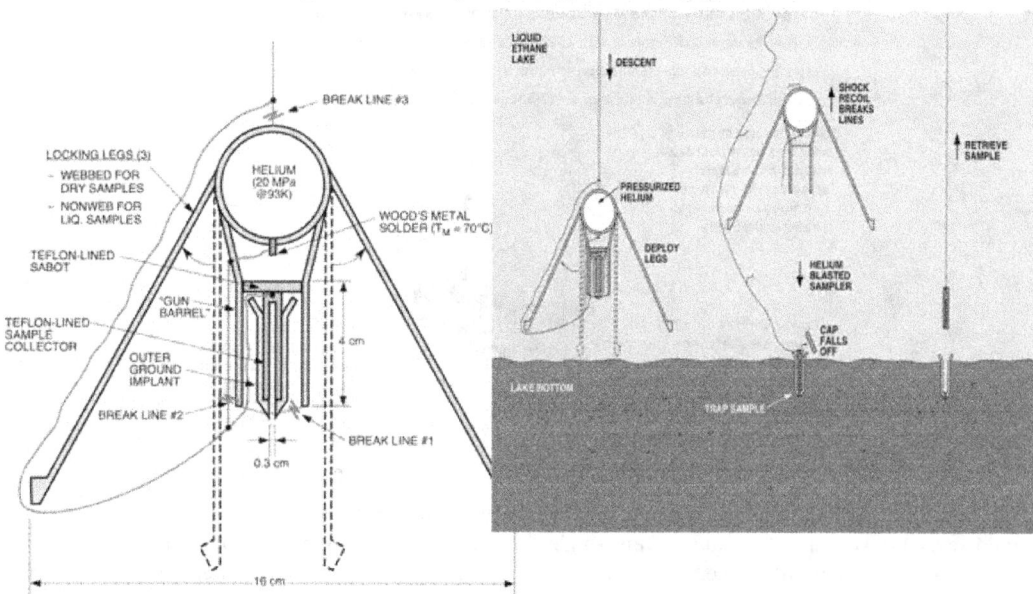

Fig. 5. Harpoon Close-up w. Retrieval System

Fig. 4 summarizes the key variables associated with the three sampling options. The harpoon's primary advantage is that it can accommodate high viscosity medias. Its disadvantage is that off-normal impacts and/or extremely hard surfaces, degrades its sample acquisition performance. Also, once the device is fired, it cannot be reused. This means that many devices, all requiring tethers, must be carried on the gondola. Both the passive and active sondes can be deployed into all environments, can be reused, but will have difficulty in viscous mediums.

The harpoon concept is shown in greater detail in Fig. 5. The device is lowered via a tether to the surface and/or subsurface. Pyros are used to drive the center sample chamber into the material where it is trapped and sealed. The umbrella shown in Fig 5 reacts rebound forces in liquid mediums so that the sample chamber thrust force stays vectored in the direction of the sampling surface. Research was done to identify cryo-liquid class pyros for both the sample chamber and for jettisoning the rebound umbrella/housing after the impact event. Once the sample is collected, the sample chamber is reeled back up to the aerial platform where it is then transferred to the instrument suite mounted in the gondola. It should be noted that the combined mass of the harpoon system was within 1-2kg of the mass of a single integrated sonde---essentially the same.

Both the passive sonde and active sonde are the same in terms of deployment by tether and internal electronics/science payload. Only the amphibious sonde is discussed here.

The amphibious sonde design was pursued first since it represented the worst case payload element in terms of design complexity. The design effort included functional design, dynamic simulation, and development/test of the first prototype. Considerable work has already been done in the area of mole penetrators. The JPL cryobot developed and tested in a glacier above the Arctic Circle on the island of Svalbard [2001, 7] was a form, fit, and functional prototype vehicle which used a cyclic passive nose heating and active water jetting approach for penetrating ice sheets and managing dust/debris in the ice. The science payload was completely integrated into the vehicle housing, where sampled melt-water was passively ingested and circulated across an ion-micro-electrode array. A 3x magnification imager was mounted normal to the long axis of the probe in the electronics bay of the probe housing. This configuration allowed the probe to image ice/debris layers as it descended. The probe was also capable of steering by selectively turning off nose heaters and turning on opposing aft shell heaters. The sonde design developed in this study drew heavily on the heritage of the cryobot while extending its capability by adding the feature of mobility in both liquid as well as on solid surfaces like the rim of a Titan crater lake.

The current science requirements only require 2-5 cm penetration depths of the solid icy surface on Titan. If liquid methane lakes exist, the probe is only required to sample meters in depth close to the crater rim, with emphasis placed on getting a small sample from the

Sonde/Mobility Platform Design

- Target Mass based on the Mars Scout cryobot flight mass = 30-33kg
- Target Power = 30W (motors on); 20W (CPU/sensors/instruments)
- Sonde upper chamber dimensions= 30cm diameter/20cm height
- Sonde lower chamber dimensions= 50-60cm length/15cm diameter

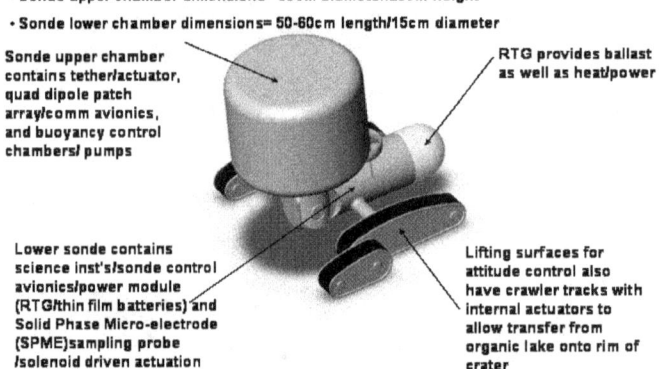

Fig 6. Planetary Autonomous Amphibious Robotic Vehicle (PAARV)

bottom of the lake. To do this with a simple system which could meet the volume constraints of the sonde (i.e., 15cm dia, 60-90cm length) we decided to look at solid phase micro-extraction (SPME) needles which adsorb sample onto a thin porous coating,, retract into a chamber where the needle can be interrogated for bulk material inorgnics, and then heated to release the organic volatiles. The SPME is mounted in the belly of the sonde where it can be deployed/retracted via solenoid or spring action close to the surface. The complete system is shown in Fig 6.

The above design has been prototyped and tested in the laboratory. This system is shown in Fig 7. The reader should note that this first version was primarily built to test its actual dynamic response against the kinematic/dynamic modeling results.

No instruments were placed on-board and the active buoyancy control system was not incorporated due to limited funding. The vehicle was teleoperated and tested in soil (the coarse Mars stimulant in the Planetary Robotic Vehicle Laboratory, JPL) and in a pool (the CalTech swimming pool). The tests showed good vehicle maneuverability in deep, heavy soil. In the pool environment, we learned that as we changed vertical orientation (pitch), subsequently changing the c.g., the vehicle was somewhat sensitive to the degree to which it could pitch and still remain stable, particularly in the presence of waves. It will be necessary to build in lower buoyancy control chambers near the nose, or provide the ability to extend a telescoping mass to maintain stable vertical equilibrium. The vehicle is shown maneuvering in the sandbox and pool in Fig 7.

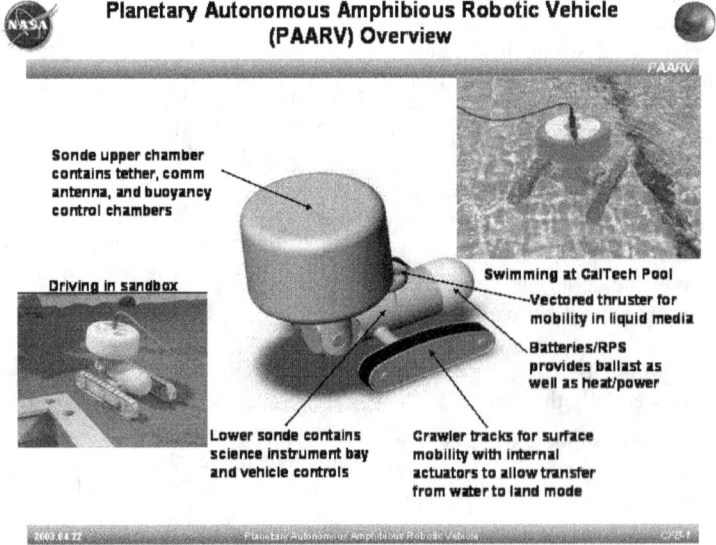

Fig 7. PAARV Being Tested in the Mars Sandbox and CalTech Pool

4. TECHNOLOGY ASSESSMENT

One of the key outputs of the Titan design analysis was to obtain a clear understanding of what we know, what we can comfortably predict in terms of launch mass/volume, and how a large scale in-situ science mission might be accomplished. Of particular importance was identification of the technology uncertainties and gaps. This first year effort identified the following critical technologies as enabling for a post-Huygens in-situ science mission:

- Ultra-light weight aerocapture/aeroshell materials;
- High efficiency SEP solar cell technology;
- High efficiency NEP/high-yield radioisotope materials;
- Extreme cold sensors/electronics/batteries
- Extremely high resolution micro-wet chemistry instruments;
- Extremely high resolution, low volume GCMS;
- Understanding the physics of cryogenic related failures/accurate failure projection models;
- GN&C w/o orbiter (surface beacons, Titan celestial nav, micro-satellites for GPS);
- Steerable phased patch array DTE comm using MEMS/nano-device technology;
- Structural/electronics packaging for extremely tight volumes;
- Cryogenic balloon materials;
- Packaging for long cruise soaks and cryogenic deployment;
- Solid Phase Micro-Extraction sample acquisition/control devices;
- Solids/residue purging technologies for instruments/transfer ports;
- Icy organic sample acquisition and transfer via controlled adsorption/desorption;
- Subsurface mobility propulsive mechanisms for dense cryo-organic liquids;
- Cryo-genic actuators/valves/seals;
- Cryo-tether materials (low shape memory/non-fracturable fiber-optics);
- Passive/amphibious mobile sondes for organic lake/crater rim sampling;
- Micro-harpoon impactors for dense icy materials (impact reaction mechanisms, cryo-liquid pyro mechanisms, low voltage/ high pressure micro tanks and valves;
- Autonomous control/fault tolerant-redundant S/W architectures;

5. SUMMARY AND CONCLUSIONS

The results of this initial study phase are significant. The following key findings were products of the design effort:

- We identified a Titan mission design that meets the science team requirements, based on a realistic payload/mass that can be injected down to the Titan surface (i.e., launch 5028kg total payload in 2011 and arrive 2017);

- We looked at near/far term launch capabilities of Adv SEP and NEP---NEP alone can deliver same payload but takes 2x as long to get to Titan (i.e., 11yrs);

- We determined that a hybrid launch/cruise system composed of an Atlas 551(Chem)+SEP+aerocapture entry looks best for the optimal:

  o Injected mass;

  o Trip time (keeping trip times ~6yrs +/-1yr per rqmt levied by science team);

- We completed the orbital insertion analysis/packaging with final mass breakdowns (total of 5028kg, 2600kg of injected mass, 100kg of science payload);

- We developed sonde and harpoon surface and sub-surface sampling designs that allow sampling of the atmospheric column, liquid lakes, and crater rim solid material;

- We developed a power and functional control architecture for the surface system (60W peak);

- We determined a way to remove the requirement for an orbiter, and developed a viable DTE communication concept using a patch array mounted on the aerial platform;

- We delineated the aerial platform/sonde design interfaces and developed dynamic models for sonde hydrodynamics;

- We developed a thermal control concept for the surface mobility/instrument delivery sondes to enable reliable surface and sub-surface operations for minimum power (20We after 5hrs).

6. ACKNOWLEDEMENTS

The research described in this paper was carried out at the Jet Propulsion Laboratory, California Institute of Technology, under a contract with the National Aeronautics and Space Administration. The authors would like to acknowledge the following Titan team members for their respective contributions to this research activity: Dr. P. Beauchamp- JPL, Dr. J. Beauchamp- CalTech, Dr. R. Hodyss- CalTech, Dr. N. Sarker- UofA, Dr. M. Smith- UofA who participated on the Titan science team. Special acknowledgment also goes to the engineering support team composed of T. Sweetser- Orbital mechanics, P. Timmerman- Power, R. Frisbee/M. Noca- Propulsion, E. Satorius/E. Archer- Communications, J. Jones/J. Hall/B. Dudik- Aerial platform/Harpoon sampling system, C. Bergh /W. Fang /E. Kulczycki- Sonde electronic/mechanical design, G. Woodward- Probe deployment/functional operation simulation, S. Chao /A. Sengupta- Survivable systems analysis, M. Quadrelli- S/W control architecture (aerial platform to sonde(s), sonde-to-sonde), and last, additional mission design support from NASA Langley (F. Stillwagen, S. Krizan, E. Dyke, orbiter/aero-capture/simulation), and NASA Glenn (M. McGuire, NEP option).

7. REFERENCES

1. Chau, Quadrelli, Zimmerman, "Search for Life Chemicals & Resources on Titan," NASA, Code R Revolutionary Aerospace Concepts, RASC, January 2003.

2. Lorenz, R. D. and Mitton, J. 2002. *Lifting Titan's Veil*, Cambridge University Press, Cambridge.

3. Yung, Y. L., Allen, M. A., and Pinto, J. P. 1984. photochemistry of the atmosphere of Titan: comparison between model and observations. *Astrophysical Journal Supplement Series* **55**, 465-506.

4. Khare, B. N., Sagan, C., Thompson, W. R., Arakawa, E. T., Suits, F., Callcott, T. A., Williams, M. W., Shrader, S., Ogino, H., Willingham, M. W., and Nagy, B. 1984. The organic aerosols of Titan. *Advances in Space Research* 4, 59-68.

5. Artemieva, N. and Lunine, J. I. 2003. Cratering on Titan: Impact melt, ejecta, and the fate of surface organics. *Icarus* 164, 471-480.

6. Lunine, J. I., Lorenz, R. D., and Hartmann, W. K. 1998. Some speculations on Titan's past, present and future. *Planetary and Space Science* 46, 1099-1117.

7. Zimmerman, W., et.al., "The Mars North Polar Cap Deep Penetration Cryo-Scout Mission," 2002 IEEE Aerospace Conference, Session 2.07, Big Sky, Montana, March 9-16, 2002.

ATTITUDE ISSUES ON THE HUYGENS PROBE: BALLOON DROPPED MOCK UP ROLE IN DETERMINING RECONSTRUCTION STRATEGIES DURING DESCENT IN LOWER ATMOSPHERE.

C. Bettanini (1), F. Angrilli (1)

[1]Centre of Studies and Activities for Space, CISAS "G.Colombo", University of Padova, via Venezia 15, 35131 Padova – Italy, Email :carlo.bettanini@unipd.it

ABSTRACT

As part of the collaboration with Italian Space Agency on HASI instrument for Huygens mission, University of Padova has been conducting since 2001 scientific activity on Stratospheric Balloon Launches from the Trapani base in Sicily. The most recent boomerang flight in July 2003 has successfully flown a mock up of the Huygens probe hosting spares of flight scientific units and extra housekeeping and scientific sensors on a parachuted descent from 33 kilometre altitude. This work presents the studies conducted on attitude reconstruction of the probe, as well as the utilisation of iterative extended Kalman filtering in investigating vanes induced spin rate and in providing a baseline for the performance evaluation of Huygens accelerometers operations. Finally some possible contributions on the reconstruction of the lower part of Titan descent for Huygens probe are suggested based on the confrontation of sensor data for 2003 flight.

1. HUYGENS PROBE OVERVIEW

Huygens is part of the Cassini-Huygens mission, a joint NASA-ESA-ASI mission for the exploration of the Saturn system. Built by an industrial consortium led by Aerospatiale, the Probe System comprises two principal elements: the 318 kg mass Huygens Probe and the 30 kg Probe Support Equipment, which remains attached to the Orbiter after Probe separation. Huygens probe will be parachuted through the atmosphere of Titan, Saturn's larger satellite, on January 14th 2005 after a dormant interplanetary journey of 7.25 years.

The scientific payload of the probe consists of a complement of six scientific instruments, which are each designed to perform a different function as the probe descends into Titan's mysterious atmosphere. The instruments are the following: Aerosol Collector and Pyrolyser (ACP), Descent Imager/Spectral Radiometer (DISR), Doppler Wind Experiment (DWE), Gas Chromatograph and Mass Spectrometer (GCMS), Surface Science Package (SSP) and Huygens Atmosphere Structure Instrument (HASI).

This last is a multi-sensor instrument that will measure the physical and electrical properties of Titan's atmosphere, which has been partially developed by University of Padova and is as a package managed by the centre for space studies of the same university. The sensors suite consists of a 3-axis accelerometer, a temperature sensor, a multi-range pressure sensor, a microphone and a electric field sensor array.

In parallel to utilising measurements from scientific packages, probe is equipped with dedicated sensor packages for health monitoring, house keeping, timing of operative sequences mainly related to shield separation and chutes deployment and attitude determination. Acceleration measures are based on a triply redundant Central Acceleration Sensor Unit (CASU) and a Radial Acceleration Sensor Unit (RASU) constituted by two equal accelerometers, all positioned on the main probe plate.

All on board data are handled by the Command and Data Management Subsystem relying on a very safe redundancy scheme, comprising two identical Command and Data Management Units which work simultaneously and are configured with hot redundancy and report data to the experiments in the so called Descent Data Broadcasts.

2. BALLOON BORNE HASI MOCK UP PROBE

HASI balloon flight campaign aim is to test the performance of scientific instruments of Huygens probe and specially the response of HASI instrument package to the thermal and fluidodynamic disturbances in an atmospheric descent.

To achieve this target University of Padova has developed a low cost mock up of Huygens probe, which hosts spare instruments of the real probe and several add on instruments, mechanically and electronically designed to be suitable for stratospheric balloon launches. A balloon mission is hence an efficient and economic way to lift a payload to a desired altitude and then drop it in a parachuted descent. The probe has been completely engineered at University of Padova and from 2001 has undergone several mechanical, electronic, thermal and fluidodynamic optimisations conducted in order to increase the available space for payloads, guarantee accessibility and maintainability of subsystems and improve performance during foreseen mission operations. The development activities have included:

- Design, development and assembly of probe mechanics

- Design, development and integration of power system
- Design, development and integration of probe electronics. Procurement of H/K sensors for probe status (attitude, temperature) monitoring.
- Design, development and coding of data acquisition system, for real-time monitoring, diagnostic and redundancy.
- Full system integration and verification.

The last evolution of the probe hosts 12 different scientific instruments with 84 different channels, acquired at different sampling rates by the on board integrated data acquisition and instrument control system during ascent, floating and descent phases.

This system is based on PC architecture and soft-real-time application allowing onboard storage and telemetry transmission satisfying all requests for real-time monitoring, diagnostic and redundancy.

Table 1. List of payload instruments hosted in 2002 and 2003 Balloon campaigns

Acronym	Sensor	provided by
HASI ACC	Triaxial accelerometer	PSSRI/Open University, UK
HASI PPI	Pressure Profile Instrument	FMI, Finland
HASI TEM	Dual Pt wire thermometers	CISAS – University of Padova, I
HASI PWA	Permittivity, Wave and Altimetry	CETP, France ESA-ESTEC/RSSD, NL IAA, Spain IWF, Austria
Huygens CASU	Single axis central accelerometer	ESA-ESTEC/RSSD, NL
Huygens RASU	Radial accelerometers (2)	ESA- ESTEC/RSSD, NL
Huygens RAU	Radar altimeter unit	ESA-ESTEC/RSSD, NL
Huygens SSP TILT	Science Surface Package tilt sensor	PSSRI/Open University, UK
Beagle2/UV sensor	UV sensor for Beagle2 on MarsExpress	PSSRI/Open University, UK
AD590	Temperature housekeeping sensors (7)	CISAS – University of Padova, I
MAG	Triaxial fluxgate magnetometer	CISAS – University of Padova, I
Inertial platform	Gyro enhanced orientation sensor	CISAS – University of Padova, I

A more challenging thermal environment is to be encountered in 2004 Antarctica flight but a new dedicated design and extensive testing have lead to a new probe configuration that can nominally operate almost in any atmospheric condition.

For data communication to ground probe relies on a dedicated ASI gondola, which provides telemetry capability for data download and sending/ receiving telecommands.

Both flights conducted in 2002 and 2003 very extremely successful, since probe was correctly launched and recovered in Sicily providing all expected scientific data.

3. HUYGENS ATTITUDE RECONSTRUCTION

Referring to [1] Huygens parachute system is designed to minimise the influence of external perturbations on probe attitude during descent thanks to a double pendulum configuration which will limit maximum probe oscillation to a 5 degree angle in presence of strong transversal winds. Furthermore the lower dome vanes configuration has been designed and tested to spin stabilise the probe and allow fast damping of aerodynamic oscillations thanks to optimised dynamic moment coefficients for angles of attack different from zero.

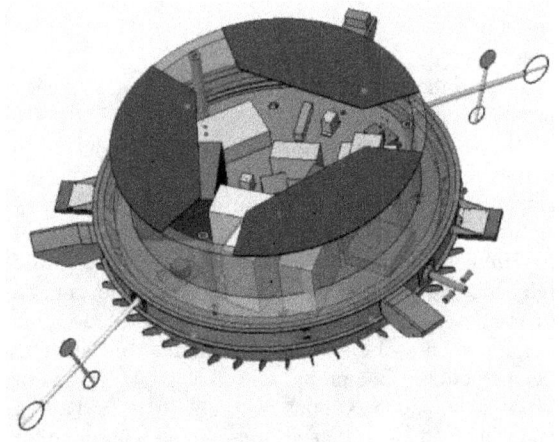

Fig. 1. 3D model of 2003 Huygens mock up.

Being attitude control of the probe mainly based on passive systems no extensive direct measurement of attitude is provided aboard the Huygens probe. Two accelerometric data sets (HASI and CASU) will be available during entry and descent to provide information on angle of attack, but the sole information about rotation will be given as spin rate around probe main axes from data of the radial accelerometer units RASU. Internal RASU electronics controls acquisition at 8 Hz sampling frequency and performs low pass filtering at 2 Hz before providing values to the CDMUs, which perform the spin calculation based on a simple radial acceleration algorithm. ($a=r\omega^2$)

Before using this algorithm two different averages are calculated, a first one every 2 second and therefore based on 16 measured values, called F1 and a second one based on the average of 64 consecutive F1 values, called G value. This last is the value used directly in the spin rate calculation algorithm which is therefore executed every 2 second basing on a average over 1024 measurements.

The two RASU accelerometers are located very close to each other on the same side of the probe main plate, so they provide redundant measurements but are also subject to several perturbing effects which should be taken in proper consideration. This effects are mainly related to the gravity acceleration which is sensed by the sensors in presence of probe oscillations and varying lateral winds which provide acceleration inputs and as well induce probe oscillations.

The entity of this disturbance on correct spin rate detection has been evaluated with a 3D dynamical model of the probe taking into consideration the most severe foreseen probe oscillations (maximum oscillation amplitude: 5 deg, maximum oscillation velocity: 2.5 deg/s) during a descent at the maximum expected spin rate (30 deg/s).[1] A 60 second long oscillation has been supposed and a gravity value of 1.35 m/s^2 as the one of Titan has also been assumed.

RASU acquisition has been simulated at 8 Hz (0.125 s time step for simulation output) and filtered with 2 Hz low pass filter. It is also notable that on Huygens probe RASU working range is software limited to 0 - 0.12 g and therefore negative readings are not considered.

Result of the simulation is shown in the following Fig. 2 where the nominal acceleration value (0.112 g at sensor radial position for expected spin rate) is perturbed by gravity because of probe oscillation.

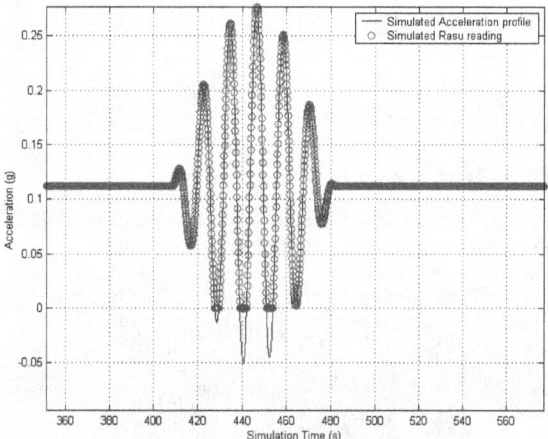

Fig. 2. Perturbing effect on RASU sensed acceleration due to gravity coupling thorough oscillation during spin controlled descent.

It can be noted that since RASU is sensing only positive accelerations, some "signal" is lost, but this does not affect much the calculations for small angles.

For larger oscillation angles (if present) the perturbation due to gravity increases and the effect on measurements can be such to not allow a accurate velocity profile reconstruction due to loss of too much information on probe dynamics.

Executing the spin rate algorithm on the simulated data produces the profiles shown in following Fig.3 and Fig.4 .

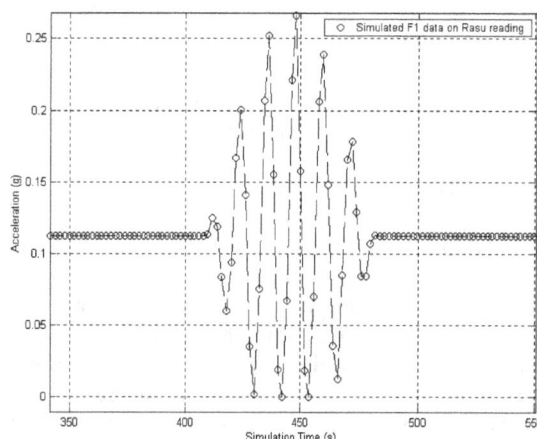

Fig. 3. F1 value on simulated RASU reading.

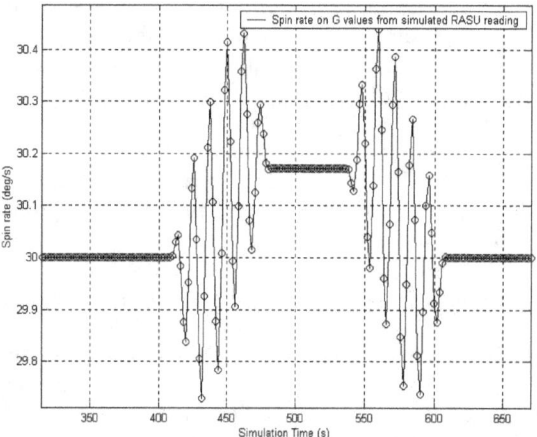

Fig. 4. Spin rates obtained through Huygens spin rate algorithm .

It can be seen that probe oscillation affects spin rate calculation for a period that is twice the extension of the oscillation due to the properties of the algorithm. A maximum 1,5% relative error on spin rate calculation can be expected for the foreseen oscillations at the maximum expected spin rate. In presence of lower spin rates a greater relative error affects measurements with values up to the same size of the expected reading. This problem can not be completed avoided through data averaging since as shown before part of the expected reading will be negative and therefore considered as zeros in data acquisition thus driving to an overestimation of the spin rate.

It should be considered that since no direct measurement of the angle of rotation is available probe rotation can be calculated only through integration of spin rate. Simulations show that in this case even a very small relative error on rotation velocity can cause the reconstructed angle value to depart in a short time from the real one. Being the knowledge of angle of rotation of great importance for some instruments like PWA improved techniques based on sensor fusion and Kalman filtering are currently under testing for reducing uncertainty in probe angle determination.

4. HUYGENS MOCK UP ATTITUDE RECONSTRUCTION

Although the mock up we realised for the balloon tests has been designed to be similar to the real Huygens some adjustments had to be conceived to adapt the configuration to a descent in earth atmosphere. This has mainly influenced the vanes configuration in order to achieve the same spin rate profile in presence of denser atmosphere but has also required a different configuration for RASU sensors. Since, as said in the previous paragraph, these accelerometers can sense gravity in case of oscillations and gravity on earth is almost eight time bigger than on Titan, the disturbance could be of the same value of the spin borne centrifugal contribution for probe inclinations around 3 degree from vertical. We also knew from previous balloon campaigns that the balloon train configuration provided by the Italian Space Agency did not have a double pendulum configuration and therefore probe oscillations can easily reach 40 degree during descent.

We had therefore to design a different configuration installing the two RASU units at the same radial distance from the center of mass but 180 degree apart facing each other. Thanks to this sensor configuration we are able to eliminate oscillation and lateral winds contributions with post mission data elaboration of units readings. A picture of the probe main plate is shown in Fig. 5.

Fig. 5. RASU sensor configuration on Huygens mock up for balloon mission.

In order to test the quality of attitude reconstruction with Huygens sensors and investigate possible perturbations we installed a commercial inertial measurement unit with high accuracy attitude determination capability.

The 2003 flight was successfully launched and recovered on June 7th 2003, starting at 6.54 AM local time and lasting around 3 hours. (130 minute ascent to 33 kilometre altitude and a 65 minutes parachuted descent).

Elaboration of IMU through quaternions shows that probe oscillations during the descent have reached values as high as 40 degree as shown in Fig.6.

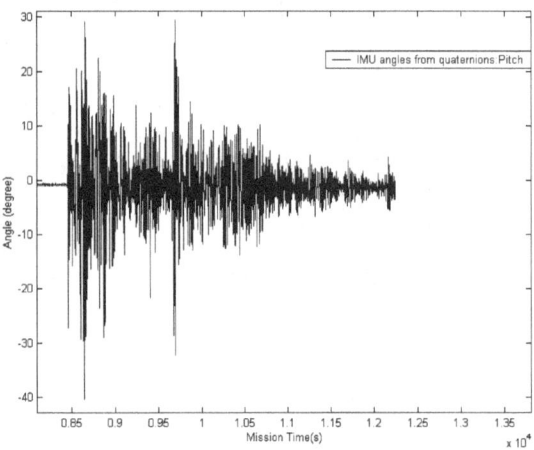

Fig. 6. Probe pitch angle during descent

Probe rotation around spin axes is also provided by IMU data elaboration as in the following Fig.7.

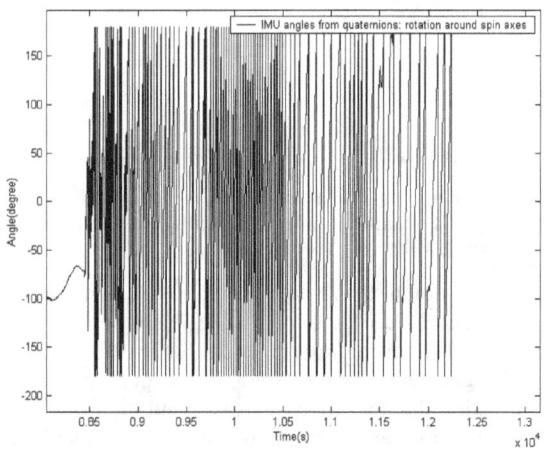

Fig. 7. Probe rotation around spin axes during descent

Since a slip ring de-coupling device has been introduced between the probe and the above telemetry gondola, probe rotation is with good confidence not perturbed by the dynamics of any parachute train element and can therefore be directly related to fluidodynamic interaction with atmosphere during the descent.

In order to calculate the descent spin rate profile and the perturbations due to lateral and vertical winds a reconstruction algorithm has been developed based on Extended Kalman filter. The Iterative Extended Kalman filter is a very good mean to address the general problem of trying to estimate the state of a discrete-time controlled process that is governed by the non linear stochastic difference equation and with non linear measurements relationship to the process. Reducing it to the essentials Kalman filter behaves as a predictor corrector where a time update based on a good knowledge of the system dynamics projects the current state estimate ahead in time. The state equation has been calculated with CFD models and relates the attitude parameters to the descent profile characteristics. The measurement update adjusts the projected estimate by an actual measurement at that time. The extended filter implies wide use of jacobian matrices with partial derivatives and for the complex parachuted descent dynamics this yields to complex symbolic calculations and long calculation time. The developed filter investigates the evolution of 7 state variables using 7 measurement data sets at different sampling rates through a so called sensor fusion algorithm.

Huygens should encounter during the Titan descent [1].

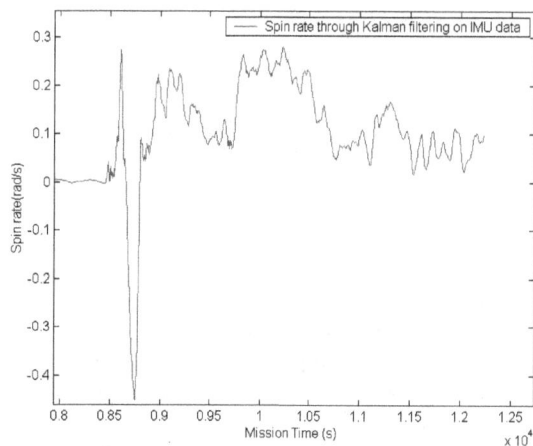

Fig. 9. Spin rate calculated with Iterative Extended Kalman on IMU data

In parallel to Kalman filtering processing of RASU data has been conducted. The previously described Huygens spin rate algorithm has been used to evaluate the spin rate of the probe to simulate an Huygens-like mission. As already described a pre processing of accelerometer data has been performed in order to disregard perturbing inputs due to undesired probe dynamics. The resulting spin rate profile based on RASU data is shown in the following Fig.10.

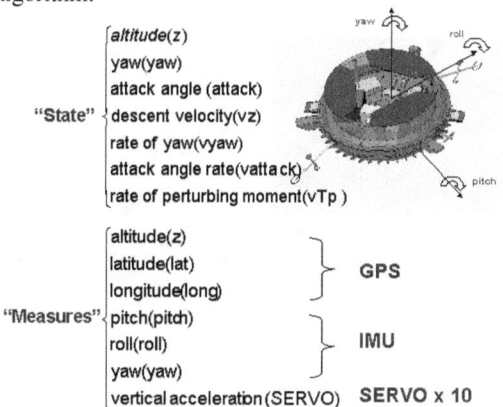

Fig. 8. Iterative Extended Kalman filter parameters

The Kalman filter has allowed the calculation of the mission spin rate profile from inertial platform measurements and its evolution is shown in the following Fig.9. It must be noted that after parachute opening the probe undergoes an unexpected counter clockwise rotation before starting the nominal clockwise rotation. This behaviour is still under investigation and is though to be caused by internal friction in the slip ring device triggering the rotation. Besides this first phase the spin rate profile is in good accordance with the expected profile[2] and the achieved rates are consistent with the ones that

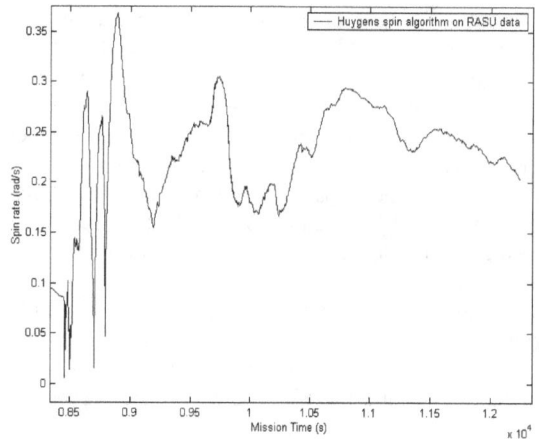

Fig. 10. Spin rate calculated with Huygens spin algorithm on RASU data

RASU derived profile seems to differ a lot from the one from Fig. 9 especially in the first phase of the descent. This is due to the fact that RASU based spin rate algorithm can not reconstruct the sign of rotation. Considering only absolute rotation values the profiles are very similar and therefore RASU based rate reconstruction provides values that are consistent with the ones derived by the more accurate inertial platform unit. The main difference is that RASU based reconstruction tends to overestimate spin rate with a maximum relative error around 20%. This happens when rotation rate is low and therefore remaining dynamic disturbances affect mostly the reconstruction as also shown in the simulations.

This effect is unlikely to be encountered on the real Huygens mission thanks to the double pendulum configuration which should consistently limit undesired motion of the probe.

Concerning rotation angle reconstruction the studies conducted on balloon flight data has shown that RASU data do not allow an accurate reconstruction through spin rate integration, either because they don't give information on probe rotation sign also because signal to noise ratio is too low to prevent drifting of state estimation. Simulation showed also that same problem is likely to be encountered in the reconstruction of the real Huygens profile angular position.

The development of new algorithm based on other fusion with other instrument data (HASI ACC or DISR) is therefore under investigation both for the balloon flight and the real Huygens mission interpretation.

5. CONCLUSIONS

Flying a mock up of the Huygens probe in a stratospheric balloon campaign has provided information on several issues related to probe attitude determination. Data elaboration has shown that radial accelerometers RASU can provide a good spin rate profile reconstruction during the descent also in earth atmosphere although several requirements must be met to relate accelerometer data to probe rotation.

Some possible perturbations on RASU readings have been underlined in case of probe oscillation and these should be considered for data investigation during the Titan descent.

Concerning the determination of probe rotation angle around the spin axis this work has underlined that RASU derived data are not sufficient for an accurate reconstruction and therefore new strategies based on sensor fusion with other on board equipment data should be implemented for Huygens and are currently under investigation.

6. REFERENCES

1. *ESA Huygens Probe Entry and Descent Analysis*, E.W.P. 1679

2. M. Antonello, C. Bettanini, *Aerodynamics Design of the Huygens Spin Vanes for the HASI*, Proceedings of International Workshop on Planetary Probe Atmospheric Entry and Descent Trajectory Analysis and Science, Lisbon 2003, ESA - SPP - 544

ATMOSPHERIC STABILITY & TURBULENCE FROM TEMPERATURE PROFILES OVER SICILY DURING SUMMER 2002 & 2003 HASI BALLOON CAMPAIGNS

G. Colombatti (1), F. Ferri (1), F. Angrilli (1), M. Fulchignoni (2) and the HASI balloon team

(1) CISAS "G.Colombo" - Università di Padova, Via Venezia, 1 35131 Padova – Italy; EMAIL: giacomo.colombatti@unipd.it
(2) Université Paris VII – LESIA, Observatoire Paris-Meudon, France

1. ABSTRACT

Experimental results and interpretation of the temperature measurements data retrieved during the balloon campaigns (in 2002 and in 2003) for testing HASI (Huygens Atmospheric Structure Instrument), launched from the Italian Space Agency Base in Trapani (Sicily), are presented.
Both ascending and descending phases are analysed; data reveal interesting features near the tropopause (present in the region between 11km-14km), where temperature cooling can be related to layers with strong winds (2002 flight); in the troposphere a multi-stratified structure of the temperature field is observed and discussed (particularly in the 2003 flight)
Finally, stability and turbulence of the atmosphere are analysed; the buoyancy N^2 parameters for both the flights show lowers value respect to standard tropospheric values corresponding to a lower stability of the atmosphere; still there is a higher stability above the tropopause. The energy spectrum of temperature data is consistent with the Kolmogorov theory: the characteristic $k^{-5/3}$ behaviour is reproduced.

2. Introduction

Up to now, no temperature profiles of the summer atmosphere from 30km altitude down to ground above Sicily have been published.
The aim of this study is to analyse the features observed during two balloon flights: the observed physical phenomena require some interpretation. Furthermore, the analysis of the temperature data will help the understanding of the temperature profiles that will be measured during the 2005 Titan mission; a training with a real set of data will help the investigation on the features that will be measured in 2005.
Scope of this work is to present the evolution of temperature profiles over Sicily during the campaigns that where conducted in the years 2002 & 2003; during the summer, two balloon tests where performed with a mock-up of the Huygens probe. On board the mock-up, HASI's temperature sensors where mounted and temperature profiles where measured both during ascent and descent phases.
In Section 3 the complete experimental set-up is described: a general presentation of the balloon flight train is followed by a description of the Huygens mock-up; the paper will focus on the instruments used for the temperature measurements.

A general description of the balloon flight is presented in Section 4 showing the balloon ground tracks and the balloon ascent and descent velocities. Horizontal displacement from launch site is measured via a GPS (Global Positioning System); horizontal velocities of the probe are assumed to be the same of the horizontal winds.
Detailed analysis of the temperature measurements is presented in Section 5; the investigation will focus on the descent phase since it's the most significant for comparison with the future real Titan profile, and is, probably, less affected by the flight train disturbances.
Finally, discussion on atmospheric stability and turbulence is carried on in Section 0.

3. Experimental set-up

The balloon flight consists in raising a specially engineered probe, a 1:1 scale mock-up of the HUYGENS probe (see), to an altitude higher than 30km with a stratospheric helium balloon and then releasing it in a parachute driven descent and collecting several scientific data measurements during the parachute drop (for a more detailed description see Fulchignoni et al., 2004; Bettanini et al., 2004).

3.1. HASI TEM & MTEM Temperature Instrument

HASI TEM sensor is a dual element platinum resistance thermometer [Ruffino et al. 1996] mounted on a fixed stem (STUB) and located outside the mock-ups' boundary layer (see), in a region where the local flow velocity is high, in order to avoid thermal contamination and promote very fast response. Resolution is less than 0.07 K and absolute accuracy less than 2 K. HASI has two redundant TEM units (TEM1 and TEM2), each one composed by 2 sensing elements (Fine and Coarse); each sensor is sampled at a frequency of 0.2 Hz; but the measurements are in sequence giving a frequency of 0.8 Hz [Fulchignoni et al., 2002]
MTEM sensor is a platinum resistance thermometer, evolution of the one mounted on the HASI experiment, designed for descent measurements in Mars' atmosphere [Angrilli et al., 1999]. In its new configuration (see Figure 1), the sensing elements (Pt wires) are suspended on a very thin non-metallic fibres truss, in order to thermally de-couple it from the supporting structure. The expected overall accuracy in the measurement is less than 0.1 K in the range 200-

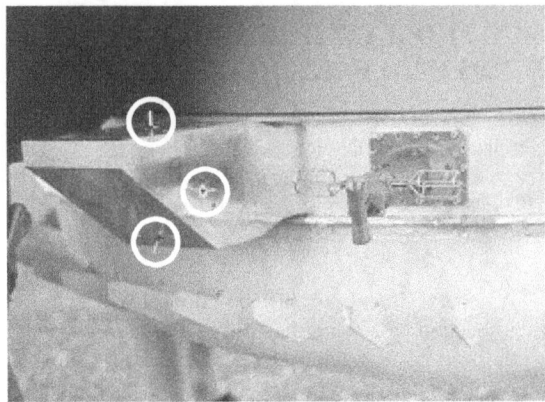 

Figure 1. left: HASI TEM close up (circles - PT100s used as housekeeping sensors); right: MTEM temperature sensors.

300 and resolution is 0.05K. The sensors are sampled at a frequency of 8 Hz.

In Figure 2 the spatial resolution as function of altitude level for the HASI TEM sensor for both ascent and descent phases is reported; after the separation phase the spatial vertical resolution is increasing because the vertical velocity of the probe is decreasing, reaching values of around 25m below height 7000m. This resolution enables us to highlight features of several tens of meters and to emphasise their evolution.

4. Balloon flight experiments

The flights were performed during two sunny days, with almost no clouds or strong winds during the launch phase; these favourable meteorological conditions ensure a local flight: maximum distance of the landing point from the launch pad was approximately 50km.

Both flight reached an altitude higher than 30km; 2003 flight reached an altitude level of 33km, requested for testing of a special operational mode of the Huygens RADAR bread-board altimeter

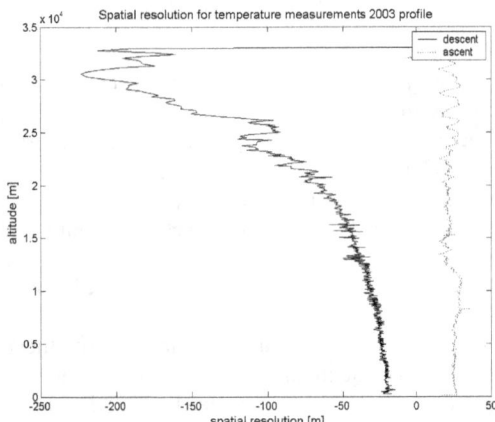

Figure 2. Spatial resolution for HASI TEM sensors for flight 2003

Ascent velocity can be considered constant with a value close to 5 m/s while descent velocity, after the transient separation phase, was measured in between 5 and 10 m/s and impact velocity was around 5-6 m/s.

The desired time at float altitude was the minimum possible compatibly with technical operations and constrains for secure landing on ground. In Figure 3 are shown the ground tracks of the balloon flight over Sicily as derived from the GPS data. For the 2002 flight it must be considered that a 2 hour floating phase at 30km was ensured for a secure prediction ground landing. During the descent phase the probe mock-up is spinning thanks to a set of vans fixed at the bottom shell of the probe so as will do Huygens probe in Titan's atmosphere.

4.1. Balloon 2002 & 2003 wind profiles

We assume that the wind, driving the probe-parachute-balloon system has only an horizontal component; so that the vertical velocity (ascending & descending) is the velocity of the probe-parachute-balloon system during ascent and descent.

Descent vertical profile for the 2002 and 2003 flights is similar, with velocities between 40 m/s and 4 m/s, for both the flights; descent velocity values above 30km altitude are relative to the transient phase after probe separation from the balloon.

As it can be seen (Figure 4 and Figure 5) the two profiles have the same trend; differences in the profiles can be observed at altitude of 25km, 14km (features present in 2003 profile but not in 2002 profile) and 3km where, locally, vertical descent velocities increase (feature present in 2002 flight).

In the 2002 profile the interesting feature occurs at around 3km: three variations in the horizontal wind velocities have been detected; in case A the velocity varies of 10m/s in 500m; this strong change is related to a layer of winds in NS direction. In B the variation is not very significant (2m/s in 180m) but there will be a good correlation with temperature data. In C the variations is smaller (6m/s in 200m) but GPS wind velocity data are incomplete due to loss of the telemetry link with the probe due to topographic configuration.

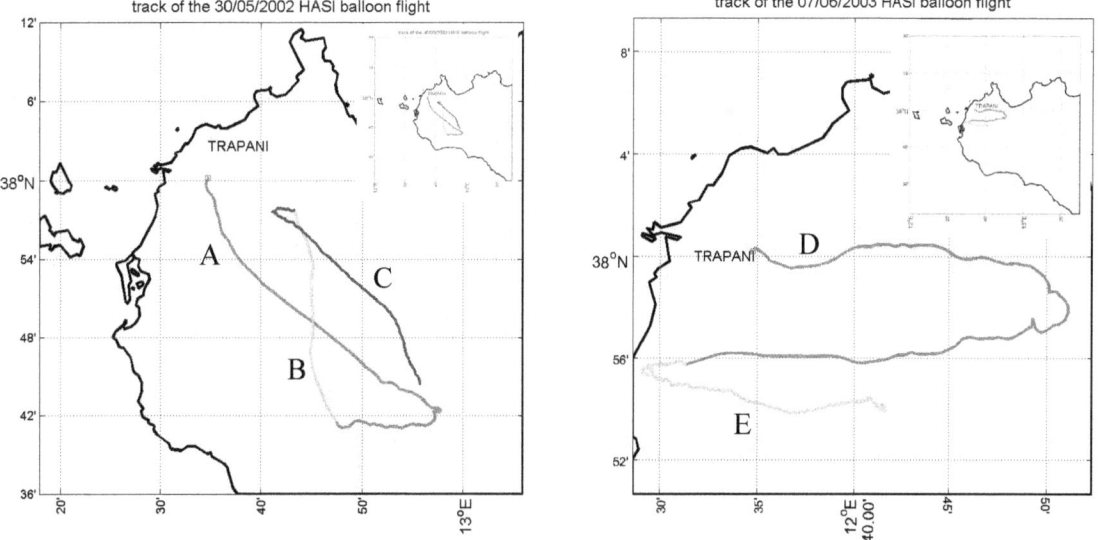

Figure 3 left: Balloon 2002 track over west Sicily; A is the ascending phase; B is the floating phase at 30km necessary for secure landing and C is the descent phase; right: Balloon 2003 track over west Sicily; D is the ascending phase and E is the descending phase

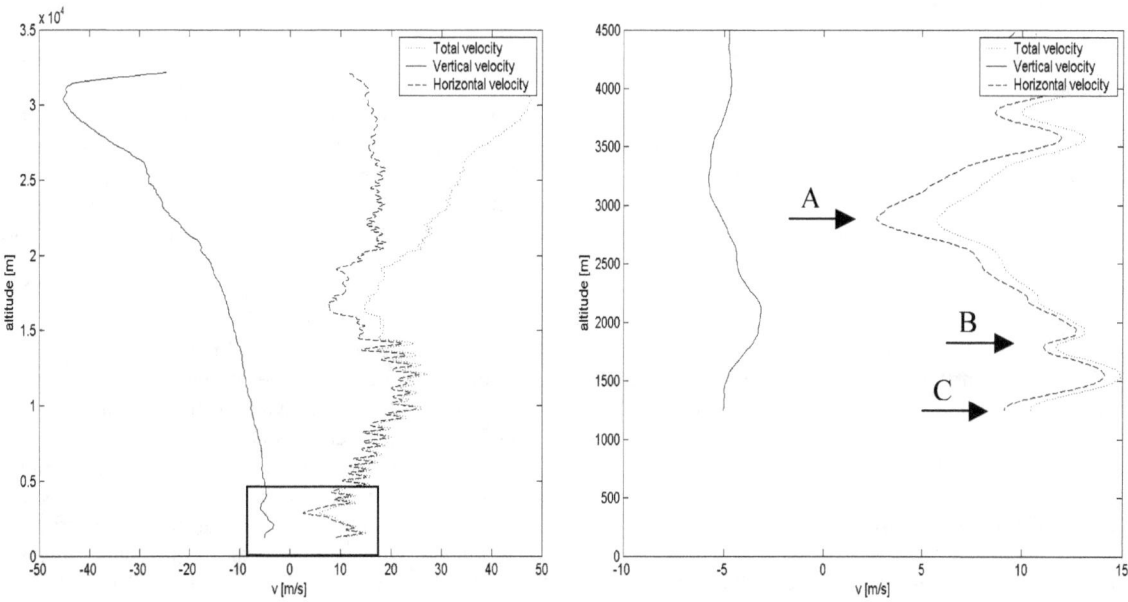

Figure 4. Wind profiles during the 2002 balloon experiment. Left: vertical velocity (line), horizontal velocity (dash) and total velocity (points)- obtained combining vertical and horizontal components-; right: close-up view of box showing features (A, B, C) at 3km altitude.

In the 2003 profile (Figure 5) at 25km the vertical velocity variation is between 2-5 m/s related to 2-3 km thick lower density layer in the atmosphere where the probe-parachute system is accelerating while the feature at 14km (see box) is due to a wind gust that accelerates the mock-up; the latter shows an oscillating velocity for 1km and an amplitude of the oscillations of around 0.2 m/s.

Different between the two horizontal wind profiles are the stronger winds that are present in 2002 near the tropopause; a Δv of more than 10m/s where the 2002 winds have values around 25 m/s reflects in a lower temperature measured by the temperature sensors.

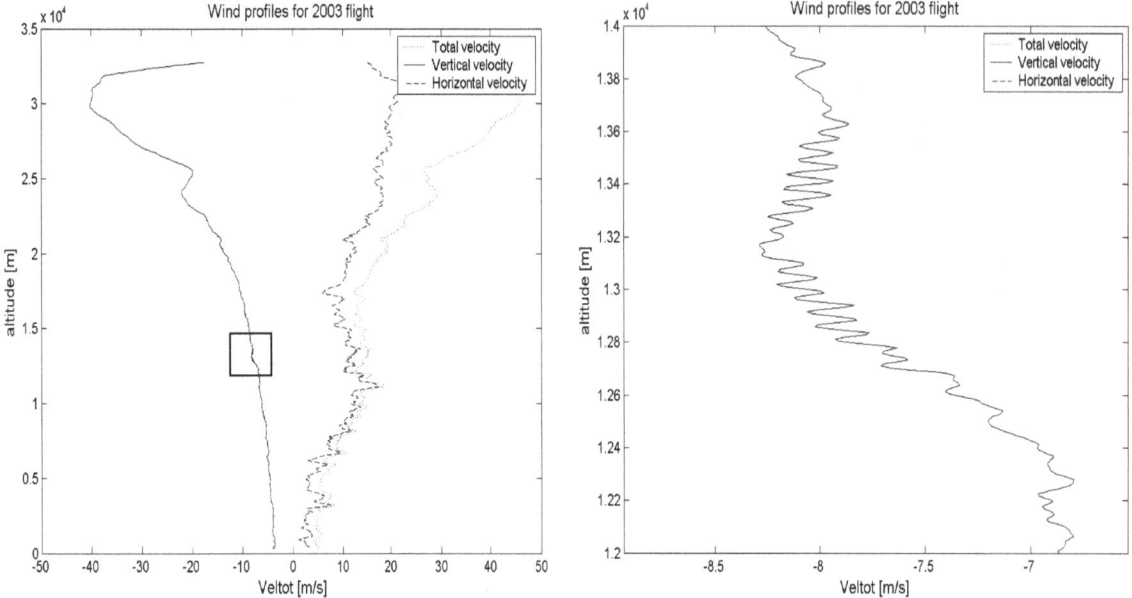

Figure 5. Wind profiles during the 2003 balloon experiment: vertical velocity (line), horizontal velocity (dash) and total velocity (points) obtained by combining vertical and horizontal components. In the close-up box the evolution of the descending velocity due to a wind gust

5. Measurements of the temperature vertical profile

During the flight campaign the temperature instruments measured data both during ascent and descent phases.
For the 2002 year campaign data recorded by HASI Tem and MTEM will be analysed; for year 2003 only data from HASI Tem will be used.
Analysis of the difference of the measured temperature respect to the static temperature shows that, since vertical velocity is less than 20-25m/s, the Mach number is less than 0.06 and the relative error is less than $4 \cdot 10^{-3}$; following this considerations no dynamic correction (refer to Fulchignoni et al. 1999, Gaborit et al. 2004) has been applied on temperature data.

5.1. 2002 balloon flight experiment: results

Focusing on the descent profiles, both the data sets present the same features; temperature at 32km are around 268K and drop down to 208K at 12.5km near the tropopause and then grow up again to 278K at 1km where the data link was lost.
In the stratosphere the temperature gradient is around 7.4 K/km for all the sensors except F1 which has a value of 7.8 K/km; the latter seems to be too high for a standard summer atmosphere above Sicily and must be disregarded (inconsistencies have been discovered in the calibration of the sensor).
Another important feature that must be analysed is the very low temperature at the inversion; in fact it can be seen that between 14km and 11.5km the temperature is not constant but decreases from 214K down to 208K
(see Figure 6 and Figure 7). This behaviour is easily correlated to the horizontal wind velocities variations measured by the GPS: winds higher than 20m/s are measured at same altitudes (see Figure 4).
The variation observed in temperature profile could be mainly due to two reasons: the mock-up passes through a layer of colder air or through a layer where stronger winds are present. From the previous observation on the wind profiles and from theoretical considerations on thermal exchange, an increase of velocity of the horizontal flow of about 5m/s (from 17m/s to 22m/s) leads to a variation, in a forced convective regime with same pressure and flow velocity values, of about 5÷10K in temperature. The conclusion is that a layer of about 3km thickness with stronger winds than the neighbours has been crossed.
At lower altitudes, below 3km, some other variations in the temperature profile are measured, respectively of 2K and 1.5K; these two variations are perfectly in accordance with feature A and B present in the wind profile (see Figure 6 & Figure 7; in MTEM profile also another feature can be observed: ΔT 1.5K correlated to variation C in wind profile, see Fig.4).
Also in this case, we argue that the mock-up is crossing layers of 200-500m thickness of stronger winds respect to upper and lower layers.

5.2. 2003 balloon flight experiment: results

In Figure 8 the HASI TEM profiles for both ascent and descent phases are presented; ascent profile is artificially shifted by 10K for graphical representation; time delay between the two profiles is decreasing at higher altitudes starting from more or less 2 hours on ground level data. The overall structure of the atmosphere is similar; inversion occurs at same altitude (see Figure 6 and Figure 7 for a similar stratification, at

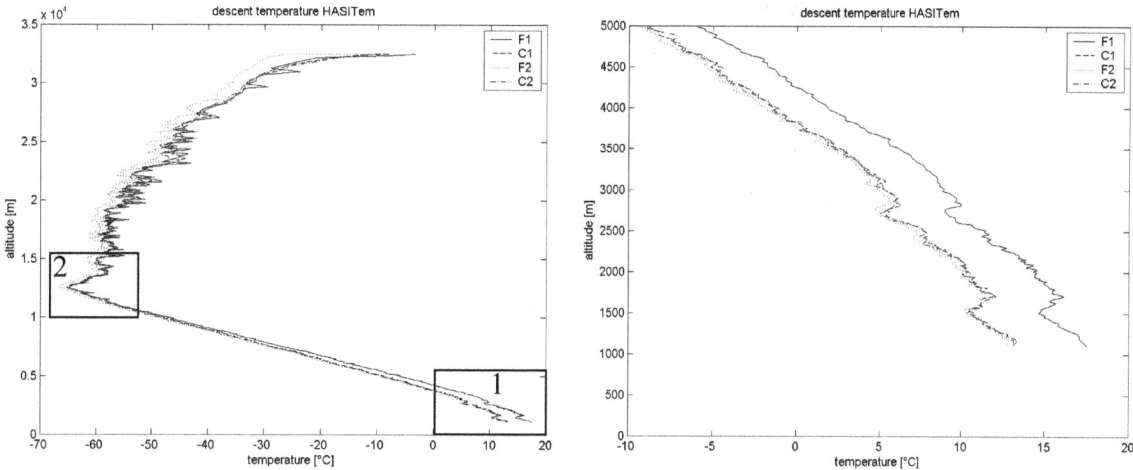

Figure 6. Balloon 2002 temperature profiles for HASI TEM; in box 2 the decrease in temperature due to increase in horizontal winds of 5m/s and on the right a close up view for box 1 showing the features below 5000m.

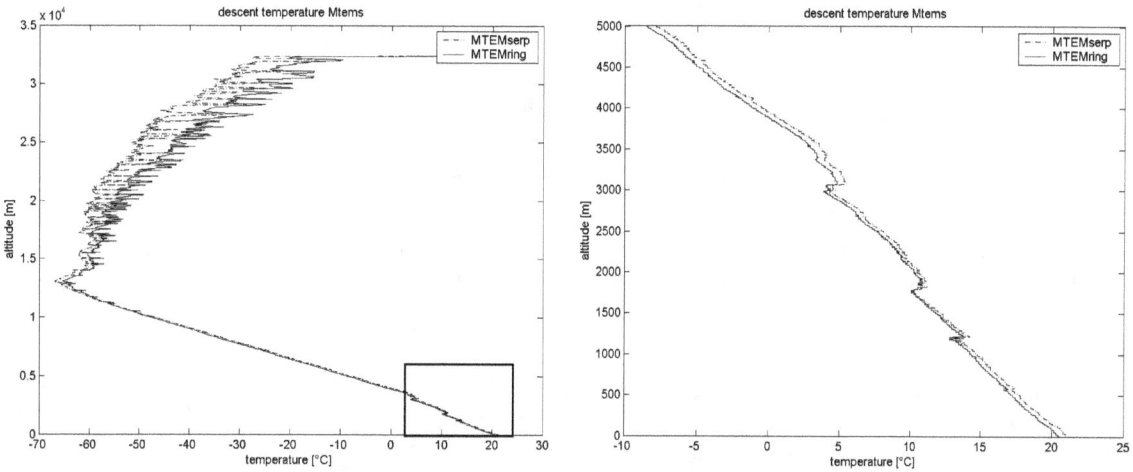

Figure 7. Balloon 2002 temperature profiles for MTEM; right: temperature variations measured when crossing layers with stronger winds.

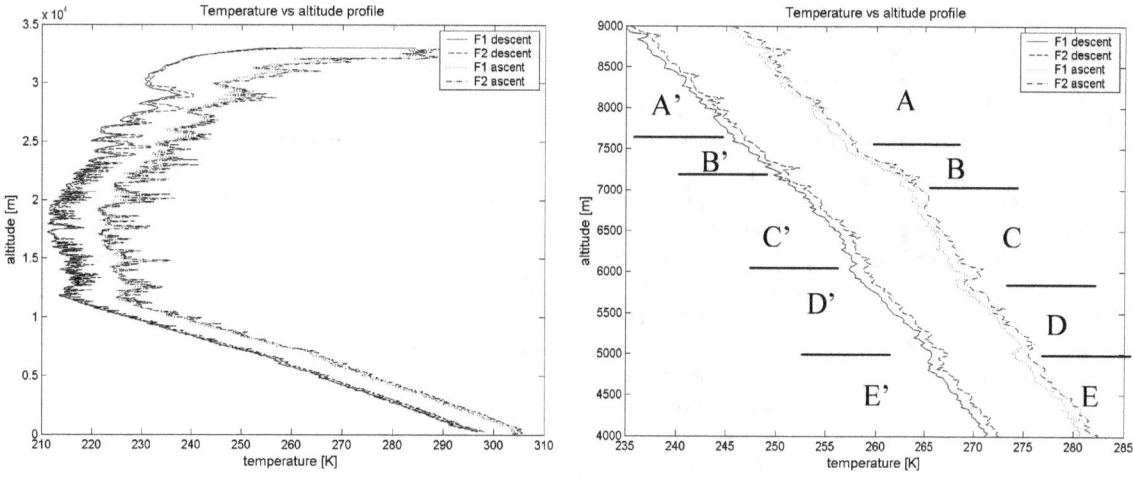

Figure 8. left: descent profile for HASI Tem; centre: temperature gradient evolution between 4000-9000m

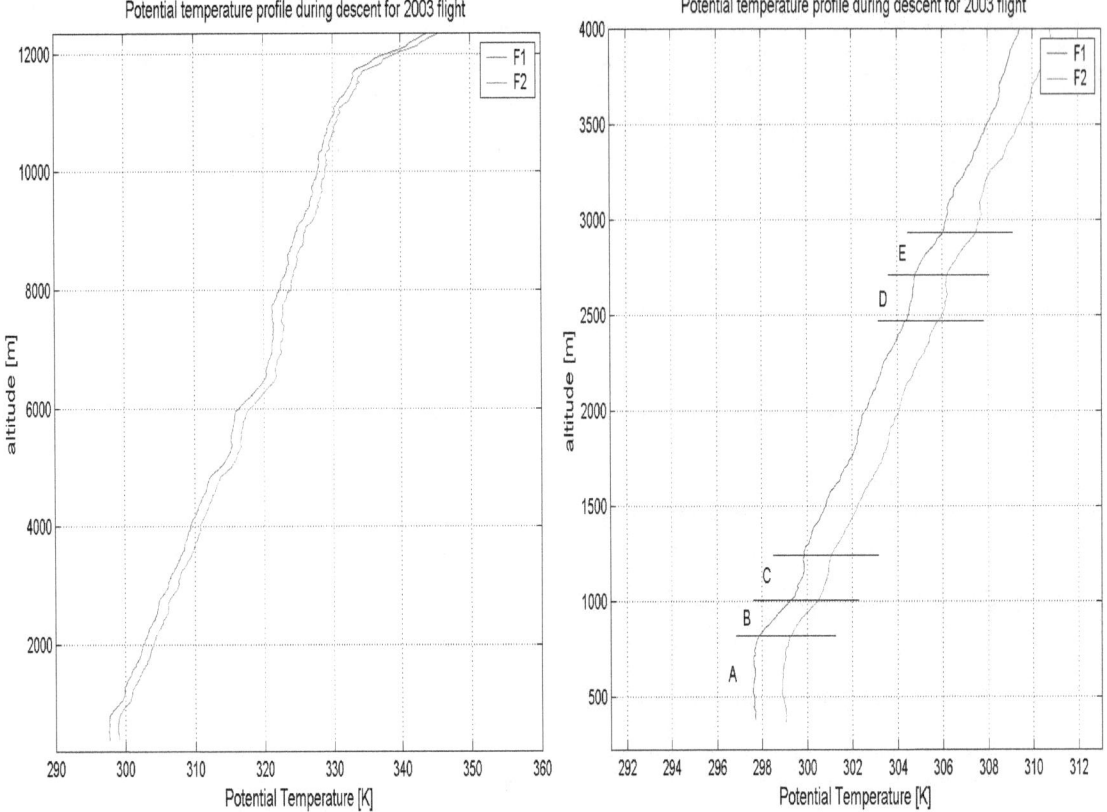

Figure 9. Balloon 2003 descent potential temperature profile between 1000m and 12000m (left) and below 4000m (right); several different regions of stability are shown (in particular B and E).

lower altitudes, measured in the 2002 flight). Above the tropopause, while the general evolution of temperature profile is uniform, the structures that are present show several changes. Some features can be recognised in both the profiles, at same altitudes while others have moved or present higher values. This is an evidence of an evolution of the temperature profile that outlines how the atmospheric structure, even in short-time windows, has a dynamic behaviour (see Figure 8; layers with prime are in descent phase). Several layers can be identified in the lower troposphere; the layers (thickness: B=400m; C=1100m; D=800m and E=5000m) are separated by a ΔT of 2K in 50m (for C/D) and 1K in 70m (for D/E); the overall structure is maintained and visible in both ascending and descending phases showing a well fixed stratification in the temperature field. In A' the gradient is −8.3K/km, in B ~ -17K/km and in C', D' and E' it's around −6.0K/km. It can be observed that during the descent phase the two layers B' and C' are not so well distinguishable but show a mixing (confirmed by the potential temperature plot showing a constant potential temperature layer between 7000-8000m; see Figure 9). Thickness of layer D' is higher than layer D confirming the evolution of the atmosphere. An overall temperature gradient of 7.1 K/km is measured in the troposphere. Other features can be observed in the upper troposphere: several sets of spikes of 3÷5K peaks in 50÷80m are present from 7000m to 11000m; these peaks are present both in ascent and in descent phases, they are not due to fluido-dynamic effects since the

configurations of the probe in the two phases are completely different and the incoming flow is different. A possible correlation between these peaks and the attitude of the probe (rotation and inclination) shows no direct correlation. Observed features must be related to the structure and configuration of the measured temperature field.

5.3. Potential temperature

The potential temperature is usually defined as the temperature a water parcel has when it is brought adiabatically to an atmospheric pressure of 1000 millibars.
The formula used for calculating the Potential Temperature (θ) is:

$$\theta = T \left(\frac{P_0}{P}\right)^{\frac{\gamma-1}{\gamma}} \quad (1)$$

where T is the measured temperature, P is the measured Pressure, P_0 is the Pressure at 1000millibars level and γ is the heat capacity ratio of the gas.

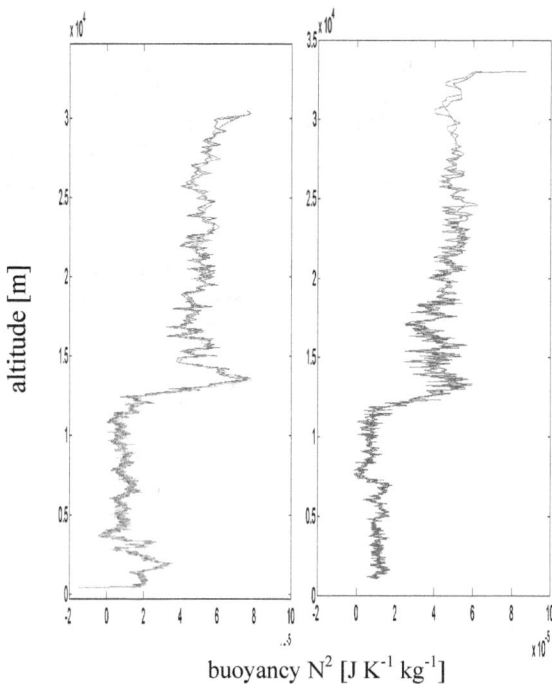

Figure 10. N^2 profiles deduced from balloon measurements from: (left) HASI TEM 2002 – mean of 40 data; (right) HASI TEM 2003– mean of 40 data

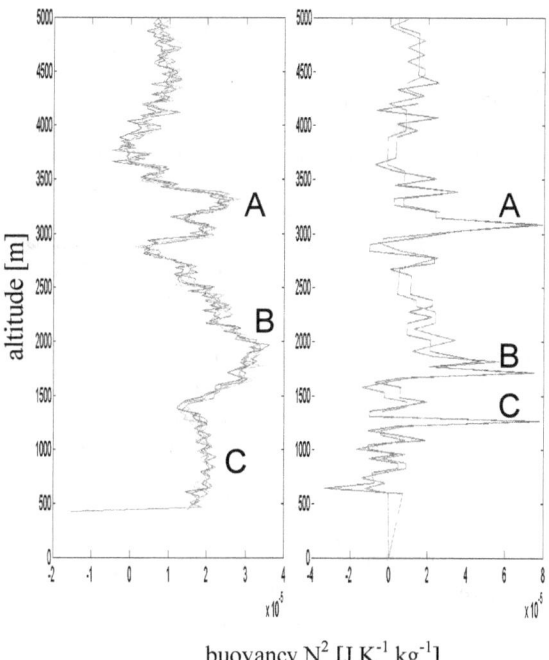

Figure 11. Close up of the N2 profiles deduced from balloon measurements below 5000m: (left) HASI TEM2002 – mean of 40 data; (right) MTEM 2002 – mean of 100 data.

Vertical stability of a dry atmosphere can be characterized by a constant variation of θ with height (9.8K/km).

The region up to 3000m shows several layers where the potential temperature is constant (see in Figure 9-right the layers A, C and D) revealing convective layers; layer A is the first layer near surface where the interactions between surface and atmosphere take place (Ekman layer); the other two layers of mixing air (C and D) can be observed at higher altitude (1100 and 2100 respectively); what is the interesting thing is that between A and C there is a layer of stable air and again above layer D; two other regions (see Figure 9 left), between 5500m-5900m and 6500m-7700m, reveal unstable regions showing the presence of two adiabatic regions where the potential temperature remains constant: this is evidence that convective mixing is taking place. Two layers, that show a more stable atmosphere are visible just below these mixing layers; these are relative thick layers (400m and 1200m respectively) that are not restricted to be in the vicinity of the tropopause revealing presence of mixing regions where density is high and where horizontal winds are mainly constant (see Figure 10).

Above 11900m the potential temperature profile reveals a stable atmosphere: above the inversion layer the stratosphere shows, as expected, a more stable behaviour.

6. Discussion

6.1. Atmospheric stability (buoyancy)

Stability of the atmosphere depends on the value of the square of the buoyancy frequency N; N^2 is defined as:

$$N^2 = \frac{g}{T}\left(\frac{\partial T}{\partial z} + \frac{g}{C_p}\right) \qquad (2)$$

where g is the gravitational acceleration, T is the atmospheric temperature (in Kelvin), C_p (=1004 J K^{-1} kg^{-1}) is the specific heat at a constant pressure and z is the altitude (in meters).

When N^2 is negative, i.e if the atmospheric lapse rate $\Gamma = \partial T/\partial z$ is bigger than the adiabatic lapse rate g/C_p, the atmosphere is unstable.

If this is the case an atmospheric air parcel will escape continuously from it's equilibrium position.

An unstable atmosphere can have an amplifying effect on small perturbations.

Stability of the atmosphere is strongly dependent on the temperature profile. Instabilities are generally present in those regions where the lapse rate Γ is high and the N^2 is low.

In Figure 10 the squared Brunt-Väisälä frequency profile N2 is shown both for the 2002 and the 2003 campaigns. It can be observed that between 12 and 14 km the stability of the atmosphere is increasing, indicating

Considering that typical stratospheric value for N2 is 4 10-5 (J K-1 kg-1), stability of the region crossed is higher in the stratosphere.

In the 2003 profile it can be observed that two layers of higher stability than the neighbour regions between 5000m and 8000m are crossed: here the potential temperature trend shows a higher stability.

Similar values of N^2 are measured in the two campaigns suggesting that no climate change has affected the Sicilian region. The N^2 profile differs a lot below the 4000m region between the 2 years: 2002 data are much more variable revealing a much more complex stratification in the lower troposphere.

It can be observed that the increased N^2 values for the features A, B and C can show a perfect correspondence to variations, at same altitudes, in the wind and temperature profiles; layers of mixing air contribute to the increase of stability of the atmosphere itself. Feature A is a 1000m layer where a corresponding drop in the wind velocity is measured and a much more stable layer is crossed and the N^2 values decrease again in less than 500m; the same happens with feature B. these profiles suggest the presence of alternate layers of stable and unstable air; in fact in the wind profiles it can be observed that there are weaker horizontal winds layers.

Same altitudes for the increased stability layers are found by the MTEM sensors for features A and B while feature C is measured at a slightly higher altitude.

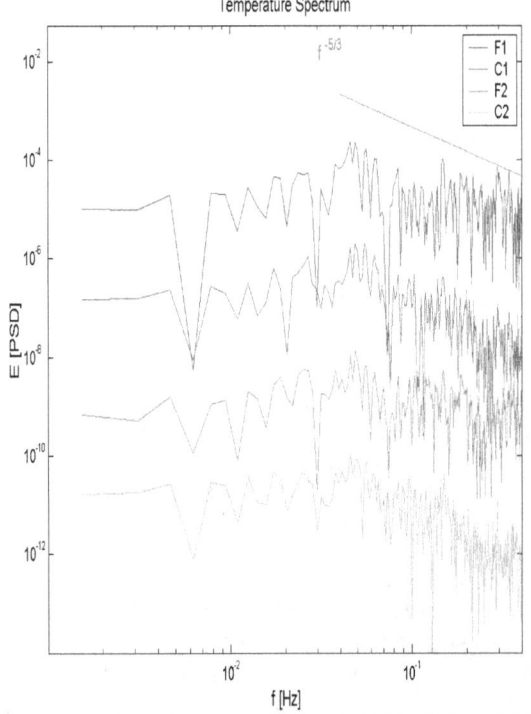

Figure 12. Temperature spectra of 2002 flight: HASI TEM (left).

6.2. Atmospheric turbulence (Kolmogorov)

Kolmogorov [1941] showed how for a 3D homogeneous and isotropic turbulence, with a downscale energy flux – positive from large to small scales, the spectrum of energy distribution follows the $k^{-5/3}$ power law and similarly Obukhov [1949] predicted a $k^{-5/3}$ spectrum for temperature fluctuations. For an analysis of global atmospheric fluxes this assumption is not realistic, horizontal dimensions are much larger than the vertical dimension, but if horizontal scales are of order of magnitude of few kilometres the Kolmogorov theory is not far from nature. With this assumption the plots (Fig. 12 and 13) show how the $k^{-5/3}$ law is followed in both the years: HASI TEM follows exactly the $k^{-5/3}$ power law; the housekeeping temperature sensor heads (PT100, see Fig. 1) show the same perfect agreement with theory.

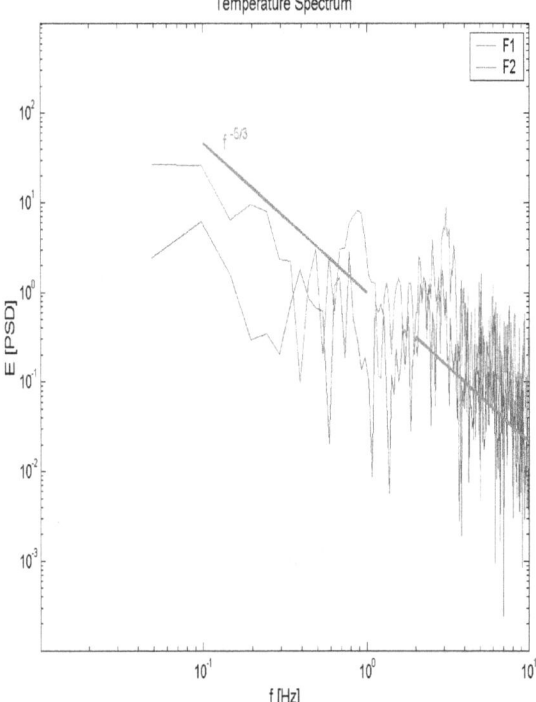

Figure 13. Temperature spectra of 2003 flight: HASI TEM (left).

7. Concluding remarks

Balloon flight experiments allowed to check performance of HASI TEM with real data sets in dynamical conditions similar to those they will experience in Titan's atmosphere. Several features and structure are observed and described.

Thermo-structure of troposphere and lower stratosphere reveal that local variations on the temperature measurements can be due to variations in the wind, both in value and in direction.

Tropopause is found to be at the same altitudes than the standard atmosphere, for the 2002 flight temperature is lower at the inversion point due to higher horizontal winds.

Analysis of both ascending and descending profiles show how several features have changed both in altitude and in intensity due to solar heating of the atmosphere.

Stability of the atmosphere is analysed and several layers of stable air are followed by unstable layers, especially in the troposphere. Atmosphere seems to be more unstable respect to standard atmosphere but still the increase in stability is found for altitudes higher than the tropopause.

Finally stability and turbulence of the atmosphere are investigated analysing the buoyancy frequency profile and the temperature spectra.

The conducted analysis has helped the understanding of the features observed in the temperature profile and will help the interpretation of the temperature measurements retrieved in the 2005 Titan mission.

8. Acknowledgements

This work has been funded by the Italian Space Agency "ASI".

References

C.Bettanini, M.Fulchignoni, F.Angrilli, P.F.Lion Stoppato, M.Antonello, S.Bastianello, G. Bianchini, G.Colombatti, F. Ferri, E. Flamini, V.Gaborit, A. Aboudan, Sicily 2002 balloon campaign: a test of the HASI instrument, *Advanc. Space Science*, **33**, 1806-1811, 2004

Fulchingoni et HASI Team, The Huygens Atmospheric Structure Instrument (HASI), ESA SP1711, 163-176,1997.

Fulchignoni, M., A. Aboudan, F. Angrilli, M. Antonello, S. Bastianello, C. Bettanini, G. Bianchini, G. Colombatti, F. Ferri, E. Flamini, V. Gaborit, N. Ghafoor, B. Hathi, A-M. Harri, A. Lehto, P.F. Lion Stoppato, M.R. Patel, J.C: Zarnecki, A stratospheric balloon experiment to test the Huygens Atmospheric Structure Instrument (HASI) *Planet.Space Scie.*,**52**, 867-880, 2004.

Kolmogorov, A.N., 1941. C.R.Acad.Sci.URSS, 30,301

A. M. Obukhov, Izv. Akad. Nauk SSSR, Ser. Geog. Geofiz. 13, 58 (1949).

Ruffino, G., A. Castelli, P. Coppa, C. Cornaro, S. Foglietta, M. Fulchignoni, F. Gori and P. Salvini, The temperature sensor on the Huygens probe for the Cassini mission: design, manufacture, calibration and tests of the laboratory prototype, *Planet. Space Sci.*, Vol. 44, Issue 10,1149-1162, 1996

Fulchignoni, M., F. Ferri, F. Angrilli, A. Bar-Nun, M.A. Barucci, G. Bianchini, W. Borucki, M. Coradini, A. Coustenis, P. Falkner, E. Flamini, R. Grard, M. Hamelin, A.M. Harri, G.W. Leppelmeier, J.J. Lopez-Moreno, J.A.M. McDonnell, C.P. McKay, F.H. Neubauer, A. Pedersen, G. Picardi, V. Pirronello, R. Rodrigo, K.Schwingenschuh, A. Seiff, H. Svedhem, V. Vanzani, J. Zarnecki, The characterisation of Titan's atmospheric physical properties by the Huygens Atmospheric Structure Instrument (HASI), *Space Science Reviews*, 104, 395-431, 2002

Angrilli, F., F. Ferri, A. Francesconi, P.F. Lion Stoppato, B. Saggin, G.Bianchini, M. Fulchignoni, New temperature sensors for descent measurements during atmospheric entry at Mars for *50th International Astronautical Congress,* Amsterdam, The Netherlands, 4 – 8 October 1999 (AIAA paper).

Gaborit, V., M. Fulchignoni, G. Colombatti, F. Ferri, C. Bettanini, Huygens/HASI 2002 balloon test campaign: probe trajectory and atmospheric vertical profiles reconstruction in press Planet.Space Scie. 2004.

PARACHUTE DYNAMICS INVESTIGATIONS USING A SENSOR PACKAGE AIRDROPPED FROM A SMALL-SCALE AIRPLANE

Jessica Dooley[(1)], Ralph D. Lorenz[(2)]

[(1)]*Lunar and Planetary Laboratory and Aerospace and Mechanical Engineering Department,
University of Arizona, 85721, USA, dooleyj@u.arizona.edu*
[(2)] *Lunar and Planetary Laboratory, University of Arizona, 85721-0092, USA, rlorenz@hindmost.lpl.Arizona.edu*

ABSTRACT

We explore the utility of various sensors by recovering parachute-probe dynamics information from a package released from a small-scale, remote-controlled airplane. The airdrops aid in the development of datasets for the exploration of planetary probe trajectory recovery algorithms, supplementing data collected from instrumented, full-scale tests and computer models.

1. INTRODUCTION

A probe-parachute system descending through a planetary atmosphere will encounter attitude motions, both due to intrinsic stability properties of the parachute and probe, and in response to wind shear. These motions must be understood, both to predict the dynamic environment which provides a context for other sensing (e.g. imaging or radar altimetry) and to permit the reconstruction of winds from on-board dynamic measurements. Attitude dynamics models e.g. [1] require experimental validation: however, full-scale tests with stratospheric balloon drops etc. pose enormous logistical difficulties and entail substantial costs [2]. We explore here what can be learned with small-scale parachute probes, since the dynamics instrumentation can now be miniaturized.

Parachute-probe drop tests of this type were previously conducted by hand-dropping small-scale sensor packages from within the atrium of the Lunar and Planetary Laboratory at the University of Arizona [3]. This testing environment was well controlled, less the modest influence of the ventilation system, but the data collection periods were very limited. The transient effects of the launch and parachute inflation left little time for data to be collected before reaching ground level, even at low descent speeds. However, this first drop test did help build a dataset of attitude motion to gain familiarity with sensor data and corresponding motions. While the drop tests focussed on low-cost and ease of use, rather than dynamic similarity, they provided the starting point for the development of a simple parachute-probe model [1]. Even more importantly, the indoor drop tests proved the usefulness of combining simple testing procedures with inexpensive electronics to explore parachute dynamics.

With a general understanding of parachute dynamics from the indoor tests, a new series of probes and alternate testing procedure was developed (See Fig. 1). This new test series aims to increase the data collection time and attempt to identify characteristics of specific parachute designs by testing a variety of parachute types such as flat, spherical, cross, and disk gap band.

Fig. 1. The two versions of the parachute-probe dynamics test-beds with a pencil for size reference. The preliminary version, shown on the right, was soon replaced with the larger version, on left, to better support the growing sensor package.

A series of Basic-X 24 micro-controllers are used to collect data from the onboard sensors. In addition, airdrops are documented with ground-based video cameras in order to document the general behaviour of the parachute-probe system for comparison with onboard sensor data.

The parachute-probe system utilizes two main sensors packages for data collection. A tri-axial, six-degree of freedom inertial measurement unit (IMU) by O-Navi provides angular rate and acceleration. The magnetic field is determined from an orthogonal triad of FGM-1 fluxgate magnetometers. Attempts were made to supplement the main sensor packages with a pressure sensor, a Global Positioning System (GPS) and a camera and processor developed by the Robotics Institute at Carnegie Mellon University (CMU), but with little success.

The airdrop tests utilize a series of parachutes manufactured for model rocket recovery. A small-scale, remote-controlled airplane is used to ferry the parachute-probe sensor package to altitudes permitting adequate descent durations. During the airplane's ascent, the probe and folded parachute are secured underneath the fuselage of the plane by rubber bands. When the desired altitude is reached, a command is sent from the pilot to release the parachute-probe package. Each data collection is initiated when the probe is separated from the bottom of the fuselage. At this time, the sensors are polled for a specified period of time and the data is stored in the EEPROM of the onboard micro controllers. After the data collection period has expired each of the onboard controllers continuously reads out the EEPROM data to a serial port until power is reset.

2. PROBE INSTRUMENTATION

The original probe design aimed to gather information about the general behaviour of parachute descent dynamics from a wide variety of sensors (see Table 1 for an approximate cost for each component). However, some sensors proved more reliable and better suited for our application than others.

2.1 Micro Controller

A series of BasicX-24 micro controllers manufactured by Net Media were used to collect and store sensor data during the parachute drops. The BX-24 is a 24-pin package containing an ATMEL micro controller running at 65,000 instructions per second, a 32K EEPROM, clock and power regulation and serial port components. As a standalone unit powered by a single 9V battery and programmed in a Basic language it can perform a range of data acquisition, communication and control tasks. We use it simply as a convenient data logger: the controller has 8 on-board 10-bit A/D converters and enough EEPROM space to store data for a usefully-long flight. (The Basic Stamp units we used previously [3], while easier to use and more robust, lack the A/D converters, run slower, and have less EEPROM space, for the same cost).

One issue we encountered with the BX-24 is that a crystal oscillator, mounted a little above the package circuit board, sheared off during some of our early tests when high impact loads were encountered. High impact loads were experienced when a significantly heavier instrumentation package was tested with an older version of the parachute-probe attachment mechanism. The more massive probe applied larger stresses on the attachment mechanism than expected, and the attachment failed during probe deployment from the ferry airplane, allowing the probe to plummet to the ground without a parachute. Rapid deceleration at ground level caused the crystals to shear off three of the four micro controllers. The attachment mechanism was upgraded with stronger metal and additional adhesive was applied to support and secure the crystal during future drops.

2.2 Primary Sensors

The Gyrocube IMU and the triad of fluxgate magnetometers provided the backbone for our data collection.

The sensor outputs (analog voltage from the Onavi, and digital pulse rate from the magnetometers) are recorded by the micro controller, scaled to an integer 0-255 and written to the controller's on-board EEPROM as 1 byte for each sample. After the flight, the EEPROM data is read out to a serial port: the data is captured by a terminal program running on a laptop computer to which the package is connected by cable. We have used similar methods in our earlier drops [3]. Construction details for similar equipment, used to measure flight dynamics of Frisbees are given in [4].

The IMU and magnetometers proved to be the most reliable sensor, as long as the probe battery voltage was monitored regularly. The monitoring of battery voltage after each drop proved important for two reasons. First, even with equipment to assist in finding the probe after landing, the probe occasionally encountered thermals that forced us on mile-long trecks through the desert to retrieve the probe. Second, even when strange weather did not carry the probe miles away, the downloading of the data stored on the micro controllers took time. The combination of long retrieval and download times and sloppy battery monitoring can drain the onboard batteries, forcing partial or full data loss. Once aware of these concerns, batteries were monitored more closely and replace often, and afternoon drops were avoided as much as possible.

2.3 Ancillary Sensors

A Global Positioning System (GPS) designed to track model rockets was added to the bottom shelf of the probe in an attempt to provide ancillary information during each drop. GPS Flight's STXe GPS receiver and radio modem and RXB2 base station receiver unit operated completely independent of the other probe sensors. Position and velocity data were transmitted directly to the ground station, while other sensor data was recorded to the micro controllers. The GPS position and velocity data was intended to be used to determine the horizontal translation, descent rate, and the approximate probe-landing site. In spite of these goals, and the ease of integrating the small receiver and radio modem into the existing probe setup, the addition of the GPS did not work as planned. The GPS antenna was mounted on the top of the probe in order to receive information from orbiting satellites during the drop. However, when the probe was mounted on the bottom of the aircraft this antenna was pointed horizontally and further attenuated by the fuselage, causing complete loss of satellite signal. This loss of signal remained through the ascent onboard the aircraft and during a majority of the drop sequence. Multiple antenna locations and probe mounting arrangements were attempted, but with little success. Future integration of a higher-quality antenna may yield better results.

A camera and processor developed by the Robotics Institute at Carnegie Mellon University (CMU) was hoped to be a novel feature of our probe airdrop tests [5]. The CMUcam was designed to provide a low-cost approach to onboard, real-time vision processing for mobile robots. The CMUcam can be programmed to track the position and size of an object that is of high contrast to its environment. For our application the camera and accompanying processor are used to track the position of the marker object in the interior of the parachute canopy in order to determine the parachute's motion relative to the probe. The small camera and processor board is mounted on the inside of the probe and the lens is directed toward the open parachute canopy. A rectangular marker made of thin paper, and of contrasting color to the parachute, is secured to the center of the canopy interior. A BasicX micro controller is wired to the CMUcam and instructs the CMUcam to capture and analyze an image for the colored marker by sending a "track color" command. The track color command is specific to the color of interest: the color of the marker inside the parachute canopy. For example, to determine the motion of the flat, orange parachute relative to the probe a black rectangle is used and the command "TC 45 60 22 35 27 45" is sent to the CMUcam. This command specifies the minimum and maximum values for red, green and blue, color of interest that the camera should track, in this case, a shade of black. The CMUcam outputs a string indicating the tracked rectangle's center of mass and corner locations in x and y coordinates, the number of pixels in the tracked region, and a value indicating the confidence the processor has that it has successfully tracked the object. In theory the CMU camera would provide information about the relative motion of the parachute-probe system by tracking the position and size of a colored marker on the inside of the parachute canopy during the decent. Ground tests using Hyperterminal and a JAVA based Graphical User Interface (GUI) provided with the CMUcam kit have yielded promising results. Using the JAVA GUI we have found appropriate maker sizes, shapes and colors that, when placed on a specific parachute, can be tracked well enough to record the spinning and translation of the parachute relative to the probe. Ground tests also revealed the influence of lighting conditions on the quality of the CMUcam tracking. Shadows cast on the chute and changing lighting conditions throughout a days worth of airdrop testing can make the marker less recognizable to the camera. However, these ground tests used Hyperterminal to view and capture the tracking data and in order to use this vision processing technique during a real drop the data must be written to the EEPROM of a micro controller onboard the probe. As of date, attempts to program the BasicX to properly store the CMUcam's string output to EEPROM have been unsuccessful. However, it is hoped that as the probe evolves, our BasicX programming abilities will also, and the CMUcam will one day provide an interesting perspective to parachute dynamics.

Experiments employing, an Omega PX139 pressure sensor to act as an altimeter have so far been unsuccessful.

Table 1. Description and approximate cost of probe instrumentation and components.

Description	Cost
BasicX Micro Controller	$50 (each)
Onavi IMU (+- 2g, +-200 degrees / second)	$510
FGM-1 Magnetometers (+- 0.7 oersted)	$43 (each)
Omega PX139 Pressure Sensor	$85
CMUcam (kit)	$109
108 db Buzzer	$5
PWM Activated Relay	$30

3. PROBE CONSTRUCTION

The outer shells of the probes are made from a plastic cardboard-like material referred to as chloraplast. Chloraplast is commonly used to make temporary signs and displays. The blue chloraplast was scored, folded, and then reinforced with epoxy filled corners to form the strong, lightweight, rectangular cases. The first probe has dimensions 20 x 10 x 8 cm, not including the foam block mounted on the bottom for shock absorbance on landing. When more sensors were added, the second probe (24 x 11 x 9 cm) was constructed and weighed about 800 grams with all sensors and power supplies. The micro controller development boards are mounted on pieces of dense foam and inserted into the probe like shelves, making the probe very modular in design. The foam shelves support the boards and are secured to the chloraplast with 4 screws, providing plug and play abilities. A downside to the multiple shelve design is that the wiring to and from the switches, batteries, micro controllers, etc can be messy and frustrating during maintenance and debugging.

Velcro straps secure the probe door in place allowing quick and easy access to each serial port on the development boards, which are continuously dumping the data stored in their EEPROM. The top of the first probe was made of chloraplast, but after absorbing the force of many deployments from the plane the chloraplast began to give way. The second version featured a hard wood top. A single piece of fishing line loops through four metal eyelets at each corner of the probe and a key ring in the center providing a well-balanced anchor point in the shape of an "X." The fishing line proved to be the strongest and least obstructive to the CMUcam's view of the open parachute canopy.

4. AIRDROP SYSTEM

The main purpose of the airdrop tests was to build a larger data set with increased data collection periods. To accomplish this we constructed an airdrop system that consists of a stable aerial platform, a reliable release mechanism, and a data collection status indicator.

4.1 Airplane Platform

A small-scale, remote-controlled (R/C) airplane called the Xtra-Easy is used to ferry the parachute-probe sensor package to the release altitude. The Extra-Easy airplane is manufactured by Hanger 9, has a wingspan of 1.75 meters and weighs approximately 3 kilograms. This platform was chosen to ferry the probe because it is a very inexpensive, stable platform, designed specifically for the beginner R/C pilot. This simplistic platform provided us with the bare minimum for flight and avoided extra complications that come with aerobatic airplanes and other "bells and whistles" available on small-scale aircraft. All the equipment needed for flight, including the complete airframe, engine, and controller, is available in a "Ready to Fly (RTF)" kit. The RTF kit provides an airframe nearly assembled for flight. Although the airframe is equipped with the standard point forty cubic centimeter motor, it was replaced with an OS .61 FX (point sixty one cubic centimeter) engine to better suit the needs of this project. This engine upgrade allowed the parachute-probe to be carried to the drop altitude with ease and speed, thereby decreasing the time between data collections. The airframe and mounting of the probe are shown in Fig. 2 and Fig. 3.

Fig. 2. The Xtra-Easy airplane taxing to the runway with the probe and parachute securely mounted under the fuselage.

Fig. 3. Upclose view of the first probe under the fuselage with the medium-sized, flat parachute.

4.2 Mounting and Release Mechanism

The probe is mounted underneath the airplane's fuselage and secured in place by a chain of rubber bands. Both ends of the rubber band chain series are permanently affixed to the plane. The middle rubber band is attached to a metallic loop that is pulled around the probe and slid over a metal rod jutting out from the opposite side of the fuselage. This metal rod is attached to the arm of a servo (a typical actuator used in remote controlled airplanes) that is connected to the same receiver box that the pilot uses to manipulate the control surfaces on the airplane. The metal rod is bent at a slight upward angle so that the loop does not slide off unintentionally before the release signal is sent. The servo is commanded to extend the metal rod outward during ascent, holding the rubber bands in place around the probe. When the pilot activates the servo from the R/C transmitter, the servo pulls the metal rod into the fuselage thereby releasing the parachute-probe system.

Data collected during the plane's ascent is of no interest to us. To ensure that we collect and record data during the descent, and not while the probe is attached to the plane, an "nc switch," or normally closed switch, was implemented on the probe. When the probe is securely mounted on the bottom of the airplane a fuselage-mounted peg depresses a push-button located on the backside of the probe opening or breaking the circuit. When the micro controllers are powered, they are instructed to monitor the status of this push-button. The micro controllers take no action while the button is depressed. When the probe is released from the plane, the peg no longer depresses the switch and the micro controllers begin to sample data from each of the sensor packages.

In preliminary airdrop tests, rough take-offs and awkward mounting of the probe under the plane made us question if the probe had begun collecting data while still attached to the plane. To clarify when data collecting or dumping was taking place we installed a 108-db buzzer and Pulse Width Modulation (PWM) activated relay. With these new components installed, the micro controller is instructed to send a PWM command to the relay to stay open while the probe is attached to the plane, closed when it detaches and collects data, and then alternate between open and closed when full data is collected and is being dumped over the serial port. When the relay is closed it allows the buzzer to sound, indicating that data is being collected. When collection is complete the buzzer pulses on and off. The buzzer and relay were very helpful in determining the status of the data collection and in helping to locate the probe after the drop.

4.3 Parachutes

Parachutes manufactured for model rocket and small Unmanned Aerial Vehicle (UAV) recovery proved ideal for our airdrop tests. Initial plans called for the testing of four types of parachute including flat, spherical, cross and disk gap band. However, time constraints have limited the research thus far to the use of flat and spherical parachutes shown in Fig. 4. Cross and disk gap band designs are planned for future tests. Specifications of the spherical and flat parachutes used in these tests are listed in Table 2. Each of the Apogee Rockets parachutes used in testing are made of 70-devier rip-stop nylon and the suspension lines are braided nylon. The spherical parachutes are made of comparable ripstop nylon and utilize similar material for suspension lines.

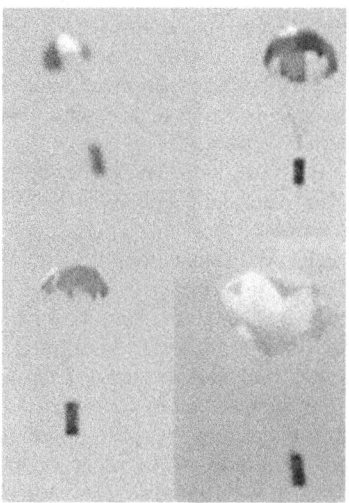

Fig. 4. The four parachutes during drop tests.

Table 2. Parachutes used for testing. The diameter, D, height, H, and suspension line length, L, are given in centimeters. Mass, M is given in grams and includes the suspension lines.

Type (Manufacturer)	D	H	L	Mass
Spherical (Spherachutes)	76	43	79	52
Spherical (Giant Leap)	132	68	112	180
Flat Octagon (Apogee)	99	48	86	65
Flat Octagon (Apogee)	152	76	140	134

5. SET-UP AND TESTING PROCEDURE

5.1 Probe Calibration

Once constructed, the sensors were calibrated by completing a series of controlled maneuvers. A small, spinning table, rotated by a servo and R/C transmitter and receiver, was constructed to calibrate the gyros on the IMU board. The accelerometers were calibrated by laying the probe on each of its 6 sides. Calibrating the magnetometers proved more difficult but was completed in a similar fashion

5.2 Test Procedure

The drop tests were conducted at the Tucson International Modelplex Park Association (TIMPA), a local remote-controlled flying field. The TIMPA facility provided ample open space and a safe, frequency controlled testing environment for our remote controlled airplane assisted drops.

During the airplane's ascent, the probe and folded parachute are secured underneath the fuselage of the plane by the previously described rubber band arrangement. The elapsed time from take-off to release altitude is less than one minute with the engine upgrade and an experienced pilot. When the desired altitude is reached, the pilot releases the parachute-probe package by retracting the rod supporting the metal loop. The probe separates from the plane and data collection is initiated when the fuselage- mounted peg no longer depresses the trigger switch. At this time, the PWM activated relay sets the buzzer to constant output and the micro controllers begin to poll the sensors for a specified period of time (just over 2 minutes for the magnetometer and IMU boards). Sensor data is stored in the onboard micro-controller's EEPROM and the drop is documented from the ground with a camcorder. Examples of camcorder footage are shown in Fig. 5. After the data collection period has expired the onboard controllers continuously reads out the EEPROM data over a serial port connection while the PWM relay alternates the buzzer signal. Upon touchdown the buzzer is deactivated and the probe is transported to a laptop computer for archiving and analysis. In contrast to real-time data transmission, the storage of data onboard during the descent avoids interruptions to the transmission and allows for more samples per second to be collected. We used a serial cable and the BasicX software to capture the data from each micro controller to files on the laptop.

6. DATA ANALYSIS

A variety of computer programs were constructed using Interactive Data Language to compute, display and compare sensor data and off-board camera footage in order to gain familiarity with the data and corresponding motions of the system and look for possible patterns due to parachute characteristics.

6.1 General Visualization with Camcorder Data

Drops were recorded from the ground in order to document the general behaviour of the probe and parachute during the descent. The drop sequence shown below is of drop 8 on July 11, 2004, beginning just 9 seconds after release from the plane. The sequence spans a mere 4.5 seconds (the images are 0.14 seconds apart) with one of the parachute lines pulled slightly shorter than the others which caused a conical pendulum motion. However, this sequence clearly shows the parachute-probe spinning clock-wise and then counter clock-wise (follow the folded parachute edge and probe edges through the sequence).

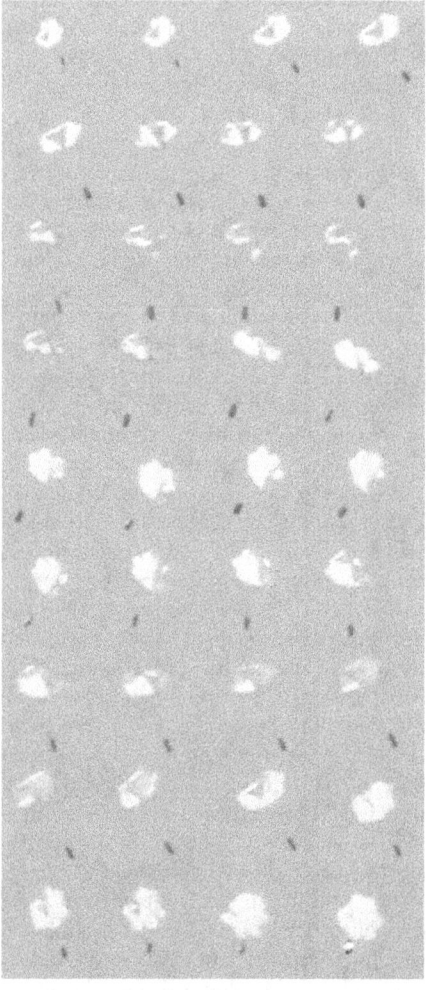

Fig. 5. Ground based imaging of the large, flat parachute during drop 8. One line is tangled causing and edge of the chute to curl over.

6.2 On-board Sensor Data

IMU data recorded for drop 8 are shown in Fig. 6. a,b,c and Fig. 7. a, b, c, respectively.

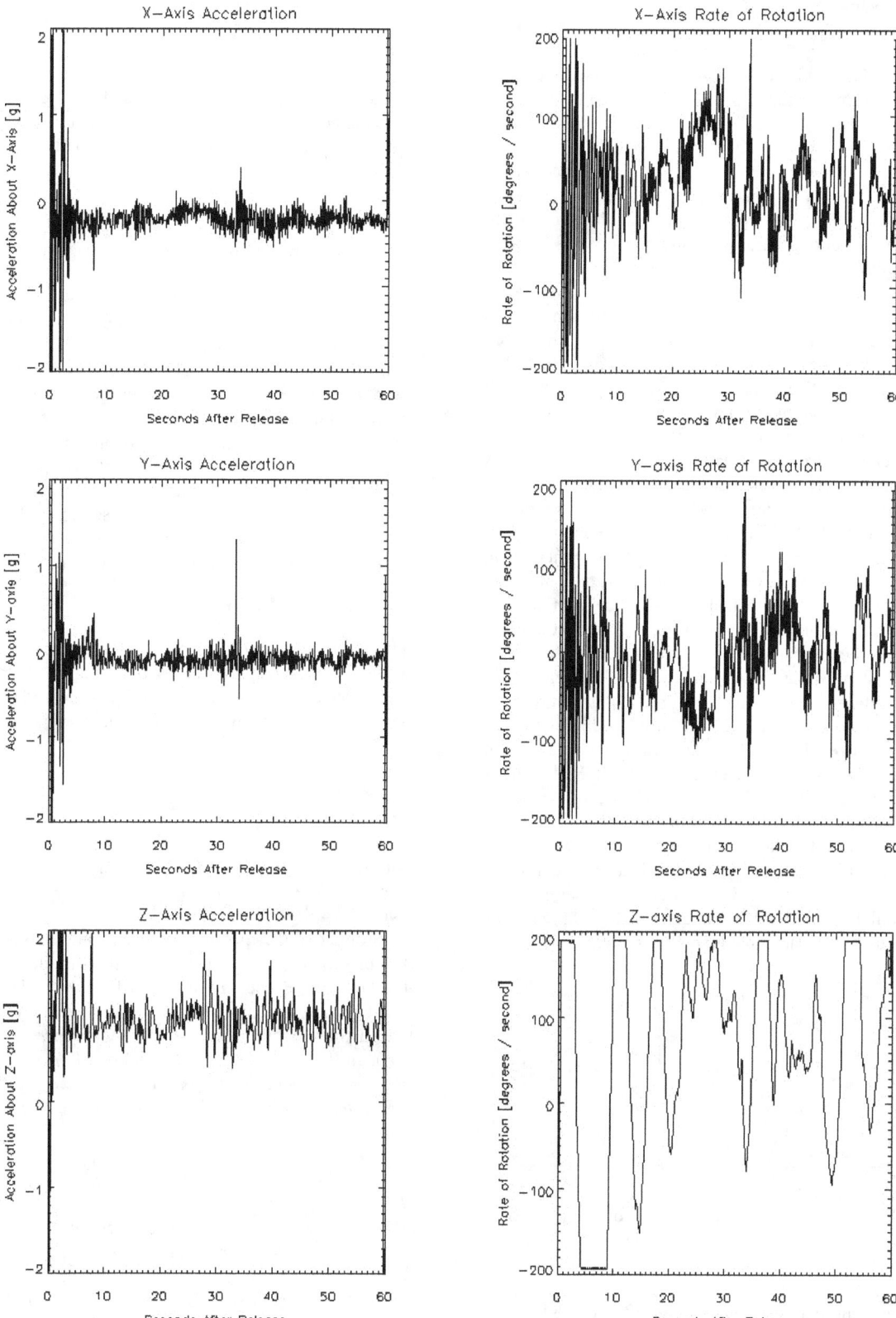

Fig. 6. a, b, c. X-, y-, and z-axis accelerometer results for drop 8 with the large, flat parachute.

Fig. 7.a,b, and c. X-, y-, and z-axis rate of rotation for drop 8 with the large, flat parachute.

The same change in rotation shown by following the parachute-probe camera sequence can be seen in the gyro plot, Fig. 7.c. It is difficult to pinpoint the true time during the drop that probe began to change rotation while looking at the image clips, but each change can be identified and matched with changes in rotation from the gyro plots, even if the timing of the two methods are not exactly in sink. Note that the flat parts of the plot are due to the gyro saturation limit of 200 degrees per second. The video record for drop 8, and others, exhibits periods of conical pendulum motion. It was suspected that this behaviour would be accompanied by persistent, rather than periodic, z-axis acceleration values greater than 1 g. Drop 8, and other drops with the flat parachute, have not confirmed this kind of result, but it may be discovered in future analysis when all instances of coning can be more closely examined. In an attempt to quantify the results from the acceleration and gyro data, a computer program was written in IDL, which calculates the mean, standard deviation, skewness, and kurtosis of 10-second segments of the sensor data collected during the drops. The results are then plotted with symbols and colours representing the general size (small, medium, large) and type of parachute (spherical or flat). We aimed to identify patterns in this data that could link the particular parachutes to specific results, but so far our attempts have been unsuccessful. Irregular flight conditions, such as tangled lines, or uncontrollable and changing weather conditions could be the cause to the irregular results. An additional approach to analysing the drop data is the use of Fast Fourier Transforms (FFT). The same, 10 second segments described above are passed through an FFT function that extracts the frequency of the oscillatory acceleration and magnetometer data. An example of how this function is applied to a segment of oscillating x-axis gyro data from drop 11 with the medium, flat parachute is shown below.

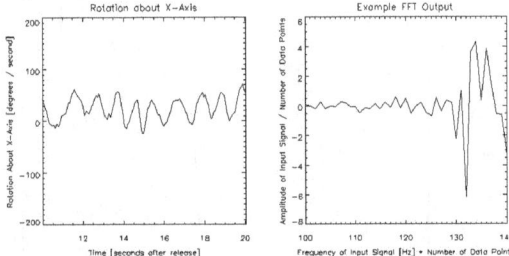

Fig. 8. Example fast Fourier transform application using x-axis gyro data.

The input function shown on the left in Fig. 8 is described with 141 data points and we can approximate the period as 1 second. The FFT command displays a spike at the value corresponding to the (frequency of the input * the number of data points used to describe the input). In this example, a series of spikes are noted around 135, therefore the FFT has identified the input pattern 0.96 Hz frequency of oscillation, or a period of 1.04 seconds. This method is used with gyro and magnetometer data to calculate the swing period of the parachute-probe system.

7. SUMMARY

Small-scale sensor packages and parachutes were airdropped from a small-scale airplane in order to gain familiarity with the descent kinematics of a planetary probe through an atmosphere and to explore interactions of the parachute system with in-situ measurements. Testing and analysis techniques used to explore the parachute-probe system have been reviewed as well as the discussion of future plans. These experiments have obvious visual appeal and are relatively inexpensive to perform. Additionally, the data acquisition equipment is in fact quite easy to assemble. This research has aided the development of a dataset for use in the evaluation of trajectory recovery algorithms and methodologies.

8. ACKNOWLEDGEMENTS

This work was supported by the Cassini project. We acknowledge and appreciate the patient assistance and outstanding flying skills of Keith Brock and the assistance provided by Frank Manning in working with the BasicX micro controller.

9. REFERENCES

1. Seiff A., Mars Atmospheric Winds Indicated by Motion of the Viking Landers During Parachute Descent, *Journal of Geophysical Research*, Vol. 98, 7461-7474, 1993.

2. Stoppato P.F. Lion et al. Stratospheric Balloon Flight Experiment Campaign for the Simulation of the Huygens Probe Mission: Verification of HASA (Huygens Atmospheric Structure Instrument) Performance in Terrestrial Atmosphere, in *Proceedings of the International Workshop of Planetary Probe Atmospheric Entry and Descent Trajectory Analysis and Science, Lisbon, Portugal, October 2003*, ESA SP-544, 303-309.

3. Dooley J., Lorenz R. D., A Miniature Parachute-Probe Dynamics Test-Bed, ESA SP-544, in *Proceedings of the International Workshop of Planetary Probe Atmospheric Entry and Descent Trajectory Analysis and Science, Lisbon, Portugal, October 2003*, ESA SP-544, 267-274.

4. R D Lorenz, Frisbee Black Box, *Nuts and Volts* Vol. 25 No. 2 (February 2004) pp. 52-55.

5. The CMUcam Vision Sensor. http://www-2.cs.cmu.edu/~cmucam/ Last updated January 14th, 2004.

SIMULATION RESULTS OF THE HUYGENS PROBE ENTRY AND DESCENT TRAJECTORY RECONSTRUCTION ALGORITHM

B. Kazeminejad[1], D.H. Atkinson[2], and M. Pérez-Ayúcar[3]

[1]Space Research Institute (IWF), Austrian Academy of Sciences, Schmiedlstr. 6, A-8042 Graz, Austria
[2]Dept. of Electrical and Computer Engineering, University of Idaho, Moscow ID-83844-1023, USA
[3]ESA Research and Scientific Support Department, ESTEC/SCI-SB, Keplerlaan 1, 2200 Noordwijk, The Netherlands

ABSTRACT

Cassini/Huygens is a joint NASA/ESA mission to explore the Saturnian system. The ESA Huygens probe is scheduled to be released from the Cassini spacecraft on December 25, 2004, enter the atmosphere of Titan in January, 2005, and descend to Titan's surface using a sequence of different parachutes. To correctly interpret and correlate results from the probe science experiments and to provide a reference set of data for "ground-truthing" Orbiter remote sensing measurements, it is essential that the probe entry and descent trajectory reconstruction be performed as early as possible in the post-flight data analysis phase. The Huygens Descent Trajectory Working Group (DTWG), a subgroup of the Huygens Science Working Team (HSWT), is responsible for developing a methodology and performing the entry and descent trajectory reconstruction.

This paper provides an outline of the trajectory reconstruction methodology, preliminary probe trajectory retrieval test results using a simulated synthetic Huygens dataset developed by the Huygens Project Scientist Team at ESA/ESTEC, and a discussion of strategies for recovery from possible instrument failure.

Key words: Huygens mission, trajectory reconstruction.

1. INTRODUCTION

1.1. Probe Mission Overview

The Huygens Probe is the ESA-provided element of the joint NASA/ESA/ASI Cassini/Huygens mission to Saturn and Titan (Lebreton and Matson, 2002). Cassini/Huygens was launched on October 15, 1997 and arrived at Saturn on July 1, 2004. Following two orbits of Saturn, the Huygens Probe will be released on December 25, 2004 and will reach Titan on January 14, 2005.

The Huygens probe carries six instruments that will perform scientific measurements of the physical and chemical properties of Titan's atmosphere, measure winds and global temperatures, and investigate energy sources important for the planet's chemistry throughout the descent mission. These instruments are the

- **Aerosol Collector and Pyrolyser (ACP)**: investigation of atmospheric aerosols in cooperation with the GCMS instrument (Israel *et al.*, 2002);

- **Huygens Atmospheric Structure Instrument (HASI)**: spacecraft acceleration measurements during the entry phase, measurement of atmospheric properties (i.e, pressure, temperature, and electric properties) during the descent phase (Fulchignoni *et al.*, 2002)

- **Descent Imager/Spectral Radiometer (DISR)**: optical/IR images and measurement of Solar Zenith Angle (SZA) (Tomasko *et al.*, 2002);

- **Doppler Wind Experiment (DWE)**: measurement of zonal wind speeds during the descent phase (Bird *et al.*, 2002);

- **Gas Chromatograph and Mass Spectrometer (GCMS)**: measurement of atmospheric composition and mole fraction of major atmospheric constituents (Niemann *et al.*, 2002);

- **Surface Science Package (SSP)**: Speed of sound, altitude, and surface properties during the descent phase (Zarnecki *et al.*, 2002);

All instruments will deliver important data containing information about the probe trajectory (and attitude). Huygens will transmit its data to the Cassini Orbiter, targeted to flyby Titan at a periapse distance of 60,000 km, during the mission. The probe data will be recorded by the orbiter's solid state recorders for later transmission to Earth.

1.2. The Probe Entry and Descent Sequence

The Huygens probe entry and descent sequence is schematically shown in Fig. 1. The probe is protected

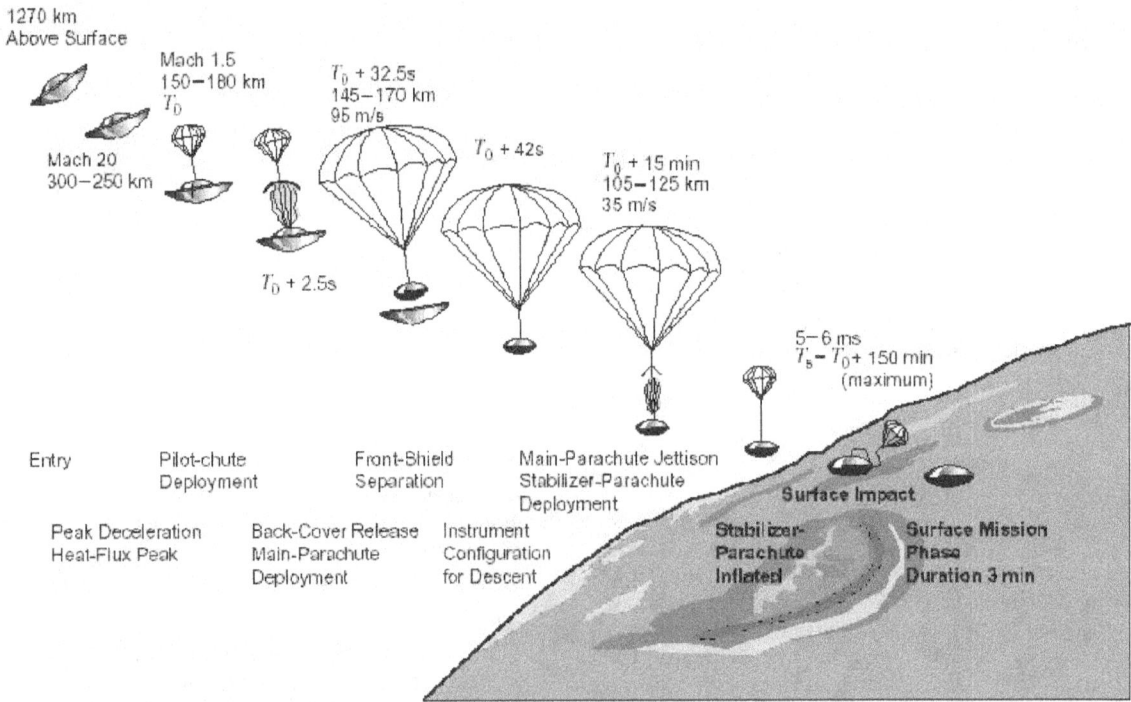

Figure 1. The Huygens probe entry and descent mission sequence;

from the atmospheric induced radiative and convective heat fluxes during entry by a 2.75 meter diameter front heat-shield as it decelerates from about Mach 22.5 to Mach 1.5 in just under five minutes. Approximately 4.45 minutes after entry the probe speed will have decreased to Mach 1.5 and the probe Central Acceleration Sensor Unit (CASU) will measure the deceleration threshold at a time designated as S_0. At S_0 the entry portion of the mission is complete and the descent mission commences.

Approximately 6.375 seconds after S_0 a parachute deployment device is fired through a breakout patch in the aft cover and a 2.59 m disk gap band (DGB) type pilot parachute is deployed. Two and one half seconds later, the probe aft cover is released and the 8.3 meter main DGB parachute is deployed. Nominally this event occurs at Mach 1.5 and an altitude of 160 km. After a 30 second delay (built into the sequence to ensure that the shield is sufficiently far below the probe to avoid possible instrument contamination), the probe speed has dropped to Mach 0.6 and the inlet ports of the probe Gas Chromatograph/Mass Spectrometer and Aerosol Collector and Pyrolyser instruments are opened and the booms of the Huygens Atmospheric Structure Instrument deployed.

The probe will descend beneath the main parachute for 15 minutes, at which time the main parachute is released and a 3.03 meter drogue parachute is deployed to carry the probe to Titan's surface. Throughout the approximately 2.5 hour parachute descent to the surface, Huygens will measure the chemical, meteorological, and dynamical properties of the Titan atmosphere. Probe experiment and housekeeping/engineering data will be transmitted to the orbiter at 8 kbit/s.

2. THE HUYGENS SYNTHETIC DATASET (HSDS)

The reconstruction of the Huygens probe will be done by the Huygens Descent Trajectory Working Group (DTWG) which is a subgroup of the Huygens Science Working Team (HSWT). To perform this task the DTWG has developed a dedicated tool, the DTWG tool, which will be described in more detail in Sec. 3.

In order to test the DTWG tool a simulated synthetic mission dataset (HSDS) was developed by the Project Scienctist Team (PST) at ESA/ESTEC (Pérez-Ayúcar et al., 2004) and was validated by the various probe instrument teams. The file format and content is fully consistent with the interface conventions between the DTWG and the instrument teams and therefore provides a perfect test case for the reconstruction capabilities of the DTWG tool.

The production and validation of the HSDS comprises the following four steps:

1. The definition of an atmosphere profile and a mission scenario (i.e., definition of initial conditions, and various simulation parameters);

2. The simulation of the Huygens probe entry and descent trajectory using the official Huygens 3DOF trajectory simulation software DTAT (Castillo and Sánchez-Nogales, 2004);

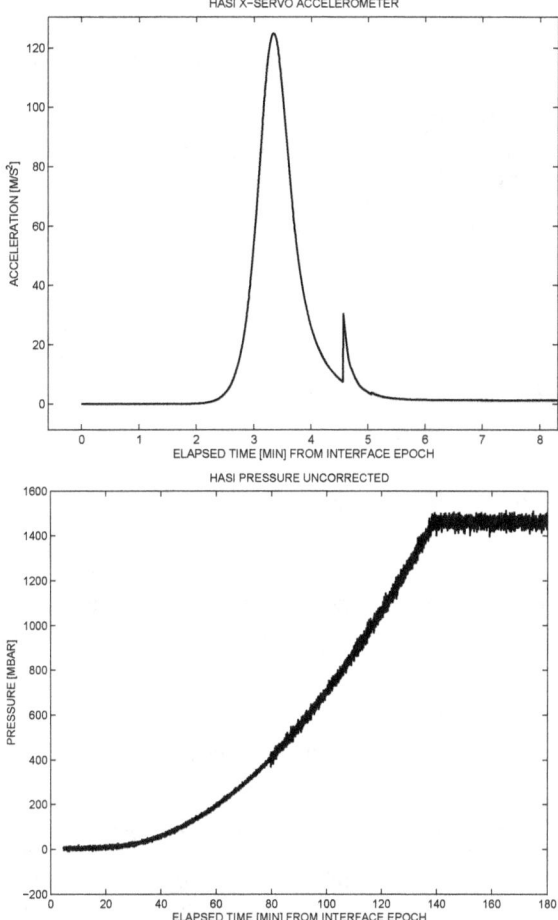
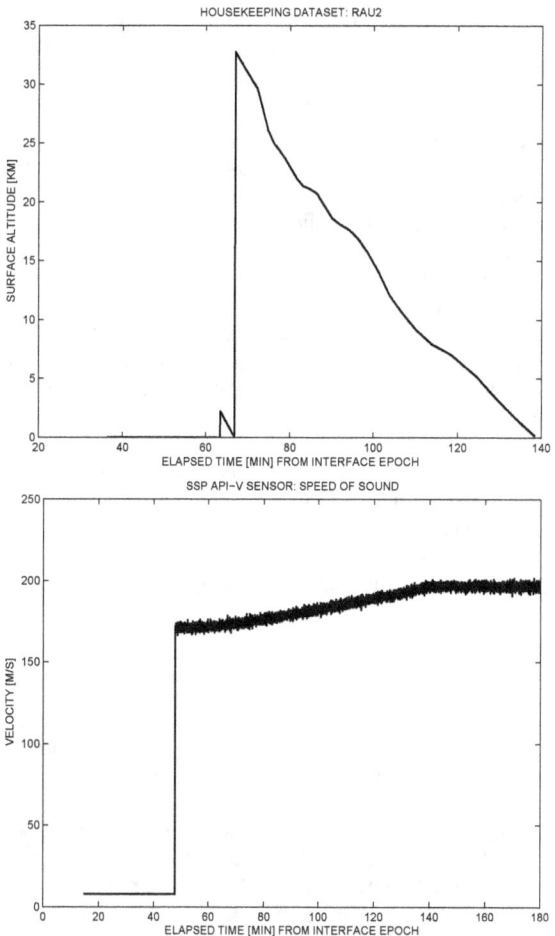

Figure 2. Examples for simulated sensor outputs of the HSDS; Upper panel: HASI X-Servo acceleration, Lower panel: HASI pressure profile;

Figure 3. Examples for simulated sensor outputs of the HSDS; Upper panel: RAU altitude profile, Lower panel: SSP speed of sound measurements;

3. The simulation of sensor outputs on the basis of the simulated trajectory;

4. The validation of the sensor outputs by the various instrument teams (PIs);

There have to date been four deliveries of the HSDS from the PST, with a continuous refinement and implementation of new features in order to better simulate the expected instrument sensor output during the actual mission in January 2005. The latest version of the HSDS (ver. 1.4) comprises the following sensor models

- HASI (3-axis) accelerometer measurements during the entry phase, pressure and temperature (corrected and uncorrected for dynamical effects) during the descent phase;

- SSP speed of sound, altitude acoustic sounder and impact time measurements;

- GCMS mole fraction measurements of the major compounds (i.e., N_2, CH_4, Ar) during the descent phase;

- DWE zonal wind measurements during the descent phase;

- DISR Solar Zenith Angle (SZA) and altitude and descent speed (derived from optical images) during the descent phase;

- Probe housekeeping data comprising engineering accelerometer, and Radar Altimeter Unit (RAU) measurements;

All sensor models were provided[1] with and without simulated prograde zonal winds and both as noise and no-noise version datasets. The four dataset versions together with a file containing the simulated trajectory (which was used for the generation of the sensor models) allow an optimized analysis of the DTWG reconstruction tool performance.

Fig. 2 and 3 show examples of the modelled sensor output from the HSDS.

---
[1] All versions of the HSDS are available online at ftp://ftp.rssd.esa.int/pub/HUYGENS/DTWG_Simulated_Data_Set/.

3. THE DTWG TRAJECTORY RECONSTRUCTION TOOL

The DTWG Trajectory Reconstruction Tool (see also Kazeminejad and Atkinson, 2004) was developed at the Space Research Institute of the Austrian Academy of Sciences in Graz, Austria under contract with the Research and Scientific Support Department of ESA. The purpose of the tool is the reconstruction of the Huygens probe entry and descent trajectory as well as the probe attitude during the entry phase (i.e., the angle-of-attack history). The tool uses the NAIF Spice toolkit and was developed in a "multi-planet" mode, i.e., it can be easily adapted for any other probe mission on any other solar system planet. In the current version (Ver.1.0) the tool is also able to reconstruct the Mars Pathfinder entry and descent trajectory and corresponding results are shown in Kazeminejad and Atkinson (2004).

The complete DTWG tool reconstruction procedure consists of the following phases:

1. **Entry Phase**: this phase comprises the reconstruction of the probe altitude and descent speed profile during the entry phase (i.e., from the interface altitude[2] down to the initiation of the parachute sequence at ~ 160 km), the reconstruction of the probe attitude (i.e., the angle-of-attack history), and the reconstruction of the upper atmosphere physical properties (i.e., density, pressure, and temperature) from the measured probe (science and/or engineering) accelerometer data;

2. **Descent Phase**: this phase comprises the reconstruction of the probe altitude and descent speed (from measured atmospheric temperature, pressure, speed of sound, atmospheric composition), the probe longitude drift (from zonal wind measurements of the Doppler Wind Experiment, and the measured Solar Zenith Angle of the DISR instrument), and the derived surface elevation (topography) with respect to the reference surface (from RAU altitude measurements) in the final portion of the descent ($\sim$30 km down to surface impact). The longitude drift reconstruction from the measured SZA is described in Allison et al. (2004).

3. **Trajectory Fitting Phase**: this phase allows an adjustment of the initial state vector at the interface altitude in order to achieve an optimum "match" of entry and descent phase by adjusting the probe initial conditions at interface altitude using a classical weighted linear least-squares fitting algorithm.

[2]The interface altitude is defined as 1270 km above Titan's reference surface and represents the official NASA/ESA handoff point where the probe initial state vector and its uncertainties (the covariance matrix) will be provided by the Cassini Navigation team to ESA.

x [km]	-1.312458638E+02
y [km]	-3.824933072E+03
z [km]	-3.697321588E+02
vx [km/s]	-2.346112519E+00
vy [km/s]	5.539336275E+00
vz [km/s]	4.588600223E-01
Titan GM [km^3/s^2]	8.978200000E+03
Saturn GM [km^3/s^2]	3.794062976E+07
Sun GM [km^3/s^2]	132712440041.940

Table 1. Huygens probe state vector at interface epoch UTC JAN 14, 2005 08:58:55.816 (inertial Titan centered EME2000 coordinate system) and primary and perturbing body gravitational constants;

4. SYNTHETIC DATASET RECONSTRUCTION RESULTS

The reconstruction results presented in this paper are based on the HSDS (V1.4) with prograde wind and no noise.

4.1. The Entry Phase

The entry phase is reconstructed by a numerically integrating the equations of motion which are outlined in detail in Kazeminejad and Atkinson (2004). The combination of the following data was used for the entry phase reconstruction effort:

- The initial conditions and physical constants taken from the *Huygens Event File* in the form of a NAIF Spice text kernel with the main parameters as specified in Table 1.

- The axial and normal accelerations derived from the HASI X-Servo, the Y-Piezo, and the Z-Piezo simulated accelerometer measurements;

- The simulated gravitational field with Titan as the primary body and Saturn and the Sun as two perturbing bodies. No flattening of the primary body was taken into account for this simulation[3].

Fig. 4 (upper and medium panels) show the reconstructed altitude and inertial velocity profiles and their respective residuals for the entry phase. One can see that the DTWG tool was able to reconstruct the descent trajectory very accurately. The lower panels of Fig. 4 show the reconstructed upper atmosphere density and temperature profiles in comparison to the Yelle et al. (1997) minimum, recommended and maximum profiles. One can readily see that an atmosphere model close to the recommended one was used for the generation of the HSDS.

[3]Note that the DTWG reconstruction tool can simulate an axisymmetric gravitational flattening field for the first zonal harmonic coefficient J_2.

4.2. The Descent Phase

The probe descent phase trajectory was reconstructed from the following datasets:

- The pressure and temperature measurements from the HASI instrument in combination with the GCMS measurements of the mole fractions (needed to infer the mean molecular mass of the gas mixture) of the major atmospheric constituents to derive altitude and descent speed;

- Optionally the SSP speed of sound measurement (in the altitude range from ∼46 km down to the surface) in connection with the HASI pressure measurements to derive altitude and descent speed;

- The DWE zonal wind measurements and the DISR Solar Zenith Angle to derive the probe longitude drift;

- The HASI and SSP accelerometer measurements at probe surface impact in order to constrain the probe impact time;

- The two RAU altitude measurements to derive the surface elevation in the final part of the descent phase (∼30 km down to 1 km);

The descent phase reconstruction was done in "reverse" mode, i.e., starting from the probe impact time (with an assumed distance to the planet center of 2575 km) upwards. This constrains the initial altitude error (which increases during the reconstruction process due to the various measurement errors of the input data) to a maximum of ±10 km (maximum estimated surface elevation with respect to the reference surface).

Fig. 5 shows the results of the descent phase reconstruction. The upper and middle panels show the direct comparison and the corresponding residuals for the altitude and the descent speed reconstructions. The lower left panel shows the reconstructed probe longitude drift from measurements of the DWE experiment (i.e., zonal wind speed measurements) and the lower right panel depicts the reconstructed surface elevation from the comparison of RAU-1 measurements with the reconstructed altitude profile from atmospheric measurements. One can see that the DTWG tool accurately reconstructed the descent trajectory.

5. ENTRY/DESCENT PHASE MERGING STRATEGY

As the entry phase and the descent phase will be reconstructed from completely different data sources (i.e., the initial state vector with corresponding uncertainties and the measured accelerations for the entry phase, and the various atmospheric properties and radar measurements for the descent phase) the following three scenarios could be envisaged:

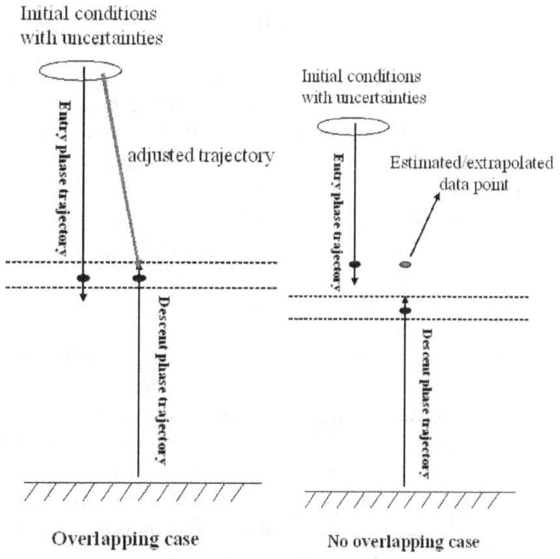

Figure 6. Entry/Descent Phase Merging scenarios; Left: overlapping case; Right: Non overlapping case;

1. The *optimum case* where the reconstructed entry and descent trajectory overlap each other and perfectly fit together. Due to the limited accuracy of the initial state vector and the noise and measurement errors of the various instruments, this case is very unlikely;

2. The *overlapping case* where the two trajectories overlap each other. In other words, for a certain time period altitude and descent speed values are available from both the entry and the descent phase trajectory reconstruction effort (see left panel of Fig. 6);

3. The *non overlapping case* where the two trajectories do not overlap each other. This scenario could happen if the actual state vector is too far away from the estimated one and the integration of the equations of motion would stop too early due to the large systematic errors (see right panel of Fig. 6).

The first case would not need any trajectory merging efforts. However, the second and third one would need to be merged in order to provide one consistent entry and descent trajectory. This merging capability is implemented into the DTWG tool in the form of a weighted linear least squares fitting algorithm, where the calculated measurement values are the altitude and/or descent speed from the entry phase and the "fitting observations" are the corresponding reconstructed values from the descent phase. The adjusting parameters are the six values of the initial state vector in numerical integration process of the equations of motion during the entry phase (the first iteration would be the state vector as delivered by the Cassini Navigation team). The testing of the trajectory merging tool capability is currently ongoing work.

6. INSTRUMENT FAILURE SCENARIOS

Any planetary probe trajectory reconstruction effort bases on a variety of instrument datasets and one needs therefore to investigate various instrument failure scenarios and their impact on the quality of the reconstruction. Part of this exercise is the definition of *critical*, *significant* and *minor* instrument datasets for both the entry and the descent phase. The difference between a critical and a significant dataset is that the lack of a critical dataset would make a trajectory reconstruction process impossible whereas the lack of a significant one would only impact the quality and reliability of the reconstruction result. A minor dataset increases the quality of the reconstruction effort but a still fairly consistent trajectory could be achieved without this input. It should be noted however that even a minor dataset could become significant or even critical if a series of input data would be missing due to a major failure of the probe system.

The entry phase reconstruction is based entirely on the measurement of the probe accelerometer data. Those are provided by the HASI instrument and to some extent (i.e., lower sampling rate and acceleration detection limits) by the probe engineering housekeeping data. The HASI accelerometer measurements are therefore considered a significant dataset for the entry phase.

The descent phase reconstruction is based on the measurement of the altitude dependent atmospheric properties like pressure and temperature. The HASI pressure and temperature measurements therefore represent a critical dataset. Note that alternative measurements which could replace one or both of these significant measurements are only available in certain parts of the descent phase (e.g., SSP speed of sound measurements from $\sim$46 km to the surface, RAU altitude measurements from $\sim$30 km down to the surface, etc.). The SSP impact sensor measurement will provide an important input for the initial epoch of the descent phase reconstruction (done in reverse mode, from the surface upwards), but could in case of failure be replaced by the measurements from the HASI accelerometers and is therefore only significant in case of a HASI accelerometer failure during the entry phase. The RAU altitude and SSP acoustic sounder datasets can be considered as minor but might however be the only reliable dataset (and therefore critical) in case of a complete HASI failure.

7. CONCLUSIONS

The Huygens Descent Trajectory Working Group has developed a dedicated tool for the reconstruction of the Huygens entry and descent trajectory on the basis of the measurements from the 6 scientific instruments and a subset of the probe's engineering housekeeping data. The tool has so far been successfully tested on the Mars Pathfinder Mission data, and a specially designed synthetic dataset that simulates the content and format of all the relevant probe sensors. The reconstruction results for the synthetic dataset are presented and discussed in this paper. The DTWG trajectory reconstruction tool was developed in the framework of a contract between the European Space Agency and the Austrian Academy of Sciences and can therefore be adapted for future planetary probe missions if required by the Agency.

8. ACKNOWLEDGEMENTS

The authors thank the European Space Agency for support of the Descent Trajectory Working Group investigation from the Huygens project.

REFERENCES

ALLISON, M., D. H. ATKINSON, M. K. BIRD, AND M. G. TOMASKO 2004. Titan zonal wind corroboration via the Huygens DISR solar zenith angle measurement. In *ESA SP-544: Planetary Probe Atmospheric Entry and Descent Trajectory Analysis and Science*, pp. 125–+.

BIRD, M. K., R. DUTTA-ROY, M. HEYL, M. ALLISON, S. W. ASMAR, W. M. FOLKNER, R. A. PRESTON, D. H. ATKINSON, P. EDENHOFER, D. PLETTEMEIER, R. WOHLMUTH, L. IESS, AND G. L. TYLER 2002. The Huygens Doppler Wind Experiment - Titan Winds Derived from Probe Radio Frequency Measurements. *Space Science Reviews* **104**, 613–640.

CASTILLO, A., AND M. SÁNCHEZ-NOGALES 2004. Huygens Descent Trajectory Analysis Tool, European Space Agency Technical Note. Technical Report HUY-ESOC-ASW-TN-1002-TOS-OFH, Deimos Space S.L., Madrid, Spain.

FULCHIGNONI, M., F. FERRI, F. ANGRILLI, A. BAR-NUN, M. A. BARUCCI, G. BIANCHINI, W. BORUCKI, M. CORADINI, A. COUSTENIS, P. FALKNER, E. FLAMINI, R. GRARD, M. HAMELIN, A. M. HARRI, G. W. LEPPELMEIER, J. J. LOPEZ-MORENO, J. A. M. MCDONNELL, C. P. MCKAY, F. H. NEUBAUER, A. PEDERSEN, G. PICARDI, V. PIRRONELLO, R. RODRIGO, K. SCHWINGENSCHUH, A. SEIFF, H. SVEDHEM, V. VANZANI, AND J. ZARNECKI 2002. The Characterisation of Titan's Atmospheric Physical Properties by the Huygens Atmospheric Structure Instrument (Hasi). *Space Science Reviews* **104**, 397–434.

ISRAEL, M. C. J.-F. B. G., S. W. H. NIEMANN, W. RIEDLER, M. STELLER, F. RAULIN, AND D. COSCIA 2002. Huygens Probe Aerosol Collector Pyrolyser Experiment. *Space Science Reviews* **104**, 435–466.

KAZEMINEJAD, B., AND D. H. ATKINSON 2004. The ESA Huygens probe entry and descent trajectory reconstruction. In *ESA SP-544: Planetary Probe Atmospheric Entry and Descent Trajectory Analysis and Science*, pp. 137.

LEBRETON, J.-P., AND D. L. MATSON 2002. The Huygens Probe: Science, Payload and Mission Overview. *Space Science Reviews* **104**, 59–100.

NIEMANN, H. B., S. K. ATREYA, S. J. BAUER, K. BIEMANN, B. BLOCK, G. R. CARIGNAN, T. M. DONAHUE, R. L. FROST, D. GAUTIER, J. A. HABERMAN, D. HARPOLD, D. M. HUNTEN, G. ISRAEL, J. I. LUNINE, K. MAUERSBERGER, T. C. OWEN, F. RAULIN, J. E. RICHARDS, AND S. H. WAY 2002. The Gas Chromatograph Mass Spectrometer for the Huygens Probe. *Space Science Reviews* **104**, 553–591.

PÉREZ-AYÚCAR, M., O. WITASSE, J.-P. LEBRETON, B. KAZEMINEJAD, AND D. H. ATKINSON 2004. A simulated dataset of the Huygens mission. In *ESA SP-544: Planetary Probe Atmospheric Entry and Descent Trajectory Analysis and Science*, pp. 343.

TOMASKO, M. G., D. BUCHHAUSER, M. BUSHROE, L. E. DAFOE, L. R. DOOSE, A. EIBL, C. FELLOWS, E. M. FARLANE, G. M. PROUT, M. J. PRINGLE, B. RIZK, C. SEE, P. H. SMITH, AND K. TSETSENEKOS 2002. The Descent Imager/Spectral Radiometer (DISR) Experiment on the Huygens Entry Probe of Titan. *Space Science Reviews* **104**, 469–551.

YELLE, R. V., D. F. STROBELL, E. LELLOUCH, AND D. GAUTIER 1997. The Yelle Titan Atmosphere Engineering Models. In *ESA-SP1177: Huygens Science, Payload and Mission*, pp. 243.

ZARNECKI, J. C., M. R. LEESE, J. R. C. GARRY, N. GHAFOOR, AND B. HATHI 2002. Huygens' Surface Science Package. *Space Science Reviews* **104**, 593–611.

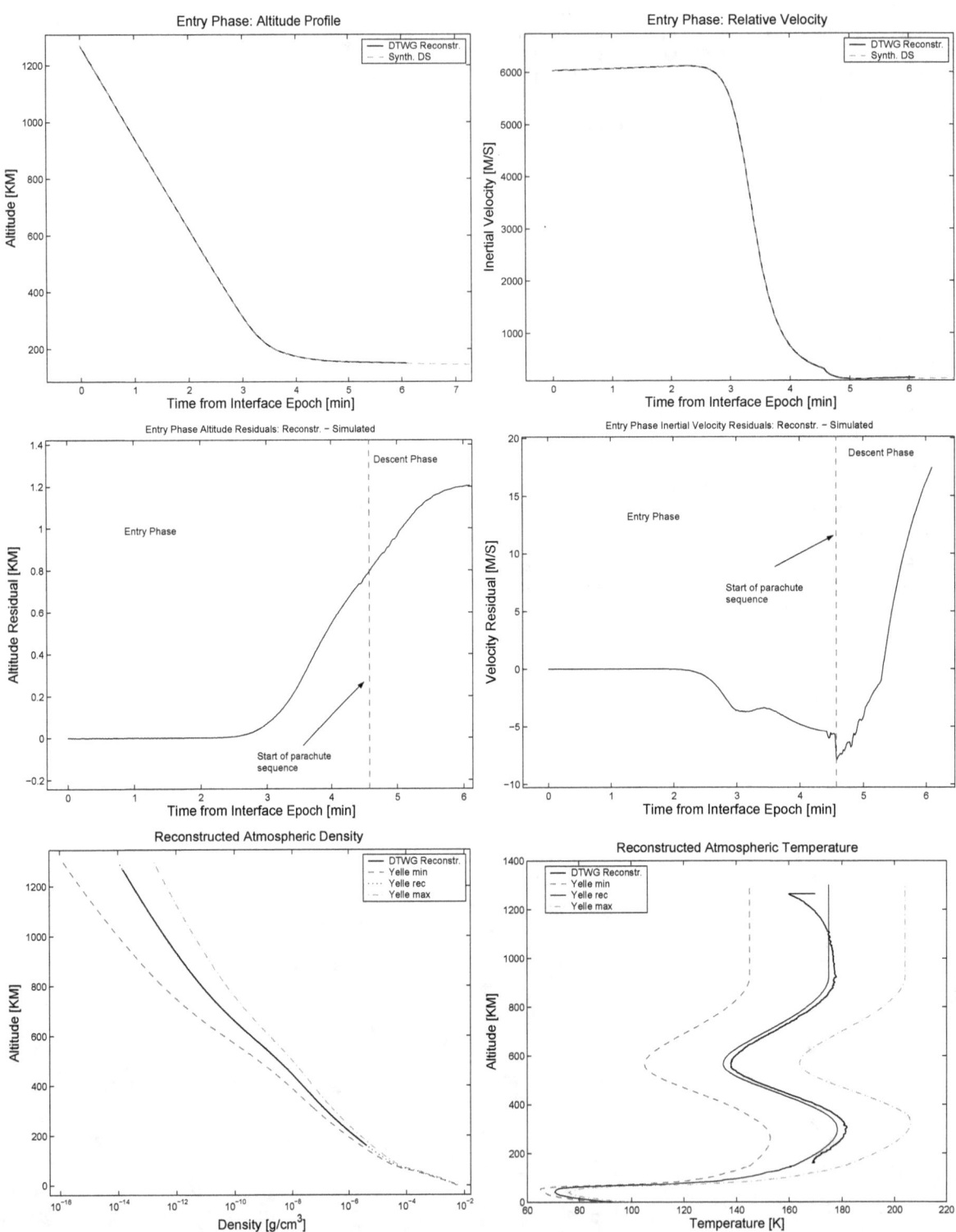

Figure 4. Entry phase trajectory reconstruction: Upper panels: comparison of reconstructed (solid) and simulated (dashed) altitude and inertial velocity profiles; Middle panels: corresponding altitude and velocity residuals (reconstructed - simulated); Lower panels: reconstructed atmospheric density and temperature profiles compared to the Yelle et al. (1997) minimum, recommended, and maximum profiles; one can see that an atmosphere model very similar to the recommended Yelle model was used for the generation of the synthetic dataset. The interface epoch is UTC JAN 14, 2005 08:58:55.816.

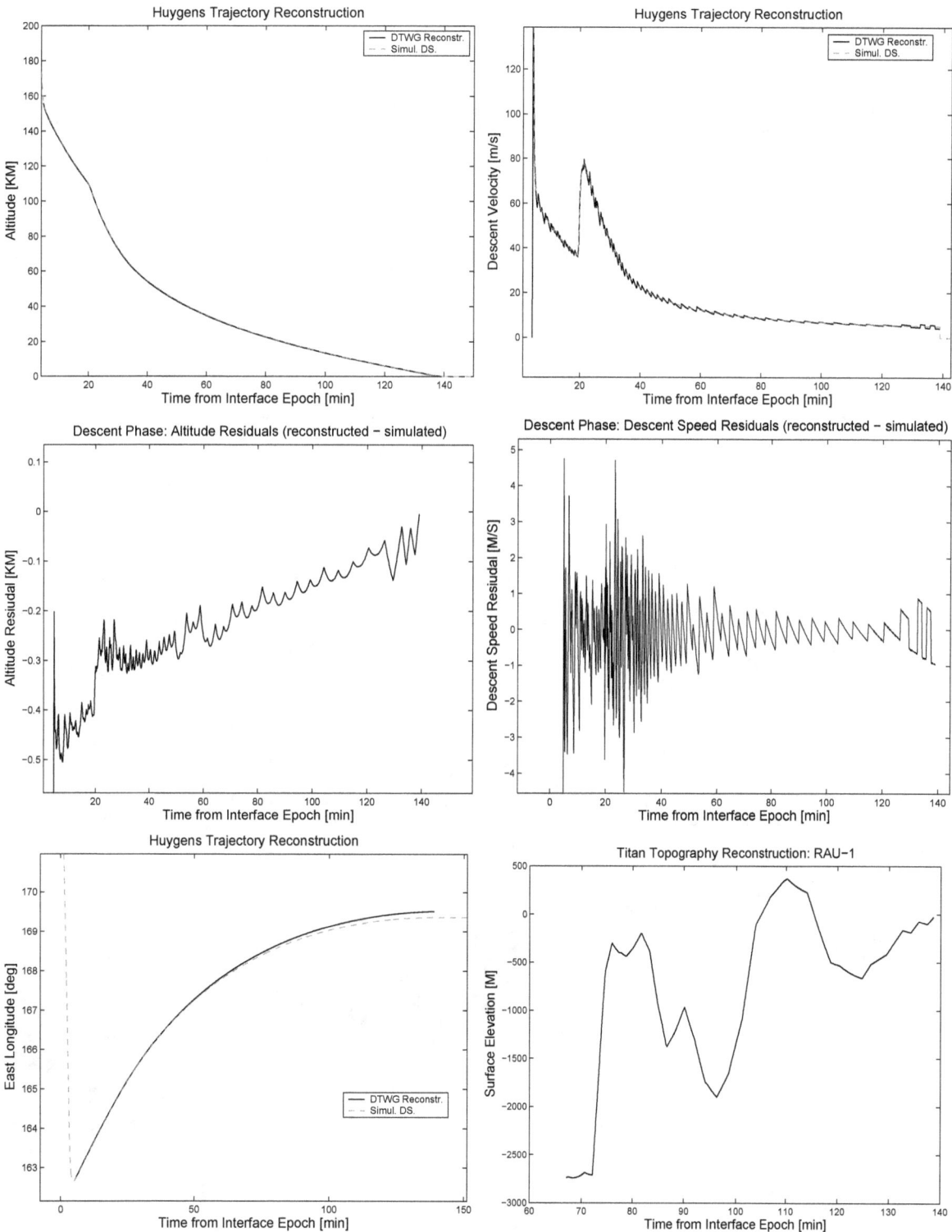

Figure 5. Descent phase trajectory reconstruction: Upper panels: comparison of reconstructed (solid) and simulated (dashed) altitude and descent speed profiles; Middle panels: corresponding altitude and descent speed residuals (reconstructed - simulated); Lower panels: reconstructed probe longitude drift due to zonal winds (left) and surface elevation from RAU-1 data (right). The interface epoch is UTC JAN 14, 2005 08:58:55.816.

STUDY OF SOME PLANETARY ATMOSPHERES FEATURES BY PROBE ENTRY AND DESCENT SIMULATIONS

P. J. S. Gil[1] and P. M. B. Rosa[2]

[1]*Instituto Superior Técnico, Secção de Mecânica Aeroespacial, Av. Rovisco Pais, 1049-001 Lisboa, Portugal.*
Email: p.gil@dem.ist.utl.pt
[2]*Instituto Superior Técnico, Secção de Mecânica Aeroespacial, Av. Rovisco Pais, 1049-001 Lisboa, Portugal.*
Email: prbrosa@mail.telepac.pt

ABSTRACT

Characterization of planetary atmospheres is analyzed by its effects in the entry and descent trajectories of probes. Emphasis is on the most important variables that characterize atmospheres e.g. density profile with altitude. Probe trajectories are numerically determined with EN-TRAP, a developing multi-purpose computational tool for entry and descent trajectory simulations capable of taking into account many features and perturbations. Real data from Mars Pathfinder mission is used. The goal is to be able to determine more accurately the atmosphere structure by observing real trajectories and what changes are to expect in probe descent trajectories if atmospheres have different properties than the ones assumed initially.

1. INTRODUCTION

Prediction and reconstruction of entry and descent trajectories of probes in planetary atmospheres is a difficult and important task as the success of a mission can depend on the correct assessment of the real conditions that probes will run into. Trajectory prediction and reconstruction have to rely on approximations that are often based on assumed knowledge of the eventual answers it tries to attain [1]. It is very important to check consistency of results and desirable to have diversity of reconstruction tools with eventually different approaches to cope with all the assumptions and phenomena involved.

ENTRAP — ENtry TRajectories in Atmospheres of Planets, is a developing software tool for precise orbit prediction and entry and descent trajectory prediction and reconstruction, capable of taking into account all kind of parameters [2]. Once it is fully developed it will be easily applied to all kind of trajectory determination in different planets and situations, allowing changes in assumptions and running all kind of tests effortlessly. Presently EN-TRAP is already in a working state although the desired flexibility and ease of use is still not achieved.

One of the aspects determining probe real trajectory is the atmospheric structure especially the density profile with altitude. Usually, an iterated procedure is used during the trajectory reconstruction from initial conditions of probe entry and previous knowledge of the atmospheric and aerodynamic properties to obtain all the atmospheric and aerodynamic information at the time of the event including the density profile.

Planetary atmosphere models can be constructed in a similar way of what is done for our planet. In the case of Earth much information is available and there are very sophisticated and complete density models [3]. For other solar system planets and satellites information is scarce and simpler models are used as a reflection of our lack of knowledge. They should however be adequate to model probe trajectories since uncertainty in other important parameters such as angle of attack or drag coefficient C_D is relatively high and more precise values can only be obtained during the trajectory reconstruction process.

When predicting an entry and descent trajectory not only an adequate atmosphere model should be considered but also how the probe will behave if atmosphere conditions are different than previously assumed since these are not in general well known. This kind of study can possibly estimate the limits of possible variations induced on the trajectory by the atmosphere local conditions and applied in mission design to foreseen undesirable situations.

In this work Mars Pathfinder (MPF) is used to assess the influence of the atmosphere density profiles used in studying probe entry and descent trajectories. From MPF data simple density profile models are derived. A comparison of simulated MPF trajectories using these models and some variations of them is performed to evaluate the dependency of some trajectory parameters regarding the atmospheric density profile with altitude. Inducing known changes in the atmosphere model parameters allow studying its effect on the simulated trajectories of

Table 1. *Mars Pathfinder entry characteristics from Spencer et al. (1999).*

Entry characteristic	Mars Pathfinder
V_e, inertial, km/s	7.264
V_e, relative, km/s	7.479 (retrograde)
Radial distance, km	3522.2
Inertial flight path angle	-14.06
Entry mass, kg	585.3
S, m^2	5.526
Angle of attack α, deg	0[a]
C_D	1.7[b]
L/D	0[a]
Guidance and control system	Spin stabilized

[a] Nominal.
[b] Nominal, for continuum flow.

probes towards a better understanding of by what extent those changes affect probe descent. This work also contributes to further test and develop our reconstruction tool with a real example.

Mars Pathfinder is a good test case for developing reconstruction tools [1] and conduct this kind of study since all information needed is available and much work has been developed that can be used for comparison (see section 2).

2. SIMULATING PATHFINDER'S ENTRY AND DESCENT

2.1. different Trajectory Reconstructions and Initial Conditions

Pathfinder entered the Martian atmosphere directly from interplanetary transfer. Direct entry led to a high entry speed. During Pathfinder's entry, descent and landing (EDL) the angle of attack between its symmetry axis and the direction of its velocity relative to the atmosphere was near-zero. The spinning about its symmetry axis was designed to be fast enough that the lift and side forces, occurring if the angle of attack was not precisely zero, were averaged to near-zero by the continuous changing direction. At 9 km altitude a parachute opened and shortly afterwards the front heatshield was released. Latter on the airbags were inflated, retrorockets fired and the lander eventually bounced on the ground more than 15 times and for longer than 1 minute, stopping ~1 km away from the impact site. A more complete description of Pathfinder's EDL can be found in [1].

Various Pathfinder trajectory reconstructions can be found in literature. The work developed by the Pathfinder scientists [4] including the accelerometer measurements and the reconstruction trajectory together with the derived atmosphere properties can be found on the Planetary Data System (PDS) which is available online [5]. An independent reconstruction by the pathfinder engineers [6, 7] was based on accelerometer, altimeter and ground-based measurements generated two more reconstructed trajectories. Both efforts used different initial conditions (i.e. different initial altitude). The reconstructed trajectories are basically identical before parachute opening. Following this event there are some differences that can be attributed to incomplete understanding of Pathfinder's aerodynamics after parachute opening [1]. MPF data was latter analyzed and used as test case in work related to the Huygens probe [8] and Beagle 2 [9] analysis tools. In this work the MPF trajectory is only simulated until parachute opening, avoiding the region where uncertainties are significative and would imply difficulty in comparing results. Table 1 summarizes the MPF entry characteristics and initial conditions considered in our work as provided by Spencer *et al.* [6, 7].

2.2. Aerodynamic Coefficients, Atmospheric Structure and Angle of Attack

Aerodynamics characteristics are necessary to design Pathfinder's trajectory and EDL control algorithms. Qualitative reasoning was used to justify the nominal zero angle of attack. To predict forces, torques and heating rates for a given atmosphere structure and probe speed, attitude aerothermodynamics studies are developed to construct an aerodynamic database in an iteration process with the nominal trajectory. If there is a suggestion during an eventual trajectory reconstruction that conditions are different than expected, additional simulations can be needed to provide relevant aerodynamic characteristics.

The ratio of the drag coefficient C_D to the lift coefficient C_L can be related to the measured ratio of axial and normal accelerations and it is proportional to the angle of attack for a given speed and atmospheric structure. The atmospheric density ρ is related to drag by

$$\rho = -\frac{2m}{C_D A} \times \frac{a_t}{v_R^2} \qquad (1)$$

where m is the probe mass, that changes along the trajectory due to the heat shield ablation, A is the probe reference area, C_D is the appropriate drag coefficient for the angle of attack and atmospheric density, temperature and composition at each instant, a_t is the acceleration along the flight path and v_R is the relative speed to the atmosphere. Atmospheric pressure is related to atmospheric density by the equation of hydrostatic equilibrium and atmospheric temperature can be obtained from the equation of state for a known atmospheric composition. An iterative procedure is then used to reconstruct the trajectory and the real atmospheric structure, and to determine the C_D and angle of attack along the EDL trajectory.

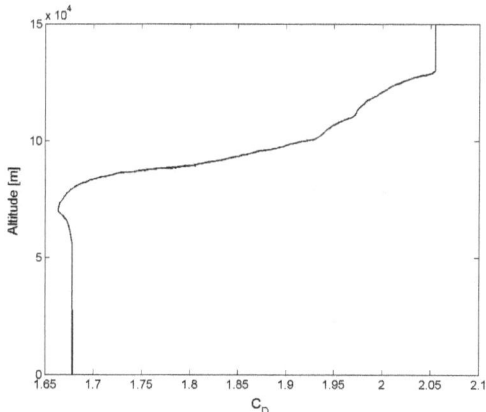

Figure 1. Considered C_D variation with altitude estimated from reconstructed values.

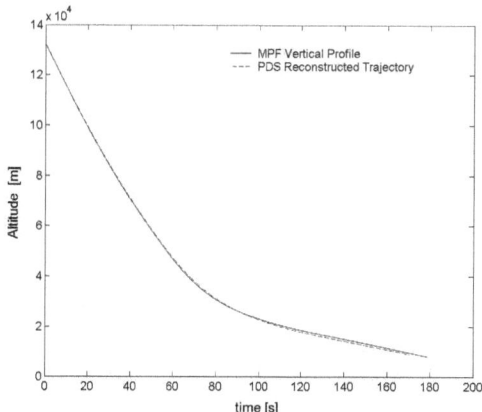

Figure 2. Comparison of the altitude profiles computed by ENTRAP *with the reconstructed from PDS. Altitude profiles are calculated with respect to the radius of the MPF landing site of 3389.72 km [11] (in [7] the altitude is given with respect to the Mars reference ellipsoid.)*

In this work the iterative process was not applied. The reconstructed PDS atmospheric density profile with altitude was adopted and the drag coefficient variation (Fig. 1) was accounted using an approximation of the values determined for the MPF reconstructed trajectory of [8] that used the aerodynamic database from [10]. The C_D values were used as reference values but when the atmospheric density profile is changed they should also vary, which was not considered. This is not a major effect although for precise calculations it should be taken into account. Lift and side forces were not considered, following the design idea that spacecraft's spin would averaged them to near-zero (see discussion of results).

2.3. Reference Atmospheric Density Profile

With all relevant parameters taken from reconstruction efforts, from the aerodynamic coefficients to atmospheric structure, it should be possible to immediately obtain a good approximation of the reconstructed trajectory without any iteration. Results should only be limited by the additional approximation of zero angle of attack. This was used to test our reconstruction tool. Comparison of the MPF vertical profile computed by ENTRAP with the reconstructed from PDS is shown in Fig. 2. They are in good agreement, with altitude residuals of less than 1 km justifying the zero angle of attack approximation. Differences are of the same order of magnitude of the found between other reconstruction efforts [8, 9] and seems to confirm the suggestion that reasonable results can be obtain using only simple aerodynamic information.

The MPF density profile with altitude from PDS was used as reference for comparison and a base to construct simple density profile models and simulating different atmospheric conditions to assess its influence in the trajectory.

3. ATMOSPHERIC DENSITY PROFILE VARIATIONS

As already indicated some variations in the atmospheric density profiles will be considered without changing the determined variation with altitude of the drag coefficient and without considering any lift. Some experimental simulations performed with different values of these parameters (not shown) suggest that differences in the results are less important than variations in density and those found between independent reconstructed trajectories. This confirms the presumption that the drag coefficient changes slowly (logarithmically) with atmospheric density [1]. Thus, MPF trajectories simulated in different atmospheric density profiles used the same determined parameters of the reference trajectory — the one obtained with the reference atmospheric density profile from PDS.

3.1. Case I: Simple Density Profile Models

To assess the influence of considering simple models for the atmospheric density profile two different models were developed: from the MPF vertical profile of the atmospheric density (the reference model) a simple exponential (one layer) and a three-layer exponential density profiles are obtained. In each layer of a model density ρ is determined by

$$\rho = \rho_{0i} e^{-\frac{h-h_i}{H_i}} \quad (2)$$

where ρ_{0i}, h_i and H_i are respectively density at the base of the layer, altitude of the base and the layer scale height. The exponential model is obtained from a simple exponential regression and similarly for the three-layer model but considering three different exponential regressions in different segments adjusted in the best way. Boundaries between layers in the three

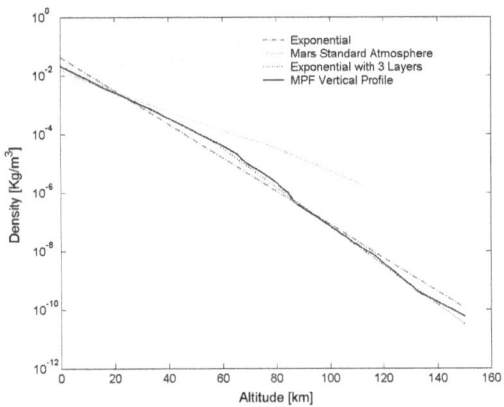

Figure 3. Comparison of simple models with the MPF profile of density with altitude (Case I).

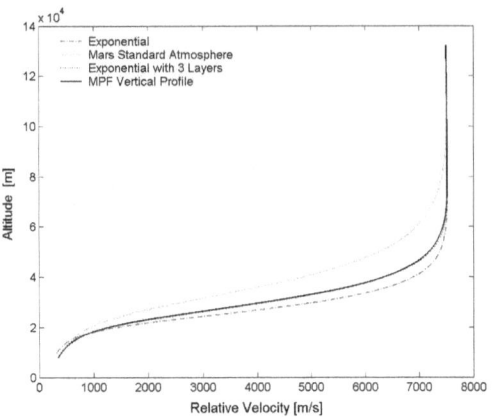

Figure 5. Comparison of relative speed with altitude for simple density profile models (Case I).

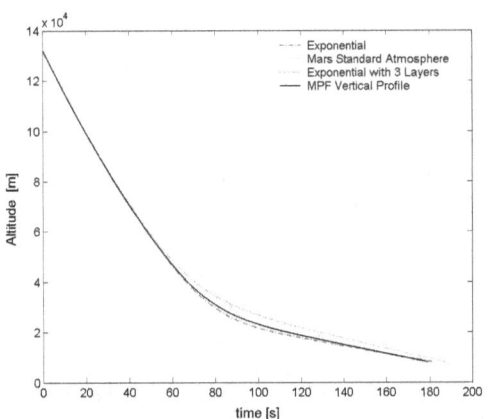

Figure 4. Comparison of altitude profiles for simple density profile models (Case I).

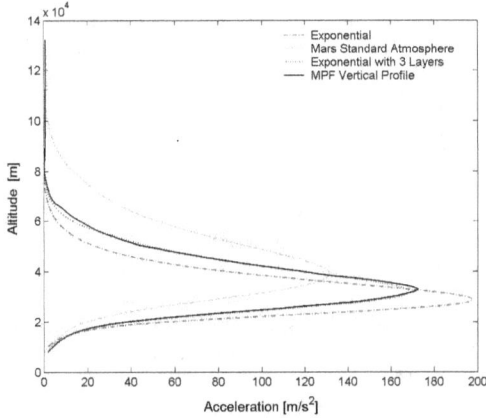

Figure 6. Comparison of acceleration with altitude for simple density profile models (Case I).

layer model are at about 20 km and 54 km altitude. A third model was considered for comparison: the Mars Standard Atmosphere (that can be found for example in http://www.grc.nasa.gov/WWW/K-12/airplane/atmosmre.html) extrapolated to much higher altitudes. Density profiles with altitude are shown in Fig. 3.

Altitude profiles for all the considered density profiles can be compared in Fig. 4. It can be seen that the three-layer exponential model is very close to the MPF profile while differences to others are significant although not large. This behavior is more pronounced in relative speed with altitude (Fig. 5) and in acceleration with altitude (Fig. 6). Aerodynamic heating (not shown) varies in a similar way as acceleration with altitude as expected. Differences in latitude and especially in longitude (also not shown) are also noticeable.

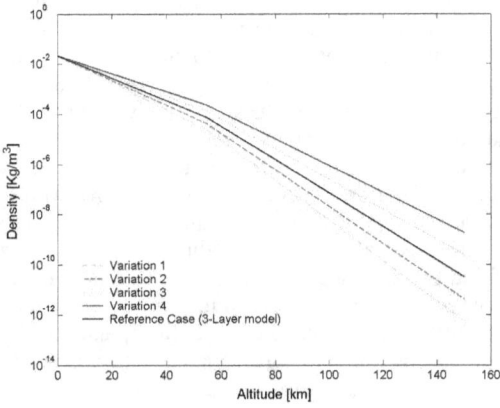

Figure 7. Case II: Comparison of the effect of varying the scale height by $\pm 10\%$ and $\pm 20\%$ in a three-layer density profile model. Variations 1 to 4 corresponds to increasing values of the scale heights.

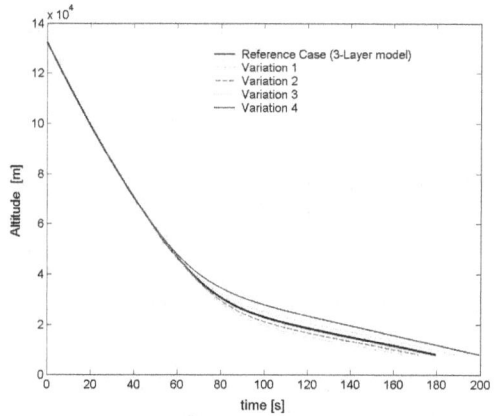

Figure 8. Comparison of altitude profiles for Case II.

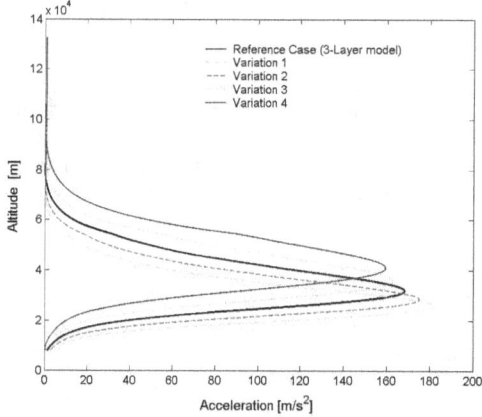

Figure 10. Comparison of acceleration with altitude for Case II.

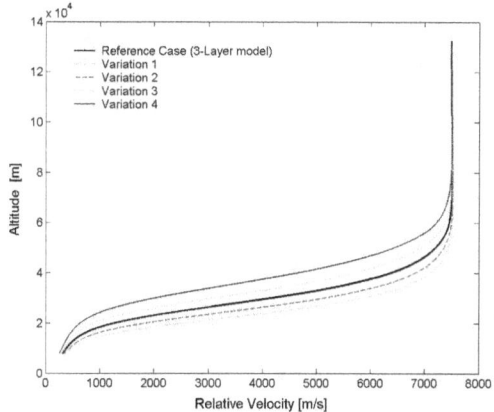

Figure 9. Comparison of relative speed with altitude for Case II.

3.2. Case II: Density Profile Model Variations

In the second set of simulations the goal was to emulate the solar cycle expansion and retraction effect in the atmosphere. The three-layer density profile model was used as reference and for simplicity the solar cycle effect was simulated by varying all scale heights H_i of the reference model in Eq. 2 by $\pm 10\%$ and $\pm 20\%$ (Fig. 7). Variations 1 to 4 of case II corresponds to increasing values of the scale heights.

As in Case I, differences in the altitude profiles (Fig. 8) are not too large but are significant and they are more compelling when relative speed and acceleration with altitude are observed (Fig. 9 and Fig. 10).

3.3. Discussion of Results

The three layer exponential model presents very small differences to the MPF profile from PDS. This is reflected in similar values for problem parameters such as acceleration and relative speed. Use of the three layer profile to model the atmospheric density profile seems to be acceptable but a precise evaluation should take into account the variation in the aerodynamic coefficients. The other models considered in Case I are not acceptable since they present huge differences in the evaluated parameters.

Although the induced variations in the scale heights in Case II were considerable, implying large variations in the evaluated parameters, results present a regularity that seems to indicate a smooth dependency on scale height.

The no lift approximation should not be a problem within the approximations used to determine the drag coefficient. It can have a positive side of separating different problems and simplifying the analysis. Since the MPF was spin stabilized, small differences found between the PDS reconstructed trajectory and the determined by EN-TRAP seem to be consistent with the possible ones being originated from small lift components induced by probe spin and not exactly zero angle of attack. This question should be further examined.

4. CONCLUSIONS AND FUTURE WORK

The three layer model for the density profile with altitude is adequate for trajectory simulations. Differences obtained are of the same order of magnitude of differences between independent reconstruction efforts (even when using relatively limited aerodynamic information). This result reinforces similar results from [9].

Density profile variations have important consequences in some of the problem parameters such as maximum acceleration; landing site can be also affected although probably less. Variations of the problem parameters with changing density profile seem to be smooth which should probably be expected because of the slow variation of some aerodynamic parameters with density.

The work developed was advantageous to confirm validation of the ENTRAP trajectory simulation tool.

This is a work in progress. Much more results can be easily obtained, from different density profiles to changing aerodynamic coefficients. One technical limitation highlighted during this work was the impossibility of applying the iterative process during simulations. This question should be addressed in the future.

Future work should point to assess dependency on the aerodynamic information and the related uncertainty. Wind can possibly have important consequences in the trajectory and to address this question more studies regarding the relations between lift, angle of attack and other aerodynamic information should be developed.

ACKNOWLEDGMENTS

This work was partially supported by Fundação Calouste Gulbenkian.

REFERENCES

[1] Withers, P., Towner, M., Hathi, B., and Zarnecki, J. Review of the trajectory and atmospheric structure reconstruction for Mars Pathfinder. In *ESA SP-544: Planetary Probe Atmospheric Entry and Descent Trajectory Analysis and Science*, pages 163–174, February 2004.

[2] Gil, P. J. S., Antunes, C. M. C., and Pedro, H. T. C. A software package for studies on spacecraft entry in planetary atmospheres. In *ESA SP-544: Planetary Probe Atmospheric Entry and Descent Trajectory Analysis and Science*, pages 335–338, February 2004.

[3] Montenbruck, O. and Gill, E. *Satellite Orbits: Models, Methods, and Applications*. Springer, 2000.

[4] Magalhães, J. A., Schofield, J. T., and Seiff, A. Results of the Mars Pathfinder atmospheric structure investigation. *J. Geophys. Res.*, 104:8943–8956, April 1999.

[5] Planetary Data System (PDS). http://atmos.nmsu.edu/PDS/data/-mpam.0001/.

[6] Spencer, D. A., Blanchard, R. C., Thurmann, S. W., Braun, R. D., Peng, C.-Y., and Kallemeyn, P. H. Mars pathfinder atmospheric entry reconstruction. Technical Report AAS 98-146, NASA, 1998.

[7] Spencer, D. A., Blanchard, R. C., Braun, R. D., Kallemeyn, P. H., and Thurman, S. W. Mars pathfinder entry, descent, and landing reconstruction. *J. Spacecraft and Rockets*, 36(3):357–366, 1999.

[8] Kazeminejad, B. and Atkinson, D. H. The ESA Huygens probe entry and descent trajectory reconstruction. In *ESA SP-544: Planetary Probe Atmospheric Entry and Descent Trajectory Analysis and Science*, pages 137–149, February 2004.

[9] Withers, P., Towner, M. C., Hathi, B., and Zarnecki, J. C. Analysis of entry accelerometer data: A case study of Mars Pathfinder. *Planet. Space Sci.*, 51:541–561, August 2003.

[10] Moss, J. N., Blanchard, R. C., Wilmoth, R. G., and Braun, R. D. Mars pathfinder rarefied aerodynamics: Computations and measurements. *J. Spacecraft and Rockets*, 36(3):330–339, 1999.

[11] Golombek, M. P., Cook, R. A., Economou, T., Folkner, W. M., Haldemann, A. F. C., Kallemeyn, P. H., Knudsen, J. M., Manning, R. M., Moore, H. J., Parker, T. J., Rieder, R., Schofield, J. T., Smith, P. H., and Vaughan, R. M. Overview of the Mars Pathfinder Mission and Assessment of Landing Site Predictions. *Science*, 278:1743–1748, December 1997.

Venus and Special Topics

LAVOISIER : A LOW ALTITUDE BALLOON NETWORK FOR PROBING THE DEEP ATMOSPHERE AND SURFACE OF VENUS

E. Chassefière[1], J.J. Berthelier[2], J.-L. Bertaux[1], E. Quémerais[1], J.-P. Pommereau[1], P. Rannou[1], F. Raulin[3], P. Coll[3], D. Coscia[3], A. Jambon[4], P. Sarda[5], J.C. Sabroux[6], G. Vitter[7], A. Le Pichon[8], B. Landeau[8], P. Lognonné[9], Y. Cohen[9], S. Vergniole[9], G. Hulot[9], M. Mandea[9], J.-F. Pineau[10], B. Bézard[11], H.U. Keller[12], D. Titov[12], D. Breuer[13], K. Szego[14], Cs. Ferencz[15], M. Roos-Serote[16], O. Korablev[17], V. Linkin[17], R. Rodrigo[18], F.W. Taylor[19], A.-M. Harri[20]

[1] Service d'Aéronomie, Pôle de Planétologie of Institut Pierre Simon Laplace (CNRS/ UPMC), Service d'Aéronomie, Université P & M Curie, Aile 45-46, 4ème étage, Boîte 102, 4 place Jussieu 75252 Paris Cedex 05, France, e-mail : eric.chassefiere@aero.jussieu.fr
[2] Centre d'Etude des Environnements Terrestres et Planétaires, Pôle de Planétologie (CETP/IPSL), France
[3] Laboratoire Interuniversitaire des Systèmes Atmosphériques (LISA, Univ. Paris 11 & 7), France
[4] Laboratoire MAGIE (Univ. Paris 6), France
[5] Groupe de Géochimie des Gaz Rares (Univ. Paris 11), France
[6] Institut de Radioprotection et de Sûreté Nucléaire (IRSN), France
[7] Laboratoire d'Electrochimie et de Physicochimie des Matériaux et des Interfaces (LEPMI), France
[8] Commissariat à l'Energie Atomique (CEA), France
[9] Institut de Physique du Globe de Paris (IPGP), France
[10] Albedo Technologies, St Sylvestre, France
[11] Observatoire de Paris, section de Meudon (LESIA/OP), France
[12] Max Planck Institut für Aeronomy (MPAe), Germany
[13] Institute für Planetology (Univ. Muenster), Germany
[14] Hungarian Academy of Science (KFKI), Hungary
[15] Space Research Group, Department of Geophysics (Univ. Eotvos), Hungary
[16] Observatorio Astronomico de Lisboa (Univ. Lisbon), Portugal
[17] Space Research Institute (IKI), Russia
[18] Instituto de Astrofisica de Andalucia (CSIC), Spain
[19] University of Oxford, United Kingdom
[20] Finnish Meteorological Institute (FMI), Finland

ABSTRACT

The in-situ exploration of the low atmosphere and surface of Venus is clearly the next step of Venus exploration. Understanding the geochemistry of the low atmosphere, interacting with rocks, and the way the integrated Venus system evolved, under the combined effects of inner planet cooling and intense atmospheric greenhouse, is a major challenge of modern planetology. Due to the dense atmosphere (95 bars at the surface), balloon platforms offer an interesting means to transport and land in-situ measurement instruments. Due to the large Archimede force, a 2 cubic meter He-pressurized balloon floating at 10 km altitude may carry up to 60 kg of payload. LAVOISIER is a project submitted to ESA in 2000 [1], in the follow up and spirit of the balloon deployed at cloud level by the Russian Vega mission in 1986. It is composed of a descent probe, for detailed noble gas and atmosphere composition analysis, and of a network of 3 balloons for geochemical and geophysical investigations at local, regional and global scales.

1. CONTEXT AND GENERAL DESCRIPTION OF THE MISSION

One of the most promising area in Solar System science is the comparative study of the three terrestrial planets (Venus, Earth, Mars). Why did the three planets evolve in such different ways, from relatively comparable initial states? The small size of Mars, favoring atmospheric escape and allowing a relatively fast cooling of the interior, certainly played a role in making the present Mars so inhospitable. Venus has almost the same size and density as Earth, and was probably initially endowed with similar amounts of volatile material. The absence of water in significant amounts, possibly explained by intense hydrogen escape at early epochs, and of molecular oxygen in the present atmosphere, which requires extremely strong primitive escape and/or massive oxidation of surface material, remains poorly constrained and understood. Because of the massive carbon dioxide atmosphere, there is a large greenhouse effect at the surface of Venus, where the temperature is ≈470°C. The way such a strong greenhouse built up and stabilized is not well understood. How did the atmosphere evolve, under the combined effects of escape and interaction with solid planet? How did the history of tectonism and volcanism influence, or was influenced by, atmospheric greenhouse, mantle convection and core solidification rate? Did the last general resurfacing event, ≈0.5-1 Gyr ago, occur in coincidence with complete freezing of the core, and vanishing of a primitive dynamo? Which processes are presently controlling the thermochemical equilibrium between the surface and the atmosphere, what is the nature of the feedback processes, what are the roles of carbon, sulfur and chlorine cycles? Is there a present volcanic activity?

The Magellan mission, through radar imaging, topography and gravity measurements, provided an enormous amount of information, allowing substantial advances in Venus geodynamics. Our knowledge of Venus geochemistry is less advanced, despite Pioneer Venus and soviet missions, which provided a good preliminary insight into atmospheric composition (and surface composition), but technological advances should allow to go much further. The Venus-Express mission will provide a wide picture of middle atmosphere thermal structure, radiative balance and circulation, and of cloud physics and chemistry. Although Venus Express, as well as the Planet C orbiter of JAXA (launch date : 2008), will also bring information on the radiative balance and composition of the deep atmosphere, this information will not be sufficient for an in-depth characterization of surface-atmosphere interaction and atmospheric geochemical cycles. In-situ measurement of the atmospheric composition, at both global and local scales, and with some vertical and horizontal sampling capability, is a necessary step for that purpose. The advantage of in-situ measurements is that they are more precise, complete, time and space resolved, than remote sensing measurements. There is a strong need for in-situ measurements of the deep atmosphere and surface material of Venus (chemical, isotopic, electromagnetic, acoustic, radioactivity probes, ...) on future missions.

The successful Venus balloon experiment of the Vega mission, near twenty years ago, strongly advocates for a deep atmosphere balloon mission, in the follow up of Venus-Express, allowing to complete atmospheric investigation, and to go further in our understanding of surface and interior states and processes. Why using balloons? In the low atmosphere of Venus, the Archimede force is large. The mass of 1 cubic meter of atmosphere at the ground level is more than 60 kg, which makes a balloon of, let say, 2 cubic meters able to carry up to 100 kg (60 kg at 10 km altitude). The low atmosphere of Venus can be compared, in terms of ambient pressure, to a region of the ocean located at 1 km below the surface. Using balloons for exploring the deep atmosphere of Venus is therefore an attractive means, provided technical challenges like thermal insulation, energy generation, tolerance of materials and components to extreme environmental conditions, can be solved. Balloons, carried by zonal winds, may travel over typically one thousand kilometers per day, even at low altitudes, and can be used as landers by deflation of their envelops. Individual balloons may therefore allow to probe the atmosphere and surface at local and regional scales, with a possible extension to a more global scale by the simultaneous use of several balloons, which can be operated in a networking approach.

A typical geochemical mission, similar to the LAVOISIER project proposed in January 2000 in response to the ESA announcement of opportunity for the flexible missions F2/F3, might be studied. It could consist of a Venus descent probe/ balloon flotilla system oriented toward : (i) measuring with a high accuracy the noble gas elemental and isotopic composition ; (ii) characterizing the main chemical cycles (carbon, sulfur, chlorine) in the low atmosphere, and their interaction with surface minerals ; (iii) helping in improving radiative transfer models through the spectral measurement of infrared fluxes, and measuring the surface temperature ; (iv) constraining small-scale and meso-scale dynamics of the deep atmosphere (winds, atmospheric mixing, meteorological events like storms) ; (v) constraining the morphological and mineralogical characteristics of the surface through visible-near infrared spectro-imaging ; (vi) searching for still hypothetical phenomena like : lightning, volcanic activity, crustal outgassing, seismic activity, remnant magnetization.

The descent probe could be equipped with ultra-sensitive and precise in-situ and optical instruments, allowing to catch a local, but precise and as complete as possible, view of the atmospheric vertical structure and surface characteristics. The descent probe could also provide a precise picture of noble gas elemental and isotopic composition. At the same time, a typical number of 3 pressurized balloons could be deployed at low altitude (10 km), and instruments onboard operated during a few days, or a few weeks, whereas balloons move around the planet, carried by winds. At the end of this period, balloons might be deflated in a controlled way and softly landed at the surface of Venus, allowing to vertically probe the lower atmosphere at different locations, and finally to perform measurements at the surface. A relay orbiter is not necessarily required, but wishable. This orbiter could also be equipped with remote sensing instruments (infra-red, microwave sounders) working synergistically with low altitude platforms.

2. SCIENCE OBJECTIVES : FOCUS ON A FEW KEY SCIENTIFIC TOPICS AND RELATED MEASUREMENT METHODS

General scientific objectives are described in [1] and [2]. A few examples of key measurements, which cannot be done from an orbiter, are given below, together with related scientific objectives. The list hereafter is far from being exhaustive. The purpose of these examples is simply to illustrate the tremendous scientific potential of a balloon mission to the deep Venus atmosphere and surface.

2.1. Noble gas elemental and isotopic composition of the global atmosphere

The primary atmospheres of the terrestrial planets may be either derived by outgassing from the solid interior or accreted from the outside (e.g. addition of cometary material). Secondary processes such as escape, outgassing from the solid planet, radiogenic contribution, solar wind addition, sputtering, chemical conversion (carbonate precipitation, photosynthesis, water decomposition, etc...) may have significantly altered their primary composition. Noble gases are ideal tracers to understand the formation and evolution of the atmosphere of terrestrial planets. Their chemical inertness and the presence of numerous isotopes among which radiogenic ones (e.g. ^{4}He, ^{40}Ar), nucleogenic ones (e.g. ^{21}Ne) and fissiogenic ones ($^{134-136}$Xe) permit to identify a number of possible sources to planetary atmospheres as well as the fingerprint of physical processes previously mentioned. Unlike chemical compounds, they can be found in the atmospheres of the terrestrial planets, in meteorites of various types and in the solar wind. Attempts at modelling the terrestrial atmospheric evolution have confirmed the high potentiality of these elements.

The composition of Venus atmosphere according to the previous measurements by Pioneer and Venera missions is : CO_2 (96.5 %), N_2 (3.5%), He (12 ppm), Ar (70 ppm), Ne (7 ppm), Kr (0.7- 0.05 ppm), Xe (?). The only element for which an isotopic composition is available is Ar but it is worth considering a confirmation because of possible interferences in the previous measurements (e.g. with HCl) especially if an accurate measurement of the ^{36}Ar/^{38}Ar is necessary. Therefore estimates of He, Ne, Kr and Xe isotopes would provide invaluable information necessary for comparative planetology and to constrain models of atmospheric formation and evolution (for both Venus and Earth).

- Helium : the abundance and isotopic composition of helium results from the competition between : (i) outgassing from the solid interior ; (ii) escape from the atmosphere ; (iii) input from solar wind. On Earth, outgassing from the solid interior consists of two well identified fluxes : one rich in ^{3}He (primordial) from the mantle and a purely radiogenic one in relation with crustal outgassing. The same components will have to be considered on Venus but with a completely different relative importance: volcanism on Earth in relation with plate tectonics is a major phenomenon. Release from the crust is a rapid process because of erosion. On Venus the much higher temperature should promote escape from deeper crustal layers but how does erosion play a role? Input from the solar wind is limited on Earth because of the magnetic field. The absence of a significant field on Venus will have an influence on the abundance and isotopic composition of He in Venus atmosphere.

- Neon : in the terrestrial atmosphere, Ne isotopes are intermediate between a planetary component (chondritic) and a solar component. The question is whether this results from mixing of these two components (mantle Ne is solar) or more simply, from mass fractionation as a result of massive escape from the atmosphere in its early stages. Mesuring Ne isotopes in the atmosphere of Venus will greatly help constraining such problems.

- Argon : ^{40}Ar is the radiogenic product of ^{40}K. On Earth, it is a major component of the atmosphere (1%), testifying of the importance of outgassing. On Venus, ^{40}Ar is less abundant than on Earth while ^{36}Ar is far more abundant, illustrating a drastic difference between the two planets. Where does ^{36}Ar come from? Its ratio to Ne precludes a significant solar contribution. Its total abundance corresponds to what is recorded in the most gas rich chondrites, a very surprising result.

- Krypton and Xenon : Krypton isotopes may provide some information on the escape processes because of the possibly mass fractionated pattern easy to interpret in the absence of radiogenic/ nucleogenic contributions. Its ratio to Ar and Xe would permit to characterize its source in Venus atmosphere. Xenon carries more information than anyone of the other noble gases because of its numerous isotopes including radiogenic/ fissiogenic isotopes from short and long lived parents. Is the terrestrial missing xenon a unique signature of our planet? Does Venus exhibit isotopic anomalies in ^{129}Xe, $^{134-136}$Xe? Are Venus atmospheric Xe isotopes mass fractionated relative to the Earth/Solar references?

It can be concluded that the behavior of planetary atmospheres, past and present, is complex. The noble gases provide much information that may greatly help constraining models of atmospheric evolution of both Earth and Venus, two planets with quite different atmosphere which otherwise look quite similar. Concerning isotopes of light elements : H, C, N and O, they exhibit highly variable isotopic compositions depending on the diverse chondrite classes and planetary objects. They result from : (i) nucleosynthetic processes ; (ii) heterogeneity in the Solar nebula ; (iii) fractionation processes upon accretion ; (iv) differenciation and evolution of planetary objects. On Venus, where the temperature is significantly higher than on the Earth, fractionation is expected to be less significant. From the isotopic composition one therefore expects mostly information dealing with comparative planetology. Under this aspect, oxygen isotopes (^{16}O, ^{17}O, ^{18}O) will be of particular significance as they permit establishing different fractionation lines for the diverse objects of the Solar system. In the future, analyses of solid samples will require an accurate knowledge of this atmospheric reference.

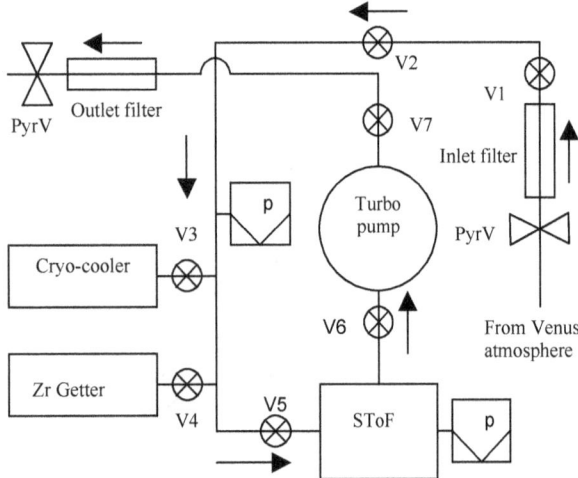

Fig. 1. Schematic view of a system composed of a static time-of-flight (StoF) mass spectrometer and separation/ purification line.

An isotopic analyzer using a time-of-flight mass spectrometer coupled with a gas separation/ purification system might be used (see Fig. 1). Noble gases are first separated from reactive gases through chemical gettering by using a zirconium getter. In a following step, noble gases are separated from each other by cryotrapping on an activated charcoal bed, cooled by a Stirling machine, yielding three separate fractions : [He, Ne], [Ar], [Kr, Xe], which may be analyzed separately in a static mode (without carrier gas). In this way, it is possible to reach high sensitivity, and to remove a few mass interferences (like $^{20}Ne/^{40}Ar$), in such a way to measure isotopic ratios with an accuracy better than 1%. This method has been extensively used in the laboratory since fifty years to analyze noble gases in terrestrial and meteoritic samples, and is presently being developed for Mars in-situ exploration [3]. Such a static ToFMS analyzer, devoted to the measurement of global atmosphere noble gas abundances and isotopic composition, could be operated on board the descent probe, by sampling atmosphere around 60 km altitude, and analyzing noble gases during the 1-2 hrs descent of the probe through the atmosphere.

2.2. Chemical composition of the atmosphere and its spatial/ temporal variability

Sulfur compounds, and their vertical structure, are insufficiently characterized, in particular below 25 km, where the uncertainty on SO_2 is nearly one decade. SO_2 is an important gas of the atmosphere, since it is very reactive. Chemical equilibrium with rocks (calcite : $CaCO_3$) calls for an equilibrium value of 100-130 ppmv of SO_2 [4]. Early measurements of SO_2 with Pioneer Venus Large Probe GC and MS and Venera 12 GC found values of 130-180 ppmv in the range 42-22 km. Later measurements with local UV spectroscopy on Vega 1 and 2 show a similar value at 40 km [5], confirmed by IR spectroscopy from the Earth [6]. However, Vega measurements show a distinct decrease of SO_2 mixing ratio for decreasing altitude, with 20 ppmv at 10 km, the lowest reliable measurement of SO_2.

COS has been measured from Earth around 35 km, S_2O is not detected but could be the "missing" strong UV absorber [7]. H_2S is measured within a factor of 2, but its vertical profile is unknown. H_2O is around 30 ppmv, but still poorly constrained. The CO mixing ratio was found to be in the range from 17±1 ppmv, measured at 12 km altitude (Venera 11/12, [8]), to 23±5 ppmv, measured at 36 km altitude (IR Earth-based observations, [9]). There is a close relationship between CO concentration and oxygen fugacity fO_2 through the thermochemical reaction :

$$CO + O_2 \rightarrow CO_2 + O$$

and oxygen fugacity may therefore be deduced from CO abundance through thermochemical models. This point is discussed in details by Fegley et al (1997).

Dual columns Gas Chromatograph

The dual columns GC is designed to perform individual qualitative and quantitative analyses of the main atmospheric species within a large dynamic range. This instrument requires a direct inlet of gaseous samples into its injection port (sampling step of 10 min., sample size of 1 µl). Each sample is then pushed by a carrier gas (Helium) through two analytical columns operating in parallel and through the detection system. The first column (type MolSieve) allows the separation of the noble gases (Ne, Ar, Kr, Xe), N_2, O_2, CO and CH_4. The second column (type Silica-PLOT or PLOT U polymer) allows the separation of SO_2, CO_2, COS, H_2S, H_2O, NH_3 and C_2 hydrocarbons.

The instrument is composed mainly by 4 sub-systems : (i) gas supply sub-system : it includes all the parts used for the injection and elution of a sample. These parts are mainly the injector, the on/off micro-valves, the carrier gas tank, the pressure regulator, the filters and the pipes ; (ii) The chromatographic columns : the two columns are metallic tubes of 10-20 m length, 0.18-0.25 mm inner diameter coated, on their inner wall, by a thin layer of stationary phase (few microns of respectively MolSieve and Silica or U phases) ; (iii) The detection sub-system : universal micro-detectors are used at each column end. A first stage of nano-TCD (Thermal Conductivity Detector) allows a non-destructive detection of all the eluted species in a range from few percent to few ppm. A second stage of miniaturised HID (Helium Ionisation Detector) could be added in order to detect species abundances in the range 100 ppm - 1 ppb ; (iv) The electronic acquisition and command sub-system.

Most of the hardware components listed above (helium tank, valves, columns, nano-TCD, electronic, ..) have already been qualified for past and present similar experiments (ACP and GCMS experiments on Cassini-Huygens mission, COSAC experiment on Rosetta cometary mission). Concerning the miniaturised HID, a R&T study is under progress to adapt commercial components to usual mission constraints.

Tunable diode laser spectroscopy

On Venus, the best altitude range for TDLAS method is the upper atmosphere, including the cloud region (from ≈80 km to ≈40 km). The accuracy of the measurements at these altitudes will be approximately constant with altitude: while descending in the atmosphere, the number of molecules in the cell in increasing, but because of line broadening the absorption remains the same. The best discrimination (isotopic measurements) occurs above 60 km. Below, where the lines broaden, the sensitivity and discrimination degrade progressively. At pressures exceeding 10 bar, the collisional broadening is no more described by Voigt profile, and is poorly parametrized, so the quantities cannot be accurately retrieved from measured absorptions. The TDLAS method is hardly applicable already at 30 km at exterior pressure. It is possible, however, to study this altitude range very important for surface-atmosphere interaction by means of active spectroscopy in a cell, if the gas will be analysed at reduced pressure. It might be possible to take advantage of the gas sampling system of a mass-spectrometer, or to employ a dedicated sampling system. The best sensitivity and discrimination is achieved at pressures of 0.1-1 bar.

Possible atmospheric targets of TDLAS in Venus atmosphere are numerous, and may be divided in two categories: (i) CO_2 and/or H_2O isotopic ratios ; (ii) molecular species (HCl, OCS, HF, H_2S, CH_4, NH_3, HBr, HI). Concerning molecular species, the typical accuracy of the measurements varies from a few ppbv (HF, CH_4) to one ppmv (HCl) and a few tens of ppmv (OCS, H_2S). Concerning isotopes of CO_2 and H_2O, which are of particular interest, the main characteristics may be summarized as follows.

Table 1 : Characteristics of molecular absorption by species of interest.

Molecule	Spectral range (cm^{-1})	Altitude (km)	Abs. Opt. Path : 10 m	Value range	Accuracy
H_2O	7295-7315	40-70	0.02-0.1	≈30 ppmv	$<10^{-3}$
		70-90	0.01-0.001		≈10^{-2}
		30-40 [1]	≈0.15		≈10^{-2}
HDO [2]		≈60	TBD	D/H≈ 150 [3]	TBD
$^{12}C^{16}O_2$	5337-5339	30-TBD	0.001		$<10^{-3}$
$^{13}C^{16}O_2$	5010-5020	TBD	0.05		$<10^{-3}$
$^{12}C^{16}O^{18}O$	5040-5050	≈60 km	0.05		$<10^{-3}$
$^{12}C^{16}O^{17}O$	5040-5050	≈60 km	0.02		≈0.01

[1] The lower atmosphere (0-30 km) is as well accessible if a gas sampling system reducing the pressure is possible.
[2] Spectroscopic data by [10].
[3] In units of SMOW value.

2.3. Oxygen fugacity

Any sound information on the oxygen mixing ratio of the troposphere of Venus would provide a definite clue for choosing between competing cosmogonic models, namely: different conditions of formation and/or different evolutionary paths. Calculations of the oxygen mixing ratio - from the CO/CO_2 ratio and the temperature- thanks to spectrometric observations of the upper Venus atmosphere from the Earth, or on the basis of mass spectrometric/gas chromatographic analysis down to the surface of the planet (*Pioneer-Venus* and *Venera 11/12*), are somewhat unsatisfactory because: chemical equilibrium may not be realized throughout the atmospheric column ; and/or uncertainties exist in atmospheric mixing rate and composition.

Given the 17±1 ppm carbon monoxide mixing ratio measured on board *Venera 11/12* in the lower Venusian atmosphere (12 km), a 50% confidence interval on such a value, extrapolated downwards, would result in an order of magnitude uncertainty on the calculated oxygen mixing ratio at the surface (735 K), a variation of 10 K yielding an error by a factor of 3 on the inferred value. The most comprehensive study of the redox state of the lower atmosphere of Venus leads to a two orders of magnitude uncertainty of the surface oxygen fugacity, centered at $10^{-20.85}$ bars [4].

On the other hand, a direct and continuous measurement of the oxygen mixing ratio from, say, 20 km down to the surface, would provide the lacking key parameter. It would allow to study the way the Venusian atmosphere chemically equilibrates, to compute the oxygen budget of the crust-atmosphere system (and, hence, deducing the surface mineralogy) and to model the origin and evolution of the planetary atmosphere. Up to now, this parameter remains an unknown, constrained by instrument sensitivities (upper limits) and by reasonable thermodynamical and geochemical considerations (lower limits), and leads to conflicting models, describing the atmosphere and lithosphere of Venus as comparatively oxidized or reduced. Moreover, if complementing the already known temperature profile, an oxygen profile would give insight into the chemical homogeneity and possible geochemical layering of the Venusian atmosphere.

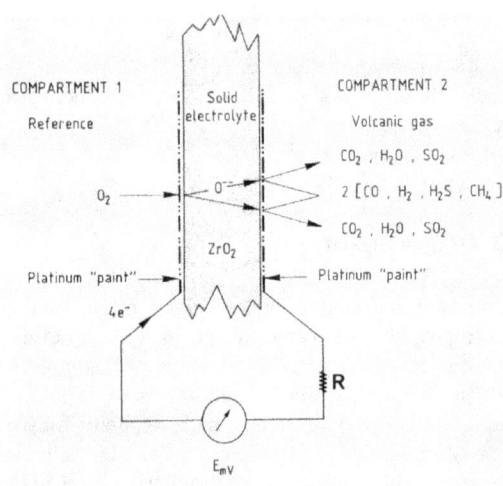

Fig. 2. Principle of a solid electrolyte oxygen sensor. Basically, it is a fuel cell "burning" the combustible atmospheric gases CO, CH_4, COS, H_2S, S_2 and H_2, into CO_2, SO_2 and H_2O, by means of oxygen ions instead of oxygen molecules. In this case, the quiet combustion is an equilibrium process. The counterpart of the flow of oxygen ions in the electrolyte is an electrical current through the external conductor. The resistance of the external circuit being very high, this current is actually negligible and electric charges equilibrate on both electrodes.

The zirconium solid-electrolyte technology offers the possibility of achieving elegantly this goal. With their high accuracy (*i.e.*, ± 0.01 in log P_{O_2} units) in the suitable range of P_{O_2} (down to 10^{-30} atm, given the high total pressures involved) and their short response-time, the ZrO_2 oxygen probes with a solid internal reference are very well adapted for continuous oxygen measurement during the descent - even very fast - of a Venus lander, or during the flight of an atmospheric vehicle. These probes are very lightweight (*ca.* 2g, including wires), do not need a power supply and deliver an electrical potential (< 1V), the radio-transmission of which to the Earth, with a ± 1 mV accuracy, is as straightforward as that of the tension of a thermocouple.

With the presently available technology, a ZrO_2 oxygen sensor (called OES, for Oxygen Electrochemical Sensor, in the following) would start delivering measurements at 12 km altitude (643 K) down to the surface of the planet (735 K) and will continue to do so until the radio equipment of the vehicle is destroyed. With some R&D efforts for increasing the ionic conductivity of the "doped" zirconium at low temperatures, the upper limit of measurement could be slightly shifted upwards, presumably up to 14 km. Heating the sensor would require a low electrical power (2 W, or so), that will add a few grams to the sensor mass (not including the power source), but would extend the range of measurements up to, at least, 30 km altitude.

2.4. Natural radioactivity

Radon (isotopes 222 and 220) is continuously exhaled by the telluric planetary surfaces. Without any scavenging process, its steady mean concentration in the lower troposphere (below the cloud deck) is directly proportional to the exhalation rate, from which the concentration of surface rocks in parent radionuclides (namely U-238 and Th-232) can be inferred (given the emanating power of the surface rocks). Moreover, the structure of the troposphere and the mixing processes driven by atmospheric dynamics will be deduced from the vertical and horizontal (if any) gradients of radon, independently from any consideration on the atmospheric chemistry : this will shed new light onto stagnation and convection processes in the Venusian lower atmosphere. Additionally, volcanic plumes, if they exist, might be detected at large distance of the location where they are formed.

The measurement of the airborne naturally occurring radionuclides of the uranium and thorium series concerns two gaseous isotopes : ^{222}Rn, ^{220}Rn, and four solid or liquid (in the condition of Venusian lower atmosphere) alpha emitting isotopes : ^{218}Po, ^{214}Po, ^{212}Po, ^{210}Po and ^{212}Bi, which are decay products of the radon isotopes.

Two different methods of measurement can be implemented for such an experiment:
- Pulse type ionization chamber, associated with spectroscopic analysis of pulse amplitudes, the ionization chamber being filled with the gas existing locally in the atmosphere.
- Gaseous scintillation techniques, based on the de-excitation of atmospheric molecular species like CO_2 and N_2 by photons emission, after excitation due to the energy transfer of alpha particles. The concerned spectral bands are located in the near UV region of the spectrum (230 to 420 nm). The measurement is then done by photon counting, using a special large aperture and wide field UV optics, coupled to an UV photo-multiplier. A spectroscopic analysis of the output pulses allows discrimination of the various alpha energies, and therefore of the different radionuclides.

The two proposed methods of measurement allow isotopic ratios between the different isotopes to be calculated, allowing study of the diffusion of these isotopes, either under gaseous form, or as aerosols, to be made. The first method described allows local information (few meters around the probe) to be retrieved, whereas the second one allows more wide spatially integrated data to be considered.

2.5. Acoustic detection onboard balloons

The acoustic signals may have three origins. The first type is related to any explosive volcanic eruption. Little is known however on the regime of volcanoes on Venus, especially the gas release during the eruption: even with the lack of water, a significant release of CO_2 and SO_2 might be expected. The second type will be related to acoustic signals generated by quakes. The lack of water in the crust will favour the brittle regime of the crust, but the seismic activity of Venus is unknown. The third type of events will be related to storms, lightning, and other atmospheric events.

Due to the short duration of the mission, the acoustic monitoring is focused for the endogenic sources on the acoustic signals which might be trapped in the atmosphere and therefore recorded at large distances from the sources. Signals from the same source might then be detected on several balloons, and the position of the source might be determined by a classical triangulation.

For atmospheric signals, short period infrasounds can be detected too, but very likely, signals detected on each balloon will be uncorrelated: we therefore propose to have the possibility to determine the propagation direction of the signal with an antenna of receivers.

In contrary to the Earth, the atmosphere of Venus shows no acoustic low velocity zone. However, the very strong decrease of the density at altitude higher than 100 km traps strongly the acoustic energy, with the maximum amplitudes of acoustic modes at altitudes of 50 km. As a consequence the quality coefficient of the fundamental and two first overtones of acoustic modes are higher than 4000 and are therefore strongly trapped in an acoustic channel between the surface and about 100 km of altitude (Figure 3a). A detailed analysis shows however that the trapping is efficient only at frequencies less than 5 mHz. In comparison, the Earth fundamental acoustic mode has a quality coefficient of about 100, and overtones, with frequencies higher than 4 mHz, are not trapped, with coefficient of about 10 (see Figure 3b for the amplitude of the fundamental mode). We can therefore expect good propagation of acoustic signals in this channel. Zonal wind will however produce a distortion of the wave front, which will decrease the precision of localisation. In addition to the acoustic signals generated by volcanoes, three normal modes associated to Rayleigh surface waves have large amplitudes in the atmosphere, as shown by Figure 4. (near 3, 4.4 and 4.9 mHz). As a consequence, the surface waves generated by any quake may produce significant pressure signals and could be detected for strong magnitudes. These signals might have strong amplitudes at periods of about 20 sec, i.e. wavelength in the atmosphere of about 7-9 km.

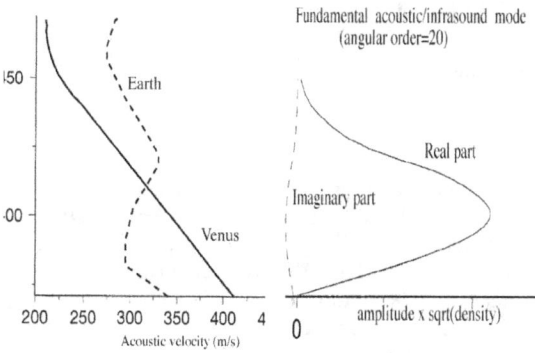

Fig. 3. 3a (left side): Acoustic velocities of Venus as compared to the Earth. 3b (right side): Distribution of the kinetic energy of the acoustic mode with angular order 20, with frequency of about 3 mHz.

The instrument is designed for the separation of the pressure changes related to the altitude variations of the balloon and the pressure changes related to an incident acoustic signal for signals in the frequency range 0.1-10 Hz, and for the determination of the direction of signal propagation in the band. We therefore propose to use a long cable of a few 100 meters, in Kapton. The mass of the cable and deployment system is expected to be close to one kg. Along this cable, 3 sensors will be mounted. The complete mass of the experiment might be 1.5 kg.

Infrasonic sensors might be similar to the infrasonic sensors developed in the frame of the Netlander European Mars network mission, and have a resolution of about 10^{-3} Pa in the infrasonic (1 mHz-0.1 Hz) frequency band.

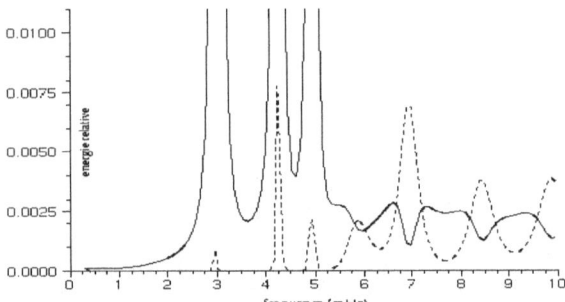

Fig. 4. Portion of the fundamental Rayleigh waves in the Venus atmosphere.

The cable can also be used for the deployment of a potential magnetometer far from the gondola of the balloon. A common instrument with both the magnetometer and the acoustic sensor may therefore be proposed and will result in a significant mass optimisation.

3. PAYLOAD QUICK LOOK

3.1. Balloon Probes

The payload (see Table 2) may be summarized as following :
- Gas chromatograph (GC), electrochemical oxygen sensor (OES), radioactive tracer detector (APID).
- Infrared radiometer (radiative transfer, composition). The VERBE experiment (Venus Energy and Radiation Budget Experiment) consists of a small radiometer operating in four spectral ranges : 1.7 µm, 2.4 µm, 3.7 µm, and an integral channel : 0.8-4 µm, looking upward and downward.
- Radio science (RS) : localization of balloons by VLBI (atmospheric dynamics).
- Radio-electric receiver (RWAEI) and optical photometer (LOD) (for lightning detection).
- Acoustic and magnetic sensor AMS (volcanoes, lightnings, meteorological events ; quakes ; remnant/ intrinsic magnetic field).
- Near infrared camera NIRI for surface characterization (morphology/mineralogy on dayside, temperature on nightside).

- Electromagnetic sounding of inner planet by magnetometry (AMS).
- Relaxation probe for conductivity measurements (RP).

Table 2 : Main budgets of balloon probe payload.

Instrument	Power (average)	Mass (kg)	Volume (Liter)	Data Rate (kbits/s)
GC[1]	2.7	1.5	1.6	0.03
OES	0	0.02	0.01	0.001
AMS[2]	0.5	1.5	<0.3	0.1
NIRI[3]	<0.1	0.12	0.15	0.4*
LOD[4]	2	0.07	0.25	<0.01
APID	0.1-1	2.3	3	0.1
VERBE	1.2	<0.5	0.1	0.2*
RWAEI[5]	1.7	0.3	<0.3	0.4*
RP	0.5	0.6-1.1	0.4-1.2	0.03
APTIV	0.5	0.7	0.5	0.36
TOTAL	10.2	8.1	7.4	1.7

[1] Assumed sampling rate of 30 mn.
[2] Cable of a few 100 meters in Kapton deployed under the gondola
[3] Assumed sampling rate of 1 image per 5 mn interval
[4] Bit rate depends on the number of spikes
[5] 3 short dipole electric antennas 1 m long each, 1 magnetic antenna 10 cm long
*Assumed data compression rate : 5

3.2. Descent Probe

The payload (see Table 3) consists of the following elements :
- Noble gas analyzer : static ToF MS with a separation/ purification line, consisting of a gettyer and a cryocooler (see above).
- Gas chromatograph (GC), electrochemical oxygen sensor (OES) and optical gas analyser (OLGA) (molecular/isotopic composition).
- Infrared spectrometer (atmospheric composition, radiative transfer), Visible/Near Infrared spectro-imager (atmospheric composition, surface mineralogy/morphology on dayside, surface temperature on night side). The VISS (Venus Imaging Solid-state Spectrometer) uses the AOTF technology (Acousto-Optic Tunable System). It may be used for reflectance spectroscopy (0.5-1.2 µm range), in order to determine the mineralogy of surface material, as well to probe atmospheric gases in the thermal infrared (1.7-4 µm and beyond). Several windows exist, where the transparency length (τ=1) is small (a few hundred meters to a few kilometers, see Table 4). It is divided into a VIS-NIR and a IR instrument. An option could be to use a grating spectrometer (Visible Near-Infrared Spectro-imager VNIR).
- Radioactive tracer detector (APID instrument, for Alpha Particle Ionization Detector).

- Atmospheric sensor package APTIV (p, T sensors, accelerometer).

Table 3 : Main budgets of descent probe payload.

Instrument	Power (average)	Mass (kg)	Volume (Liter)	Data Rate (kbits/s)
MS[1]	5.6	5	1	0.1
GC[2]	8	1.5	1.6	0.15*
OES[3]	2	0.02	0.01	0.001
OLGA/TDLAS[4]	1.3	0.4	0.2	1.6*
OLGA/UV[5]	5	1	1	0.1
VISS/VIS-NIR[6]	2.5	1	1	5.3*
VISS/IR[7]	6	1.1	1	0.4*
APID[8]	0.1-1	2.3	3	0.1
APTIV	0.5	0.7	0.5	0.36*
TOTAL	31.9	13.0	9.31	8.1
Alternative to VISS/VIS-NIR[9] :				
VNIS	7	3	4.5	3

[1] Assumed time analysis of 1 hr.
[2] data rate : 80 kb per 10 mn analysis.
[3] Power : 2 W if heating required (z>15 km), 0 W if not.
[4] 1 measurement per 5 mn interval assumed..
[5] 1 measurement per 10 s interval assumed.
[6] 1 spectro-image per 5 mn interval assumed.
[7] 1 linear image per minute assumed.
[8] Option 1 : pulse-type ionization chamber.
[9] 1 spectro-image per minute assumed.
*Assumed data compression rate : 5

Table 4 : Characteristics of visible-near infrared transparency windows.

Spectral interval (μm)	Transparency length [1] ($\tau=1$)	Nature of the signal
0.6 - 0.9	1.5-10 km	Solar flux
0.96 - 1.035	15 km	Solar flux
1.09 - 1.11	20 km	Solar flux
1.16 - 1.195	10 km	Solar flux
1.27-1.28	1 km	Solar flux
1.72-1.75	1 km	Thermal emission
2.21-2.46	100 m	Thermal emission
3-3.7	100 m	Thermal emission

[1] Without Rayleigh diffusion

4. INDICATIVE SYSTEM MASS BUDGET

The spacecraft at launch is composed of : (i) The bus, equipped with the telecommunication system, which may be similar to the platform developed for the Mars-Express mission (mass : 440 kg) ; (ii) The descent probe, which is an heritage of the Huygens probe (mass : 190 kg) ; (iii) 3 Balloon Probes (mass : 135 kg each), which can benefit of the heritage of the Mars balloon ; (iv) Hydrazine for maneuvers (mass : ≈50 kg). The total mass of the spacecraft at launch is ≈1100 kg (see table 5 below).

Table 5 : Tentative mass budget of a descent probe-3 balloons system (unit : kg).

Bus	Mars-Express platform	439
	Propellant	50
	Total	489
Descent probe	Structure	70
	Thermal shield	40
	Thermal insulation	40
	Instruments	15
	Batteries	10
	Electronics	10
	Telecom	2
	Total	187
Balloon probe	Cocoon gondola (structure and thermal insulation)	42
	Thermal shield	10
	Instruments	10
	Batteries	10
	Electronics	5
	Telecom	2
	Pressurized vessels	25 x 2
	Helium	6
	Total	135 x 3
TOTAL		1081

5. LAUNCH AND OPERATIONAL SEQUENCE

The spacecraft is put on an Earth-Venus transfer orbit by a Soyuz-Starsem launcher. The arrival velocity at Venus, 3 months later, is about 4 km/s. At the pericenter, the velocity of the fly-by bus increases up to nearly 11.5 km/s. The date of arrival at Venus of the bus is denoted by t_0. The sequence of operations is as following :

- $t_1 = t_0 - 20$ **days** : release of the the 3 Balloon Probes (BPs). By adequately spinning the bus, the release velocities of the BPs, that is their velocities relative to the bus, are adjusted in such a way to distribute them on a great circle of typically 1500-2000 km radius at Venus level (initial required relative velocity of 1 m/s), centered on a point, to be defined, near Venus disk center. BPs are expected to sample latitudes and local times (e. g. (i) one in dayside southern hemisphere, (ii) one in dayside northern hemisphere, (iii) one in nightside). The mass of the Bus + Descent Probe (DP) system is now 680 kg.

- $t_1 + 1$ **hr** : chemical deceleration of the Bus+DP system by 200 m/s. By assuming a specific impulsion Isp of 310 s, the required mass of hydrazine for deceleration is about 45 kg. The mass of the Bus+DP system drops down to 635 kg.

- $t_2 = t_0 - 19$ **days** : release of the DP, placed on an adequate trajectory toward the target point on Venus. The mass of the bus system is now 445 kg.

- $t_2 + 1$ **hr** : chemical deceleration of the Bus system by 30 m/s and reorientation of the bus on a fly-by trajectory. Required mass of hydrazine : 5 kg. The mass of the Bus system is now about 440 kg, that is the weight of the Mars Express platform.

- **Arrival at Venus** in the following order: BPs at t_0 - 24 hr, DP at t_0 - 2 hr, Bus at t_0. Assuming a launch in January 2009 (as in the proposal to ESA), the parameters at arrival are: (i) Venus-Earth distance: 0.52 AU ; (ii) Sun-Earth/Earth-Venus angle : 43° ; (iii) Bus-Venus/Earth-Venus angle : ≈30°. The fly-by Bus and the Earth are facing nearly the same sector of the Venus surface, allowing using both Bus/Probes and Earth/Probes radio links (for localization purpose in the last case, by using VLBI technique) ; (iv) The asymptotic trajectory of the Bus, not strictly parallel to the ecliptic plane, points towards north ecliptic, with an angle of ≈30° with respect to the ecliptic plane. The sub-Bus point on Venus is therefore located near 30° south latitude. The night-side sector represents about 2/3 of the full disk, and is located to the east. The day side-sector is therefore in the morning (due to the retrograde rotation of Venus).

- **t_0 - 24 hr** : deployment of balloons at 10 km altitude and beginning of balloon observation phase. The Bus/BP distance is about 350,000 km. During the 24 hours of their operational phase, the balloons, carried by the wind (a few meters per second, up to 20 m/s), cover a distance smaller than 1500 km. Because the rotation of Venus is very slow (period : 117 days), the bus may be considered as facing the same sector of the Venus disk during the final one-day approach. If BPs are deployed at a distance of 1500-2000 km from the center of the disk, they are in constant radio visibility from the bus. BP-Earth and BP-Bus data transmission, localization of balloons by VLBI.

- **t_0 - 2 hr** : (i) Progressive deflation of balloons and descent down to the surface, reached at t_0, 24 hours after deployment (duration of the descent phase : 2 hours). (ii) Injection of the DP in the Venus atmosphere for a 1 hr to 2 hrs (2 hrs assumed in the present scenario) descent phase. The descent phases of the DP (from 100 km to 0 km) and the 3 BPs (from 10 km to 0 km) are therefore synchronized, the 4 probes reach the surface at the same time. At t_0 – 2 hr, the Bus/probes distance is about 40,000 km, and the velocity of the bus is about 7 km/s. High data transmission rate between the probes and the Bus.

- **t_0** : (Hard) landing of probes and end of operations. The Bus/probes distance is of the order of 5000 km. The velocity of the bus is about 10 km/s.

- **$t > t_0$** : The bus is occulted by the planet. Data are transmitted to Earth in a subsequent phase, after de-occultation of the bus.

6. SYSTEM CONCEPT AND THERMAL CONTROL

The bus is designed to take the Descent and Balloon Probes to Venus and also to carry a data relay system for communicating with Earth. An off-the-shelf technology is planned to be used wherever possible. In particular, the bus concept and structure should be based on the Mars Express spacecraft. In this hypothesis, the total mass of the bus, including propellant requires for maneuver and braking phases, is about 485 kg. Up to 3 Balloon Probes can be accommodated on the bus and then released to Venus. Each probe is made of a gondola protected by a thermal shield during the entry phase. The gondola has a hot and a cold compartment. The hot compartment is the outer balcony of the gondola and is dedicated to the sub-systems which can withstand the atmospheric environment (in particular the thermal conditions). These sub-systems are mainly the power s/s and the Helium pressurized vessels. The cold compartment is similar to a thermal cocoon. This tighten spherical structure, containing mainly the scientific payload, ensures a thermal control of its internal volume within an acceptable range (up to 40°C). The thermal shield is designed to absorb the thermal flux due to the entry of the probe from the upper Venusian atmosphere down to the altitude measurement. The shield concept could be based on the technology retained for the Huygens probe front shield sub-system. Its weight can be evaluated to 15 kg.

At an altitude of about 20 km, slightly higher than the measurement altitude, the thermal shield is released, the balloon envelope is deployed from the gondola and begins to be inflated with Helium. The balloon envelope is designed to receive a total Helium mass of 6 kg, corresponding to an inflated volume of about 2 m^3, in order to stabilize the 60 kg gondola at the measurement altitude of 10 km. The Helium gas is contained in pressurized vessels made of metallic insert (Titanium thin wall sphere) reinforced by an external carbon composite thermo structure. Two vessels of 40 cm diameter, pressurized at 60 MPa, are required per balloon probe. The estimated mass of a single vessel is 20 kg. After complete inflation of the balloon envelope, the vessels are released from the gondola in order to remove their 40 kg mass and to reduce the total mass to 60 kg. After altitude stabilization of the balloon probes, the operational phase can start. The objective is to have 20 hours of scientific operations at 10 km altitude, and then to allow the descent of the gondola down to the surface by venting the Helium content of the envelope. This phase will take about 2 hours, the estimated science operation phase of the balloon probes is then close to 24 hours.

The descent probe is designed to perform scientific measurements during its 1 hr (or 2 hrs) descent time through the Venusian atmosphere from 80 km altitude to the ground. The concept is similar to the balloons probes, but with increased system dimensions (hemisphere of 160 cm diameter instead of 80 cm diameter sphere) and weight (250 kg instead of 115 kg) in order to accommodate the specific payload. The structure is an hemispherical thermal cocoon and an outer balcony protected by a thermal shield. The shield is used both to absorb the thermal flux due to the atmospheric descent and to allow aerobraking of the probe. A parachute could be eventually added to reach the best compromise between scientific return optimization and system constraints such as system thermal behavior and power resources. Some of the instruments require windows to look towards the ground. A first option could be to integrate these optical windows to the shield. A second option could be to integrate the windows on the bottom part of the probe and to release the shield during the descent. The outer balcony "hot compartment" is dedicated to sub-systems which can be in isothermal conditions with the probe environment (mainly telecom parts such as antenna, batteries, ..). The thermal insulation of the cocoon is designed to limit the internal temperature to acceptable value (about 40°C) during the 1 hr to 2 hrs lifetime. The internal cold compartment receive the complete payload and the other sub-systems (electronics, ..).

Both the balloons and descent probes have a complex and variable thermal environment during their operational phase. All the parts which can withstand this environment are not thermally protected in order to reduce the size of the temperature controlled area. This is the case in particular for the hardware components of the telecom s/s and for the batteries (described below). The temperature controled area is designed to keep all the thermally sensitive equipments below 40°C. This task is more critical on the balloon probes, because of their 24 hours lifetime, than in the descent probe. Taking into account the limited power resources, a concept of a passive thermal cocoon made of concentric layers of PCM (Phase Change Materials) have then been validated for the balloon probes.

The thermal dynamic simplified model is based on a spherical gondola. A preliminary thermal study has been led. The concept of the gondola consists of a spherical cocoon with a series of concentric layers having the following characteristics (from the outer edge to the inner edge) : insulating layer, liquid water layer, insulating layer, paraffin layer, inner compartment with the payload. The thermal conductivity of the insulator has been assumed to be 0.04 W m^{-1} K^{-1}. For the paraffin, a calorific capacity of 1000 J kg^{-1} K^{-1} is assumed, for a volumic mass of 800 kg m^{-3}. Melting temperature is 40°C, and the latent heat of fusion is 260,000 J kg^{-1}. About 4000 configurations have been analyzed. The better compromise consists of a 80 cm diameter cocoon, with successive layers, from the outer to the inner, of thickness : 8 cm (insulator), 2 cm (water), 8 cm (insulator), 4 cm (paraffin) (see Fig. 5). The volume of the central compartment is 24 l and the time of survival about 32 hours. This time has been obtained with conservative assumptions, in such a way to account for the necessary weakening of performances due to thermal "leaks" (entry gas and optical lines, structure). The weight of the cocoon, including insulator, water and paraffin, and payload (20 kg) is 60 kg.

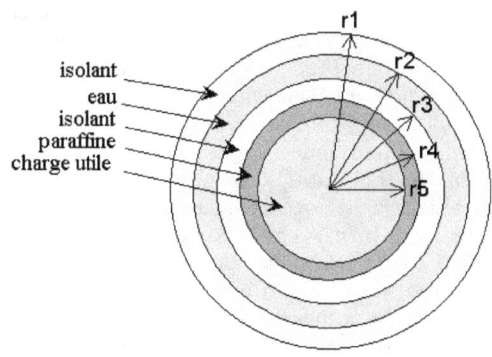

Fig. 5. Schematic view of the cocoon with concentric layers consisting of (from the outer to the inner) : insulator, water, insulator, paraffin, central compartment for payload.

The use of high temperature batteries aboard the balloon and descent probes is an important key to solve the thermal control requirement by reducing the mass located in the cold compartment and the volume of this compartment. It means that the power sub-system should work at about 700 K. Both lithium and sodium batteries are under development. These batteries allow a stable energy storage at ambient temperature, it means that no significant energy loss is expected after integration of the power s/s and during the cruise to Venus. They can then be operated in the range 600-800 K. The lithium-sulfur batteries are commonly operated in the range 620-670 K. Practical energy densities in the range 100-150 Wh/kg are reported on the optimized couple Li/FeS$_2$ batteries. The sodium-sulfur (Na/S$_2$) batteries have a similar operating temperature range and practical energy densities of 100 Wh/kg are reported. A technological trade-off is required between these technologies but a mean energy density of 100 Wh/kg, compatible with the Venus environmental constraints, can be retained for the power s/s mass and performance estimation.

7. TELECOMMUNICATIONS

A detailed scenario of the telecommunication sequence has been established.
- Phase 1: When the bus is still at large distance from Venus (> 100,000 km), BPs directly transmit to the Earth. For an emitted power of 10 W, which requires an input power of the order of 30 W (30% efficiency), the small, omnidirectional, antenna of the BP allows transferring 10 to 100 Bits per second (Bps) to the Earth.
- Phase 2: When the bus/ probe distance falls below 100,000 km (last 6 hrs), the data transmission rate from the probes to the bus becomes significantly larger than the probe/ Earth data transmission rate,

increasing from 0.25 kBps to about 10 kBps at 5,000 km, in the final stage of data transmission. Note that the emitted power can be 10 times greater on the DP, because of the larger amount of energy available (due to its shorter journey in Venus atmosphere: 2 hrs versus 24 hrs), allowing emitting up to 100 kBps in the final stage. With such a high data rate, all DP data can be transmitted in real time. For BPs, it is possible to transmit a substantial fraction of data collected during the one-day mission in the last hours of operations by using an onboard memory of a few hundred Mbits. Note that the totality of BP data (about 100 Mbits with a compression factor of 5) may hardly be recovered, which therefore requires onboard electronic intelligence (compression, detection of rare events...) in order to select data to be transmitted.

8. CONCLUSION

Deploying balloons at low altitude (typically 10 km) rather than at middle altitudes (30-60 km), would be of high interest for several reasons. First, at 10 km altitude, the probe has a direct optical access to the surface in the visible range ($\tau<1$ in the 4 visible windows) for morphology/ mineralogy purpose. Second, at only 1/2 atmospheric scale height above the surface, the surrounding atmosphere is at chemical equilibrium with the surface, which is of great interest for geochemical measurements. Furthermore, the probability to detect chemical, radioactive, thermal, acoustic signatures of surface activity (volcanism, quakes, ...) is increased near the surface. Finally, if it is possible to design a balloon system able to sustain these harsh environmental conditions, this system can be operated, without further substantial modification, during descent and landing if balloons are deflated in a final stage. Using balloons as descent probes in a multi-site perspective is an attractive possibility that deserves to be studied. Because temperature sharply decreases with altitude, it might be possible to relax constraints by deploying balloons at higher altitude (i.e. 20 km). A trade-off will have to be found, in terms of deployment altitude, between science return and technical difficulties (and risk).

The proposed flotilla of balloons may provide interesting network science, in the case a single event may be detected simultaneously in different regions of Venus: acoustic/chemical/ radioactive signal due to surface activity (volcanism, quakes,...), electromagnetic signal produced by storms (lightning), chemical horizontal gradients due to atmospheric circulation etc... Balloons, used as local probes and/or as a network, and complemented by a descent probe (possibly also by an orbiter), will bring unprecedented information about deep atmosphere, surface mineralogy, geochemical cycles and, ultimately, Venus history and evolution. Note that a detailed knowledge of deep atmosphere and surface is required for preparing a Venus sample return mission on the long term. A number of R&T studies have to be led about sensors, onboard data treatment, thermal insulation, heat-tolerant components,... From preliminary studies, a 1-day-lived balloon seems thermally and energetically possible. Note that active cooling (using radioactive thermonuclear generators) might allow to increase the lifetime of balloons, provided thermal insulation, and tolerance to pressure and chemical environment, are sufficiently performing. For the descent probe, the heritages from Russian Venera missions, NASA Pioneer Venus mission and ESA Titan Huygens probe may certainly be used with benefit.

9. REFERENCES

[1] Chassefière, E., D. Coscia, et al. A mission to Venus : LAVOISIER. Answer to ESA call for mission proposals for two fleximissions (F2 and F3), January 2000

[2] Chassefière, E., et al. The Lavoisier Mission : a system of descent probe and balloon flotilla for geochemical investigation of the deep atmosphere and surface of Venus, *Adv. Space Res.,* Vol. 29, N°2, 255, 2002.

[3] Jambon, A., E . Quémerais, E. Chassefière, et al. PALOMA : in-situ measurements of the isotopic composition of Mars atmosphere, *Concepts and Approaches for Mars Exploration*, LPI Contribution N° 1062, p. 160-161, 2000.

[4] Fegley, B., M. Yu. Zolotov, and K. Lodders, The Oxidation State of the Lower Atmosphere and Surface of Venus. *Icarus* 125, 416-439, 1997.

[5] Bertaux, J.L. et al., VEGA-1 and VEGA-2 entry probes: an investigation of local UV absorption (220-400 nm) in the atmosphere of Venus (SO_2, aerosols, cloud structure). *J. Geophys. Res.* 101, 12709-12745, 1996.

[6] Bezard, C. de Bergh, L. Giver, Q. Ma, and R. Tipping, Near-infrared light from Venus' nightside: a spectroscopic analysis. *Icarus* 103, 1-42, 1993.

[7] Na, C.Y., and L.W. Esposito, Is Disulfur Monoxide a Second Absorber on Venus?, *Icarus* 125, 364-368, 1997.

[8] Marov, M.Ya., A.P. Galtsev, and V.P. Shari, Radiation transfer and the water vapor contents in the Venus atmosphere. *Astron. Vestnik* 19, #1, 15-41 (in Russian), 1985.

[9] Pollack, J.B., B. Dalton, D., Grinspoon, R. Watson, R. Freedman, D. Crisp, D. Allen, B. Bezard, C. de Bergh, L. Giver, Q. Ma, and R. Tipping, Near-infrared light from Venus' nightside: a spectroscopic analysis. *Icarus* 103, 1-42, 1993.

[10] Bykov, A.D. et al., The infrared spectra of H_2S from 1 to 5 µm. *Canadian J. of Phys.* 72, 989-1000, 1994.

ESA VENUS ENTRY PROBE STUDY

M.L. van den Berg [1], P. Falkner [1], A. Phipps [2], J.C. Underwood [3], J.S. Lingard [3],
J. Moorhouse [4], S. Kraft [4], A. Peacock [1]

[1] *Science Payload & Advanced Concepts Office, European Space Agency, ESTEC,
P.O. Box 299, 2200 AG Noordwijk, The Netherlands, E-mail: MvdBerg@rssd.esa.int*
[2] *Surrey Satellite Technology Ltd., Guildford, Surrey GU2 7XH, United Kingdom*
[3] *Vorticity Ltd., Chalgrove, Oxfordshire OX44 7RW, United Kingdom*
[4] *Cosine Research B.V., 2333 CA Leiden, The Netherlands*

ABSTRACT

The Venus Entry Probe is one of ESA's Technology Reference Studies (TRS). The purpose of the Technology Reference Studies is to provide a focus for the development of strategically important technologies that are of likely relevance for future scientific missions. The aim of the Venus Entry Probe TRS is to study approaches for low cost in-situ exploration of Venus and other planetary bodies with a significant atmosphere. In this paper, the mission objectives and an outline of the mission concept of the Venus Entry Probe TRS are presented.

1. INTRODUCTION

The Venus Entry Probe is an ESA Technology Reference Study (TRS) [1]. Technology reference studies are model science-driven mission studies that are, although not part of the ESA science programme, able to provide a focus for future technology requirements. This is accomplished through the study of several technologically demanding and scientifically meaningful mission concepts, which have been strategically chosen to address diverse technological issues.

Key technological objectives for future planetary exploration include the use of small orbiters and in-situ probes with highly miniaturized and highly integrated payload suites. The low resource, and therefore low cost, spacecraft allow for a phased strategic approach to planetary exploration, thus reducing mission risks compared to a single heavy resource mission.

2. VENUS EXPLORATION IN CONTEXT

More than twenty missions have been flown to Venus so far, including fly-bys, orbiters, and in-situ probes. These past missions have provided a basic description of the planet, its atmosphere and ionosphere as well as a complete mapping of the surface by radar. The upcoming comprehensive planetary orbiters, ESA's Venus Express (launch 2005)[2] and Planet-C from ISAS (launch 2007)[3], will further enrich our knowledge of the planet. These satellite observatories will perform an extensive survey of the atmosphere and the plasma environment, thus practically completing the global exploration of Venus from orbit. For the next phase, detailed in-situ exploration will be required, expanding upon the very successful Venera atmospheric and landing probes (1967 - 1981), the Pioneer Venus 2 probes (1978), and the VEGA balloons (1985).

3. MISSION OBJECTIVES

The objective of the Venus Entry Probe Technology Reference Study is to establish a feasible mission profile for a low-cost in-situ exploration of the atmosphere of Venus. An extensive literature survey has been performed in order to identify a typical set of scientific objectives for such a mission. From this survey, the following set of key issues has been derived (with references to review articles):

[SR1] *Origin and evolution of the atmosphere*

A major question is to understand why and how the atmosphere has evolved so differently compared to Earth. This can only be investigated by in-situ measurements of the isotopic ratios of the noble gases [4, 5].

[SR2] *Composition and chemistry of the lower atmosphere*

Accurate measurements of minor atmospheric constituents, particularly water vapour, sulphur dioxide and other sulphur compounds, will improve our knowledge of the runaway greenhouse effect on Venus, atmospheric chemical processes and atmosphere-surface chemistry, and will address the issue of the possible existence of volcanism [4, 5].

[SR3] *Atmospheric dynamics*

Venus has a very complicated atmospheric dynamical system. The driving force behind the zonal supperrotation, the dynamics of the polar vortices and the meridional circulation as well as the cause of temporal and spatial variations of the cloud layer opacity are all rather poorly understood [4, 5, 6].

[SR4] *Aerosols in the cloud layers*

Measurements of the size distribution, temporal and spatial variability as well as the chemical composition of the cloud particles is of interest for better understanding the thermal balance as well as the atmospheric chemistry [4]. Furthermore, it has been suggested that the unidentified large (~ 7 μm diameter) cloud particles might contain microbial life [7, 8].

[SR5] *Geology and tectonics*

Key outstanding questions on the surface of Venus are the mineralogy, the history of resurfacing as well as of volcanism [5]. Resolving the global tectonic structure and (improved) topographical mapping will improve our understanding on these issues.

4. MISSION DESIGN

4.1 Mission requirements

In order to address the science objectives, the following mission requirements have been imposed on the Venus Entry Probe TRS:

[MR1] In-situ scientific exploration at an altitude between 40 and 57 km at all longitudes by means of an aerobot [SR1-4].

[MR2] Vertical profiles of a few physical properties of the lower atmosphere at varying locations across the planet by means of atmospheric microprobes [SR3].

[MR3] Remote atmospheric sensing to provide a regional and global context of the in-situ atmospheric measurements (also concurrent with the aerobot operational phase) [SR2-4].

[MR4] Remote sensing of the polar vortices with a large field of view and a repeat frequency less than 5 hours [SR3].

[MR5] Remote sensing of the Venus atmosphere at all longitudes and latitudes [SR2-4].

[MR6] Remote sensing of the Venus surface by means of a ground penetrating radar and radar altimeter [SR5].

4.2 Mission concept

The mission concept that is able to fulfil all requirements consists of a pair of small-sats and an aerobot, which drops active ballast probes. A two-satellite configuration is required in order to commence

Table 1. Mission baseline scenario.

S/C Module	Measurements	Strawman payload	Requirements
Venus Polar Orbiter (VPO)	- Atmospheric composition - Atmospheric dynamics - Atmospheric structure	- Microwave sounder - Visible-NIR imaging spectrometer - UV spectrometer - IR radiometer	- Large FOV - Resolution ~ 5 km - Operational before aerobot deployment - Aerobot communications
Venus Elliptical Orbiter (VEO)	- Subsurface sounding - Topographical mapping	- Ground penetrating radar - Radar altimeter - Entry probe	- Low periapse (radar) - Entry probe deployment - Data relay to Earth
Aerobot	- Isotopic ratios noble gases - Minor gas constituents - Aerosol analysis - Pressure, temperature etc. - Tracking and localization of microprobes	- Gas chromatograph /Mass spectrometer with aerosol inlet - Nephelometer - IR radiometer - Meteorological package - Radar altimeter	- Long duration (different longitudes) - Microprobe deployment - Altitude 40 - 57 km (aerosols)
Atmospheric microprobes	- Pressure, temperature - Light level (up and down) - Wind velocity	- P/L fully integrated with probe	- Operational down to 10 km or less

the remote sensing atmospheric investigations prior to the aerobot deployment (MR3).

One satellite will be in a polar Venus orbit. The Venus Polar Orbiter (VPO) contains a remote sensing payload suite primarily dedicated to support the in-situ atmospheric measurements by the aerobot and to address the global atmospheric science objectives. The second satellite enters a highly elliptical orbit, deploys the aerobot and subsequently operates as a data relay satellite, while it also performs limited science investigations of the ionosphere and the surface (after lowering the apoapse).

The aerobot consists of a long-duration balloon, which will analyse the scientifically interesting Venusian middle cloud layer. During flight, the balloon deploys a swarm of active ballast probes, which determine vertical profiles of pressure, temperature, flux levels and wind velocity in the lower atmosphere.

The concept of a long-duration balloon with ballast probes is not new and has been proposed before, see e.g. [9, 10, 11]. The focus of the Venus Entry Probe study is to identify the critical technologies associated with such a concept with the aim to successfully support the technology development of a miniaturized aerobot system with atmospheric microprobes.

Table 1 gives an overview of the mission baseline scenario, including a strawman payload suite. Because atmospheric science investigations (large field of view and high polar revisit frequency, see MR4) and surface radar investigations (low periapse) pose different requirements on the operational orbit, the Venus Polar Orbiter will carry the atmospheric remote sensing instrumentation and the Venus Elliptical Orbiter the radar instrumentation. The tentative operational orbits for both spacecraft are listed in Table 2.

4.3 Launch and transfer to Venus

A Soyuz-Fregat 2-1B launch from Kourou has been selected as the baseline for the Venus Entry Probe TRS because it is a cost-efficient and highly reliable launch vehicle. The mass capability for direct escape to Venus is about 1400 kg. The Earth departure phase can be optimized by launching the Soyuz-Fregat into a highly elliptical Earth orbit, with the spacecraft providing the delta-V for Earth escape [12].

A standard high thrust heliocentric transfer from Earth to Venus is envisaged, because this is the most cost-efficient and flexible option for a mission to Venus. The launch opportunities are primarily driven by the Earth-Venus synodic period of 1.6 years. The 3.4° inclination of Venus' orbit to ecliptic causes a variation in the Earth-Venus distance, so that the delta-V requirements vary at successive optimum launch windows.

The typical transfer time for a half solar revolution transfer is between 120 and 160 days, with a delta-V requirement for Venus orbit insertion (250 km × 66,000 km) typically less than 1.4 km/s [12]. Depending on Earth departure strategy and planetary geometry, a high-thrust chemical propulsion system can typically bring into Venus orbit a spacecraft mass between 900 kg and 1150 kg.

The VPO and VEO spacecraft can travel as a composite or individually. As the composite configuration is more mass and cost-efficient (mission operations), this is currently selected as the baseline, with the VEO providing the propellant for departure and Venus orbit insertion.

4.4 Venus Polar Orbiter spacecraft

The 3-axis stabilized Venus Polar Orbiter spacecraft is based on a thrust tube structural concept, because of its low mass and simplicity of design. The propulsion system consists of a conventional dual mode bipropellant system, using Hydrazine and Nitrogen Tetroxide for high thrust manoeuvres and Hydrazine monopropellant thrusters for low thrust.

Table 2. Operational orbits for the Venus Polar Orbiter and the Venus Elliptical Orbiter spacecraft.

	VPO	VEO
Periapse (km)	2000	250
Apoapse (km)	6000	7500 – 20000
Period (hr)	3.1	3.1 – 6.2
Inclination	~ 90°	~ 75 - 90°

Table 3. Venus Polar Orbiter mass budget.

Item	Mass(kg)
Science instruments	30
Communications	22
Structure	51
Propulsion	63
ACS	10
OBDH	4
Power	21
Thermal control	14
Subtotal	**215**
System margin (20%)	43
Total dry mass	**258**

Table 3 shows the top level mass budget for the Venus Polar Orbiter. A mass budget of 30 kg (including margins) has been allocated for the remote sensing atmospheric science instruments (see Table 1). The payload instruments will be integrated into a highly integrated payload suite. By merging individual instruments onto one platform and sharing resources on a system architecture level, considerable mass and power reductions can be achieved without sacrificing the scientific performance. The science data obtained by the Venus Polar Orbiter will be relayed to the Venus Elliptical Orbiter through an X-band link.

4.5 Venus Elliptical Orbiter spacecraft

Table 4 lists the top-level mass budget for the Venus Elliptical Orbiter. For cost reduction purposes, the commonality of platform and subsystems between the VPO and VEO will be exploited as much as possible. As a consequence, the VEO spacecraft also uses a similar thrust tube concept and a dual mode propulsion system.

The Venus Elliptical Orbiter will stay in a highly elliptical orbit until deployment of the entry probe, which is initiated after the VPO has reached its final orbit and the instrument calibration phase has been completed. During this first phase, the VEO primarily acts as a relay station to Earth for data from the VPO as well as from the aerobot, possibly via the VPO. The Ka-band has been selected for communications to Earth, whereas X-band communication is the baseline for the inter-satellite communications.

After the operational phase of the aerobot has ended, the VEO will progress to its final low elliptical orbit (250 km × 7,500 – 20,000 km) in order to start the detailed (sub)surface radar investigations. The current preliminary mass budgets allow for an apoapse of 20,000 km using chemical propulsion. Further work is in progress to assess whether aerobraking or spacecraft mass reduction are viable routes towards a lower apoapse, and consequently a larger surface coverage. A mass of 20 kg has been reserved for a ground penetrating radar and a radar altimeter. Possibly a wide field camera will be included as well.

4.6 Entry vehicle

Fig. 1 shows a conceptual drawing of the entry vehicle. The aeroshell has a 45° sphere-cone geometry, which provides a good packaging shape and aerodynamical stability. Most of the volume of the entry probe is taken up by the spherical gas storage tank, which is surrounded by the ring-shaped gondola. For storage of the balloon inflation gas, a conventional gas tank has been baselined, though alternatives such as cold gas generators or chemical storage of hydrogen are being considered.

In Table 5 the tentative top-level mass budget for the Venus entry vehicle is summarized. Because the design study is still in an early phase, a 25% design maturity margin has been added.

4.6.1 Probe release

The entry probe will be released from the VEO spacecraft, while it is in a highly elliptical orbit with an orbital period in excess of 24 hours. To keep the entry vehicle design simple, the VEO spacecraft will provide the required velocity and orientation for the probe entry. After release of the probe, the spacecraft will perform a re-orbit burn.

Deployment from orbit has been chosen as the baseline because direct entry from the interplanetary transfer hyperbola would require a complicated interplanetary transfer trajectory or orbit insertion scenario in order to fulfil the requirement of starting the remote sensing atmospheric science investigations with the VPO prior to aerobot deployment (MR3).

Table 4. Venus Elliptical Orbiter mass budget.

Item	Mass (kg)
Science instruments	20
Entry Probe	85
Communications	32
Structure	65
Propulsion	135
ACS	10
OBDH	4
Power	19
Thermal control	5
Subtotal	**375**
System margin (20%)	75
Total dry mass	**450**

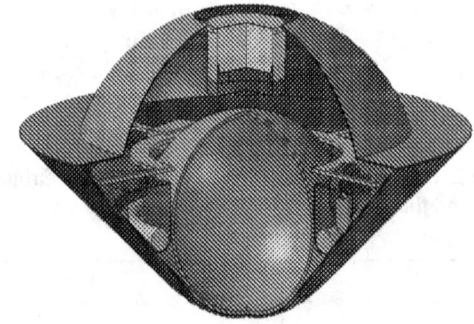

Fig. 1. Conceptual drawing of the entry probe.

4.6.2 Entry, descent and deployment

The probe will enter the dense Venus atmosphere with a velocity of 9.8 km/s and a flight path angle between 30° and 40°, as this scenario yields the best overall system mass. A steep entry angle will cause the probe to penetrate deep within the atmosphere quickly, leading to high accelerations and heat fluxes. However, since the deceleration to subsonic velocities occurs very quickly, the total absorbed heat is relatively low. Additionally, the short entry duration enables a quick release of the aeroshell (~20 seconds), thus minimizing the time for the absorbed heat to soak through the heat shield.

The heat shield material consists of Carbon-Phenolic, which is capable of withstanding very high heat fluxes (~ 300 MW/m^2), much higher than the peak heat flux of ~20 MW/m^2 for a 40° entry angle. The maximum entry flight path angle is set by the 200 g acceleration capability of the payload.

The deployment sequence is depicted in figure 2. The 45° sphere-cone entry probe is designed to be stable in the hypersonic and supersonic regimes, so that no active control is required. Just above Mach 1.5, a disk-gap-band or a ribbon parachute will be deployed by a pyrotechnic mortar. The parachute stabilizes the probe as it decelerates through the transonic regime. The front aeroshell will be released a few seconds after parachute deployment when the subsonic regime has been reached. To prevent heating from the back cover, the rear aeroshell will be distanced from the aerobot by a tether. At a velocity of ~20 m/s and altitude of ~55 km, the balloon will be deployed. The parachute and rear aeroshell are released and the inflation of the balloon is started. The parachute will be designed with a small amount of glide to ensure lateral separation between the parachute and the balloon. The inflation time of the balloon is a trade between the minimum altitude and the aerodynamic loads on the balloon. Currently, an inflation duration of 20 seconds and a minimum altitude of 54 km is foreseen. The gas storage system will be released after inflation of the balloon, and the aerobot will gradually rise to cruise altitude.

4.6.3 Aerobot

The balloon will stabilize at an altitude of 55 km. At this altitude all the scientific objectives outlined in section 3 can be addressed, while the environment is relatively benign (30 °C and 0.5 bars [13]).

The goal for the aerobot operational mission duration is to travel at least twice around Venus. Taking the average speed of 67.5 m/s from the VEGA balloons that flew at a similar altitude [14], one obtains a minimum flight duration of 14 days.

A light gas balloon with slight overpressure is considered the most suitable candidate for the Venus aerobot, because such a balloon complies best with the operational requirements for a long duration mission. As the gas leaks out of the super pressure balloon, the float altitude will increase until there is insufficient gas for positive buoyancy (and the balloon sinks to the surface). A carefully selected microprobe drop scenario could partially compensate for the loss of balloon gas and thus maximize the operational lifetime. Gas release mechanisms and gas replenishment systems are also being considered in order to compensate for

Table 5. Venus entry vehicle mass budget.

Item	Mass (kg)
Gondola in-situ science instruments	4.0
Atmospheric microprobe system	4.0
Aerobot-VEO communications	1.6
Gondola structure and separation system	6.9
Gondola OBDH	0.6
Gondola power	5.6
Gondola environment	0.3
Subtotal (Gondola)	**23.0**
Balloon (including gas, envelope and deployment system)	5.4
Gas storage system	14.9
Entry and descent system	24.8
Total mass entry vehicle	**68.1**
Design maturity margin (25%)	17.0
Mass Entry Vehicle (with margin)	**85.1**

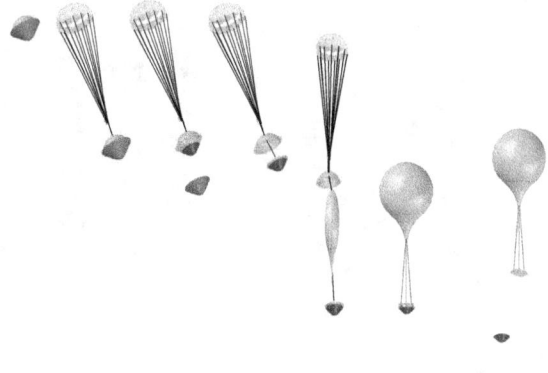

Fig. 2. A schematic of the Venus aerobot deployment.

temperature changes in the balloon gas due to gradients in solar radiation at the day/night and night/day terminators.

Hydrogen has been selected as the baseline for the balloon inflation gas, with helium as a backup option. Though the mass of gas storage systems for hydrogen and helium are similar, the main advantage of hydrogen is that it generally has a lower gas leakage rate compared to helium, which is a monatomic gas. The main disadvantage of using hydrogen is its hazardousness.

The balloon envelope material should have an extremely low leakage rate, possibly requiring welded seams. Additionally, the deployment will have to be carried out in a controlled manner to avoid the slightest damage to the envelope.

4.6.4 Gondola

Figure 3 shows a conceptual drawing of the gondola layout. A strawman payload suite has been defined, which can fulfil the mission objectives. It consists of a gas chromatograph/gas spectrometer (with aerosol inlet), a nephelometer, solar and IR flux radiometers, a meteorological package, a radar altimeter and the atmospheric microprobe system. An assessment study is currently in progress to integrate all instruments, except the atmospheric microprobe system, into two highly integrated payload suites with a total mass of 4 kg and an average power consumption of 5 W.

Electrical power will be provided by amorphous-silicon solar cells, which are mounted on the gondola surfaces, yielding sufficient power during the day.

During the night, primary or secondary batteries will be used.

Currently, Lithium-thionylchloride primary batteries have been selected as the baseline, as this is the most mass-efficient solution due to their high energy density (~590 Wh/kg). The important drawback of soluble cathode lithium cells is that they are less safe than the more common solid electrolyte Lithium cells. As an alternative, Li-polymer secondary batteries are considered which can be recharged during the day. As the energy density is significantly lower (~170 Wh/kg), the mass penalty for using rechargeable batteries is about 4 kg.

In order to save mass, the payload and communication duty cycles will be substantially lower during the night, resulting in an average night-time power consumption of 5 W, compared to 11 W during the day.

4.6.5 Atmospheric microprobes

The fifteen atmospheric microprobes on board of the aerobot serve a twofold purpose:

- Perform scientific meaningful measurements
- Drop ballast in order to increase the operational lifetime of the aerobot

The atmospheric microprobes measure in-situ vertical profiles of selected properties of the lower atmosphere from the aerobot float altitude down to at least 10 km altitude. Due to the stringent mass limitations of the aerobot, they should be as low-weight as possible. This limits the choice of sensors that can be carried with the microprobes. Currently, the following measurements are foreseen: pressure, temperature, and solar flux levels. The horizontal wind velocity will be deduced from the trajectory of the microprobes. This set of measurements, performed at different longitudes, will provide new insights in the atmospheric dynamics and the heat balance on Venus (see MR2).

In order to investigate both the local weather patterns on Venus as well the global atmospheric dynamics, the 15 microprobes will be dropped in 5 separate drop campaigns, spaced equally over the mission lifetime. The three probes in a drop campaign will be released with an interval of 5 minutes.

Localization and communication of the small microprobes is a challenging task and is therefore subject of a separate technology development activity [15]. A preliminary assessment by Qinetiq indicated a

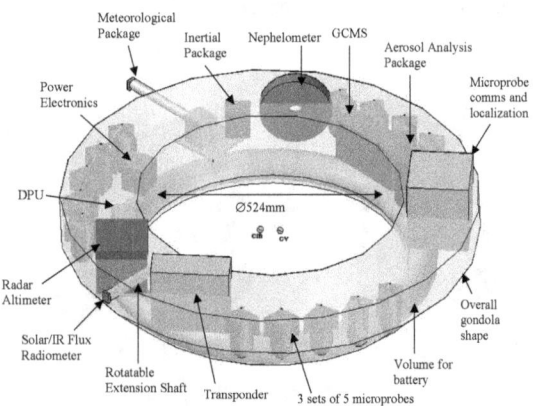

Fig. 3. The layout of the gondola.

mass of 1.4 kg for the communication and localization system and 104 g for a fully functional microprobe, assuming a 5-year technology development horizon.

5. SUMMARY

The Technology Reference Studies are a tool to identify enabling technologies and to provide a reference for mid-term technology developments that are of relevance for potential future scientific missions. Early development of strategic technologies will reduce mission costs and shorten the mission implementation time. As the enabling technologies mature and mission costs reduce, the scientific community will benefit by an increased capability to perform major science missions possible at an increased frequency.

The Venus Entry Probe Technology Reference Study concentrates on in-situ exploration of Venus and other planetary bodies with a significant atmosphere. The mission profile provides a reference for the development of enabling technologies in the field of atmospheric entry systems, micro-aerobots, atmospheric microprobes and highly integrated miniaturized payload suites.

ACKNOWLEDGEMENTS

The support of the study team at Surrey Satellite Technology Ltd., in particular Adrian Woodroffe, Alex Cropp, Nadeem Ghafoor and Jeff Ward, is gratefully acknowledged. Julian Harris from Swiss Space Technology is acknowledged for assessing the miniaturization of the electronics for the payload instruments.

REFERENCES

1. See http://sci.esa.int/science-e/www/object/index.cfm?fobjectid=33170.

2. *Venus Express*, Mission Definition Report, ESA-SCI(2001)6, 2001. Available at www.rssd.esa.int/SB-general/Missions.html.

3. Oyama K.-I., Imamura T. and Abe T., Feasibility study for Venus atmosphere mission, *Advances in Space Research*, Vol. 29, 265-271, 2002.

4. Moroz V. I., Studies of the Atmosphere of Venus by Means of Spacecraft: Solved and Unsolved Problems, *Advances in Space Research*, Vol. 29, 215-225, 2002.

5. Titov D.V. et al. Missions to Venus, *Proc. ESLAB 36 Symposium*, ESA SP-514, 13-20, 2002.

6. Taylor F.W., Some fundamental questions concerning the circulation of the atmosphere of Venus, *Adv. Space Res.*, Vol. 29, 227-231, 2002.

7. Cockell C.S., Life on Venus, *Planetary and Space Science*, Vol. 47, 1487-1501, 1999.

8. Schulze-Makuch D. and Irwin L.N., Reassessing the possibility of life on Venus: Proposal for an astrobiology mission, *Astrobiology*, Vol. 2, 197-202, 2002.

9. Blamont J., The exploration of the atmosphere of Venus by balloons, *Advances in Space Research*, Vol. 5, 99-106, 1985.

10. Kerzhanovich V., et al. Venus Aerobot Multisonde Mission: Atmospheric relay for imaging the surface of Venus, *IEEE Aerospace Conference Proceedings*, Vol. 7, 485-491, 2000.

11. Klaasen K.P. and Greeley R., VEVA Discovery mission to Venus: exploration of volcanoes and atmosphere," *Acta Astronautica*, Vol. 52, 151-158, 2003.

12. Kemble S., Taylor M.J., Warren C. and Eckersley S., Study of the Venus microsat in-situ explorer, TRM/IP/TN1, EADS Astrium Ltd., 2003.

13. Seiff A., et al. Models of the structure of the atmosphere of Venus from the surface to 100 km, *Advances in Space Research*, Vol. 5, 3-58, 1985.

14. Andreev R.A., et al. Mean zonal winds on Venus from Doppler tracking of the Vega balloons, *Sov. Astron. Lett.*, Vol. 12, 17-19, 1986.

15. Microprobe localization and communication prototype system under development by Qinetiq (ESA TRP contract 17946/03/NL/PA).

ROTARY-WING DECELERATORS FOR PROBE DESCENT THROUGH THE ATMOSPHERE OF VENUS

Larry A. Young[(1)], Geoffrey Briggs[(2)], Edwin Aiken[(3)], Greg Pisanich[(4)]

[(1)] Army/NASA Rotorcraft Division, MS 243-12, NASA Ames Research Center, Moffett Field, CA, USA 94035, larry.a.young@nasa.gov
[(2)] Center for Mars Exploration, MS 239-20, NASA Ames Research Center, Moffett Field, CA, USA 94035, gbriggs@mail.arc.nasa.gov
[(3)] Army/NASA Rotorcraft Division, MS 243-10, NASA Ames Research Center, Moffett Field, CA, USA 94035, eaiken@mail.arc.nasa.gov
[(4)] QSS Group, Computational Sciences Division, MS 269-3, NASA Ames Research Center, Moffett Field, CA, USA 94035, gp@ptolemy.arc.nasa.gov

SUMMARY

An innovative concept is proposed for atmospheric entry probe deceleration, wherein one or more deployed rotors (in autorotation or wind-turbine flow states) on the aft end of the probe effect controlled descent. This concept is particularly oriented toward probes intended to land safely on the surface of Venus. Initial work on design trade studies is discussed.

1. INTRODUCTION

A NASA-sponsored NRC "Decadal Study" was recently completed [1] wherein solar system exploration priorities were assessed by a broad survey of planetary science requirements. One of the outcomes of this study was the high priority assigned to a probe/lander mission to the surface of Venus to gain an improved understanding (above that attained by the USSR Venera lander missions in the 1980s and the more recent Magellan radar orbiter) of the history of the planet through measurements of the elemental and mineralogical composition of the surface and of surface-atmospheric interactions. Given the young age of most of Venus' surface, special interest focussed on gaining access to the oldest terrains, namely, the highland *tessera*.

To be presented at the 2nd International Planetary Probe Workshop, NASA Ames Research Center, Moffett Field, CA, August 23 - 27, 2004.

In response to the Decadal Study, NASA is initiating the P.I.-led *New Frontiers* Program and at least one Venus atmospheric probe/lander mission is under study in a collaboration between academia, industry and NASA-JPL and NASA-ARC. This mission would ideally build upon the science, and to some degree the technology derived, from the Soviet missions in the 1960's (the Pioneer Venus probes were not designed for landing). The Venera technology -- using bluff-body (flat-plate) decelerators -- provides passive control of the probe descent rate with altitude and thus allows for neither surface hazard avoidance nor precision landing capability (Fig. 1). The Venera technique – and ideally other passive aerodynamic decelerators – are acceptable for lowland sites. The Magellan radar images of the highland tessera indicate that such passive technology will make landing on the tessera very risky because of terrain roughness and steep slopes.

Fig. 1. Venera Flat-Plate Decelerator

Future Venus lander missions call for an active controlled-descent decelerator. In many respects such control is easier for Venus than for Mars because Venus has a thick atmosphere with a surface pressure of about 90 bars (comparable to pressures a kilometer beneath the surface of our oceans). Such a dense atmosphere makes the use of active aerodynamic decelerators a potentially ideal solution for the descent over highland *tessera*. (The high surface temperatures of Venus do represent a challenge for mission lifetime and for mechanical device actuation and need to be accounted for in the later stages of the design process.)

One active aerodynamic controlled-descent concept is the rotary-wing (RW) decelerator (Fig. 2), wherein the autorotating rotors can precisely control both the rate and angle of descent so that hazards can be detected (by optical imaging and laser altimetry) and avoided and so that touchdown can be gentle. These probe autorotating rotors are capable of being slowed down by braking action as well as potentially being able to perform a collective pitch-angle step input for the final soft-flare landing maneuver.

2. CONCEPT DESCRIPTION AND PAST WORK

In general, use of active aerodynamic control to perform enhanced planetary probe entry and descent is a very desirable characteristic. In particular, use of active aerodynamic control is an essential entry probe attribute to avoid surface hazards during the final stages of landing in unknown and uncertain territory, when there is a high probability of encountering extremely rough terrain. The problem is further compounded with probe thermal management issues for Venus, i.e., it is necessary to provide for high descent speeds through regions of lower-priority interest -- to minimize overall descent time and corresponding heat build-up in the probe's interior -- and to provide for low-speed descent, a soft landing, and more time on the surface and in the lower atmospheric regions of high interest. Rotary-wing decelerators potentially promise a satisfactory solution to these problems.

Fig. 2 is an illustration of one approach to implementing a three-rotor RW-decelerator for a Venus probe. Fig. 2 also sequentially depicts (left to right, top to bottom) the release of the probe from the aeroshell, the deployment and full extension of the rotor booms and rotors, and the deployment of landing gear.

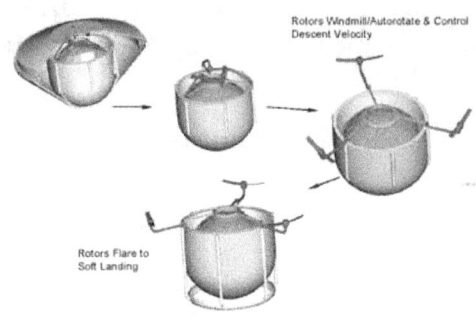

Fig. 2. Rotary-Wing Decelerator for Venus Probe

3. NOTIONAL MISSION & MAXIMIZING SCIENCE RETURN

Researchers [2-9] have previously examined rotor entry decelerators for space mission applications. But none of this past work specifically examined the feasibility of applying this technology to Venus missions. This work does, however, build upon earlier planetary aerial vehicle work by the Army/NASA Rotorcraft Division and the Center for Mars Exploration [10]. The thick surface atmosphere of Venus allows for the usage of very small rotors for deceleration. On the basis of pure aerodynamic deceleration potential, RW-decelerators can at best only match a flat-plate, or bluff body, decelerator – the real advantage of the concept is in the ability to effect a controlled descent (both rate and trajectory angle), soft flare landings, and possibly electrical power generation during descent. Note that the folding support arms shown in the conceptual sketch of Fig. 2 are perhaps an unnecessary design feature; with typical aeroshell shapes, and the compact rotor sizes of a RW-decelerator, rigid (always deployed) support arms are likely feasible instead.

4. GENERAL TECHNICAL APPROACH

The overall objective of the work is to establish the feasibility of RW-decelerators in terms of performance and cost in comparison to proven Venera-class decelerator technology in the context of providing Venus probes with hazard avoidance and safe landing capability on the ancient Venus highlands.

The problem being pursued is envisioned to have three components:

• First, engineering analysis to refine the RW-decelerator conceptual design and to identify key technologies that need to be matured/developed.

• Second, proof-of-concept prototyping of small-scale underwater "test articles" employing a multi-rotor RW-decelerator (as a terrestrial surrogate for a Venus atmosphere probe) to demonstrate trim-control laws.

• Third, feasibility demonstrations with a larger underwater surrogate probe (release/submergence of the prototype in a large body of water) of various active controlled-descent, hazard avoidance, and precision "landing" strategies (i.e. implementation of information and control system technologies).

This paper focuses on the preliminary engineering design analysis.

5. DESIGN SPACE AND SIZING ANALYSIS

A subset of the design space for the engineering trade studies for the Venus probe RW-decelerator concept is shown in Table 1. All RW-decelerators incorporating one or more rotors are capable of descent rate control. Only decelerator systems with three or more rotors are capable of descent angle/trajectory trim control. All RW-decelerators must incorporate rotor collective pitch-angle step input control to be able to perform a soft flare landing (decelerating to net zero vertical velocity). If some form of rotor collective pitch-angle control is not provided for then some moderate level of landing-gear impact (nonzero vertical velocity) upon surface contact will occur.

Table 1. Design Space

# Rotors	Descent Rate Control	Descent Trajectory Control	Soft Flare Landing	Pitch Control
1	X		X	X
2	X		X	X
3	X	X	X	X
4	X	X	X	X

Employing first-order quasi-steady analysis, Fig. 3 illustrates the first-order influence of rotor size (and number) on probe descent speed, as a function of altitude. For example, a simple estimate of rotor size for a Venus RW-decelerator, for a near-surface design descent speed of 8.5 m/sec for a 200kg probe (without aeroshell), is 0.42 meters diameter for an individual rotor in a three-rotor decelerator system operating in ideal autorotation (pre-touchdown rotor "flare"). The probe pressure-vessel diameter is assumed to be approximately 0.7 meters. Note, that for the single-rotor case, rotor blade-root cutout is assumed to be equal to the probe pressure vessel diameter, i.e., $r_c=D$; for all other cases, it is assumed that $r_c=0$.

As noted earlier, Fig. 3 rotor size estimates were based upon a simple analysis; the details of the analysis are as follows. From [11], for ideal autorotation, the descent speed, V, is given by the approximate expression

$$V \approx bv_h \qquad (1)$$

Where the constant $b \approx -1.71$.

Correspondingly, the ideal hover induced velocity is given by the expression

$$v_h = \sqrt{\frac{T}{2\rho A}} \qquad (2)$$

Where T is the required rotor thrust, A is the rotor disk area ($A = \pi(R^2 - r_c^2)$), and ρ is the atmospheric density at the prescribed probe altitude.

Now, given Eqs. 1 and 2, the rotor size (in terms of R, the rotor radius) can be given in terms of the required (ideal) autorotation descent velocity, V.

$$R = \sqrt{\left(\frac{b^2}{2\pi}\right)\left(\frac{T}{\rho V^2}\right) + r_c^2} \qquad (3)$$

Where, again, r_c is the blade-root cut-out for the rotor(s)

Each rotor will have to provide the following amount of Thrust, T, during descent, recognizing that the entry body in itself will have a drag coefficient of C_D and a frontal area of S.

$$T = \frac{1}{N}\left[(m - \Delta m_B)g - \left(\frac{1}{2}\rho V^2\right)C_D S\right] \qquad (4)$$

$$\Delta m_B = 2\rho \left[\sqrt{\frac{S}{\pi}} \left(f - \frac{1}{2} \right) S + \frac{\pi}{3} \left(\sqrt{\frac{S}{\pi}} \right)^3 \right] \quad (5)$$

Note that Eq. 5 accounts for the buoyancy effects of the probe in the thick lower atmosphere of Venus, assuming the probe is a rounded-nose finite cylinder of fineness ratio, f. (Fineness ratio is the ratio of probe longitudinal axis length to the maximum radial axis dimension.) Buoyancy is a small, but nontrivial, contribution to the Venus probe descent speed profile; buoyancy, of course, is a substantial contributor for the proposed surrogate submersible probe testing.

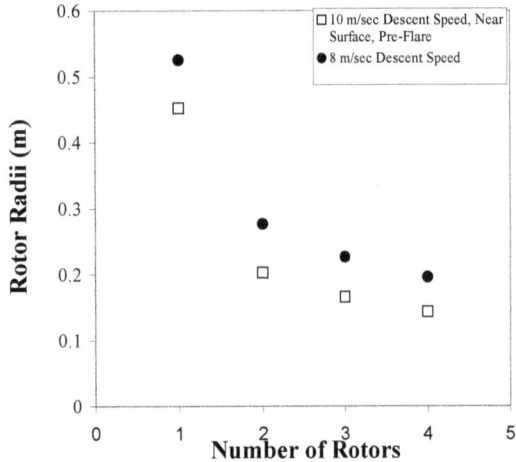

Fig. 3. Rotor Decelerator Size Relative to Autorotation Descent Speeds

Trim control (to vary descent-angle and trajectory) for a four-rotor RW-decelerator is fairly straightforward. Trim for a four-rotor-system simply entails differential rotor braking between the four rotors, note that opposing pairs of rotors spin in opposite directions. Fig. 4 illustrates the connection between the application of rotor differential braking torque and the subsequent (in sequence) reduced rotor thrust, probe bluff-body angular displacement, and the resulting bluff-body normal- and side-force generation. Symmetry considerations for the four-rotor decelerator system allow for yaw control to be effected in the same manner as the pitch control shown in Fig. 4, with an orthogonal pair of rotors. For the four-rotor decelerator system, pitch and yaw control are de-coupled from each other. Descent rate is influenced to a slight degree by pitch and yaw control braking-torque inputs, as they reduce the overall thrust by $2\Delta T$.

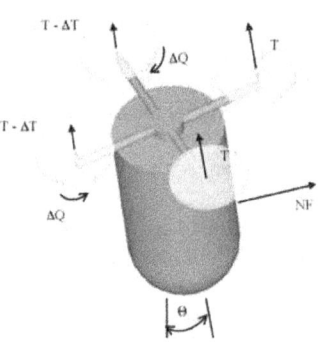

Fig. 4. Four-Rotor (Pitch) Trim Control

Trim control can still be implemented on a three-rotor decelerator system, but it entails a more complex approach. Overall, the three-, versus four-rotor, decelerator design has better volume/packaging characteristics while stowed in the entry aeroshell.

6. QUASI-STATIC DESCENT PROFILES

There are four phases of probe descent with rotary-wing decelerators: 1. release from the entry vehicle aeroshell and initial rotor spin-up and high-speed deceleration of probe, 2. transition phases where the rotor passes through the turbulent and vortex-ring states, 3. low-speed and low-altitude terminal descent, and 4. rotor flare and soft landing. The engineering analysis work to date focuses on the last two stages of probe descent. Future work will couple probe RW-decelerator control laws with a high-level closed-loop controller to validate the viability of hazard avoidance and precision landing using a variety of hypothetical sensors and terrain feature-recognition techniques as applied to Venus-representative simulated terrain.

Figure 5 shows the ideal autorotation descent speed profile with altitude, for the lower extremes of Venus's atmosphere, using a quasi-static aerodynamic analysis based in part on Eqs. 1-5; Venus atmospheric properties were taken from [14]. Figure 5 also illustrates how higher descent speeds result from RW-decelerator configurations having higher disk loading (ratio of rotor thrust to rotor disk area, T/A, N/m^3). This holds true for conventional helicopters as much as it does for the Venus probe rotary-wing decelerators. Therefore, a careful design balance must be maintained between the compactness of the rotary-wing decelerator package -- with correspondingly higher disk loading -- and achieving low probe descent speeds. Also shown in

Fig. 5 is the estimated terminal velocity profile for the probe body without the RW-decelerators or, alternatively, with the decelerators providing no effective braking action. This can be considered as being the maximum descent speed of probe through Venus' lower atmosphere.

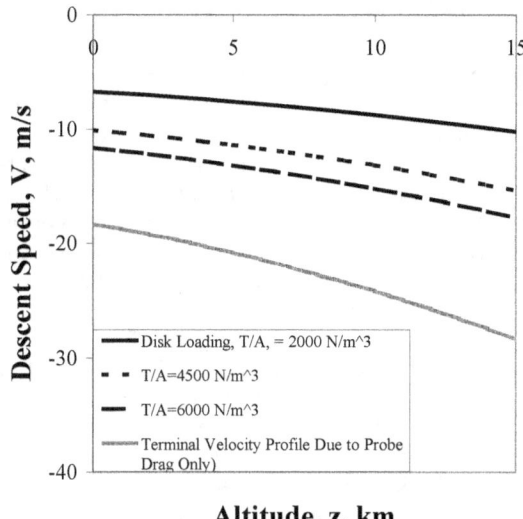

Fig. 5. Probe Descent Speed as a Function of Altitude (Ideal Autorotation)

Figures 6a-b are representative plots of rotor operating conditions, in terms of tip Mach and Reynolds numbers, during descent through the lower extremes of Venus' atmosphere. Note that blade solidity is the ratio of total blade planform area to rotor disk area. As can be readily seen in Fig. 6a, the lower the blade solidity the higher the tip speed required to provide for adequate lift for probe autorotation. Correspondingly, the higher the blade solidity – and therefore the mean effective blade airfoil chord length – the higher the tip Reynolds number. In both cases, though, the RW-decelerator tip Mach and Reynolds numbers fall within the range of engineering experience for conventional/terrestrial rotary-wing aerodynamics.

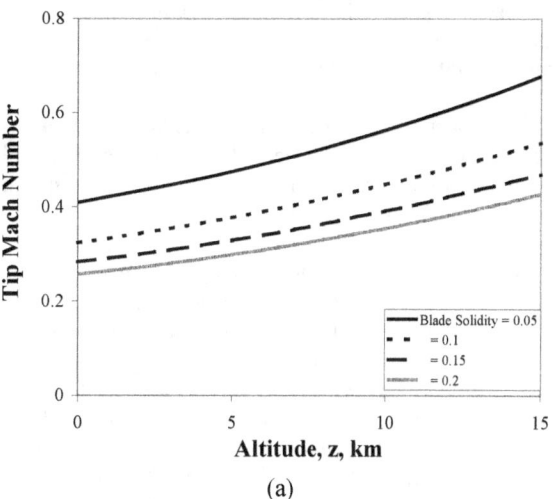

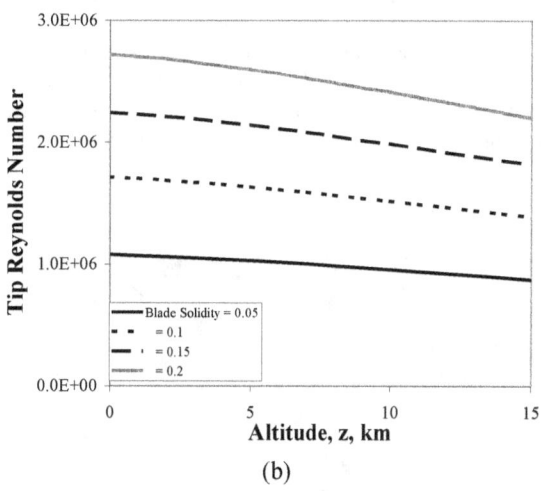

Fig. 6. Rotor Operating Conditions During Descent: (a) Tip Mach and (b) Tip Reynolds Numbers as a Function of Altitude (and Blade Solidity)

7. FUTURE SURROGATE PROBE TESTING

The use of underwater submersibles to demonstrate and evaluate teleoperation and robotic technologies for NASA planetary science missions is not a new technical approach. Previous work has been conducted, such as the Ames TROV project [12].

Though Venus's lower atmosphere has pressure levels comparable to the ocean depths on Earth, the analogy between the two is only of limited aerodynamic value. However, on the other hand, there is considerable value in the possible test and evaluation of surrogate underwater probes for the proof-of-concept testing of descent trim-control laws and terminal stage guidance and navigation and autonomy technologies.

The majority of demonstrations will entail use of small-scale probes that will be released in an artificial pool/tank of water. The test and evaluation team will place artificial hazards (orange markers) at the bottom of the pool (Fig 7). Control of the probe hazard avoidance and precision landing guidance will be provided by using simple optical imagers, existing vision-system software, a pool-side lap-top computer, and radio-frequency (RF) or ultrasound I/O for telemetry and control inputs. The proposed simple vision-system initially to be used in the demonstrations has been previously used for other, similar vehicle guidance projects [12-13]. Additionally, other sensors and systems will be based in part on experience gained in the development of small robotic underwater vehicles. In the final demonstrations the probe will be of larger size and capability and there will be increased realism of terrain hazards at the bottom of the natural body of water (a lake such as Tahoe or Mono) where the underwater landscape can be conveniently evaluated prior to field trials.

Fig. 7. Small-scale Surrogate Probe Testing

8. CONCLUDING REMARKS

Preliminary work related to the use of rotary-wing decelerators for application to Venus entry-probes/landers has been found to be very promising. A considerable amount of work remains to be performed – including work in the areas of control law development, hazard avoidance strategies, and surrogate probe testing.

10. ACKNOWLEDGMENTS

The contribution of Jim Kennon, Projects Division, to this work is gratefully acknowledged.

11. REFERENCES

1. "New Frontiers in the Solar System: An Integrated Exploration Strategy," NRC Decadal Study.
2. Levin, A.D. and Smith, R.C., "Experimental Aerodynamics of a Rotor Entry Vehicle," 3rd AIAA Aerodynamic Decelerator Systems Technology Conference, El Centro, CA, September 23-25, 1968, Technical Report NO. 69-11 April 1969.
3. Iverson, J.D., "The Magnus Rotor as an Aerodynamic Decelerator,"3rd AIAA Aerodynamic Decelerator Systems Technology Conference, El Centro, CA, September 23-25, 1968, Technical Report NO. 69-11 April 1969.
4. Smith, R.C. and Levin, A.D., "The Unpowered Rotor: A Lifting Decelerator for Spacecraft Recovery," 3rd AIAA Aerodynamic Decelerator Systems Technology Conference, El Centro, CA, September 23-25, 1968, Technical Report NO. 69-11 April 1969.
5. Levin, A. D. and Smith, R. C., "Experimental aerodynamic performance characteristics of a rotor entry vehicle configuration. 1 – Subsonic," NASA-TN-D-7046, 1971.
6. Levin, A. D. and Smith, R. C., "Experimental aerodynamic performance characteristics of a rotor entry vehicle configuration. 2 – Transonic," NASA-TN-D-7047, 1971.
7. Levin, A. D. and Smith, R. C., "Experimental aerodynamic performance characteristics of a rotor entry vehicle configuration. 3 – Supersonic," NASA-TN-D-7048, 1971.
8. Levin, A. D. and Smith, R. C., "An analytical investigation of the aerodynamic and performance characteristics of an unpowered rotor entry vehicle," NASA-TN-D-4537, 1968.
9. Giansante, N. and Lemnios, A. Z., "The dynamic behavior of rotor entry vehicle configurations. Volume 1 - Equations of motion," NASA-CR-73390, 1968.
10. Young, L.A., Chen, R., Aiken, E., and Briggs. G., "Design Opportunities and Challenges in the Development of Vertical Lift Planetary Aerial Vehicles," American Helicopter Society (AHS) Vertical Lift Aircraft Design Conference, San Francisco, CA, January 2000.

11. Johnson, W., *Helicopter Theory*, Princeton University Press, 1980.
12. Hine, B., Stoker, C., Sims, M., et al, "The Application of Telepresence and Virtual Reality to Subsea Exploration," 2nd Workshop on Mobile Robots for Subsea Environments, Proceedings ROV'94, Monterey, CA, May 1994.
12. Plice, L., Pisanich, G., Lau, B., and Young, L.A., "Biologically Inspired 'Behavioral' Strategies for Autonomous Aerial Explorers on Mars," IEEE Aerospace Conference, Big Sky, MT, March 2003.
13. Pisanich, G., Young, L.A., Plice, L., Ippolito, C., Lau, B., Lee, P., "Initial Efforts towards Mission-Representative Imaging Surveys from Aerial Explorers," SPIE Electronic Imaging Conference, San Jose, CA, January 2004.
14. Lodders, K. and Fegley, Jr., B., *The Planetary Scientist's Companion,* Oxford University Press, 1998.

A Search for Viable Venus and Jupiter Sample Return Mission Trajectories for the Next Decade

Jason N. Leong and Dr. Periklis Papadopoulos
Department of Mechanical and Aerospace Engineering
San José State University
One Washington Square
San Jose, CA 95192-0080, USA

ABSTRACT

Planetary exploration using unmanned spacecraft capable of returning geologic or atmospheric samples have been discussed as a means of gathering scientific data for several years. Both NASA and ESA performed initial studies for Sample Return Missions (SRMs) in the late 1990's, but most suggested a launch before the year 2010. The GENESIS and STARDUST spacecraft are the only current examples of the SRM concept with the Mars SRM expected around 2015. A feasibility study looking at SRM trajectories to Venus and Jupiter, for a spacecraft departing the Earth between the years 2011 through 2020 was conducted for a university project. The objective of the study was to evaluate SRMs to planets other than Mars, which has already gained significant attention in the scientific community. This paper is a synopsis of the study's mission trajectory concept and the conclusions to the viability of such a mission with today's technology.

1. INTRODUCTION

The Apollo Program's lunar landing missions represent the only successful attempts to date in which a specimen from a celestial body has been return to the Earth for study. It has been over thirty years since the Apollo Program. The concept of a SRM affords the scientific community the opportunity to closely study material from another planet. This opportunity can help answer question such as the chemical composition of a planet's surface and the atmospheric composition of a planet of interest. This type of information assists scientists in better understanding the Earth and its future.

While not a new concept, SRMs previously have not been in the forefront of planetary exploration. Because of the requirement to "return a sample," considerable energy is required to successfully complete such a mission. The energy requirement for SRMs equates to propellant mass and ultimately the launch costs of the mission. Previous efforts have concluded multiple launches are necessary because of the large required propellant for the SRM [1].

The current state of rocket and propulsion technology, however, warrants a feasibility evaluation for the SRM occurring in the next decade. The Evolved Extended Launch Vehicle (EELV) represents the backbone of US launch vehicles in the near future.

Since Mars is the current focus of NASA, alternative planets are used to evaluate the feasibility of SRM. The planet Venus, which was explored by the Magellan spacecraft from 1990 until it was de-orbited into the planet in October 1994, was chosen for the inner planet study case. Venus represents an alternative to the currently popular Mars exploration missions. Venus is also closer to Earth than Mars thereby increasing the chances for a feasible mission. The outer planet case utilizes the planet Jupiter, which was explored by the Galileo spacecraft until it was de-orbited into Jupiter's atmosphere in September 2003. Jupiter's moon Europa is of particular scientific interest because of its icy surface, which is believed to hold the building blocks of life, warrants a comprehensive "Mars-like" exploration may be in the future [2]. Jupiter itself is the closest of the outer planets, which makes it a suitable bounding case.

To narrow the scope of the architecture study, three constraints were observed. The first constraint is the Earth departure date. NASA has speculated the year 2011 as the earliest launch of a MSRM. Other SRMs, which utilize reuse of the MSRM technology, can therefore occur no earlier than 1 January 2011. The second constraint is the ten-year MET for the operational phase of the mission. This constraint was chosen for two reasons. Ten-years is just short of the orbital period of Jupiter, the closest of the outer planets. Ten-years also reduces the required computational simulation time to a manageable level.

2. BACKGROUND

A Matlab math model is utilized to compute the transfer orbit trajectories and thus the ΔV for this study. The ΔVs are essential to perform the viability assessment of each case study mission. To solve for the trajectories, two fundamental orbit mechanics methods are used:

- The patched-conic Approximation
- Lambert's Problem

2.1. The Patched-conic Approximation

The model makes use of the patched-conic approximation to determine the interplanetary trajectories. The patched-conic method breaks the interplanetary trajectory into three small distinct problems, and then solved using the two-body system. The patched-conic approximation is an industry-accepted method when making a first-ordered analysis for interplanetary trajectories. The method has the advantage of shorting the computational time without sacrificing the integrity of the generated data [3].

The simplifying assumption of the patched-conic is based on the utilization of the two-body system to approximate the motion of the spacecraft through its trajectory. The idea of the two-body system is when considering only two bodies, the spacecraft and the celestial body it orbits, the gravitational attraction of the celestial body is the dominant effect on the spacecraft. Because of the dominating gravitational effect of the celestial body such as a planet, the spacecraft is said to be within the "sphere of influence" of the planet. Fig. 1 illustrates the various trajectory geometries. For all three parts of the patched-conic approximation, the perturbation effects by all other bodies including the spacecraft itself are considered small because of the sphere of influence concept.

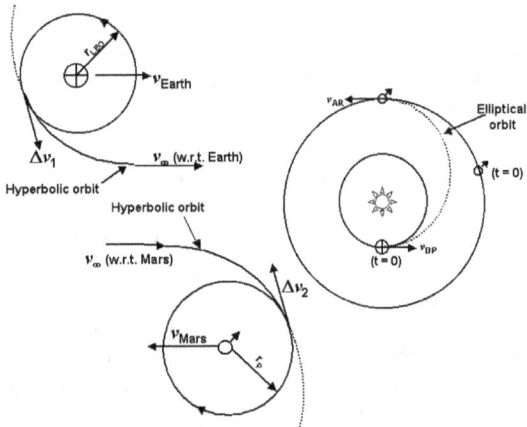

Fig. 1: Patched-Conic Departure, Interplanetary, and Arrival Trajectories

Additional assumptions are made for trajectory simulation purposes. The first assumption is that a Launch Vehicle (LV) places a spacecraft into an initial circular Low Earth Orbit (LEO) altitude of 300 km. This assumption is used as an initial condition for the model. For the interplanetary transfer orbit, a Hohmann transfer is assumed, which yields the theoretical minimum ΔV. Since the atmosphere of Venus begins at approximately 100 km. altitude, a circular orbit at this altitude is assumed for the insertion and corresponding departure orbits for the Venus SRM (VSRM) study case. Because of the general scientific interest, the orbital altitude of Jupiter's moon, Europa, is assumed for the Jupiter SRM (JSRM) study case. No attempt is made to model the orbit about Europa itself.

2.2. Lambert's Problem

Although the patched-conic method provides the information for the spacecraft to perform the ΔVs, the orbital elements still need to be determined. For the model, the user selects the departure and the arrival dates for the transfer orbit. This provides the Time of Flight (TOF) for the interplanetary trajectory. The departure date also fixes a departure position of the departure planet and thus the spacecraft with respect to the sun. Likewise, the arrival date fixes the arrival position of the target planet also with respect to the sun. Combining these three elements together, the two position vectors and the time between them, uniquely defines the transfer orbit. The problem of two position vectors and the TOF between them is known as "Lambert's Problem." There have been several methods for solving Lambert's problem. Because of its robustness, the algorithm developed by Battin [4] is ideal for general-purpose use and is implemented for this study.

2.3. Methodology

Because of the ten-year departure period and ten-year mission life, over 26-million trajectories are generated per case for this study. To solve for the transfer orbit trajectories and the mission ΔVs, a systematic mission analysis process is developed. Fig. 2 illustrates the mission analysis process, at a high level, as implemented in the Matlab software code. The model takes a departure date then steps through each arrival date increasing TOF. Once all arrival dates have been iterated upon, the departure date is incremented and the process repeats itself. After all the departure dates have been iterated, the model reduces the data by searching for the minimum ΔV for that particular case.

Once the user inputs the range of departure dates and derives the last arrival date, the model converts the dates to Julian days for analysis. A matrix of arrival dates verses departure dates is then created with each value representing an elapsed Julian day from the departure date. The planetary ephemeris of the target planet is updated using Lambert's problem based on reference planetary ephemeris form observational data and the arrival date. Similarly, the planetary ephemeris of the departure planet is updated based on the departure date. This gives enough information, the two position vectors and the time between them, to solve Lambert's problem for the interplanetary transfer orbit.

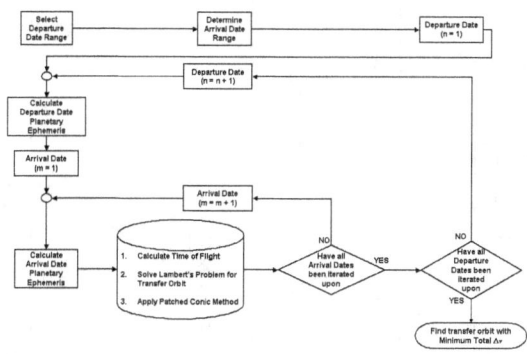

Fig. 2: Matlab Mission Analysis Model

3. RESULTS

To evaluate the trajectories for both ΔV_{Total} and TOF a figure of merit (FOM) [5] analysis is performed with the goal of finding the trajectory, which maximizes the total FOM. Since minimizing the ΔV_{Total} equates to a savings in propellant mass, the ΔV parameter is weighted twice as important than TOF. The scoring for the ΔV_{Total} is simply a ten to one scale with the smallest ΔV_{Total} as ten and the largest ΔV_{Total} as one. The TOF is also scored on a ten to one scale with the shortest TOF as ten and the longest TOF as one. Following this evaluation criterion, an equation is developed to score each candidate trajectory:

$$2 \times \Delta V_{Total} \text{ score} + \text{TOF score} = \text{FOM}_{Total}$$

The results show the minimum ΔV_{Total} transfer orbit opportunities for each year of the next decade. For each departure date, ten years of arrival dates were evaluated to determine the minimum ΔV_{Total} for that year.

3.1. Venus Sample Return Mission Trajectories

Tables 1 and 2 show the Earth to Venus velocity and flight time characteristics for the minimum ΔV transfer orbits per Earth departure year respectively. Table 1 provides the required ΔV maneuvers to depart Earth and insert the spacecraft into Venus orbit. The table also includes the velocity at infinity (V_∞) of the hyperbolic departure orbit, and the corresponding escape energy (C_3). Table 2 provides the departure and arrival dates and the associated TOF of the trajectory.

Table 1: Velocity characteristics for Earth to Venus transfer

Year	ΔV_1 (km/sec)	ΔV_2 (km/sec)	V_∞ (km/sec)	C_3 (km²/sec²)
2011	4.1554	3.1367	4.6678	21.7884
2012	3.3039	3.8206	1.5093	2.2780
2013	3.3349	3.7323	1.7213	2.9629
2014	3.6720	3.3514	3.2456	10.5339
2015	3.5653	3.4376	2.8486	8.1145
2016	3.2167	4.1041	0.6028	0.3633
2017	3.3580	3.6342	1.8643	3.4755
2018	3.3854	3.5954	2.0208	4.0836
2019	4.0909	3.1399	4.5008	20.2576
2020	3.2585	3.9054	1.1306	1.2783

Table 2: Time of Flight for Earth to Venus transfer

Year	Departure Date	Arrival Date	TOF (Days)
2011	12/29/11	8/25/12	240
2012	12/31/12	11/20/13	324
2013	1/8/13	11/27/13	323
2014	12/28/14	9/22/15	268
2015	1/1/15	9/26/15	268
2016	1/1/16	12/18/16	352
2017	12/28/17	10/20/18	296
2018	1/4/18	10/26/18	295
2019	12/31/19	8/26/20	239
2020	12/26/20	11/15/21	324

Tables 3 and 4 show the Venus to Earth velocity and flight time characteristics respectively for the minimum ΔV transfer orbits per Venus departure year. The parameters in Table 3 are similar to that of Table 1 with the exception of excluding V_∞ and the corresponding C_3. Although these values are calculated in the Matlab model, their main use is for LV sizing. In the case of the return trajectory, there is no LV because the spacecraft itself performs the ΔV required to insert the payload into a hyperbolic Venus departure orbit. The values of V_∞ and the corresponding C_3 are thus not applicable to the return trajectory in the context of this project. The data in Table 4 is similar to that in Table 2.

Table 3: Velocity characteristics for Venus to Earth transfer

Year	ΔV_1 (km/sec)	ΔV_2 (km/sec)	V_∞ (km/sec)	C_3 (km^2/sec^2)
2012	4.4792	3.4601	N/A	N/A
2013	3.8434	3.3575	N/A	N/A
2014	3.2888	3.9291	N/A	N/A
2015	4.3108	3.4127	N/A	N/A
2016	3.3875	3.6384	N/A	N/A
2017	3.1493	4.4450	N/A	N/A
2018	4.0564	3.3329	N/A	N/A
2019	3.5294	3.6128	N/A	N/A
2020	4.4987	3.5093	N/A	N/A
2021	3.8439	3.3414	N/A	N/A

Table 4: Time of Flight for Venus to Earth transfer

Year	Departure Date	Arrival Date	TOF (Days)
2012	5/12/12	7/7/13	421
2013	8/5/13	7/8/14	337
2014	10/29/14	7/9/15	253
2015	6/9/15	7/6/16	393
2016	12/21/16	1/4/17	14
2017	11/26/17	7/10/18	226
2018	7/8/18	7/9/19	366
2019	10/2/19	7/10/20	282
2020	5/12/20	7/8/21	422
2021	8/4/21	7/8/22	338

Table 5 shows the results of the FOM scoring for the VSRM. The results show planning a mission utilizing the highest score for both the departure and return trajectories yields no solution since the return trajectory occurs before the departure trajectory. The dilemma for mission planning is which trajectory, departure or return, to select. Utilizing the FOM evaluation criteria based on departure trajectory yields TOF to Venus as 295 days but, limits the return trajectories to after 26 October 2018. Basing the mission on the return trajectory criteria yields TOF to Venus as 268 days. In the course of this study, a 14-day return trajectory in late 2016 was found which seems to take advantage of a favorable planetary alignment between Venus and Earth.

Table 5: Trajectory Figures of Merit for Venus SRM

Departure Trajectories				Return Trajectories			
Year	ΔV Score	TOF Score	FOM Score	Year	ΔV Score	TOF Score	FOM Score
2011	2	9	13	2012	2	2	6
2012	5	3	13	2013	7	6	20
2013	6	4	16	2014	6	8	20
2014	7	8	22	2015	3	3	9
2015	8	8	24	2016	10	10	30
2016	1	1	3	2017	4	9	17
2017	9	5	23	2018	5	4	14
2018	10	6	26	2019	9	7	25
2019	3	10	16	2020	1	1	3
2020	4	3	11	2021	8	5	21

3.2. Jupiter Sample Return Mission Trajectories

Tables 6 and 7 show the Earth to Jupiter velocity and trajectory characteristics for the minimum ΔV transfer orbits per Earth departure year respectively. Tables 8 and 9 show the Jupiter to Earth velocity and trajectory characteristics respectively for the minimum ΔV transfer orbits per Jupiter departure year. The velocity and trajectory characteristics in all the JSRM tables are similar to their VSRM counterparts.

Table 6: Velocity characteristics for Earth to Jupiter transfer

Year	ΔV_1 (km/sec)	ΔV_2 (km/sec)	V_∞ (km/sec)	C_3 (km^2/sec^2)
2011	6.1493	7.7272	8.5524	73.1430
2012	6.4899	10.1822	9.0945	82.7106
2013	7.5551	14.3265	10.6831	114.1287
2014	7.2125	10.7404	10.1870	103.7759
2015	7.1223	10.7077	10.0544	101.0916
2016	7.2544	5.6990	10.2485	105.0314
2017	7.1542	5.7088	10.1014	102.0386
2018	7.0373	5.7384	9.9284	98.5723
2019	6.8993	5.8030	9.7220	94.5172
2020	6.7369	5.9309	9.4759	89.7929

Table 7: Time of Flight for Earth to Jupiter transfer

Year	Departure Date	Arrival Date	TOF (Days)
2011	7/11/11	6/17/14	1072
2012	7/15/12	8/15/14	761
2013	10/18/13	12/31/23	3726
2014	9/22/14	12/31/24	3753
2015	9/21/15	12/31/24	3389
2016	7/13/16	4/10/26	3558
2017	7/13/17	4/10/26	3193
2018	7/13/18	4/10/26	2828
2019	7/13/19	4/10/26	2463
2020	7/11/20	4/10/26	2099

The differences between the candidate departure orbits are seen in the ΔV data in Table 6. Between 2016 and 2020 the ΔV_1 magnitude has a decreasing trend while the ΔV_2 magnitude has a corresponding increasing trend. The change in the ΔV's are directly related to the changes in the TOF in Table 7 because of the differences in the calculated trajectories despite the similar departure dates and the same arrival date. The trend also suggest a lower ΔV_{Total} in the years just beyond 2020, which is beyond the scope of this study.

Table 8: Velocity characteristics for Jupiter to Earth transfer

Year	ΔV_1 (km/sec)	ΔV_2 (km/sec)	V_∞ (km/sec)	C_3 (km²/sec²)
2020	7.2233	5.7540	N/A	N/A
2021	7.3096	5.8867	N/A	N/A
2022	7.5485	6.2501	N/A	N/A
2023	7.5431	6.2420	N/A	N/A
2024	7.4279	6.0674	N/A	N/A
2025	7.3931	6.0144	N/A	N/A
2026	7.4131	6.0449	N/A	N/A
2027	10.4812	10.3805	N/A	N/A
2028	20.5148	22.4696	N/A	N/A
2029	32.4785	35.5349	N/A	N/A

Table 9: Time of Flight for Jupiter to Earth transfer

Year	Departure Date	Arrival Date	TOF (Days)
2020	7/5/20	1/8/23	917
2021	1/1/21	12/17/22	715
2022	2/4/22	11/4/24	1004
2023	12/31/23	9/22/26	996
2024	12/31/24	8/26/27	968
2025	6/17/25	8/15/27	789
2026	1/26/26	7/18/28	904
2027	1/1/27	7/31/29	942
2028	1/1/28	8/28/29	605
2029	1/1/29	9/27/30	634

Table 10 shows the results of the FOM scoring for the JSRM. The results show a similar mission planning dilemma as the VSRM results.

Table 10: Trajectory Figures of Merit for Jupiter SRM

Departure Trajectories				Return Trajectories			
Year	ΔV Score	TOF Score	FOM Score	Year	ΔV Score	TOF Score	FOM Score
2011	5	9	19	2020	10	5	25
2012	4	10	18	2021	9	8	26
2013	1	2	4	2022	4	1	9
2014	2	1	5	2023	5	2	12
2015	3	4	10	2024	6	3	15
2016	6	3	15	2025	8	7	23
2017	7	5	19	2026	7	6	20
2018	8	6	22	2027	3	4	10
2019	9	7	25	2028	2	10	14
2020	10	8	28	2029	1	9	11

4. DISCUSSION

A major concern for any space mission is the minimization the total mass of the spacecraft, which includes the mass of the propellant necessary to perform the mission. For SRMs in general, a significant amount of propellant is required since the spacecraft must perform a ΔV to place itself onto a return hyperbolic trajectory. A viable mission must also have mass allocated for the spacecraft subsystems such as electrical power, guidance and navigation, thermal control, and communications in addition to the physical structure which amount to the "dry mass" of the spacecraft. The "wet mass" of the spacecraft is the dry mass with the addition of the required propellant mass. The wet mass represents the gross mass of the spacecraft injected into space by the LV. The LV "throw" capability is therefore the parameter used for evaluation of mission feasibility.

For this study, the Delta IV is assumed for mission feasibility evaluation. Fig. 3 shows the Earth escape energy performance for the various configuration of the Delta IV. From Fig. 3 the Delta IV Heavy configuration can "throw" approximately 7800 kg of gross spacecraft mass if the required energy to escape Earth's gravity is 10 km^2/s^2 or less.

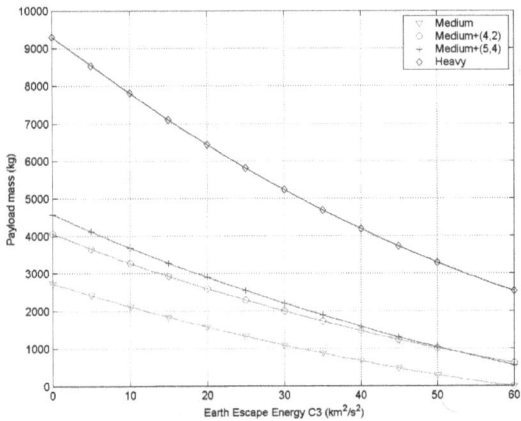

Fig. 3: Predicted Planetary Mission Performance of Delta IV Launch Vehicles [6]

4.1. Case Selection

To determine the feasibility of the two mission cases, a representative mission plan from each study case is selected. In both study cases, the representative mission plan uses the FOM criteria for the return trajectory then matches the "best" departure trajectory, based on FOM score. Although no JSRM mission plan meets the ten-year MET constraint, the representative case is evaluated for completeness.

Table 11 shows the selected VSRM evaluation case. The total MET for the VSRM is 734 days or two years, three days. Comparing the transfer time to Venus with the Magellan mission, the selected VSRM plan arrives 195 days sooner than the Magellan spacecraft, which had a transfer time of 463 days.

Table 11: Selected Mission Plan for VSRM

Parameter	Earth to Venus	Venus to Earth
Departure Date	1/1/15	12/21/16
Arrival Date	9/26/15	1/4/17
Time of Flight (Days)	268	14
ΔV_1 (km/sec)	3.5653	3.3875
ΔV_2 (km/sec)	3.4376	3.6384
ΔV_{Total} (km/sec)	7.0029	7.0259
Departure V_∞ (km/sec)	2.8486	N/A
Launch Vehicle Escape Energy C_3 (km^2/sec^2)	8.1145	N/A

Table 12 shows the selected JSRM evaluation case. The total MET for the JSRM is 4,177 days. Comparing the JSRM departure transfer time to that of the Galileo mission to Jupiter, the JSRM arrives at Jupiter in just under three years while Galileo took six years of transfer time using planetary gravity assists.

Table 12: Selected Mission Plan for JSRM

Parameter	Earth to Jupiter	Jupiter to Earth
Departure Date	7/11/11	1/1/21
Arrival Date	6/17/14	12/17/22
Time of Flight (Days)	1072	715
ΔV_1 (km/sec)	6.1493	7.3096
ΔV_2 (km/sec)	7.7272	5.8867
ΔV_{Total} (km/sec)	13.8765	13.1963
Departure V_∞ (km/sec)	8.5524	N/A
Launch Vehicle Escape Energy C_3 (km^2/sec^2)	73.1430	N/A

4.2. Mission Feasibility Evaluation

Examination of the C_3 from both missions and comparing them to the curves in Fig. 3, a maximum spacecraft mass is determined. For the VSRM, an interpolated payload mass of 8085 kg is found. By assuming direct injection, the first ΔV performed at the 300 km altitude is ignored. An extrapolated mass of 1918 kg is determined for the JSRM. The spacecraft mass for the JSRM is nearly 25% that of the VSRM because of the large C_3 term. Given these two mass figures and the ΔV information from both mission plans, wet and dry mass estimates are derived using the rocket equation, with specific impulse (Isp) as the variable. Current bi-propellant systems have and Isp in the range of 200 to 450 seconds.

Fig. 4 illustrates the wet/dry mass estimates for both missions assuming direct injection of the spacecraft into the transfer orbit by the Delta IV Heavy LV. The spacecraft performs the remaining three ΔV maneuvers outlined in the mission plans.

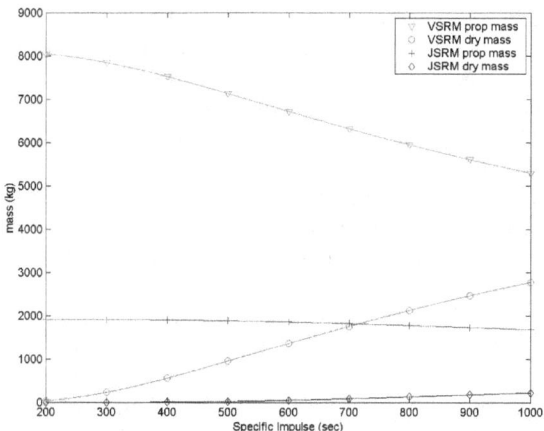

Fig. 4: Spacecraft mass breakdown (three-ΔV case)

For the VSRM case, the 8085 kg gross mass breaks down into 7329.4 kg of propellant and 755.6 kg of dry mass for an Isp of 450 seconds. More challenging, from a mass perspective, is the JSRM with only 16.8 kg of dry mass to allocate for the same Isp as the VSRM. Even with an Isp of 1000 seconds, the JSRM could only allocate 227.3 kg of dry mass to the spacecraft subsystems. To approach the 755.6 kg dry mass of the VSRM with an Isp of 450 seconds, the Isp required for the JSRM is 2300 seconds.

Because of the mass challenges, a two-ΔV scenario is conceived for study. A ballistic return trajectory is considered to eliminate the need to perform a ΔV to insert the spacecraft into Earth orbit. A ballistic trajectory assumes the mission is planned well enough such that the returning sample capsule will re-enter Earth's atmosphere on a precise trajectory to land on a predetermined spot on the Earth for recovery. Fig. 5 shows the mass estimates for the two-ΔV scenario.

For the VSRM two-ΔV case, the 8085 kg gross mass breaks down into 6362.2 kg of propellant and 1722.8 kg of dry mass for an Isp of 450 seconds. Similar analysis for the JSRM shows a propellant mass of 1854.4 kg and a dry mass of 63.6 kg. Although the dry mass for the JSRM has increased almost four times that of the three-ΔV case, the JSRM still has significant challenges when it comes to dry mass.

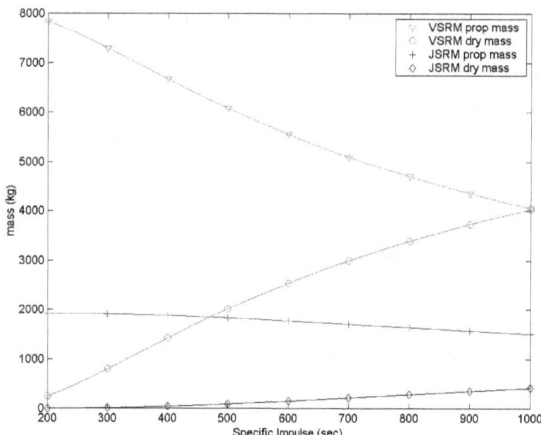

Fig. 5: Spacecraft mass breakdown (two-ΔV case)

Since Magellan and Galileo did not have a requirement to return to Earth, a strait comparison of dry mass is somewhat biased. The above two-ΔV and three-ΔV cases assume the total dry mass of the spacecraft including the lander returns to Earth. An actual SRM would most likely abandon the lander after that portion of the mission was over. Abandoning any unnecessary dry mass has the advantage of reducing mass of the spacecraft for the return trajectory. Each mission phase has it own unique dry mass configuration leading to a form of mass staging, which is common in multi-stage rockets. Table 13 shows the estimated dry mass for the various pieces of hardware form a JPL study. Assuming these are the required dry masses for any SRM, the gross mass of the spacecraft with propellant can be determined for both SRM study cases.

Table 13: JPL VSRM study dry mass estimate [1]

Subsystem	Departure Trajectory dry mass	Return Trajectory dry mass
Orbiter Vehicle	680 kg	680 kg
Earth Entry Vehicle	20 kg	20 kg
Lander	931 kg	N/A
Ascent System	476 kg	N/A
Sample	N/A	5 kg
Dry Mass Totals	2107 kg	705 kg

The table shows two mass configurations, the departure trajectory and the return trajectory configuration. For the departure trajectory configuration, the LV must lift the Orbiter Vehicle and Earth Entry Vehicle as well as the Lander and Ascent System. For the return trajectory, the Lander and the Ascent System are abandoned since they were only required to collect the five-kilogram sample, rendezvous with the Orbiter Vehicle, and transfer the sample to the Earth Entry Vehicle. Once sample transfer is complete, the Ascent System is jettisoned. The Lander itself remains on the surface of the planet surface. The dry mass of the return trajectory is approximately one-third that of the departure trajectory. Since the change in dry mass affects the required propellant mass, a savings in total gross mass is realized.

Fig. 6 demonstrates the advantages of mass staging technique for the VSRM case. It shows, that for an Isp of 450 seconds and the dry mass configurations for the different mission stages shown in Table 13, the total spacecraft gross mass at launch is 6363 kg. This gross mass includes 4256.0 kg of propellant mass for the two-Dv maneuvers. The propellant required to insert the spacecraft into Venus orbit is 3442.4 kg. To perform the Dv maneuver for a ballistic return trajectory to Earth requires 813.6 kg of propellant.

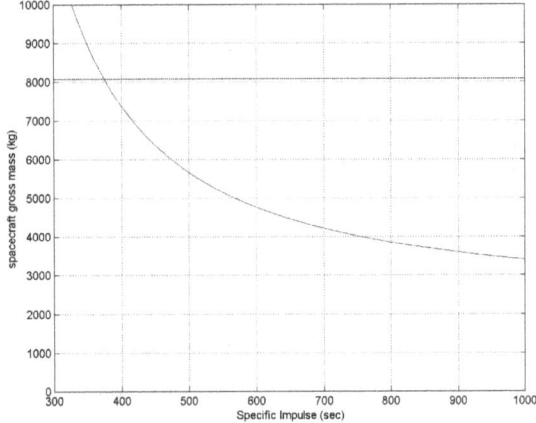

Fig. 6: VSRM gross spacecraft mass estimate (two-ΔV case) using staging

The staging technique indicates a VSRM is feasible with current bi-propellant technology because the gross mass at launch is less than the reference 8085 kg LV capability. The remaining lift capability of 1722 kg equates to a 21.3% margin. Margin of this magnitude is essential during the developmental phase of any program to cover any unforeseen contingency situations.

Since the dry mass of the JPL proposed SRM is greater than the 1918 kg lift capability of the Delta IV Heavy for the JSRM, the mission is automatically not feasible. Other techniques in mission design could be employed for the JSRM such as the gravity assist, but that technique has the detriment of significantly increasing the TOF. In the case of To make the JSRM possible, a LV with significantly more capability needs to be developed or a different mission design other than direct injection, such as gravity assist trajectories, is required.

5. CONCLUSIONS

This study has shown that an inner planet SRM and specifically a VSRM is feasible in the 2011 – 2020 timeframe. The selected mission plan meets the key parameters of minimizing both ΔV and MET for the mission. The mission duration of two years and three days meets the less than ten-year objective of the study. The spacecraft's gross mass estimate of 6363 kg meets the lift capability of the Delta IV Heavy LV with 21.3% margin. The current state of the art propulsion technology is sufficient to meet the required ΔVs of the mission. This study has also shown that a JSRM is not feasible in the 2011 – 2020 timeframe. The selected mission plan does minimize ΔV and MET, but the MET exceeds the ten-year parameter of this study. Significant LV development is required to increase the lift capability for a direct injection. This type of LV development is currently not planed to meet this study's timeframe. Even with improvements to LV lift capability, the JSRM spacecraft requires significant advances in propulsion technology to reduce the required propellant. Since Jupiter is the closest outer planet, study of SRM's to any of the other outer planets is not recommended given the current state of technology.

6. REFERENCES

1. Rogers, David et al. 1999. *Venus Sample Return: A Hot Topic*. JPL report 99-1991.

2. Nilsen, Erik N. 2000. Future Mission Concepts for the Exploration of the Solar System. In the *Bio-Inspired Engineering of Exploration Systems 2000 Workshop* held Pasadena, CA. 4-6 December 2000.

3. Prussing, John E. and Bruce A. Conway. 1993. *Orbital Mechanics*. Oxford University Press, Inc. pp. 124-128.

4. Battin, Richard H. 1999. *An Introduction to the Mathematics and Methods of Astrodynamics*. American Institute of Aeronautics and Astronautics, Inc. pp. 325-342.

5. Larson, Wiley J. and James R. Wertz, Eds. 1992. *Space Mission Analysis and Design, Second Edition*. Microcosm, Inc. pp. 59-61.

6. The Boeing Company. 2000. *Delta IV Payload Planner's Guide*. The Boeing Company.

SYNERGY BETWEEN ENTRY PROBES AND ORBITERS

Richard E. Young[1]

[1]NASA Ames Research Center, MS 245-3, Moffett Field, CA 94035 USA, Email: Richard.E.Young@nasa.gov

ABSTRACT

We identify two catagories of probe-orbiter interactions which benefit the science return from a particular mission. The first category is termed "Mission Design Aspects". This category is meant to describe those aspects of the mission design involving the orbiter that affect the science return from the probe(s). The second category of probe-orbiter interaction is termed "Orbiter-Probe Science Interactions", and is meant to include interactions between oribter and probe(s) that directly involve science measurements made from each platform. Two mission related aspects of probe-orbiter interactions are delivery of a probe(s) to the entry site(s) by an orbiter, and communication between each probe and the orbiter. We consider four general probe-orbiter science interactions that greatly enhance, or in certain cases are essential for, the mission science return. The four topics are, global context of the probe entry site(s), ground truth for remote sensing observations of an orbiter, atmospheric composition measurements, and wind measurements.

1. INTRODUCTION

The principal distinguishing measurement feature of atmospheric entry probes/surface landers, as compared to observations from orbit or flyby spacecraft, is that probes/landers typically make in-situ measurements. Conducting remote sensing of a planetary atmosphere or surface in order to obtain composition, cloud information, thermal characteristics, or winds, usually involves the inversion of spectra obtained with various forms of spectrometers or radiometers yielding results that are model dependent. Particular key measurements can be identified that cannot adequately be made remotely, either because the sensitivity of measurement is insufficient to measure the desired quantity to the required accuracy, or because there is no feasible remote sensing observation that can return the desired information.

On the other hand it is often desired to know the global distribution of key quantities, and this is only feasible from an orbiter. Entry probes/surface landers give essentially point measurements at the entry location, either by providing vertical profiles of atmospheric quantities at one horizontal location, or a measurment from a particular surface location. The optimal program is a balanced set of in-situ and remote sensing observations that complement each other.

We identify two catagories of probe-orbiter interactions which benefit the science return from a particular mission. The first category is termed "Mission Design Aspects". This category is meant to describe those aspects of the mission design involving the orbiter that affect the science return from the probe(s). The second category of probe-orbiter interaction is termed "Orbiter-Probe Science Interactions", and is meant to include interactions between oribter and probe(s) that directly involve science measurements made from each platform. Examples from each category are discussed below.

2. MISSION DESIGN ASPECTS

Two mission related aspects of probe-orbiter interactions are delivery of a probe(s) to the entry site(s) by an orbiter, and communication between each probe and the orbiter. Delivery of probes from an orbiter has the potential to allow access to desirable probe entry sites that otherwise could not be reached. Communication between probe(s) and orbiter has the potential to allow access to desirable probe entry sites that would not be available by direct communication to Earth, as well as the potential of direct science collaboration. Examples using past missions to Venus and Jupiter will be discussed below to illustrate how such probe-orbiter interactions can pay off in the future.

Fig. 1 (adapted from [1]) shows the distribution of the Pioneer Venus probes in a coordinate system fixed with respect to the subsolar point on Venus. Also shown is the distribution of the Venera series of probes. The four Pioneer Venus probes and the Venera probes up through Venera 8 were delivered by a dedicated bus spacecraft and communicated directly to Earth. The PV small probes were released almost simultaneously from the probe bus, with the large probe having been released a few days earlier from the same bus. Note that because of the communication constraint directly to Earth, all these probes entered either on the night side of Venus or in the early morning, local time. On the other hand, Veneras 9-12 all communicated either with a flyby parent spacecraft (V11-V12) or an orbiter (V9-V10). These probes were able to descend in the noon and afternoon regions of the atmosphere.

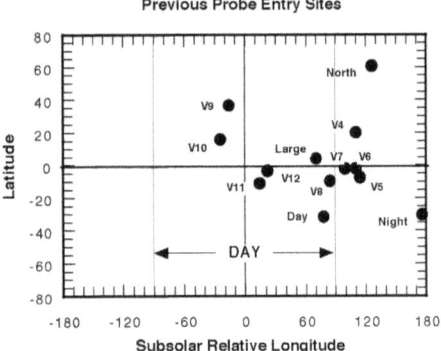

Fig. 1. Locations of previous probe entry sites at Venus, in subsolar longitude. Adapted from [1]. Subsolar point is at center of figure. The probes designated Large, North, Day, and Night are the PV probes. The other probes are Veneras 8-12.

There are meteorological reasons why reaching local noon and later meridians is desirable. For example, there is evidence of convective cells having large horizontal scales (500-1000 km) occurring at cloud levels in the mid to late afternoon local time ([2] and references therein). This could be evidence that the thermal structure of the atmosphere differs significantly in the afternoon from what has been measured in the early morning and night regions. If so, there are implications for understanding the Venus superrotation and related circulation patterns, as well as mixing of trace species from the surface to cloud levels, which in turn affects cloud microphysics and composition. But, as Fig. 1 illustrates, accessing afternoon regions of the atmosphere depends on communicating with an orbiter (or possibly flyby spacecraft) and not directly to Earth.

Of even higher current science priority is reaching particular regions of the Venus surface. It has been established by the science community [3] that it is imperative to determine the elemental and mineralogical composition of the Venus surface at a variety of sites, including especially the highland tessera. Such information is essential in trying to understand how Solar System terrestrial planet formation may have been similar or different among the planets, and in what ways differences may have occurred.

Fig. 2 is a topography map (originally in color) produced by the Pioneer Venus Orbiter [4]. The tessera are located in the high regions, but the lowland plains are also of interest. Each type of region should be sampled, and preferably at multiple sites. In order to so will almost surely require both delivery by, and communication with, an overflying spacecraft for each landed science package. It would be highly unlikely that all desired landed sites would be accessible by probe/landers launched from a single carrier on a flyby trajectory, nor is it likely that probes at such diverse sites would each be able to communicate directly to Earth. Therefore, an orbiter component to a Venus mission investigating the surface and/or atmosphere would seem an essential part of the mission.

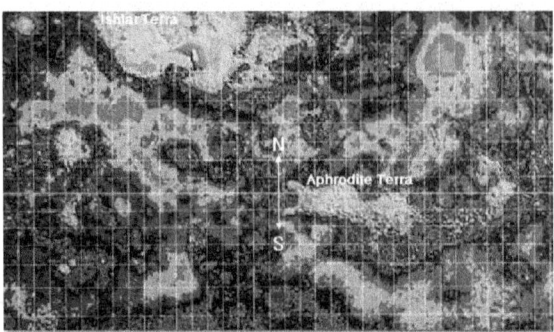

Fig. 2. Venus topography as determined from the Pioneer Venus Orbiter. Figure adapted from [4]. In translation to black and white from original color figure, most relative altitude information is lost. Certain regions, for example, Ishtar Terra or Aphrodite Terra, contain very high topography. However, the figure illustrates the diversity of terrain on Venus.

Turning now to Jupiter, the Galileo Mission illustrates the need for probe-orbiter interactions in a number of ways. Here we will consider only probe delivery and communications, but later, various other science aspects also will be discussed.

The Galileo probe was released from the orbiter about 5 months and 80 million kilometers from Jupiter. There were a number of mission trajectory constraints [5], but the mission was designed such that the orbiter overflew the probe entry site near the Jovian equator during the probe descent through the atmosphere. This allowed the probe telemetry to be received by the orbiter in real time. In future Jupiter probe missions there are strong scientific reasons for targeting probes to mid and high latitudes [3]. There are a number of technical challenges associated with entry of probes in regions other than the Jovian equator, and probe delivery and communication will be a central mission design issue.

The optimum communication strategy for a probe entering Jupiter's atmosphere, in terms of probe telemetry data rate, is communication with an overflying spacecraft. The reasons are as follows.

First, such a scenario minimizes the communciation path length. Since probe telemetry signal amplitude is inversely proportional to distance squared from the probe for a given antenna radiation pattern, minimizing the path length minimizes the attenuation of the probe signal, and hence maximizes the amount of science data that can be returned.

Second, because of thermal protection and trajectory considerations, probe mission arrival scenarios will likely dictate that probes enter in the late afternoon or early evening, local Jovian time. For example, the Galileo probe entered Jupiter's atmosphere near the equator but at a solar zenith angle of 67°. Therefore, the probe's initial descent was very near the evening terminator on Jupiter, such that by the time the probe reached the 15 bar pressure level (about 45 minutes into the mission), Jupiter's rotation caused it to be descending on the night side of the planet.

Communication losses through the atmosphere due to absorbers and clouds are minimized if the comminication path is vertical or nearly so, as would be the case to an overflying spacecraft. For probes entering in the late afternoon, or at mid to high Jovian latitudes, communication to Earth would have a long slant path through the atmosphere, causing more probe telemetry attenuation. In addition, probe communication would be rather limited in time before no signal could be received at Earth due to Jupiter's rotation, combined with rotation of the receiving station on Earth due to Earth's rotation. For these reasons it seems likely that optimal probe missions to the outer planets will involve associated orbiters, or at least flyby spacecraft.

Beside the probe entry site accessiblity issue, there may be direct scientific gain to be had because of a probe-orbiter telemetry link. For example, one of the very significant, but unanticipated, scientific benefits of the Galileo probe-orbiter telemetry link was derivation of the vertical distribution of the abundance of NH_3 by inversion of the probe-orbiter radio signal amplitude as a funtion of depth [6]. This turned out to be one of the most important results from the probe mission, first, because the abundance of N in the form NH_3 is central to understanding the evolution of Jupiter; second, there was no other method that was capable of deriving the vertical distribution of NH_3; and finally, the observed NH_3 abundance profile behaved in a totally unanticipated manner below the upper NH_3 ice cloud deck (NH_3 vapor condenses directly to ice to form the upper cloud layer on Jupiter). Prior to the Galileo probe mission, the Jovian C/N ratio was thought to be about 2 times solar, but the probe results showed that C/N ≈ solar. This result necessitated a major change in thinking about how Jupiter acquired its inventory of heavy elements (see [7] and references therein for discussion of all these aspects of the NH_3 abundance).

In summary, there are at least three mission design aspects involving orbiters that have significant potential for enhancing science return from a probe mission: a) delivery of probes to desirable entry sites, b) communication between probe(s) and orbiter, thereby enabling access to desirable probe entry sites, and c) science measurements that directly take advantage of a probe-orbiter telemetry link. The previous discussion of the Pioneer Venus and Galileo missions has illustrated examples in each of these areas.

3. ORBITER-PROBE SCIENCE INTERACTIONS

We identify four general probe-orbiter science interactions that greatly enhance, or in certain cases are essential for, the mission science return. Each will be illustrated below for Venus and Jupiter as was done before, but each is applicable to any probe mission. The four topics are, global context of the probe entry site(s), ground truth for remote sensing observations of an orbiter, atmospheric composition measurements, and wind measurements. More topics can probably be considered, but we limit the discussion here to these four.

3.1 Global context

The importance of obtaining the global context of entry probe site(s), usually from an orbiter, can be illustrated by the experience of the Galileo probe mission. Planned high resolution approach images of the Galileo probe entry site, to be taken by the Galileo orbiter just before probe entry, were canceled in the mission sequence because of the failure of the Galileo orbiter high gain antenna and the occurrence of other orbiter spacecraft complications. This had the potential of leaving unknown the particular cloud and atmospheric features through which the probe descended, thereby leaving the global context of the probe measurements uncertain.

As it turned out and as described below, ground based measurements were able to identify the atmospheric feature into which the probe entered, and this identification has been crucial for trying to understand and interpret various aspects of the probe data. However, ground based observations cannot always be counted on to provide the appropriate contextual information, and therefore, such information obtained from an orbiter (or possibly flyby spacecraft) is almost essential.

Fig. 3, taken from Fig. 3 in [9], illustrates the probe entry and descent trajectory, projected on NASA IRTF 4.78 µm false color images of the probe entry site. Several points are apparent from the figure. First, the probe apparently descended in the southern region of a 5 µm hot spot. These are regions located slightly north of the Jovian equator that correspond to local clearings in the clouds. They are bright near the 5 µm region of the spectrum because thermal emission from deeper atmospheric levels near 4-5 bars is being observed. The southern location of the probe entry site in the hot spot is significant for interpreting the probe wind observations (cf. [8]).

Second, the probe was within the hot spot (at least as far as horizontal position) throughout the entire descent portion of the mission, descent portion meaning that part of the mission where the probe was making direct atmospheric measurements. Immersion in the hot spot was an important factor for understanding the vertical profiles of condensible species. In fact, had we not known that the probe descended in a hot spot, interpretation of the composition measurements would have been extremely difficult, if not impossible (cf. [7] and references therein for detailed discussion).

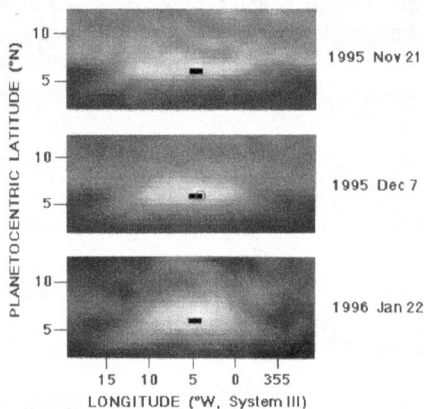

Fig. 3. Location of the Galileo probe entry site within the entry hot spot, taken from [9]. Images were taken by the facility near-infrared camera at the NASA IRTF at the summit of Mauna Kea in Hawaii. The dark bar depicts the longitudinal extent of the probe entry path starting from 450 km above the 1 bar pressure level. A 1-σ circle shows the effects of pointing uncertainty of the telescope on the location of the final portion of the probe entry. The panels from different dates were aligned together using a drift rate of 103 ms^{-1} relative to System III.

Third, the hot spot maintained its integrity for the two month period illustrated in the figure, and actually did so for much longer [9]. This is a crucial property of hot spots that must be matched by theoretical models attempting to simulate conditions at the probe entry site.

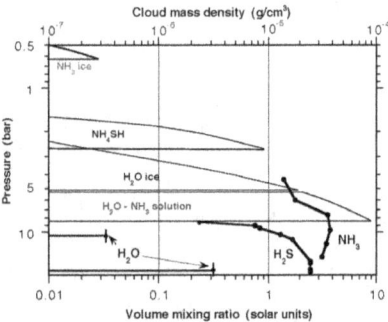

Fig. 4. Condensible species abundance profiles measured by the Galileo probe, together with a thermochemical equilibrium cloud model (see [7] for details).

Fig. 4 illustrates the vertical profiles of the condensible species NH_3, H_2S, and H_2O as observed by the Galileo probe [10]. Prior to the probe mission it was expected that the abundances of the above species would correlate with their cloud condensation levels. So that, for example, NH_3 abundance would have a constant mixing ratio below the NH_3 ice clouds, and follow close to a saturation mixing ratio for some distance above the cloud bottom. Fig. 4 shows that that is not at all what was observed for NH_3, nor for any of the other condensible species with regard to their respective cloud condensation levels (note H_2S combines with NH_3 to form the NH_4SH cloud).

Furthermore, each species increased with depth below its condensation level at a different fractional rate than the others. Although NH_3 and H_2S were observed to eventually reach constant mixing ratios at depth, H_2O did not at any depth sampled by the probe.

Based on knowledge that the Galileo probe descended in a 5 μm hot spot, models of hot spots have been proposed that can at least qualitatively, if not completely quantitatively, explain the observed vertical abundance profiles of the three condensible species (e.g., [8]). It is now believed that the unusual vertical distributions of NH_3, H_2S, and H_2O are the result of the peculiar atmospheric dynamics associated with the hot spot through which the probe descended.

Had we not known the global context of the Galileo probe entry site, the above situation would have been almost impossible to comprehend, and the science return from the probe mission would have been significantly degraded. Similar conclusions can be reached for probe missions to Venus in which understanding the dynamic meteorology or spatial variations in composition are important goals. An orbiter giving the global context of probe entry sites is extremely valuable, especially since we will either not always be able to obtain such information from Earth, or not be able to obtain it from Earth with sufficient spatial resolution to be useful.

3.2 Ground truth

As was mentioned previously, a great strength of having both probes and an orbiter in a planetary mission is that both local *in-situ* and global remote sensing science measurements can be accomplished. Orbiter measurements have the capability to extend probe measurements over global scales, place the probe measurements in global context, and remotely sense regions not accessible by probes. On the other hand, probe measurements can be a great aid to orbiter measurements by providing calibration for orbiter remote sensing observations, which by necessity, involve model dependent inversion of the remote sensing data to obtain desired physical quantities.

The Galileo probe mission again illustrates the advantages of having both kinds of spacecraft in this context. As discussed in [7], prior to the Galileo probe encounter certain Earth based and Voyager spacecraft remote sensing observations of Jupiter's atmospheric composition were considerably in error with respect to particular key species.

For example, the Galileo probe measurements of helium abundance showed that the Jovian helium abundance as derived from Voyager was about 30% too low [11, 12]. Based on this result, a reassessment of the Voyager He mixing ratio for Saturn indicated that the Voyager value there was too low by a factor of 3-4 [13]. Voyager Jovian water abundance values, which pertained to regions within 5 μm hot spots, were found by the Galileo probe measurements to be in error by 1-2 orders of magnitude. This error is now thought to be due to a calibration problem with the Voyager IRIS instrument in the spectral region having wavelengths shorter than 5 μm [14]. In order to fit the Voyager IRIS data, an additional opacity somewhere between 3 and 8 bars is required. As another example, ground based and Voyager determinations of the NH_3 abundance indicated a C/N ratio about twice the solar value in Jupiter's atmosphere (cf. [7] and references therein), whereas the probe measurements indicated a C/N ratio less than or near solar [6, 10, 12]. This result represented a considerable change and a major surprise, one that affects proposed scenarios of Jupiter's formation and evolution. Each of these examples demonstrates the value of ground truth measurements.

On the other hand, the Galileo probe experience shows that a single vertical profile of measurements of particular quantities can be hard to generalize to the whole planet. Because of the probe entry into a 5 μm hot spot, the condensible species NH_3, H_2S, and H_2O behaved in very unexpected ways as a function of depth, as discussed earlier (see Fig. 4). If the Galileo orbiter had had instrumentation, such as a microwave radiometer to sound NH_3 and H_2O, the probe data could still have provided the necessary ground truth for retrieval of NH_3 and H_2O, and the orbiter could then have reliably sounded the deeper atmosphere and other latitudes and longitudes to obtain a comprehensive picture of the NH_3 and H_2O abundances.

3.3 Composition

We have already discussed how the Galileo orbiter-probe telemetry signal was used to derive the abundance of the key species NH_3 in the Jovian atmosphere. There are other important instances where measurements by both orbiter and probe(s) would pay handsome dividends for determining the composition of a planetary atmosphere. A few examples are given below.

One of the key questions regarding the composition of the Venus atmosphere is the abundance distribution of CO, because it is generally accepted that the oxidation state of the lower atmosphere of Venus is controlled by the net thermochemical reaction [15]

$$2CO + O_2 = 2CO_2$$

Fig. 5 illustrates the CO concentration as a function of height as implied by various remote sensing and in-situ observations. The CO mixing ratio evidently decreases with decreasing altitude below cloud levels near 65 km, although there is considerable uncertainty in the observations. As noted in the figure, cloud level CO concentrations have been derived entirely from Earth based infra-red remote sensing observations. Every probe to Venus has started taking in-situ measurements below 65 km because of entry considerations. If that is also the case in future probe missions, then remote sensing of CO at cloud levels from an orbiter becomes a very desirable objective, and is necessary if the global distribution of CO is to be obtained above the clouds where it is produced. Lower in the atmosphere, at altitudes between 35-45 km, a combination of Earth based remote sensing and in-situ measurements were used to obtain the CO mixing ratio. At the lowest altitude levels, only in-situ measurements of CO concentration exist [15].

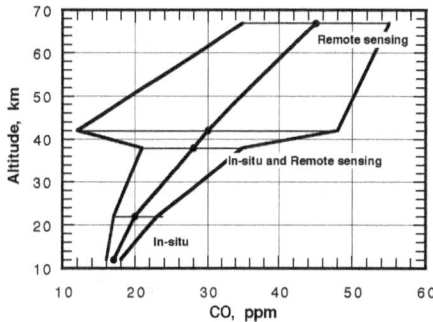

Fig. 5. Vertical distribution of CO in the atmosphere of Venus as determined from ground based and probe measurements. The middle curve shows the nominal velues, the range consistent with the measurement errors is given by the left and right curves. Based on data given in [15].

The abundance profile of CO is a good example of where the combination of in-situ and remote sensing observations can be used together to establish an important result. In future Venus missions involving both probes and an orbiter, it should be possible to completely nail down the CO distribution, both with respect to height and global position, but especially in the lowest atmospheric scale height. Once the entire CO distribution is accurately known, the chemistry of the atmosphere, and especially the chemistry between the atmosphere and solid surface, will be much better constrained. Clearly, remote sensing from an orbiter coupled with in-situ measurements from probes will be necessary to completely characterize CO and its chemistry.

Instrumentation that would be required to measure CO abundance would be an IR spectrometer on the orbiter, coupled with a GCMS and perhaps an IR spectrometer on the probe(s).

As another example of the benefit of simulatneous measurments from both an orbiter and probe(s), we consider the distributions of NH_3 and H_2O in the Jovian atmosphere. Referring again to Fig. 4, it can be seen that the Galileo probe was not able to determine the deep equilibrium mixing ratio of H_2O, although it did do so for NH_3 and H_2S within the error limits. The global abundance of H_2O is critical for developing understanding of giant planet formation and evolution (cf. [7] for discussion). Had the Galileo orbiter been equipped with a microwave radiometer, then using the probe measurements for ground truth, the radiometer data could have been used to derive the H₂O abundance deep in the atmosphere, as well as give a global picture of the H₂O distribution. Thus, in future Jupiter missions aimed at measuring atmospheric composition, entry probes coupled with an orbiter are highly desirable, and this is in fact the scenario recommended by the SSE Decadal Survey [3].

3.4 Winds

Obtaining the global distribution of winds in a planetary atmosphere is an area which requires close collaboration between an orbiter (or possibly flyby spacecraft) and probe. There are two aspects of this collaboration. First, sufficiently accurate tracking of probes necessary to determine winds to a resolution of about 1 ms^{-1} usually involves an orbiter or at least flyby spacecraft. Winds to this accuracy are usually necessary if one wants to understand the overall circulation. Second, in order to fit vertical profiles of wind determined from probe tracking into the context of the global circulation, orbiter measurements of global scale winds are required. These points can be illustrated by both the Pioneer Venus and Galileo experiences.

Fig. 6 illustrates the vertical profiles of westward and northward wind as determined from ground based differential very long base line interferometry (DVLBI) for each Pioneer Venus probe [16]. In this case the bus delivering all four probes was used to determine a reference trajectory which could be used to eliminate certain systematic errors in the probe wind determinations. Had this not been done, the accuracy of the winds from tracking the probes from Earth would have been significantly degraded. For example, eddy and mean meridional wind amplitudes at pressures greater than 1 bar, thought to be important for maintaining the superrotation, are of this magnitude in the deep Venus atmosphere.

Fig. 7 taken from [18], which shows the cloud level meridional winds obtained from tracking of cloud features by the Pioneer Venus orbiter, as well as the flybys of Mariner 10 and Galileo, illustrates the point that the global wind patterns at cloud levels, in which the probe measured winds were imbedded, could only be determined from global remote sensing.

Thus, in order to obtain good quantitative resolution on wind vertical structure (from probes) as determined within the context of global scale winds (from remote sensing), combining probes and orbiter measurements represents an optimal measurement strategy.

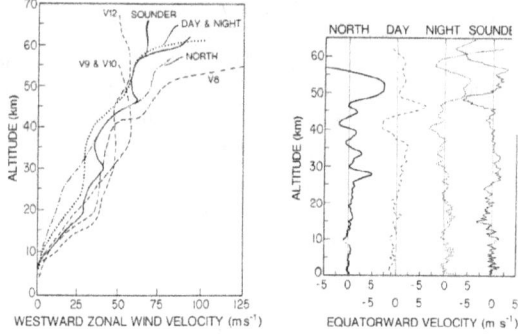

Fig. 6. Horizontal winds in Venus atmopshere as determined from tracking of the PV probes and Veneras 8-12. Taken from [16], [17].

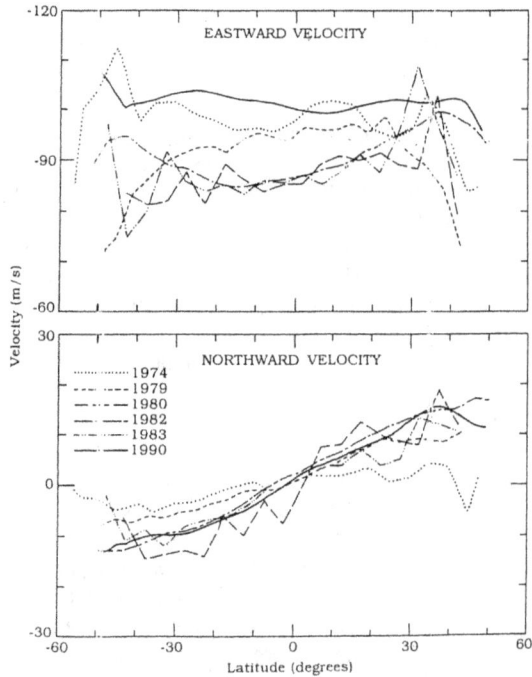

Fig. 7. Mean horizontal winds at cloud levels as determined from the PV orbiter, Mariner 10, and Galileo flyby. Taken [18].

The Galileo probe wind measurements illustrate why both probe and orbiter are required to adequately measure winds on the outer planets. Fig. 8 shows the winds measured by tracking the probe using two completely different tracking platforms. The upper curve gives the wind profile derived from tracking of the probe carrier frequency using the Very Large Array (VLA) set of radio telescopes [19]. As it turned out, the probe was visible from the VLA, and the probe carrier frequency (though not the full telemetry string) was detectable by the VLA. Thus, it was possible to obtain an independent determination of the Jovian winds from that obtained using the orbiter, a very valuable addition to the probe mission. However, by the time the probe reached about 4-5 bars pressure, absorption of the probe signal by ammonia through the long path length in the atmosphere caused loss of signal detectability from the VLA.

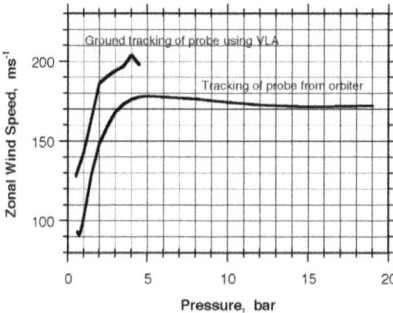

Fig. 8. Jovian zonal wind profiles at the Galileo probe entry site as derived from Doppler tracking of the probe. Data taken from [19, 20]. See text for discussion of differences between the curves and uncertainties in the measurements.

The lower curve in Fig. 8 shows the wind as determined from Doppler tracking of the probe from the Galileo orbiter [20]. The two curves are qualitatively the same. The offset of 30-40 m s^{-1} between the VLA and orbiter-tracked winds would probably be significantly reduced if the VLA analysis was redone to incorporate the most recent determinations of probe descent velocity, which the VLA analysis uses at the beginning to derive the zonal wind profile. On the other hand, the winds near 1 bar are subject to significant error in the orbiter Doppler tracking method because of the almost vertical orientation of the orbiter-probe geometry, and derived values range from about 80 to 120 ms^{-1}.

The primary questions regarding the Jovian winds prior to the Galileo mission were how deep did the winds extend below cloud levels, and did they increase with depth. These are questions that apply to all the outer planets. The fact that the Galileo orbiter was able to track the probe to much deeper levels in Jupiter's atmosphere than possible from the VLA illustrates the advantage of tracking probes from orbiters when deriving winds for the outer planets. The geometry, in terms of long slant paths in the atmosphere for the probe signal to reach Earth, and the timing of having the ground based receiving station in view of the probe telemetry transmission at the right time to measure winds, conspire to make probe tracking from the ground for wind measurements rather limited. Long telemetry slant paths through the atmosphere limit wind tracking to shallow depths because of atmospheric attenuation due to clouds and signal absorbing trace species such as ammonia.

4.0 SUMMARY

We have discussed two catagories of probe-orbiter interactions which benefit the science return from a probe mission. The first category, "Mission Design Aspects", describes those aspects of mission design concerning an orbiter that can affect the science return from probe(s). The second category of probe-orbiter interaction is termed "Orbiter-Probe Science Interactions", and is meant to include interactions between oribter and probe(s) that directly involve science measurements made from each platform. We have shown, using the Pioneer Venus and Galileo missions as examples, how two mission related aspects of probe-orbiter interactions, delivery of a probe(s) to the entry site(s) by an orbiter, and communication between each probe and the orbiter, can considerably enhance the mission science return.

We also considered four general probe-orbiter science interactions that greatly enhance, or in certain cases are essential for, mission science return. The four topics are, global context of the probe entry site(s), ground truth for remote sensing observations of an orbiter, atmospheric composition measurements, and wind measurements. For each case particular examples drawn from Pioneer Venus or Galileo were identified that demonstrated the advantages of having probes and orbiters interact during a mission.

Future missions to Venus or the outer planets will probably have more ambitious goals than either Pioneer Venus or Galileo. Combining probes and orbiters in a mission design, and using each as observing platforms, seems to offer the greatest mission flexibility and science return to address these more ambitious goals.

5.0 REFERENCES

1. Colin, L. The Pioneer Venus Program, *J. GEOPHYS. RES.*, Vol 85, 7575-7598, 1980.
2. Baker, R.D. and Schubert, G. Cellular Convection in the Atmopshere of Venus, *NATURE*, Vol 355, 710-712, 1992.
3. Solar System Exploration Decadal Survey: Space Studies Board, National Research Council, "New Frontiers in the Solar System: An Integrated Exploration Strategy", July 8, 2002.
4. Pettengill, G.H. et al. Pioneer Venus Radar Results: Altimetry and Surface Properties, *J. GEOPHYS. RES.*, Vol 85, 8261-8270, 1980.
5. D'Amario, L.A., et al. Galileo Trajectory Design, *SPACE SCI. REV.*, Vol 60, 23-78, 1992.
6. Folkner, W.M., and Woo, R. Ammonia Abundance at the Galileo Probe Site Derived from Absorption of its Radio Signal, *J. GEOPHYS. RES.*, Vol 103, 22847-22855, 1998.
7. Young, R.E. The Galileo Probe: How It Has Changed Our Understanding of Jupiter, *NEW ASTRON. REV.*, Vol 47, 1-51, 2003.
8. Showman, A.P., and Dowling, T.E. Nonlinear Simulations of Jupiter's 5-micron Hot Spots, *SCIENCE*, Vol 289, 1737-1740, 2000.
9. Orton, G., et al. Characteristics of the Galileo Probe Entry Site from Earth-based Remote Sensing

Observations, *J. GEOPHYS. RES.*, Vol 103, 22791-22814, 1998.

10. Wong, M.H. et al. Updated Galileo Probe Mass Spectrometer Measurements of Carbon, Oxygen, Nitrogen, and Sulfur on Jupiter, *ICARUS*, Vol 171, 153-170, 2004.

11. Von Zahn, U. et al. The Helium Mass Fraction in Jupiter's Atmosphere Found to Match That in the Sun's Convective Zone, *J. GEOPHYS. RES.*, Vol 103, 22815-22829, 1998.

12. Niemann, H.B., et al. The Composition of the Jovain Atmosphere as Determined by the Galileo Probe Mass Spectrometer, *J. GEOPHY. RES.*, Vol 103, 22831-22845, 1998.

13. Conrath, B.J., and Gautier, D. Saturn Helium Abundance: A Reanalysis of Voyager Measurements, *ICARUS*, Vol 144, 124-134, 2000.

14. Roos-Serote, M., et al. Constraints on the Tropospheric Cloud Structure of Jupiter from Spectroscopy in the 5 μm Region: A Comparison Between Voyager/IRIS, Galileo/NIMS, and ISO/SWS Spectra, *ICARUS*, Vol 137, 315-340, 1999.

15. Fegley, B. et al. Geochemistry of Surface-Atmosphere Interactions on Venus, *VENUS II*, Eds. S.W. Bougher, D.M. Hunten, R.J. Phillips, Univ. of Arizona Press, 591-636, 1997.

16. Counselman, C.C. et al. Zonal and Meridional Circulation of the Lower Atmosphere of Venus Determined by Radio Interferometry, *J. GEOPHYS. RES.*, Vol 85, 8026-8030, 1980.

17. Schubert, G. General Circulation and the Dynamical State of the Venus Atmosphere, *VENUS*, Eds. D.M. Hunten, L. Colin, T.M. Donahue, V.I. Moroz, Univ. of Arizona Press, 681-765, 1983.

18. Gierasch, P.J. et al. The General Circulation of the Venus Atmosphere: An Assessment, , *VENUS II*, Eds. S.W. Bougher, D.M. Hunten, R.J. Phillips, Univ. of Arizona Press, 459-500, 1997.

19. Folkner, W.M., et al. Earth-based Radio Tracking of the Galileo Probe for Jupiter Wind Estimation, *SCIENCE*, Vol 275, 644-646, 1997.

20. Atkinson, D.H. et al. The Galileo Doppler Wind Experiment, *J. GEOPHYS. RES*, Vol 103, 22911-22928, 1998.

Cross-Cutting Topics

DEVELOPMENT OF A SOLID STATE THERMAL SENSORS FOR AEROSHELL TPS FLIGHT APPLICATIONS

Ed Martinez[1], Tomo Oishi[2], Sergey Gorbonov[3]

[1] NASA Ames Research Center, Moffett Field CA 94035
[2] Ion America, Moffett Field CA 94035
[3] ELORET Corporation, Moffett Field CA 94035

1. INTRODUCTION

In-situ Thermal Protection System (TPS) sensors are required to provide verification by traceability of TPS performance and sizing tools. Traceability will lead to higher fidelity design tools, which in turn will lead to lower design safety margins, and decreased heatshield mass. Decreasing TPS mass will enable certain missions that are not otherwise feasible, and directly increase science payload. NASA Ames is currently developing two flight measurements as essential to advancing the state of TPS traceability for material modeling and aerothermal simulation: heat flux and surface recession (for ablators). The heat flux gage is applicable to both ablators and non-ablators and is therefore the more generalized sensor concept of the two with wider applicability to mission scenarios.

This paper describes the continuing development of a thermal microsensor capable of surface and in-depth temperature and heat flux measurements for TPS materials appropriate to Titan, Neptune, and Mars aerocapture [1,2], and direct entry. The thermal sensor is a monolithic solidstate device composed of thick film platinum RTD on an alumina substrate. [3,4] Choice of materials and critical dimensions are used to tailor gage response, determined during calibration activities, to specific (forebody vs. aftbody) heating environments. Current design has maximum operating temperature of 1500K, and allowable constant heat flux of $q=28.7$ W/cm^2, and time constants between 0.05 and 0.2 seconds. The catalytic and radiative response of these heat flux gages can also be changed through the use of appropriate coatings. By using several co-located gages with various surface coatings, data can be obtained to isolate surface heat flux components due to radiation, catalycity and convection. Selectivity to radiative heat flux is a useful feature even for an in-depth gage, as radiative transport may be a significant heat transport mechanism for porous TPS materials in Titan aerocapture.

2. TEST RESULTS IN A FOREBODY TPS APPLICATION

In December 2003 a series of arc jet tests (Table 2.2) were conducted to evaluate thermal microsensor performance mounted in-situ in a 3-inch diameter blunt cone model of Fibrous Reinforced Ceramic Insulation (FRCI-12 (Shuttle tile) coated with RCG. The test series exposed 5 models to cold wall heat fluxes of 42 and 60 W/cm2 in 13 total exposures. This configuration of tile and coating is extremely well characterized as it is used for Shuttle acreage TPS. The configuration replicates a forebody heatshield application where the sensors were mounted on a plug [Fig 2.1], and inserted from behind into the model. A top view picture of 5 different models can be seen in Fig 2.2. The sensors were mounted 1/16 inch below the surface of the coating, and a thermocouple was mounted 0.01 inches below the thermal sensor to provide a comparison. The objectives of this test series were to demonstrate the ability to record RTD data during an arc jet run; demonstrate thermal shock survivability beyond 900 C of the sensors; obtain sensor performance versus independent measurements; and, obtain data on performance limits (T, delta-T, Q, and Q-dot). This was the first test series which used the thermal microsensors with TPS in an arc jet test. All objectives were met with complete success, as demonstrated in the data below.

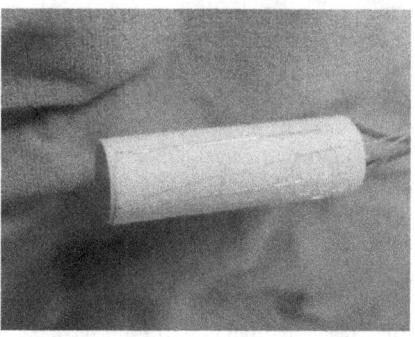

Fig 2.1. Thermal Sensor Plug

Fig 2.2. Blunt Cone Models

Model Name	Sensor location Depth [in]	# of runs	Date	Duration (sec)	Cold Wall Heat flux (W/cm^2)	Sensor current (mA)
S1	1/16	4	12/04/03	128	41.3	1
			12/08/03	183	60.5	1
			12/18/03	169	59.6	10
			12/18/03	167	60.4	15
S2	1.468	1	12/04/03	30	41.3	1
S3	0.3	2	12/05/03	302	42.2	1
			12/08/03	242	41.0	1
S4	1/16	4	12/05/03	303	42.2	1
			12/08/03	136	41.0	1
			12/18/03	168	59.6	10
			12/18/03	140	60.4	20
S5	0.3	2	12/05/03	304	42.2	1
			12/08/03	243	60.5	1

Table 2.2. Arc Jet Test Series for Heatshield simulation

A material response simulation was performed at NASA Ames using the Fully Implicit Ablation and Thermal Response Simulation Code (FIAT). The predictions and data presented in Figs. 2.3 and 2.4 are for 42 W/cm^2 and 60 W/cm^2 cold wall heat flux, respectively. Figure 2.3 shows the comparison of in-depth heat flux history between prediction and the thermal sensor reading. The in-depth heat flux is calculated at 0.1587 cm (1/16") below the surface, where the sensor is located. As seen in the graph the comparison is excellent. The peak heat fluxes match within 5%, and the shapes of the curves almost overlay. The transition at 400 seconds happens when the model is removed from the arc jet flow, at this point the surface temperature is much higher than the surrounding environment and heat radiates away from the model instead of into it, as seen from 100 to 400 seconds. Also important is the almost overlay of the calculation and measurement of the return to baseline from 400 to 1400 seconds. This is a strong indicator that the sensor is functioning correctly with no baseline or calibration shifts. This is the first known direct measurement of heat flux in-situ of a TPS material.

A comparison of temperature between the thermal sensor and the thermocouple mounted 0.01 inches below the thermal sensor can be seen in Fig 2.4. As seen by the dashed lines the peak temperatures are within +/- 2%, and within +/-5% on the cold soak after the model is removed from the arc jet stream. A similar result is seen for a different test at the higher heat flux of 60 W/cm2 in Fig 2.5. These experimental comparisons between a traditional method of the thermocouple versus the thermal microsensor verifies the ability of the sensor to make accurate long duration temperature measurements to at least 1000 C. Figure 2.6 shows the repeatability of the heat flux measurements for 4 different tests at 60 W/cm2. These results are typical for all 13 exposures.

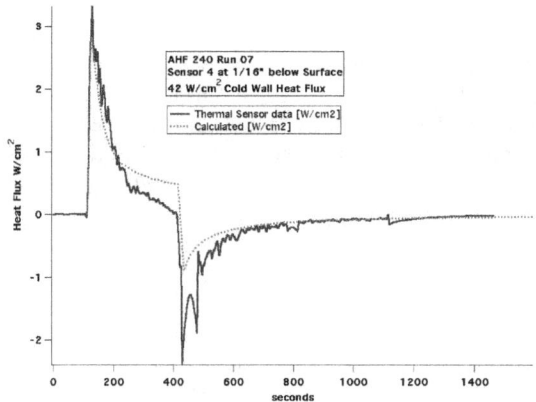

Figure 2.3. In-depth heat flux (42 W/cm^2)

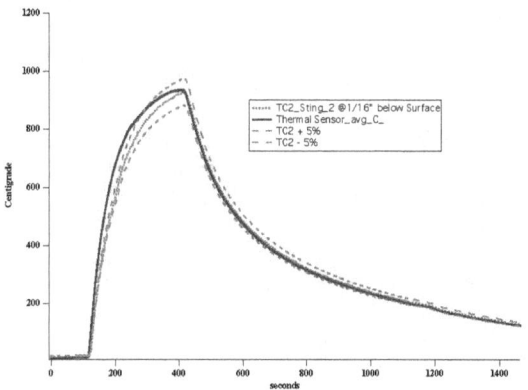

Fig 2.4. In-depth temperatures at 42 W/cm^2

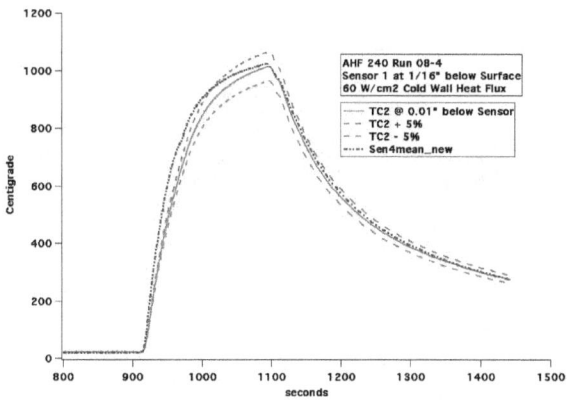

Fig 2.5. Comparison of in depth temperatures at 60 W/cm^2

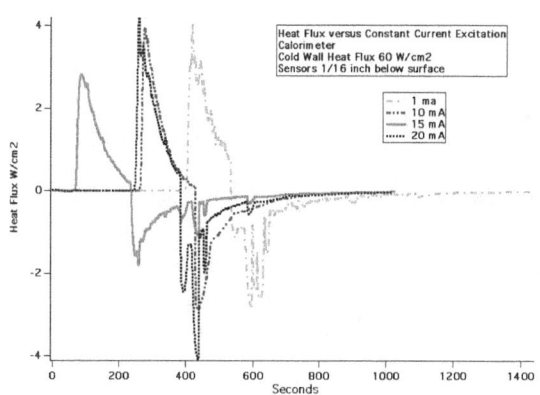

Fig 2.6. Repeatability of in-situ heat flux at 60 W/cm^2

3. TEST RESULTS IN AN AFT SHELL APPLICATION

Pre and post-dictions were performed using the FIAT code. The measured surface heat flux, using a foil calorimeter (Gardon gage) [5,6] was used as the input for the FIAT computation. To match the measured maximum surface temperature of 725 C at t = 254 sec, the input maximum heat flux was scaled to 5.7 W/cm^2. Figure 2.7 presents the comparison between computation and data at the model surface and bond-line. The predicted temperature profiles (symbols) generally agree with TC data. However, the sensor temperature reading (colored in black) responds as expected versus the thermocouple data. When the materials get close to steady state conditions both the sensor and material temperatures converge, as seen in the plot.

This is due to the mismatch in diffusivity between the ceramic thermal barrier and the low-density silica tile. Also, there are significant differences in emmisivity between the sensor (_ ~ 0.3) and the uncoated tile (_ ~ 0.5), causing temperature variations between the two as energy is reradiated away from the surface of each. This is an example of the data necessary to calibrate response specific to materials, locations, and times of interest during reentry.

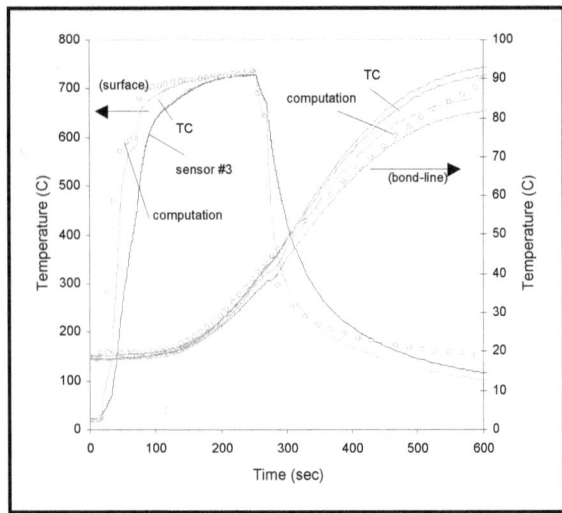

Figure 2.7. Comparison between prediction and TC data

4. Conclusions

The key quantities in the design of the TPS are the surface heat flux (convective, catalytic, and radiative), integrated heat load, stress (both pressure and shear), and material response (including ablation rates and/or surface recession). The fidelity of the physical models would be vastly improved through validation against accurate measurements of these quantities throughout a flight trajectory and consequently, the uncertainties/margins in the TPS design would be reduced. The key requirements for sensor development are that the sensors be lightweight and robust (able to withstand the launch and flight conditions and the elevated temperatures during measurement) and that they be equally applicable in both ground-based facilities and in flight (for calibration purposes).

5. REFERENCES

[1] Wercinski, P.F., Henline, W., Tran, H., Milos, F., Papadopoulos, P., Chen Y.K., and Venkatapathy, E., "Trajectory, Aerothermal Conditions, and Thermal Protection System Mass for the Mars 2001 Aerocapture Mission," AIAA Paper 97-0472, January 1997.

[2] Throckmorton, D.A., "Benchmark Determination of Shuttle Orbiter Entry Aerodynamic Heat Transfer Data," *Journal of Spacecraft*, v. 20, p. 219-224, 1983.

[3] Noltingk, BE (ed.) Instrumentation Reference Book, Butterworths & Co. Ltd. London, 1988.

[4] Julian W. Gardner, "Microsensors: principles and applications," Wiley, 1994.

[5] Gardon, R., "An Instrument for the Direct Measurement of Intense Thermal Radiation," *Rev. Sci. Instrum.*, 24, No. 5, pp. 366–370, 1953.

[6] Jones T V "The thin film heat transfer gauges - a history and new developments," *4th National UK Heat Transfer Conf., Manchester* (London: ImechE) pp 1–12, 1995.

[7] Simmons S G, Hager J M and Diller T E, "Simultaneous measurement of time-resolved surface heat-flux and free stream turbulence at a stagnation point," *Heat Transfer* v. 2, ed G Hetsroni (New York: Hemisphere), 1990.

ULTRALIGHTWEIGHT BALLUTE TECHNOLOGY ADVANCES

Jim Masciarelli [1] and Kevin Miller [2]

Ball Aerospace & Technologies Corp., Boulder, CO
[1] *jmasciar@ball.com, Co-Principal Investigator*
[2] *klmiller@ball.com, Principal Investigator*

ABSTRACT

Ultralightweight ballutes offer the potential to provide the deceleration for entry and aerocapture missions at a fraction of the mass of traditional methods. A team consisting of Ball Aerospace, ILC Dover, NASA Langley, NASA Johnson, and the Jet Propulsion Laboratory has been addressing the technical issues associated with ultralightweight ballutes for aerocapture at Titan. Significant progress has been made in the areas of ballute materials, aerothermal analysis, trajectory control, and aeroelastic modeling. The status and results of efforts in these areas are presented. The results indicate that an ultralightweight ballute system mass of 8 to 10 percent of the total entry mass is possible.

[*keywords*: aerocapture, inflatable, thin-film structures, rarefied flow, Direct Simulation Monte Carlo (DSMC)]

NOMENCLATURE

ΔV   Delta-V, Change in Velocity
Kn Knudsen number
M Mach number
p pressure
q heating rate

1. INTRODUCTION

Traditional entry technology relies on an aeroshell or heat shield to provide aerodynamic deceleration and protect the spacecraft from high entry heating rates. The innovative concept behind using ballutes for entry and aerocapture missions centers on deployment of a large, lightweight, inflatable aerodynamic decelerator (ballute) whose large drag area allows the spacecraft to decelerate at very low densities high in the atmosphere. This "fly higher, fly lighter" concept provides the required aerodynamic deceleration with relatively benign heating rates. The low heating rates experienced during atmospheric entry and deceleration enable the use of lightweight construction techniques for the ballute, resulting in revolutionary mass performance compared to traditional entry technologies. For purposes of evaluating specific performance of ballute technology in a flight like scenario, this paper addresses aerocapture at Titan. The Titan aerocapture mission requirements used for the analyses described in this paper are summarized in Table 1.

Table 1. Titan Aerocapture Mission Requirements

Parameter	Titan
Entry Mass (kg)	1000
Ballute Mass (kg)	41
Area with Ballute (m^2)	751
Area without Ballute (m^2)	3.8
Entry C_D	1.7
S/C C_D	2.3
Entry Speed (km/sec)	6.5
Max. Allowable Heating (W/cm^2)	3.0
Entry Flight Path Angle (°)	-39
Entry Altitude (km)	1000
Atmospheric Scale Height (km)	41
Pass duration (s)	~3600
Ballistic Ratio	~150+

The ballute would be stowed in a small volume attached to the spacecraft, then deployed and inflated prior to entry into the atmosphere. After entry, the vehicle decelerates due to drag high in the atmosphere. Once the velocity has been sufficiently reduced, the ballute is separated. For aerocapture missions, the spacecraft exits the atmosphere and continues to apoapsis, where a small propulsive maneuver is used to place the vehicle in the desired final orbit. For entry or landing missions, the vehicle can transition to the use of a traditional parachute or propulsion system for terminal descent and landing.

As mentioned above, compared with chemical propulsion or rigid aeroshell solutions to orbit capture, ballutes offer substantial mass performance benefits. In addition, due to the characteristics of inflatable design, ballutes also offer significant benefits in flight system level packaging and other critical areas. Whereas more traditional systems introduce significant design constraints, such as propellant management, slosh, nutation and control of the center of gravity (c.g.) in the case of chemical orbit capture, or packaging envelopes, thermal control, cruise stage functionality and c.g. in the case of rigid aeroshells, a "fly higher, fly lighter" self contained package of about 0.5 m^3 constitutes a ballute system prior to inflation. This frees the spacecraft configuration from constraints imposed by an aeroshell and enables flexible and reconfigurable entry system designs. Spacecraft components do not have to be packaged within the wake of an aeroshell. The spacecraft center of gravity does not have to be strictly controlled to maintain aerodynamic stability, as the large ballute provides a very stable aerodynamic configuration with large performance margins. Because the ballute is only deployed when needed, there is no aeroshell to impede spacecraft functionality (e.g. heat rejection or obstruction of antenna and instrument views from the spacecraft).

In addition, due to its compact packaging, a ballute solution precludes the tight operational constraints, such as heat rejection, that arise from flying a spacecraft within an aeroshell. Finally, "fly higher, fly lighter" ballutes expand the envelope of orbits that can be reached without incurring a large propellant penalty since capture is achieved at a relatively high periapsis. Thus, ballutes enable aeroassist to be used even where constraints imposed by aeroshells would be prohibitive.

The In Space Propulsion (ISP) program, managed at Marshall Space Flight Center, is conducting development on propulsion technologies. Our team, consisting of Ball Aerospace, ILC Dover, NASA Langley, NASA Johnson, and the Jet Propulsion Laboratory, is working to develop two variants of the "Fly higher, fly lighter" ballute technologies. These concepts are illustrated in Fig. 1a and 1b. As shown, the trailing ballute (Fig. 1a) technology implementation concept uses a large toroid tethered in the wake of the primary spacecraft to decelerate. Critical technology development status on trailing ballutes has been summarized previously [1]. The clamped ballute concept (Fig. 1b) also uses a trailing ballute attached to the aft end of the primary spacecraft, but it uses a conic thin film web to attach the spacecraft to the toroid. There are advantages and disadvantages to each approach, and these will be addressed in this paper. Performance characteristics of the two technology approaches are summarized in Table 2. Our systems analyses on the two technology implementation concepts include trajectory, aerothermal, structural, and thermal analyses, along with materials testing and hypersonic wind tunnel testing. This paper will specifically address recent advances in:

- Aerothermal Analysis and Testing
- Materials Evaluation and Testing
- Trajectory Control
- Aeroelastic Modeling
- System Mass Performance

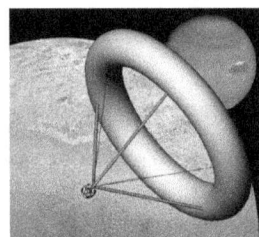 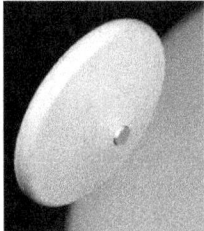

Fig. 1 a and b. Trailing Ballute and Clamped Ballute Technology Implementation Concepts

Table 2. Ultra lightweight ballute performance

	Trailing	Clamped
Peak dynamic pressure	21 Pa	52 Pa
Peak convective heating rate	1.8 W/cm^2	2.1 W/cm^2
Peak temperature on ballute	403 C	435 C
Ballute system mass	97 kg	74 kg
Aerocapture mass fraction	10%	8%

2. AEROTHERMAL ANALYSIS AND TEST

We are performing aerothermal analysis of ballute configurations to define the aerodynamic loads and heating environment for the ballute throughout the aerocapture trajectory. Because the trajectory spans free molecular, transitional, and continuum flow regimes, a suite of tools is required. For Computational Fluid Dynamics (CFD) Navier-Stokes solutions, we are employing the Langley Aerothermodynamic Upwind Relaxation Algorithm (LAURA) code [2]. We are using the Direct-Simulation Monte Carlo (DSMC) technique for high-fidelity free molecular and transitional flow modeling with the DSMC Analysis Code (DAC) [3]. We also make use of the DAC Free-Molecular (DACFREE) tool, which is an engineering level code that implements free-molecular and

Newtonian aerodynamics methods for quick evaluation of ballute configurations.

For the trailing ballute, we have completed quasi-dynamic modeling and analyses for Titan aerocapture. This has included stepwise DSMC and CFD analysis of the entry system over many points in the trajectory, including the peak dynamic pressure and peak heating points. For the clamped ballute, we have completed analysis at the peak heating and peak dynamic pressure points. The results from these analyses (aero loads and heating distribution over the ballute as a function of time) were used as inputs to the structural and thermal analyses. Sizing of the ballute system elements is then based on the structural and thermal analyses. Element sizing includes the number of tethers, cross section of tethers, and tailored film thicknesses for the ballute.

There is a potential for unsteady flow around the spacecraft and the trailing ballute. Therefore, we are also using CFD models of the flow field to examine the trailing ballute two-body flow stability and interactions. Our objective is to identify how changes in geometric design parameters such as ballute trailing distance, ballute aspect ratio, and ballute diameter influence the flow stability.

Hypersonic tests have been conducted on representative ballute configurations at facilities at NASA Langley Research Center and at the University of Virginia. These tests have provided an empirical basis for the DSMC and CFD models. The University of Virginia tests provide data on the three dimensional flow fields and ballute surface response under the rarefied conditions representative of the "Fly Higher, Fly Lighter" trajectory. Results have correlated well with DSMC models. The Langley tests, examining features such as angle of attack, and ratio of trailing distance to ballute diameter over various flow conditions, have provided a basis for understanding of geometries and their effects on shock-shock interaction. Schlieren images from some of the tests, shown in Fig. 2, indicate the shock interfaces as a function of the flight system geometry. In addition, these tests have provided a basis for correlation of models to predict the geometry and conditions for unsteady flow.

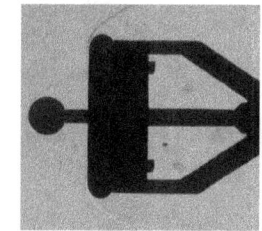
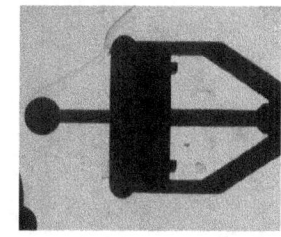

Fig. 2. NASA Langley hypersonic testing of trailing ballute configurations

3. MATERIALS EVALUATION AND TESTING

The objective of our ballute materials efforts is to investigate the availability and applicability of candidate materials, evaluate their mechanical properties, and determine how well they can survive the predicted operating environment within the constraints of our current ballute design concepts. Our current efforts have focused on material performance at elevated temperatures, seaming ability, strength of seams, and impacts on material performance due to folding and packaging.

We have completed a broad survey of potential thin film materials for ballute construction, including several varieties of Kapton, Upilex, and PBO. We have examined PBO and Kevlar tensile materials for tethers. In many instances, material performance data does not exist for the temperature ranges of interest. Therefore, we are conducting tests to determine material properties at operational conditions, plus margin. Our thin film testing has included raw material samples and seamed samples. To examine effects of packaging and storage, we have tested both pristine and creased samples. In all cases, we determined material strength and elongation at room temperature and up to 500°C (see Fig. 3). The results of these tests are incorporated into our nonlinear finite element models for ballute structural design, sizing, and analysis. Based on this work, we have selected Upilex film for the inflatable torus and PBO for the tethers for our reference trailing ballute design concept.

Fig. 3. Thin film material testing has identified candidate materials compatible with "Fly Higher, Fly Lighter" Environments

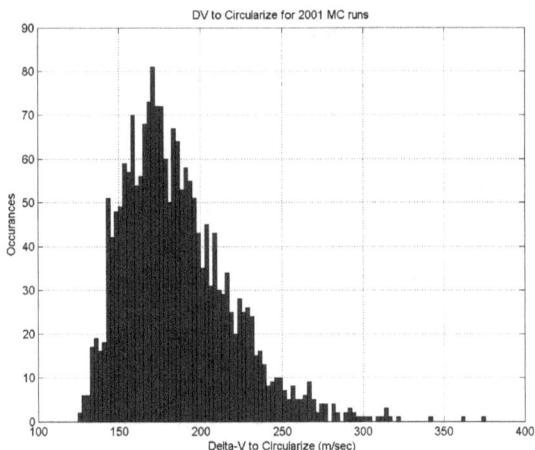

Fig. 4. Monte Carlo trajectory results demonstrate robust aerocapture performance

4. TRAJECTORY CONTROL

The effectiveness of using a drag-only device to capture into orbit relies on critical timing of the ballute separation under navigation, atmospheric, and design uncertainties. We have developed an on-board algorithm that uses measured deceleration due to drag and a numerical trajectory predictor to issue the ballute separation command during flight and achieve the target orbit.

We have tested our ballute separation algorithm using Monte Carlo trajectory simulations of aerocapture at Titan. The simulations include dispersions in entry conditions, uncertainty in navigation knowledge, variability in aerodynamics, and variation in atmospheric density with random perturbations using the Titan GRAM atmosphere model [4]. The results of these simulations, illustrated in Fig. 4, show that the separation algorithm provides excellent performance, with 100 percent of 2000 Monte Carlo cases capturing. A small propulsive maneuver is required after atmospheric exit to achieve the final orbit. Comparing our Monte Carlo results to those for aerocapture using an aeroshell with lift modulation [5] shows that the post-aerocapture propulsive ΔV required for aerocapture with a ballute is the same magnitude as that required for an aeroshell.

5. AEROELASTIC MODELING

One of our primary technology development efforts involves the methodology and tools to analyze the aeroelastic problem for ultralightweight ballute concepts. This problem involves the combination of aerodynamics, dynamics, elasticity, and thermal analysis due to the heating encountered in the entry environment (see Fig. 5). There is no known aeroelastic modeling capability for the nonlinear, hypersonic, rarefied flow regime associated with our ultralightweight ballute concepts. Currently, we are using an uncoupled static analyses to evaluate the change in aerodynamic loads and aerothermal performance due to structural deflections, thus necessitating the use of large margins to account for dynamic load factors. A coupled analysis capability will provide better model of the dynamic response of the ballute system, enabling a more robust design with reasonable margins and optimum mass performance.

We are currently working on developing a ballute aeroelastic modeling capability that couples existing nonlinear structures, thermal, and aerodynamic codes (e.g., MSC MARC, LS-DYNA, LAURA, DACFREE, DAC). We have identified an initial tool architecture, are evaluating analysis tools, and developing interface codes. The analytical architecture that we have defined provides a fully coupled solution including aerothermal, structural, and thermal elements. The aeroelastic modeling toolset is specifically designed to provide dynamic performance analysis of non-linear, thin film systems under rarefied, hypersonic conditions. Our plan is to develop a complete working implementation of this tool set, and then test its capabilities through wind tunnel test of flexible ballute models.

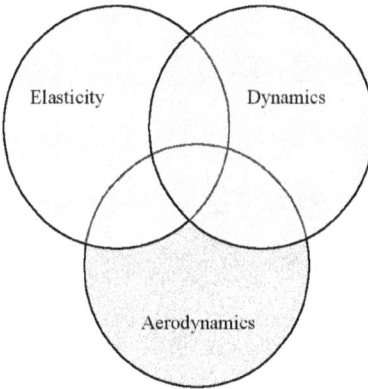

Fig. 5. Aeroelastic modeling approach will provide coupled analyses for non-linear inflatable structures in a rarefied, hypersonic flowfield.

6. SUMMARY

Our ultralightweight ballute technology development efforts have shown that existing thin film materials are compatible with the ultralightweight ballute mission concept for aerocapture at Titan. Aerothermal analysis and testing has been completed to define the design environment and understand the flow field dynamics. Monte Carlo trajectory analysis shows that the aerocapture trajectory can be controlled under dispersions. Based on the analysis completed to date, an ultralightweight ballute system mass of 8 to 10 percent of the total entry mass is possible.

Ultralightweight ballutes offer a revolutionary performance benefit compared to other orbit insertion technologies. Using our current estimate of ballute system mass and the ΔV it provides for aerocapture at Titan, ultralightweight ballutes provide an equivalent specific impulse (I_{sp}) of 5000 sec. This compares very favorably with other orbit insertion technologies as shown in Table 3.

Table 3. Comparison of Equivalent Specific Impulse (I_{sp}) for Various Orbit Insertion Technologies

Technology	I_{sp}
Ultralightweight Ballutes	5000 sec
Rigid Aeroshell	1000 sec
Ion Propulsion	3100 sec
Bipropellant Propulsion	330 sec

Based on our systems analysis, testing, and development efforts, ultralightweight ballute system Technology Readiness Level (TRL) is currently at 3 to 4. Our team will continue to advance ultralightweight ballute technology with the goal of achieving system TRL of 6, so that the benefits of this technology can be applied to space science and exploration missions.

7. ACKNOWLEDGMENTS

Funding for this development effort was provided by the NASA Office of Space Science under contract NAS8-02130. The authors would like to thank Les Johnson, Bonnie James and Erin Richardson at the MSFC In Space Propulsion Program office for their sponsorship and direction of these efforts. The authors would also like to thank Jim Stein, Denise Lawless, Jody Ware and Gumersindo Rodriguez, ILC Dover; Peter Gnoffo, Tom Horvath, and Greg Buck, NASA Langley Research Center; Duane Hill, Kristen Damiani, Dave Waller, and Duffy Morales, Ball Aerospace & Technologies, Corp., Reuben Rohrschneider and Bobby Braun at Georgia Tech, and Jim McDaniel and Eric Cecil at the University of Virginia, who made significant and substantial contributions to the body of work referenced in the development of this paper.

8. REFERENCES

1. Miller, K.; Gulick, D.; Lewis, J.; Trochman, B.; Stein, J.; Lyons, D.; Wilmoth, R.; "Trailing Ballute Aerocapture: Concept Feasibility and Assessment," AIAA/ASME/SAE/ASEE Join Propulsion Conference, Huntsville, AL, July 2003.
2. Gnoffo, P. A.; "An Upwind-Biased, Point-Implicit Relaxation Algorithm for Viscous, Compressible Perfect-Gas Flows," NASA TP-2953, Feb. 1990.
3. LeBeau, G. J.; Lumpkin, F. E.; "Application Highlights of the DSMC Analysis Code (DAC) Software for Simulation Rarefield Flows," Computer Methods in Applied Mechanics and Engineering, Vol. 191, Issue 6-7, 2001, pp. 595-609.
4. Justus, C. G.; Duval, A.; Johnson, D. L.; "Engineering-Level Model Atmospheres for Titan and Neptune," AIAA-2003-4802, AIAA/ASME/SAE/ASEE Joint Propulsion Conference, Huntsville, AL, July 2003.
5. Way, D.; Powell R.; Edquist, K.; Masciarelli, J.; Starr, B.; "Aerocapture Simulation and Performance for the Titan Explorer Mission," AIAA 2003-4951, AIAA/ASME/SAE/ASEE Joint Propulsion Conference, Huntsville, AL, July 2003.

New Approach for Thermal Protection System of a Probe During Entry

Boris Yendler[1], Nathan Poffenbarger[2], Amisha Patel[2], Ninad Bhave[2], Periklis Papadopoulos[2]

[1] LMTO. Lockheed Martin, Sunnyvale, CA.94089
(2) SJSU, San Jose State University, One Washington Square, San Jose, CA 95112

ABSTRACT

One of the biggest challenges for any thermal protection system (TPS) of a probe is to provide a sufficient barrier for heat generated during descent in order to keep the temperature inside of the probe low enough to support operational temperature of equipment. Typically, such a goal is achieved by having the ceramic tiles and blankets like on the Space Shuttle, silicon based ablators, or metallic systems to cover the probe external surface.

This paper discusses the development of an innovative technique for TPS of the probe. It is proposed to use a novel TPS which comprises thermal management of the entry vehicle. It includes: a) absorption of the heat during heat pick load by a Phase Change Material (PCM), b) separation of the compartment which contains PCM from the rest of the space vehicle by a gap with a high thermal resistance, c) maintaining temperature of the internal wall of s/c cabin temperature by transfer heat from the internal wall to the "cold" side of the vehicle and to reject heat into the space during the flight and on a ground, d) utilization of an advanced heat pipe, so called Loop Heat Pipe to transfer heat from the cabin internal wall to the cold side of the s/c and to reject the heat into environment outside of the vehicle. A Loop Heat Pipe is capable of transferring heat against gravity.

INTRODUCTION

An important element of planetary missions is the design of TPS capable of shielding the vehicle from aero-heating during the atmospheric entry. Currently, depending on the aero-thermodynamic heat loads, Shuttle ceramic tiles, ablators, or metallic thermal protection systems are used.

The new proposed approach for thermal management of probe TPS is to use a Phase Change Material (PCM) to store the incident heat generated during an intense transient heating environment of descent.

This stored heat is then rejected back into the environment, once the high heat load has dissipated. Due to the magnitude of the expected heat loading and the likely operating environments, it would be desirable to use a PCM with a high melting temperature. This, however, creates heat transfer from the PCM to the probe's payload compartment. Management of this heat flux is accomplished through the use of insulation, a secondary PCM, and a use of Loop Heat Pipe (LHP).

CONCEPTUAL DESIGN

Three heat loads, low, medium and high, are considered in the current effort (see Table 1). The total heat load was calculated assuming 1 minute of a heat flux which profile is depicted in Fig.1.

	Peak Stagnation Heat flux.	Total Heat Load
Low	50 (w/cm^2)	59.8 Mega Joules
Medium	100 (w/cm^2)	120 Mega Joules
High	200 (w/cm^2)	239 Mega Joules

Table 1: Minimum, Nominal, and Maximum heat load cases

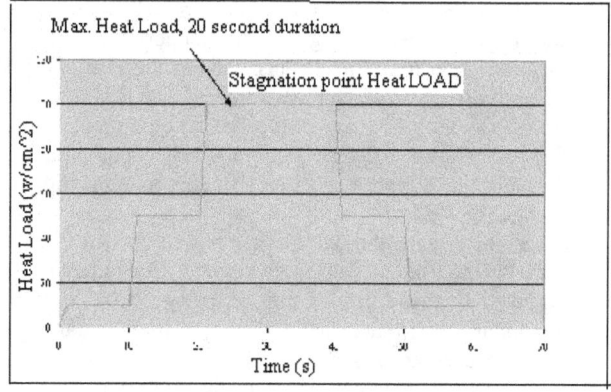

Figure 1: Normalized heat load vs. time

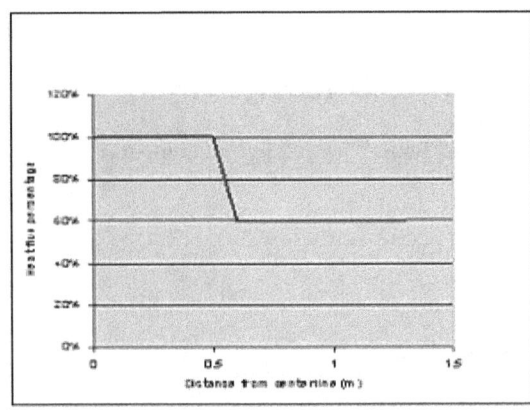

Figure 2: Heat flux distribution over the heat shield surface, as a percentage of the heat flux at the stagnation point.

Figure 2 shows a distribution of the heat flux over the surface of the fore body that presents simplified version of an actual distribution of the heat flux over the heat shield.

The current study assumes that the stagnation point temperature of the outer skin of the fore body does not exceed 825 K, due to the effect of the Phase Change Material, discussed below.

The major goals of two proposed designs (see Fig. 3 and Fig. 4) are: a) absorb coming heat during heat pick load by a PCM, b) separate the compartment which contains PCM from the rest of the probe by a gap with a high thermal resistance, c) maintaining temperature of the internal wall of the probe cargo compartment.

Design I and II consists of the same components. The incoming heat is conducted through the outer skin of the fore body (1) and gets absorbed by the Primary Phase Change Material (2) with a high melting point. The Primary PCM (2) is separated from the probe cargo compartment by a gap with a support structure (3). The gap limits heat transfer from the Primary PCM to the probe cargo compartment. The heat leak from Primary PCM into the cargo compartment is due to radiation between gap walls and conduction through the support structures. The role of the secondary PCM (4) with low melting point is to absorb heat leaks through the gap and to control the cargo compartment wall temperature. It should be noted that the roles of the Primary PCM and Secondary PCM are distinct. While the Primary PCM is used to absorb heat conducting from the heated outer skin thus maintaining the outer skin at a

Design I

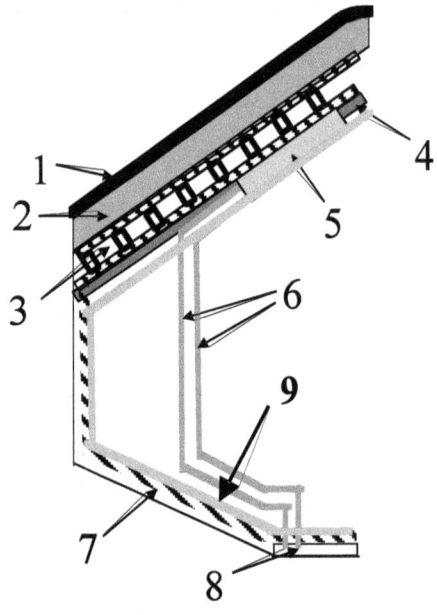

Figure 3: Design I. 1 – heat shield, 2 – primary PCM, 3- gap with support structure, 4 – secondary PCM, 5 – LHP evaporator, 6 – LHP transport lines, 7 – cargo inner wall, 8 – LHP condenser with radiator, 9 – LMI

desirable temperature, the secondary PCM exists to regulate the temperature of the wall of the cabin. The LHP evaporator (5) takes in heat accumulated by the Secondary PCM (4) and transfers the heat via transport lines (6) to the LHP condenser (8) shown together with a radiator situated on the afterbody. The internal wall of the cargo compartment is covered with MLI (9) for better protection of the cargo compartment.

Environment temperature should be less then secondary PCM temperature in order to remove heat from the secondary PCM. Design I is suitable for a cold environment with temperatures around 273 K or below. This could be a limitation for some mission requirements, like landing during a hot season.

Design II

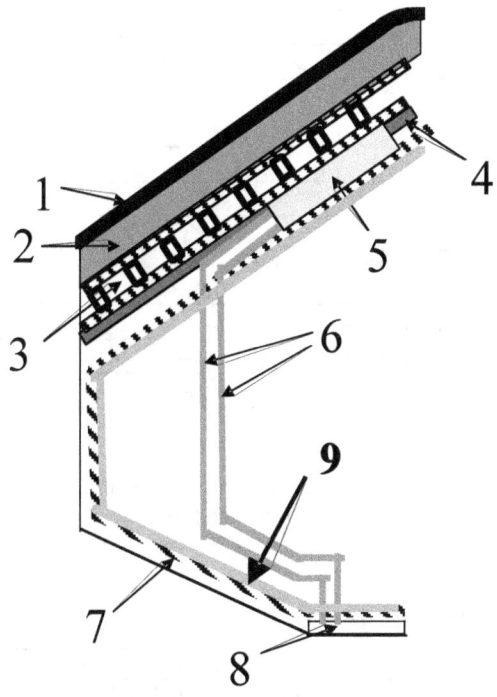

Figure 4: Design II. 1 – heat shield, 2 – primary PCM, 3- gap with support structure, 4 – secondary PCM, 5 – LHP evaporator, 6 – LHP transport lines, 7 – cargo inner wall, 8 – LHP condenser with radiator, 9 – LMI

This limitation is circumvented in Design II where the secondary PCM with elevated melting point is employed. It allows rejecting heat from the Secondary PCM into the hot environment. Since the operating temperature of the LHP evaporator is higher then 100 C, an additional layer of protection in the form of a gap is inserted between the LHP and the inner wall. Thus, in the Design II, the incident heat is absorbed by the Primary PCM (2). The heat leak through the gap (3) is absorbed by the Secondary PCM(4) which is not longer adjacent to the cabin wall. A gap is inserted between the Secondary PCM (4) and the cabin wall (7). The LHP evaporator (6) is embedded in the Secondary PCM (4). Heat transfer from the secondary PCM to the condenser/radiator assembly (9) via LHP evaporator (6) and transport lines (7).

LHP

The Loop Heat Pipe has been designed for operation against gravitational/acceleration forces [9]. Since its creation in 1970's, LHP moved from R&D stage into commercial applications. LHP is standard equipment now for some communication satellites.

The operating temperature for LHP has been relatively low in the past, less than 450 K. However, experiments have shown that LHP's can transfer heat at higher temperatures, using materials like Cesium as the working fluid [9]. LHPs are good choices for heat transfer in a planetary probe since LHPs capability of transfering heat is much less sensitive to the orientation relative to the acceleration vector then conventional heat pipes.

The LHP considered here is made out of Titanium with a copper wick. The choice of wick is based on the wetting properties of the fluid and its interaction with the wick. Results of the modeling using the SINDA/FLUINT pre-built LHP [11] model suggest that a LHP can transfer up to 2KW of heat from the Secondary PCM to the condenser at medium to high temperatures. For the two designs described here (Design I & Design II), intended to operate at an evaporator temperature of 373 K and 523 K respectively, the working fluids proposed are water, and Diphenyl-Diphenyl Oxide Eutectic, also known as "Thermex", respectively.

PCM

A good PCM candidate the TPS considered here should be of a high density, high specific heat, high heat of fusion and appropriate difference between melting and boiling points. A PSM with a high specific heat and a high heat of fusion will have high heat storage capability leading to more heat absorbed. It is also important to choose PCM with a high density since a higher density of material translates into lesser volume of PCM required for heat absorption.

PCM LiF with melting point of 848.2 C was chosen as the Primary PCM for Designs I and II. Material AlBr3 with melting point of 97.5 C is suitable for Design I as the Secondary PCM. TlNO3 with melting point of 206 C is used as the Secondary PCM in Design II. All three PCMs are safe and don't present hazard to the payload. However, the PCM compatibility with probe structural elements and payload is beyond of the scope of presented effort and is not considered here.

MLI

Multi-layer insulation (MLI) blankets provide heat-resistant transfer in or out of the body. It is proposed here to cover the internal wall of the cargo compartment with MLI in order to prevent heat from leaking into the cargo bay and harming the occupants or sensitive electronics. Several types of readily available MLI blankets can be used. A good example is Aluminized Kapton, which exhibits a desirable α/ε ratio for the exposed outer surface, and can comfortably sustain temperatures around 550 K for extended periods.

ANALYSIS

It is assumed that the probe landed with the heat shield down so that the gravitational vector is directed from after body towards the heat shield (Fig.5). The heat stored in the probe is rejected by radiation and convection (Fig. 5). The heat from the Primary PCM is removed from the heat shield by radiation and by natural conduction. The heat transferred from the secondary PCM to the radiator is also removed by radiation and natural conduction.

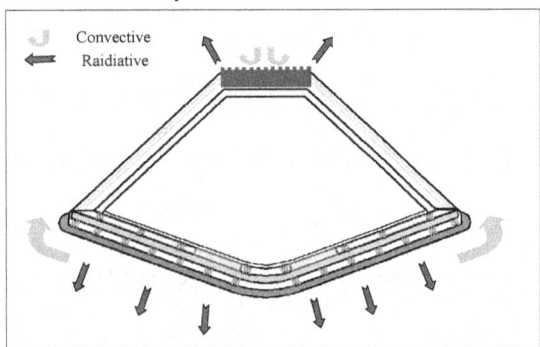

Figure 5: Cooling of the probe

No force convection is considered here in order to understate the convective cooling rates, thus maximizing the stress on the thermal management system. Only radiation between the fore body and the environmental surface is included into the model. The effect of the probe on the environment is neglected.

The convective and radiative cooling rates were calculated for a cold, diffuse CO_2 atmosphere which environmental conditions are listed in Table 2.

Table 2 Environment conditions during probe cooling

Environment	
	Assumed Surface Conditions
CO_2 Density g/m^3	1.21E-02
Temperature (K)	217
Pressure (Pa)	496
Prandlt Number	7.73E-04
Speed of sound (m/s)	230
specific heat (j/g k)	0.75443
Thermal Conductivity (W/m*k)	0.010647
Kinematic Viscosity (m^2/s)	9.01E-04
Thermal Diffusivity (m^2/s)	1.17E+00
Coef. of thermal expansion	0.004608295
Gravity (m/s^2)	3.7

Using LHP pre-build model of SINDA/FLUINT Version 1.0, the LHP has been sized for maximum of 2kW of heat rejection. The LHP parameters are shown in Table3.

Table 3. Parameters of 2kW LHP

Evaporator size	0.25 m length, 0.03 m OD
Transport line sizes	0.5 m length, 0.005 m ID
Condenser size	1.5 m length, 0.005 m ID
Radiator dimensions	0.59 m Ti base diameter, 1500 0.01 m diameter, 0.1 m length copper fins.

Heating and cooling of the probe for three levels of the heat load (see Table 1) were analyzed. Typical results of conducted analysis are show in Fig. 6 and Fig. 7.

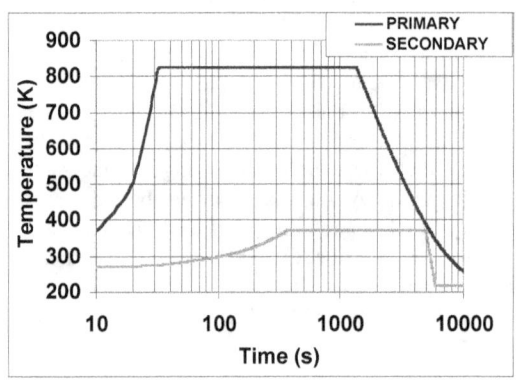

Figure 6. Temperature of Primary and Secondary PCMs for high heat load (200 w/cm^2)

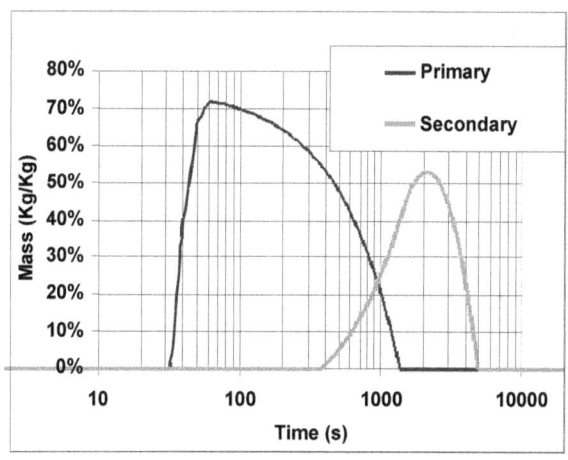

Figure 7. Melted mass as percent of PCM mass for high heat load (200 w/cm^2)

Table 4 contains the results summary for all three heat loads considered here.

Table 4. Simulation Results Summary

	low	medium	high
	50 W/cm^2	100 W/cm^2	200 W/cm^2
Primary PCM			
mass [kg]	43.7	87	173.9
time to reach melting point [sec]	33	33	33
Time for completed solidification [sec]	382	704	1352
melted mass maximum [%]	71.2	71.9	71.9
Secondary PCM			
mass [kg]	5	5	10
time to reach melting point [sec]	188	187	370
Time for completed solidification [sec]	1186	2526	4910
melted mass maximum [%]	22.1	55.1	52.8

As presented results indicate, the proposed TPS is capable of absorbing incident heat and dissipating it in reasonable time. The Primary PCM reaches the melting point before the heat load ends. After this point, the incoming heat melts the Primary PCM. An accumulated in the Primary PCM heat is rejected by radiation and convection. When all accumulated heat of fusion dissipates and PCM is fully solidified, the Primary PCM temperature starts to drop. As data in Fig.6 and Fig. 7 shows, the Secondary PCM temperature reaches melting point at the time when the Primary PCM is almost solidified.

As data in Table 4 indicates, the chosen amount of the Primary PCM provides almost 30 % safety margin. It means that amount of the Primary PCM could be reduced if the safety margin is considered too high. The Secondary PCM margin is even higher (see Table 4). It allows the reduction of an amount of the Secondary PCM in particular for the low heat loads.

One of the additional advantages of using PCM is ability to maintain the temperature of the heat shield at or below the required temperature level, which can be quite lower the existing TPS. It creates an opportunity to use a metal for the outer skin of the fore body which could lead to change of characteristics of a boundary layer and extend the transition point from laminar to turbulent flow.

CONCLUSION

Conceptual designs of integrated multiple layers of PCM, complemented by an LHP, were proposed as possible layouts of the PCM-LHP based re-usable TPS.

It was shown that the proposed TPS is capable of absorbing incoming heat and maintaining the temperature of the probe cargo compartment. One of the advantages of proposed TPS design is ability to maintain a required temperature of the outer skin of the fore body.

PCM technology was studied and determined possible candidates for the Primary and Secondary PCM. The amount of PCM, which provides at least 30% of safety margin, was determined. It was shown that all components of the proposed TPS are available or under development now.

The current effort presents the conceptual design of the re-usable TPS. Future work is required to determine all parameters of the system including but not limited to:
compatibility of the PCMs and structural elements of the probe; performance envelop of LHP during flight and after landing; structural strength of the system; etc.

Cost analysis also needs to be done.

REFERENCES

1) Anderson, W.G., "Sodium-Potassium (NaK) Heat Pipe", *Heat Pipes and Capillary Pumped Loops*, Ed A. Faghri, A.J. Juhasz, and T. Mahelky, ASME HTD-vol. 236, pp. 47-53, 29th Annual National Heat Transfer Conference, Atlanta, Georgia, August 8-11, 1993

2) Maidanik, Y.F., Feshtater, Y.G. and Goncharov, K.A., *Capillary-Pump Loop for the Systems of Thermal Regulation of Spacecraft*, 4th European Symposium on Space Environmental and Control Systems, Florence, Italy, October 1991.

3) Levy, E.K, "Theoretical Investigation of Heat Pipes Operating at Low Vapor Pressures" J. Engineering for industry, pp.547-552, November 1968

4) Cytrinowicz, D., Hamdan, M., Medis, P., Shuja, A., Henderson, H.T., Gerner, F.M., and Golliher, E., *MEMS Loop Heat Pipe Based on Coherent Porous Silicon Technology*.

5) Spitler, J.D., C. Yavuzturk, S.J. Rees. 2000. *In Situ Measurement of Ground Thermal Properties*. Proceedings of Terrastock 2000, Vol. 1, Stuttgart, August 28-September 1, 2000, pp. 165-170.

6) Yun, J., Kroliczek, E., *Operation of Capillary Pumped Loops and Loop Heat Pipes*, TAS PCB, Vol. 2, No. 6, June 2002

7) Bystrov P. I., Kagan D. N., Krechetova G. A., and Shpilrain E. E., *Liquid-metal coolants for heat pipes and power plants*, Hemisphere Publishing Corporation, Washington, 1990.

8) Dunn, P.D. and Reay, D.A., *Heat Pipes* Pergamon Press, Elmsford, New York, 1976

9) Anderson W. G., et al. High Temperature loop heat pipes, IECIC Paper No. 95-18, August 1995

10) Incropera F.P. and DeWitt D.P. *Fundamentals of Heat and Mass Transfer 4th Edition*, John Wiley & Sons, New York, 1996

11) Baumann, J, Rawal, S. Viability of Loop Heat Pipes for Space Power Solar Applications, AIAA 2001-3078, 2001

HyperPASS, a New Aeroassist Tool

Kristin Gates, Angus McRonald, Kerry Nock

Global Aerospace Corporation
711 West Woodbury Road, Suite H
Altadena, CA 91001
USA

ABSTRACT

A new software tool designed to perform aeroassist studies has been developed by Global Aerospace Corporation (GAC). The Hypersonic Planetary Aeroassist Simulation System (HyperPASS) [1] enables users to perform guided aerocapture, guided ballute aerocapture, aerobraking, orbit decay, or unguided entry simulations at any of six target bodies (Venus, Earth, Mars, Jupiter, Titan, or Neptune). HyperPASS is currently being used for trade studies to investigate (1) aerocapture performance with alternate aeroshell types, varying flight path angle and entry velocity, different g-load and heating limits, and angle of attack and angle of bank variations; (2) variable, attached ballute geometry; (3) railgun launched projectile trajectories, and (4) preliminary orbit decay evolution. After completing a simulation, there are numerous visualization options in which data can be plotted, saved, or exported to various formats. Several analysis examples will be described.

1 BACKGROUND

The Hypersonic Planetary Aeroassist Simulation System (HyperPASS) has been an ongoing project at Global Aerospace Corporation (GAC) for the past three years. Its beta version was completed in May 2004 and is currently undergoing validation. The validated version, HyperPASS 1.0, is set to be released sometime in Fall 2004.

2 VALIDATION

HyperPASS has been validated using a 2 degree-of-freedom (2DOF) system. This system is been used by one of the authors for contract work at NASA's Jet Propulsion Laboratory (JPL) for aeroassist and launch approval studies.

2.1 Titan Aerocapture

Titan Aerocapture was simulated using the Hunten [2] atmosphere model along with the following vehicle parameters and entry conditions:

Table 1 Titan Aerocapture Parameters

PARAMETER	VALUE
L/D	-0.242
m/CDA (kg/m^2)	61.279
Atmospheric interface altitude (km)	1000.000
Entry velocity (km/s)	6.000
Entry flight path angle (deg)	-33.300

Fig. 1 compares altitude vs. velocity at the level-off or periapsis point of each trajectory. The trajectories reach periapsis about 2 seconds apart with an altitude difference of only 62 m and a velocity difference of less than 25 m/s. During the entry phase, the velocities of the two trajectories agree within 1 m/s at any given altitude and vary by less than 8 m/s during the exit phase. This reflects a remarkable agreement between the HyperPASS and 2DOF simulations.

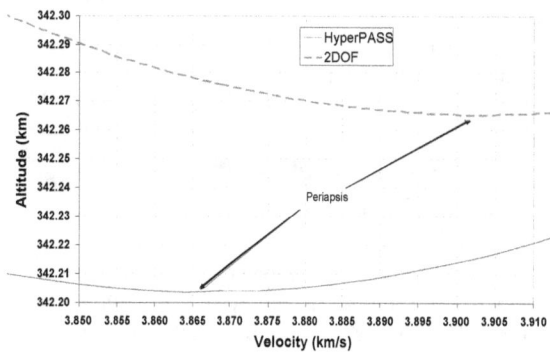

Fig. 1. Trajectory comparison at periapsis for Titan aerocapture

The flight path angles (FPA) are in close agreement throughout the entire simulation. It is seen in Fig. 2 that, upon reaching the exit altitude (1000 km), the FPA differs by less than 0.2 deg, which is the maximum FPA divergence seen in this simulation.

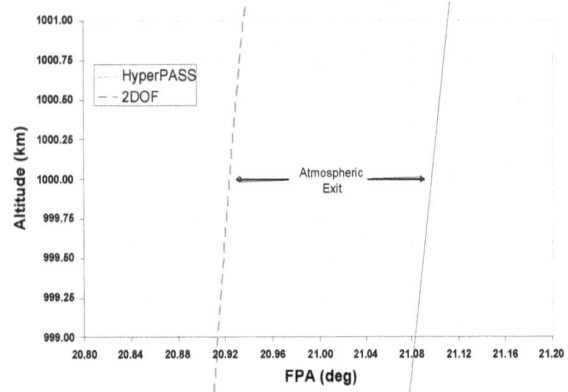

Fig. 2. FPA comparison at atmospheric exit for Titan aerocapture

The total deceleration force is likewise in agreement, with a maximum divergence of 0.015 gees.

2.2 Mars Landing

The Mars landing validation case was performed using the COSPAR90 [3] atmosphere model and the following parameters:

Table 2 Mars Landing Parameters

PARAMETER	VALUE
L/D	0.000
m/CDA (kg/m^2)	56.420
Atmospheric interface altitude (km)	125.000
Entry velocity - planet relative (km/s)	5.763
Entry flight path angle - planet relative (deg)	-11.300

Both simulation systems propagated the trajectory until a velocity of 500m/s was achieved. HyperPASS took about 0.5 seconds longer to reach this stopping condition. Looking at altitude as a function of time as shown in Fig. 3, it is found that above 60 km the altitude variance between the two simulations does not exceed 100 m. The greatest altitude divergence (~ 250 m) occurs during the last 10 seconds of simulation and is depicted in the figure below.

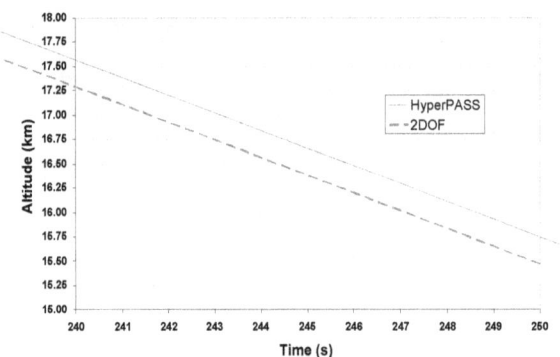

Fig. 3. Maximum altitude divergence for Mars landing

Next, attention was paid to the altitude versus velocity profiles. When each trajectory reached a 100 km altitude, the velocity difference was less than 2 m/s. At an altitude of about 38 km, the velocity divergence was at its peak (~55 m/s), as shown in Fig. 4. Upon reaching the target velocity of 500 m/s, the two trajectories showed a 162 m altitude difference.

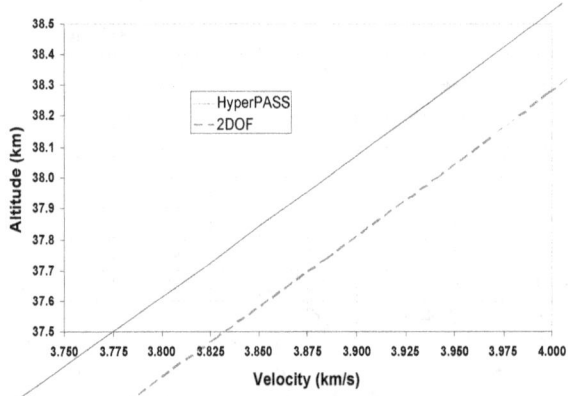

Fig. 4 Maximum velocity divergence for Mars landing

When a comparison was made between the FPA profiles, it was found that the values were extremely consistent between the two simulations, with a maximum divergence of less than 0.02deg.

2.3 Neptune Aerocapture

The Neptune aerocapture validation case uses the Hall [4] exponential atmosphere model. This run had an initial altitude of 440 km and was propagated to a 1200 km exit altitude. Parameters and initial conditions for the simulation can be viewed in Table 3.

Table 3 Neptune Aerocapture Parameters

PARAMETER	VALUE
L/D	0.632
m/CDA (kg/m^2)	208.030
Initial altitude (km)	440.000
Initial velocity - planet relative (km/s)	27.553
Initial flight path angle - planet relative (deg)	-7.440

The two simulations reached periapsis within 1 second of each other. Fig. 5 shows that HyperPASS achieved a periapsis altitude 2.08 km higher than that of the 2DOF system, but that the velocities at periapsis had less than 2 m/s difference.

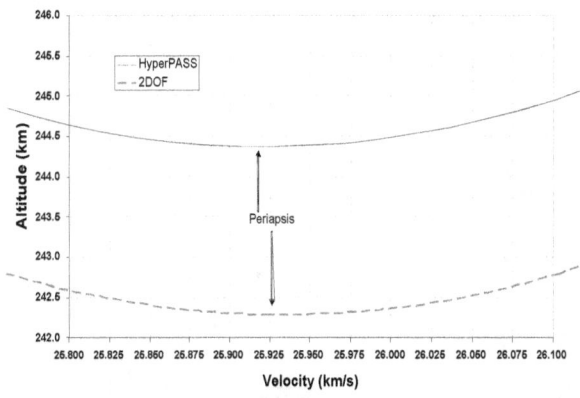

Fig. 5. Trajectory comparison at periapsis for Neptune aerocapture

The greatest variance in velocity occurred between altitudes of 450-500 km, which can be seen in Fig. 6. The systems reached exit (1200 km) less than 2 seconds apart, with a velocity difference of only 11 m/s.

The flight path angles at exit, differed by 0.03 deg and the total g-load experienced by the vehicle did not vary significantly between the two simulations. The greatest divergence occurred at periapsis, with a difference of 0.14 gees.

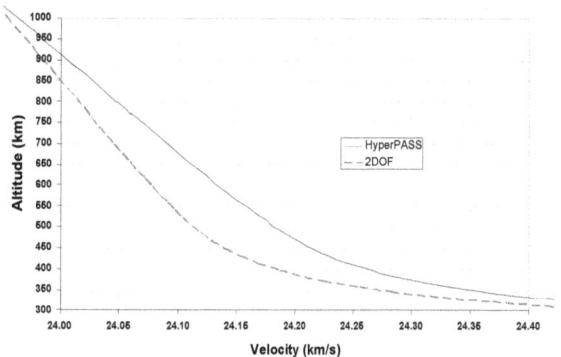

Fig. 6. Maximum velocity divergence during Neptune aerocapture

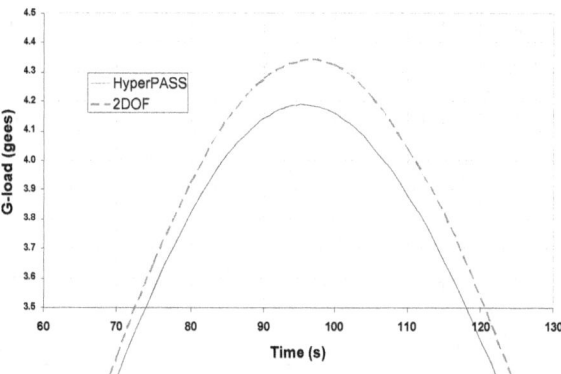

Fig. 7. Maximum g-load variance during Neptune aerocapture

3 HYPERPASS GUIDANCE CAPABILITIES

While the previous cases were performed with no implemented guidance (for validation purposes), HyperPASS possesses the ability to output the optimal trajectory, given a set of user entered initial and target conditions. The following examples include guided aerocapture at Mars, guided ballute aerocapture at Neptune, and aerobraking at Venus.

3.1 Aerocapture

HyperPASS performs guided aerocapture simulations by choosing an optimal entry FPA and guiding the vehicle through a bank-modulated aerocapture, in order to achieve the desired exit conditions. The following simulation parameters were entered into HyperPASS.

Table 4 Mars Guided Aerocapture Input Parameters

PARAMETER	VALUE
L/D	0.632
m/CDA (kg/m^2)	208.293
Entry altitude (km)	125.000
Entry velocity (km/s)	10.180
Target altitude (km)	125.000
Target velocity (km/s)	4.200

HyperPASS selected an entry FPA of -12.00 deg and was able to achieve the desired exit conditions using the angle of bank profile shown in Fig. 8. Fig. 9 displays the state parameters for the Mars guided aerocapture case.

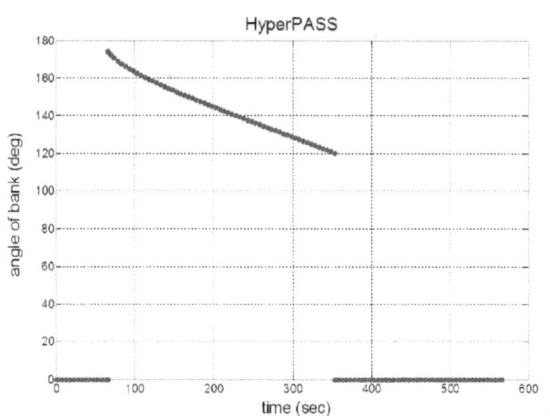

Fig. 8. Bank Angle Profile for Mars Guided Aerocapture

3.2 Ballute Aerocapture

To perform a guided ballute aerocapture simulation, HyperPASS chooses the optimal entry FPA and then determines the proper ballute cut time necessary to meet the user entered target conditions. The user can also enter ballute specifications such as shape, dimensions, and aerial density. Table 5 lists the Mars ballute case input parameters.

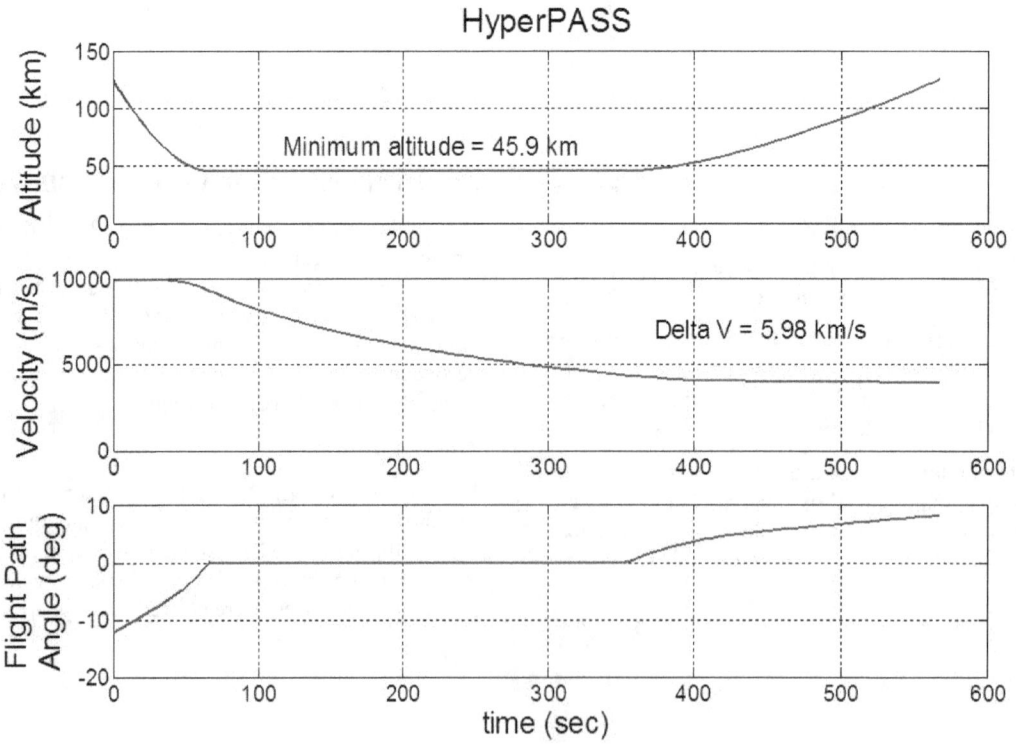

Fig. 9. State Vector for Mars Guided Aerocapture

Table 5 Mars Ballute Aerocapture Input Parameters

PARAMETER	VALUE
Ballute type	Sphere (CD = 0.9)
L/D	0.632
m/CDA - without ballute (kg/m^2)	54.072
m/CDA - with ballute (kg/m^2)	1.000
Entry altitude (km)	150.000
Entry velocity (km/s)	5.748
Target altitude (km)	150.000
Target velocity (km/s)	4.200

For the Mars ballute case, HyperPASS selected an entry FPA of -8.86 deg and cut the ballute after 195 seconds, in order to exit at the indicated target conditions. Fig. 10 displays the altitude, velocity, and FPA as a function of time.

3.3 Aerobraking

Aerobraking is simulated by performing consecutive atmospheric passes until the desired apoapsis altitude is reached. HyperPASS automatically implements raise periapsis maneuvers if the user entered heating limit is exceeded during aerobraking. Also, HyperPASS will perform orbit insertion and orbit circularization maneuvers if so desired. Table 6 gives the parameters used for the simulation presented here.

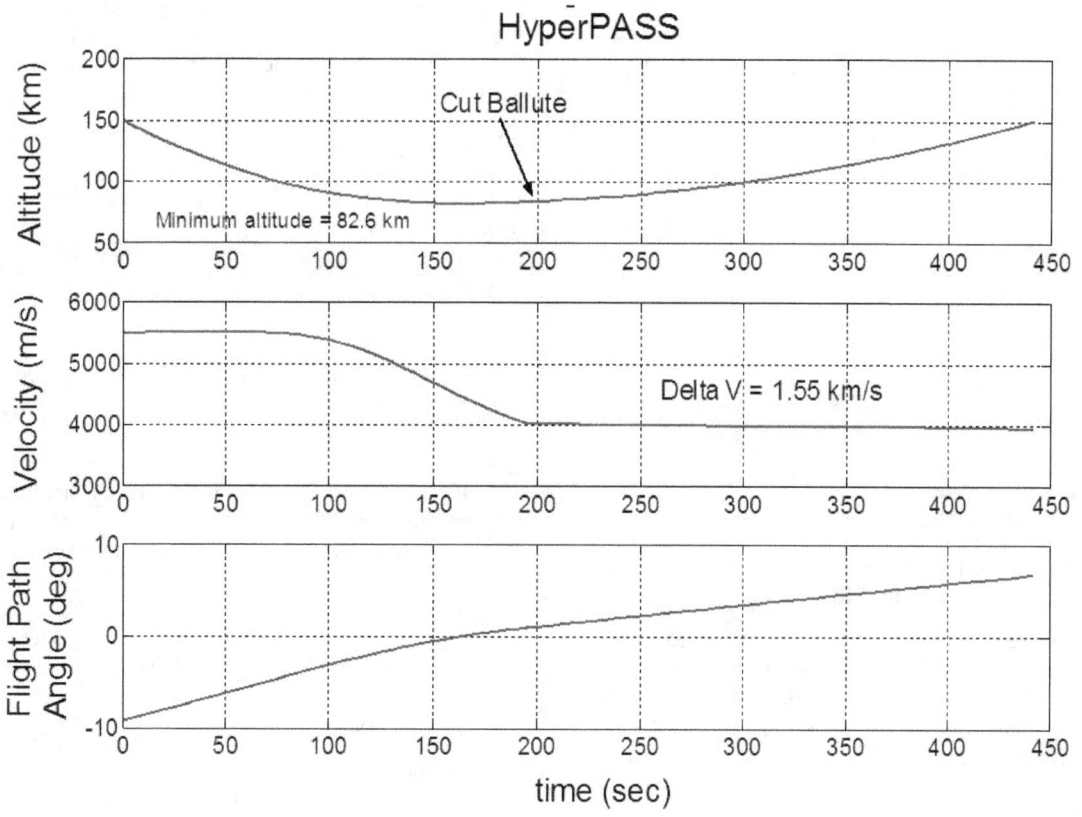

Fig. 10. State Vector for Mars Ballute Aerocapture

Table 6 Venus Aerobraking Input Parameters

PARAMETER	VALUE
L/D	0.000
Drag coefficient	2.200
m/CDA (kg/m^2)	21.739
1st periapsis altitude (km)	129.500
1st periapsis velocity (km/s)	8.586
Desired apoapsis altitude (km)	1500.000
Free Molecular Heating Limit (W/cm^2)	0.300
Raise periapsis altitude (km)	1.000

The desired apoapsis altitude was achieved in 689 atmospheric passes. Orbit period is given as a function of periapsis pass in Fig. 11.

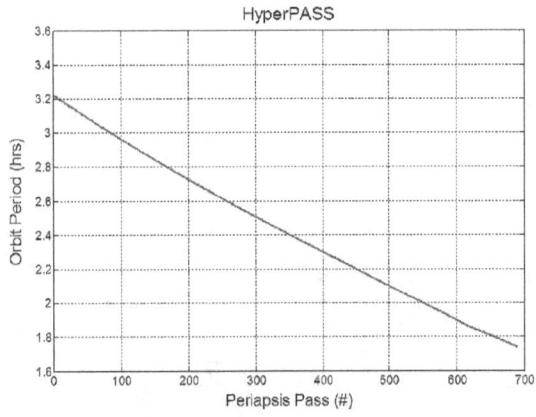

Fig. 11 Orbit period over time for Venus aerobraking

The free molecular heating limit was exceeded only once, which is apparent from the heating profile given in Fig. 12. At this point in the simulation, a maneuver was implemented to raise the periapsis altitude by the user specified "raise periapsis altitude".

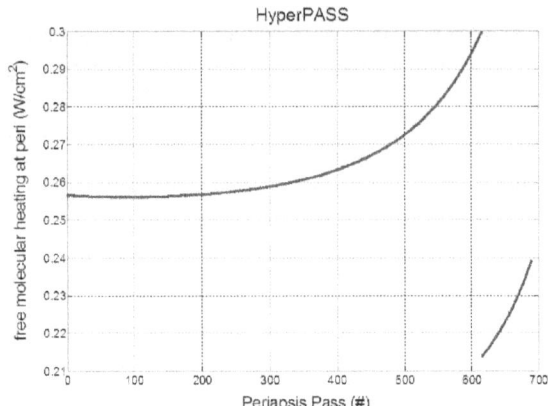

Fig. 12. Heating at periapsis for Venus aerobraking

4 CONTINUING DEVELOPMENT

HyperPASS is currently awaiting validation from a high fidelity simulation system. Planned improvements for future versions of HyperPASS include the generation of atmospheric data from Global Reference Atmospheric Models (GRAM) [5] and automated vehicle parameterization. Also, further bank modulated aerocapture development will include the added capability to maintain orbit inclination as opposed to the simple one-sided profile currently being used.

5 REFERENCES

1. Gates K. L. and Nock K. T. *HyperPASS User and Installation Manual*, http://www.gaerospace.com/projects/HyperPASS/HyperPASS_Manual.pdf.

2. Hunten, Titan Atmosphere Model, Prepared for NASA AIMES RC Preliminary Draft, 1981.

3. Pitts D. E., et al. The Mars Atmosphere: Observations and Model Profiles for Mars Missions, NASA Johnson Space Center report JSC-24455, 1990.

4. Hall J. L. and Lee A. K., Aerocapture Trajectories for Spacecraft with Large Towed Ballutes, AAS 01-235.

5. Justus C. G., "A Mars Global Reference Atmosphere Model (Mars-GRAM) for mission planning and analysis, *AIAA Paper No. 90-0004*, 28th Aerospace Sciences Meeting, 1990.

Performance of a Light-Weight Ablative Thermal Protection Material For the Stardust Mission Sample Return Capsule

M. A. Covington

ELORET Corporation, 690 W. Fremont Ave., Suite 8, Sunnyvale CA 94087
acovington@mail.arc.nasa.gov

ABSTRACT

New tests and analyses are reported that were carried out to resolve testing uncertainties in the original development and qualification of a lightweight ablative material used for the Stardust spacecraft forebody heat shield. These additional arcjet tests and analyses confirmed the ablative and thermal performance of low density Phenolic Impregnated Carbon Ablator (PICA) material used for the Stardust design. Testing was done under conditions that simulate the peak convective heating conditions (1200 W/cm^2 and 0.5 atm) expected during Earth entry of the Stardust Sample Return Capsule. Test data and predictions from an ablative material response computer code for the in-depth temperatures were compared to guide iterative adjustment of material thermophysical properties used in the code so that the measured and predicted temperatures agreed. The PICA recession rates and maximum internal temperatures were satisfactorily predicted by the computer code with the revised properties. Predicted recession rates were also in acceptable agreement with measured rates for heating conditions 37% greater than the nominal peak heating rate of 1200 W/cm^2. The measured in-depth temperature response data show consistent temperature rise deviations that may be caused by an undocumented endothermic process within the PICA material that is not accurately modeled by the computer code. Predictions of the Stardust heat shield performance based on the present evaluation provide evidence that the maximum adhesive bondline temperature will be much lower than the maximum allowable of 250°C and an earlier design prediction. The re–evaluation also suggests that even with a 25 percent increase in peak heating rates, the total recession of the heat shield would be a small fraction of the as-designed thickness. These results give confidence in the Stardust heat shield design and confirm the potential of PICA material for use in new planetary probe and sample return applications.

1. INTRODUCTION

The renewed interest in space missions to explore other planets has created a need for new advanced heat shield materials capable of efficiently protecting spacecraft under very high heating conditions. Such conditions may be experienced both during entry into the atmospheres of planets of interest and during reeentry into Earth's atmosphere for return missions. Very little development of new, efficient ablative materials has been pursued in the past two decades (since the Apollo and Viking spacecraft) due partly to the lack of missions requiring such materials.

The Stardust mission, as part of NASA's Discovery Program in 1995, created a requirement for new ablative heat shields as an enabling technology to meet the spacecraft mass goals. The Stardust mission [1] was designed as a mission to fly by the comet, Wild 2, at close range for the collection of cometary debris as well as to obtain interplanetary dust samples and return them to Earth within a Sample Return Capsule (SRC). The success of the mission requires that this Sample Return Capsule protect the collected samples during Earth atmospheric entry at an inertial velocity of 12.6 km/sec by keeping the SRC internal structure at temperatures that meet a science requirement to keep the sample materials below 70°C. These conditions result in nominal values for stagnation point heating flux of 1200 W/cm^2, peak surface pressures of 0.5 atm, and an integrated heat load of 36.5 kJ/cm^2 for the baseline entry.

To meet the requirements for the Stardust mission, one of a family of lightweight ceramic ablator materials developed at NASA Ames Research Center was selected for the forebody heat shield of the Stardust Sample Return Capsule. This material, Phenolic Impregnated Carbon Ablator (PICA), consists of a commercially available low density carbon fiber matrix substrate impregnated with phenolic resin. Some characteristics of this family of lightweight ablator materials and processing methods are given in [2]. The Stardust program resulted in intensive material development, modeling, and testing efforts [3] to provide a heat shield for the high convective heating conditions expected during Earth entry while under constraints of limited time and funding. Because of uncertainties in the heating rate calibrations carried out under the original test activities, a second project was initiated to reexamine the arcjet test conditions, the PICA ablative and thermal performance, and the modeling used to design the Stardust flight heat shield. Details of this project are reported in [4], and the summarized results are presented in this paper.

2. TESTS AND ANALYSES

2.1 PICA Material Description

The material used for the Stardust forebody heat shield is one of a class of low density, charring ablative materials recently developed at the NASA Ames Research Center. The PICA material is made from a fibrous carbon matrix insulation (Fiber Materials, Inc. under the trade name Fiberform®) impregnated with a commercial phenolic resin. The phenolic-formaldehyde resin (Borden Chemical SC1008®) used in the Starudust formulation creates a porous thermoset material after polymerization that has final bulk densities ranging from 0.22 to 0.27 g/cm^3, depending on the processing employed. More extensive details of the processing of PICA materials are given in [2].

2.2 Arc Jet Tests

The tests and related analyses were carried out to investigate the performance of PICA under conditions appropriate to the Stardust SRC entry environment. The test program utilized tests in a high energy arc jet to obtain needed data on both the ablative performance and the thermal performance of PICA material by varying the model size and the arc jet operating conditions.

The NASA Ames 60 MW Interaction Heating Facility [5] was used to provide the aerothermal test environment required to simulate Stardust SRC entry conditions as it was for the earlier Stardust development and qualification testing. Sixteen PICA flat-faced cylindrical models of 2.54 cm and 5.08 cm diameters were tested to obtain ablative performance data at the approximate conditions expected at the SRC peak convective heating flux and for heating rates at a required heat shield design margin above this. To measure thermal performance, sixteen flat-faced cylindrical models of 10.16 cm diameter were tested at lower convective heating rate conditions. Radiation heating from the entry shock layer previously had been found to be unimportant for the Stardust mission [6] as was the case for these arc jet tests. A summary of the configurations of these models and their test conditions are given in Tables 1 and 2. The stream enthalpy values in Tables 1 and 2 were deduced from laminar flow heat transfer relationships [7] using the measured pitot pressure and stagnation point heat flux to both copper heat sink calorimeters and water-cooled calorimeters.

Table 1. 2.54 cm and 5.08 cm diameter PICA models and test conditions

Model No.	Run No.	Model Diameter (cm)	Test Time (sec)	Heating Rate (W/cm^2)	Heat Load (kJ/cm^2)	Stagnation Pressure (atm)	Enthalpy (MJ/kg)	PICA Thickness (cm)	PICA Thickness (inch)
23	12E	2.54	15	1630	24.5	0.65	29.5	5.72	2.252
22	12W	2.54	10	1630	16.3	0.65	29.5	5.72	2.252
26	14E	2.54	20	1630	32.6	0.65	29.5	5.72	2.252
24	14W	2.54	10	1630	16.3	0.65	29.5	5.72	2.252
28	15E	2.54	15	1630	24.5	0.65	29.5	5.72	2.252
27	15W	2.54	6	1630	9.8	0.65	29.5	5.72	2.252
30	16E	2.54	22	1630	35.9	0.65	29.5	5.72	2.252
29	16W	2.54	17	1630	27.7	0.65	29.5	5.72	2.252
10	9E	5.08	30	1150	34.5	0.65	29.5	5.66	2.228
11	9W	5.08	20	1150	23.0	0.65	29.5	5.66	2.228
12	10E	5.08	35	1150	40.3	0.65	29.5	5.66	2.228
13	10W	5.08	25	1150	28.8	0.65	29.5	5.66	2.228
15	11E	5.08	40	1150	46.0	0.65	29.5	5.66	2.228
14	11W	5.08	20	1150	23.0	0.65	29.5	5.66	2.228
17	17E	5.08	39	1150	44.9	0.65	29.5	5.66	2.228
16	17W	5.08	37	1150	42.6	0.65	29.5	5.66	2.228

Table 2. 10.16 cm diameter PICA models and test conditions

Model No.	Run No.	Flat-Face Model Diameter (cm)	Test Time (sec)	Heating Rate (W/cm^2)	Total Heat Load (kJ/cm^2)	Stagnation Pressure (atm)	Enthalpy (MJ/kg)	PICA Thickness (cm)	(inch)
1	15E	10.16	69	580	40.0	0.45	29.5	6.05	2.380
2	15W	10.16	86	580	49.9	0.45	29.5	6.05	2.380
3A	14E	10.16	20	580	11.6	0.45	29.5	2.24	0.880
3B	13E	10.16	40	580	23.2	0.45	29.5	3.25	1.280
4A	17W	10.16	30	400	12.0	0.20	29.5	2.24	0.880
4B	13W	10.16	20	580	11.6	0.45	29.5	3.25	1.280
5A	17E	10.16	30	400	12.0	0.20	29.5	2.24	0.880
5B	18E	10.16	40	400	16.0	0.20	29.5	3.25	1.280
6A	14W	10.16	20	580	11.6	0.45	29.5	2.24	0.880
6B	18W	10.16	60	400	24.0	0.20	29.5	3.25	1.280
7A	12E	10.16	15	580	8.7	0.45	29.5	2.74	1.080
7B	12W	10.16	15	580	8.7	0.45	29.5	2.74	1.080
8A	11E	10.16	10	580	5.8	0.45	29.5	2.74	1.080
8B	11W	10.16	20	580	11.6	0.45	29.5	2.74	1.080
9A	16E	10.16	15	400	6.0	0.20	29.5	2.74	1.080
9B	16W	10.16	29	400	11.6	0.20	29.5	2.74	1.080

Experimental test data were compared with computed response results to develop and refine an analytical model that would satisfactorily predict both the ablative and thermal performance of PICA heat shields. These comparisons and results for the prediction of Stardust entry performance are given in more detail in following sections.

2.2.1 Test Models

Drawings of the PICA model configurations are shown in Figs. 1 and 2. Typical 2.54 cm and the 5.08 cm diameter models are illustrated in the drawing of Fig. 1 with model and graphite adapter dimensions proportionally scaled depending on the model diameter. The 10.16 cm diameter models are illustrated in Fig. 2. Details of the instrumented 10.16 cm models are shown in Fig. 3. The flat-faced 2.54 cm and 5.08 cm models had a corner radius of 0.239 cm and 0.476 cm, respectively. The 10.16 cm diameter models had a radius of 0.953 cm. All models were fabricated from flight-qualified PICA material from the same processing lot as that used for the Stardust flight heat shield. The average density of the PICA billet used for the models was 0.266 g/cm^3 as determined from small samples taken from multiple locations throughout the billet. The sidewalls of 2.54 and 5.08 cm models were uncoated but nearly all of the 10.16 cm models were coated with a graphite-based slurry (Graphi-Bond®) to minimize the escape of internally-generated pyrolysis gases out the sides.

The 2.54 cm and 5.08 cm models were retained in a graphite adapter using a graphite pin as shown in Fig.1. These graphite adapters were, in turn, attached to a facility model support arm with a stainless steel threaded mounting tube and a boron nitride insulation sleeve. This insulating sleeve was necessary to electrically isolate the model from the grounded support arm and reduce noise on the instrumentation signals.

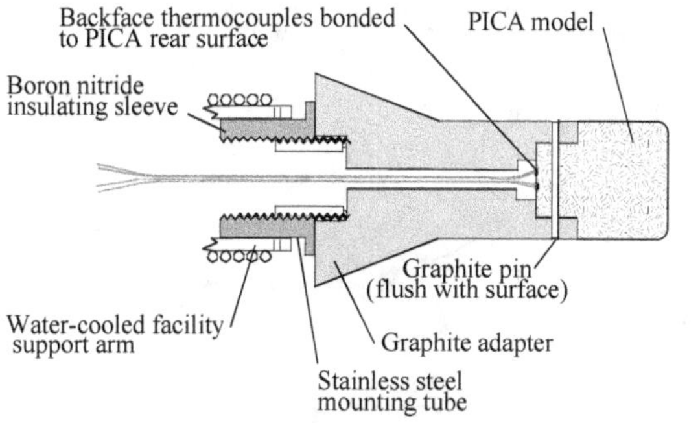

Fig. 1 5.08 cm diameter PICA model

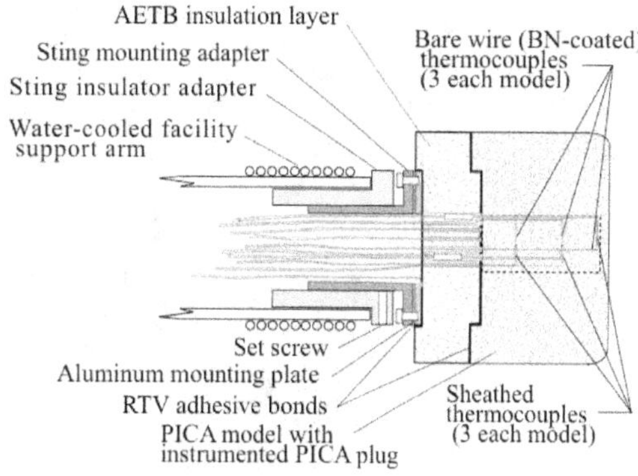

Fig. 2 10.16 cm diameter model

The 10.16 cm diameter models were constructed as shown in Fig. 2 with a 2.54 cm thick layer of Alumina Enhanced Thermal Barrier (AETB) material behind the PICA layer for thermal isolation and approximation of an adiabatic back wall condition. The PICA samples, AETB layers, and aluminum mounting plates were attached to each other with silicone adhesive as indicated in Fig. 2.

2.2.2 Test Model Instrumentation

The high heating rates and resulting high material temperatures used in the arc jet tests resulted in limitations on the type and number of material performance measurement sensors that could be incorporated. Because of the high rate of temperature increase and the high maximum temperatures (>3000°C) expected in the 2.54 cm and 5.08 cm diameter models, only backface temperature and surface temperature measurements were made on these models. Backface temperatures were obtained using 0.254 mm diameter Type R thermocouples attached to the model rear face with a graphite-based cement (Graphi-Bond®) as illustrated in Fig. 1. Two of these backface thermocouples were attached to each model for redundancy.

The 10.16 cm diameter models were instrumented using multiple thermocouple probes and bare wire thermocouples to measure in-depth, bondline, and backface temperatures. All in-depth sensors were mounted into a 2.54 cm diameter cylindrical PICA core that was subsequently inserted into the larger PICA model.

Only temperature measurements using the sheathed thermocouple probes are reported in this paper; a comparison of the sheathed thermocouple and the bare wire thermocouple measurements will be published separately.

The thermocouple probes were constructed of Type S thermocouple wire of 0.127 mm diameter encased in a 0.508 mm diameter platinum sheath and insulated with MgO powder to prevent electrical shorting to the sheath wall. These sheathed probes were bent at a 90° angle 1.27 cm from their tips to provide a configuration that allowed insertion into the test material along a constant depth line assumed to be along an isotherm and normal to the heat flux on the front face of test material. Such a temperature sensing configuration with the sensor wires or sheaths aligned along an isotherm and having a sheath length to diameter ratio of at least 25:1 (as in this case) minimizes measurement error due to conduction losses [8].

Accurate placement of both the sheathed thermocouple probes and the wire thermocouples was assured by insertion into carefully drilled holes at the specified depths measured from the unablated front face of the models. An insulative coating was applied to the wire thermocouples by dipping into a boron nitride slurry and then drying prior to insertion into the models. It was noted, however, that this coating was unevenly removed when the wire was pulled through the models during insertion so that the wire was probably not electrically insulated from the PICA in either the initial virgin or in the charred state.

Thermocouple In-Depth Locations
Distance from plug front face
(All dimensions in cm)

T/C	x	y	z	T/C to be used
1	0.20	-	-	Platinum sheath - Type S
2	-	1.60	-	Platinum sheath - Type S
3	-	-	3.60	Platinum sheath - Type S)
4	0.20	-	-	0.025 D bare wire BN-coated - Type S
5	-	1.60	-	0.025 D bare wire BN-coated - Type S
6	-	-	3.60	0.025 D bare wire BN-coated - Type S
7	In PICA/AETB bondline			Type K
8	In PICA/AETB bondline			Type K
9	In AETB/Al mounting plate bondline			Type K

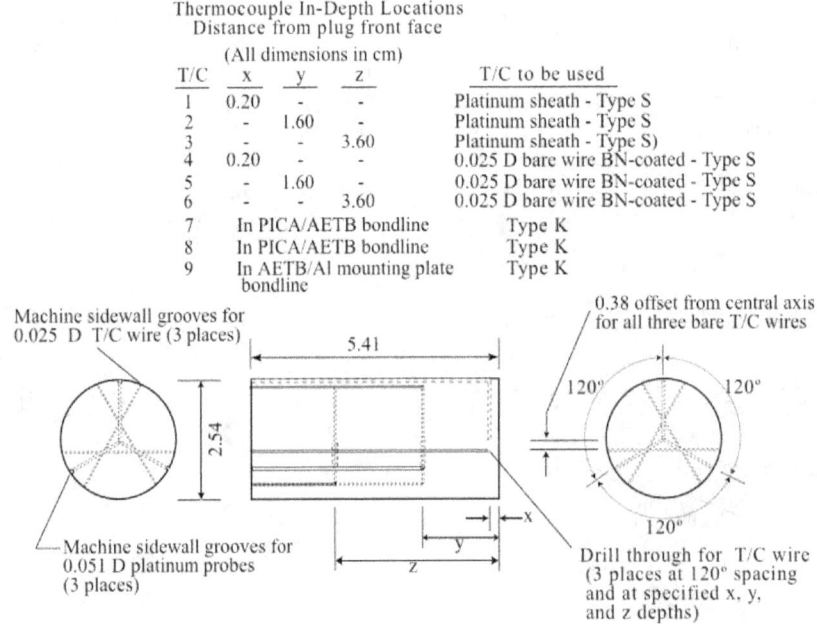

Fig. 3 Typical 10.16 cm diameter model instrumentation details

Bondline temperatures were measured by thermocouples (Type K) mounted within the silicone adhesive bondline between the rear face of PICA models and the AETB layer. Backface temperatures were sensed with thermocouples (Type K) attached to the rear face of this AETB material. Two bondline thermocouples were used on each model for redundancy.

Surface temperature data were obtained using two different single-wavelength optical pyrometers as well as a dual-wavelength (two-color) optical pyrometer. An imaging infrared video pyrometer system also was used to measure temporal temperature distributions on models during tests.

2.1.3 Stream Calibrations

Stream measurements were performed to set the heating rate and pressure conditions for these tests. Stagnation pressure for all arcjet conditions was measured using water-cooled pitot probes. For conditions used with the 2.54 cm and 5.08 cm diameter models, two different hemisphere-cylinder copper heat sink calorimeters were used to measure the cold wall convective heating flux. One calorimeter had a diameter of 3.05 cm and a nose radius of 5.84 cm and the other had a diameter of 3.05 cm and a nose radius of 10.16 cm. Both calorimeters had a corner radius of 0.152 cm. The data from a series of calibration runs with these two calorimeters were used to select two test conditions. One selected condition gave a cold wall, fully catalytic heating rate of 1630 W/cm^2 for tests of the 2.54 cm diameter models. This same condition provided a cold wall, fully catalytic convective heating rate of 1150 W/cm^2 for the 5.08 cm diameter models. The measured stagnation pressure at this test condition for both smaller models was 0.65 atm. The actual measured heating rate values of these non-flat faced calorimeters were corrected using the geometric correlation factors of [9] to provide the assumed heating flux to the flat faced PICA models actually tested. The front surfaces of the copper heat sink mass in these calorimeters were carefully cleaned before each run to assure that a highly catalytic surface for dissociated gas species recombination was present to fulfill the assumption of a fully catalytic wall.

Calibration runs for the 10.16 cm diameter models used 7.62 cm diameter water-cooled hemisphere calorimeters with Gardon-type thin foil heat flux sensors mounted at the stagnation point to define two test conditions. One selected nominal condition for tests of the 10.16 cm diameter PICA models was a cold wall, fully catalytic heating rate of 400 W/cm^2 and a stagnation point pressure of 0.20 atm, and the other was at a heating rate of 580 W/cm^2 and stagnation pressure of 0.45 atm.

2.1.4 Test Environments

The arc jet test conditions and test times are shown in Table 1 for the 2.54 and 5.08 cm diameter models and in Table 2 for the 10.14 cm diameter models. The exposure times for the smaller models varied from 6 sec to 40 sec and resultant total heat loads were from 9.8 kJ/cm^2 to 44.9 kJ/cm^2 (see Table 1). For the 10.16 cm diameter models, the two different arc jet operating conditions provided model exposure times from 10 to 86 sec, and total heat loads from 5.8 kJ/cm^2 to 49.9 kJ/cm^2 on PICA models of varying thickness as shown in Table 2. The arc jet operating conditions for all the tests was at a nominal stream total enthalpy of 29.5 MJ/kg. Radiation heating to the models from the shock layer at all of these conditions was negligible.

2.3 Material Performance Modeling

Modeling of the ablation and thermal performance of the PICA material used the FIAT (Fully Implicit Ablation and Thermal) computer code described in [10]. This code was used in a mode that models in-depth conduction, kinetically-controlled pyrolysis, blowing due to pyrolysis gases, and surface recession as a function of time in a one-dimensional porous ablative material. The PICA properties used with this code were a combination of measured thermophysical properties and polymer pyrolysis kinetics, and adjusted property values based on thermal response data from these tests. The measured specific heat and thermal conductivity of virgin material were taken from [11]. The initial values from [11] for char thermal conductivity and specific heat were iteratively adjusted to give the best fit to thermal response data over the range of test results. The Arrhenius kinetic constants for phenolic pyrolysis from [12] were used. Pyrolysis gas enthalpy values for the ablation products were calculated using an equilibrium thermochemistry program [13]. A PICA virgin and char surface emissivity of 0.9 was assumed that is consistent with the value [2] measured for PICA and that has been used for other carbonaceous ablators. The PICA material ablation model was validated using the arcjet surface recession and thermal response data from these tests as discussed in following sections.

2.4 Data Analysis and Computational Model Comparisons

2.4.1 Ablation Performance

The 2.54 cm and 5.08 cm diameter models were tested at the highest heating rates and stagnation pressures as previously described to measure surface ablation rates at conditions approximating those for a nominal Stardust entry (1200 W/cm^2) and at least a 25% higher heating rate. The surface recession rate is taken as the best measure of ablative performance in this study. The recession rate data for these two smaller models are listed in Table 3, and recession rate data for the 10.16 cm models are given in Table 4. These data are plotted and compared to the steady state surface recession rate calculated by the FIAT code in Figs. 4 and 5. A least squares fit of data at both the 1150 and 1630 W/cm^2 calculated by the FIAT code using revised properties are given in Fig. 5. The plot shows that the average measured recession rates for both the 400 W/cm^2 and 580 W/cm^2 levels are higher than that predicted by FIAT code with the predicted steady state rate being 11% low at the 400 W/cm^2 level and 7% low at the 580 W/cm^2 condition. This agreement between measured and predicted recession rates using the FIAT model is satisfactory considering the range of high heating fluxes the model attempts to cover and the test and model parameter uncertainties. The curves for the calculated FIAT response show that, even at the 400-600 W/cm^2 heating range, there is a reasonably long initial period of non-steady ablation of at least 40 seconds until steady state values of surface recession and temperature are

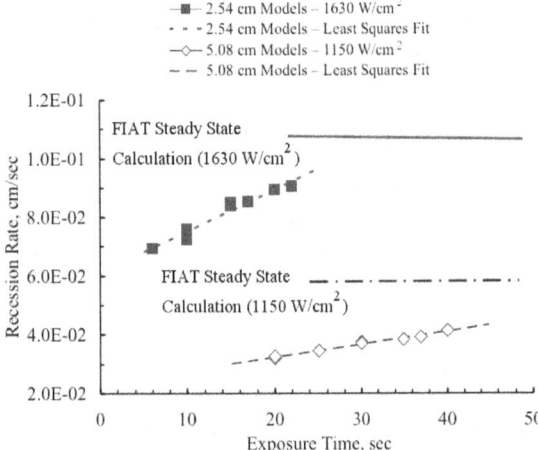

Fig. 4 Surface recession for 2.54 and 5.08 cm models

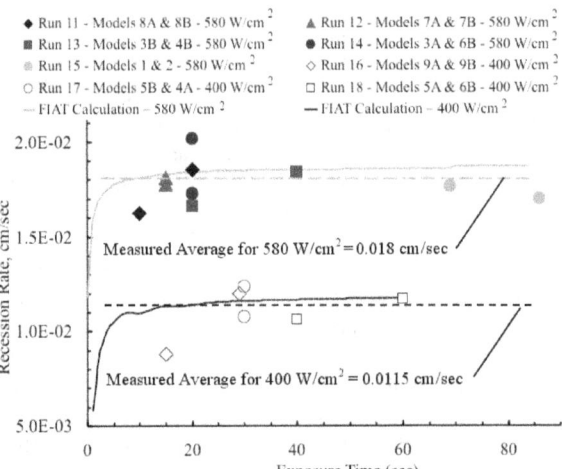

Fig. 5 Surface recession for 10.16 cm models

heating levels show a clear trend of increased recession caused by one or a combination of effects. One effect may be due to the increasing depletion of the phenolic resin at these high heating conditions, thus reducing blowing by pyrolysis gases from the front face and resulting in reduced convective heating blockage. High sidewall heating to the small diameter models under these conditions invalidates the assumption of one-dimensional slab heating inherent in the FIAT calculations. Also, the progressive rounding of the model front face with increasing exposure time and a resulting decrease in the effective nose radius would increase the convective heating. The data in Fig. 4 show that the recession rates for the smaller models approach that predicted by the FIAT code and, at both heating levels, the recession is less than the calculated steady state rate over the range of test times experienced. The data for the surface recession of the 10.16 cm models and comparison with the transient recession rate as reached. Recession asymmetries developed on the 2.54 and 5.08 cm models that are thought to be due to misalignment with the peak heating profile in the arc heater stream at the high heating rate conditions, and recession measurements were only made at the center for these models. On the 10.16 cm diameter models, asymmetric recession was not observed and recession measurements were made at the center, 1.0 cm away from the center, and at the edge of the model. The results are shown in Table 4. The front surface roughness on all models tested was greater post-test than on the pre-test machined surfaces; however, the surfaces exhibited no evidence of large scale spallation and visually appeared reasonably smooth and uniform at all conditions.

2.4.2 Thermal Performance

Surface and in-depth temperature measurements from the 10.16 cm diameter models were used to define the thermal response and to derive the analytical response model as previously discussed. The temperature data from a selected number of tests on these models were used to revise the thermophysical properties for use in the FIAT response code. None of the recession data from either the 2.54 cm, 5.08 cm, or 10.16 cm models was used for defining the properties since changes in these properties over ranges of interest have minimal effect on the recession rates. The temperature response data used were those from Models 3B, 4B, 5B, 6A, 6B, 7B, and 9B. These were selected because they were the most complete sets of data, had the best instrumentation signal reliability, and included a representative range of PICA layer thicknesses from 2.24 cm to 3.25 cm and model diameter to thickness ratios from 4.55 to 3.125. The approach used to revise the modeling parameters was to modify only the char conductivity and char specific heat, and re-run the FIAT code for a new set of predictions that was compared with the experimental in-depth temperature profiles for the 8 sets of data from the models selected. This process was then iteratively repeated until it was judged that the revised model predictions were in reasonable agreement with the sets of measured data. The char thermal conductivity and specific heat were chosen as the properties to vary since they are the two with the greatest uncertainty.

Figs. 6 through 11 show representative in-depth and surface temperature data and compare these data with FIAT code predictions using the revised property set that gave the best agreement. In general, the comparison of the agreement is based on maximum temperature reached at a given in-depth location because of an observed temperature rise lag that did not match the predicted monatomic temperature rise of the computer calculations. This failure to predict the observed lag in in-depth and bondline temperatures was found in all data for this and other tests of PICA material, and is discussed more fully later in the paper. The maximum temperature was chosen for this reason as the basis of comparison between measured and predicted results. For each of the temperature plots of Figs. 6 through 10, the legends show in parenthesis the depth of the installed thermocouple probes from the original unablated surface.

The data from model 9B are typical for the lowest heating rate of 400 W/cm^2. Model 9B had a test time of 29 sec and an integrated heat load of 11.6 kJ/cm^2. In-depth thermocouple and pyrometer-measured surface temperature data are presented in Figs. 6a, 6b, and 6c for this model. It is seen that temperatures calculated

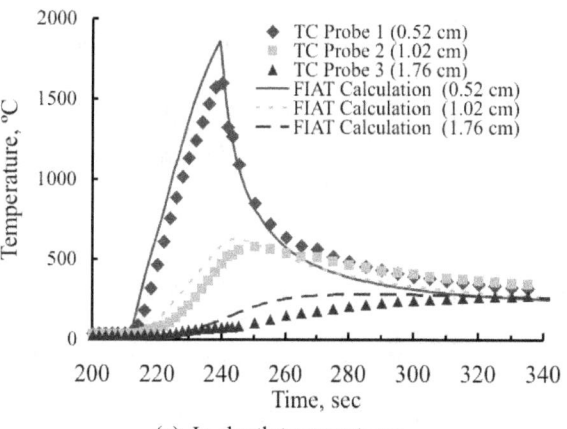

(a) In-depth temperatures

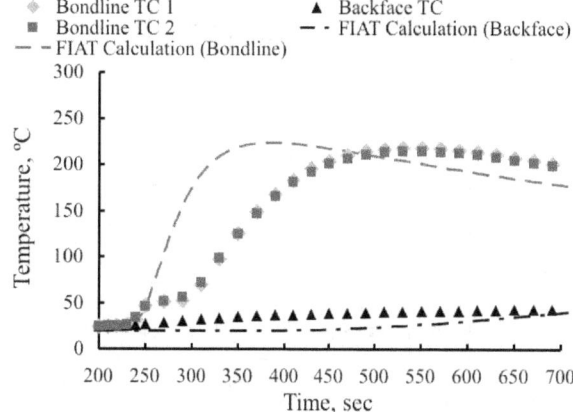

(b) Bondline and backface temperatures

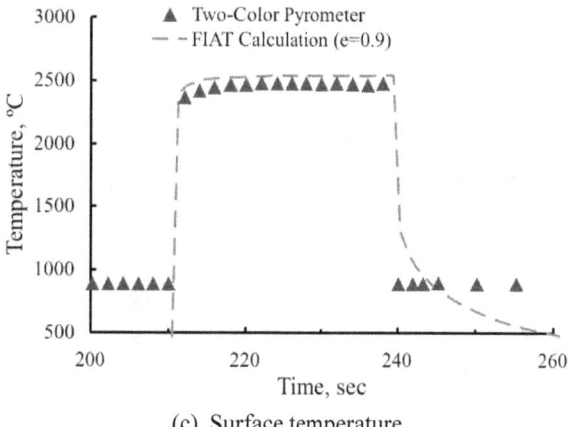

(c) Surface temperature

Fig. 6 Comparison of experimental and calculated thermal response for Model 9B. Heating rate=400 W/cm^2; stagnation pressure=0.20 atm; heating time 29 sec.

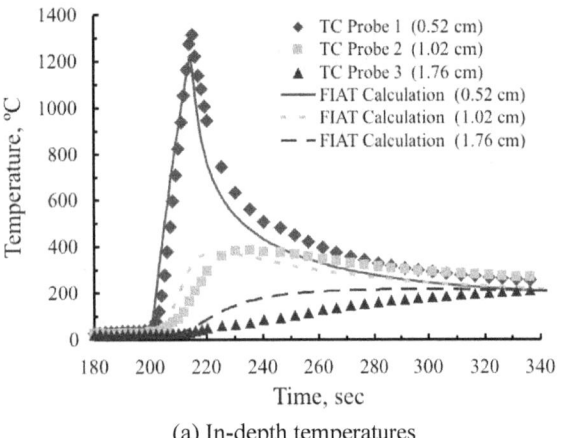

(a) In-depth temperatures

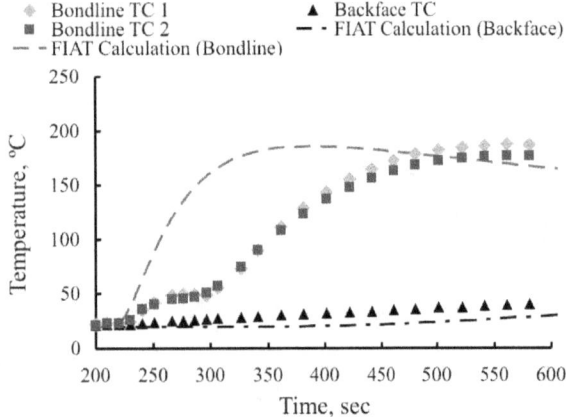

(b) Bondline and backface temperatures

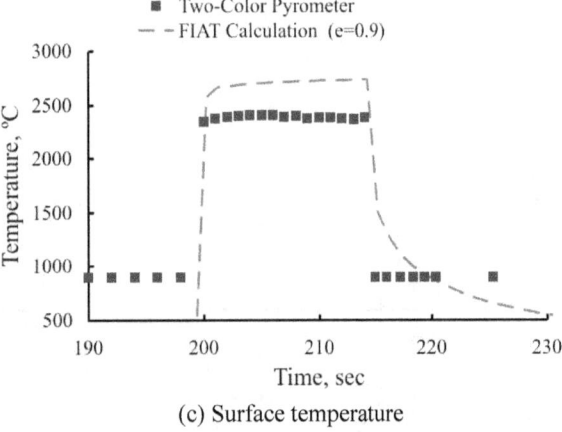

(c) Surface temperature

Fig. 7 Comparison of experimental and calculated thermal response for Model 7B. Heating rate=580 W/cm^2; stagnation pressure=0.45 atm; heating time=15 sec.

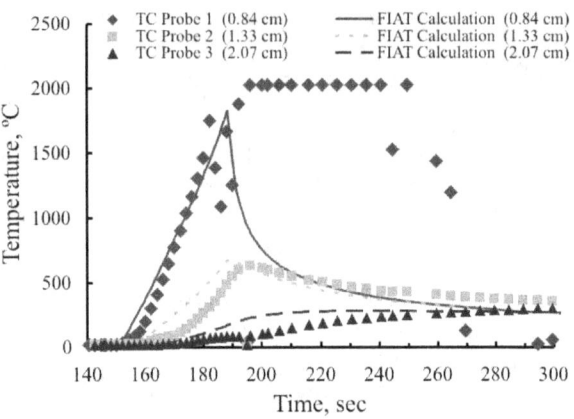

(a) In-depth temperatures

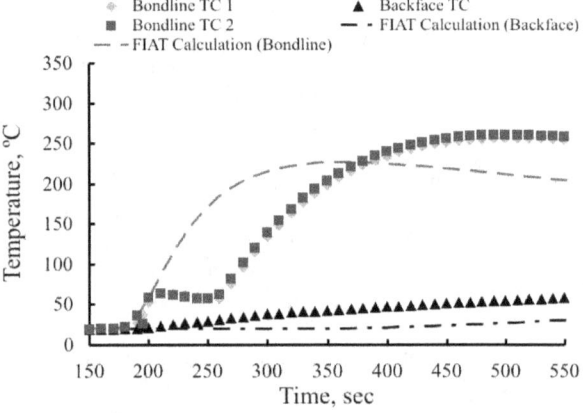

(b) Bondline and backface temperatures

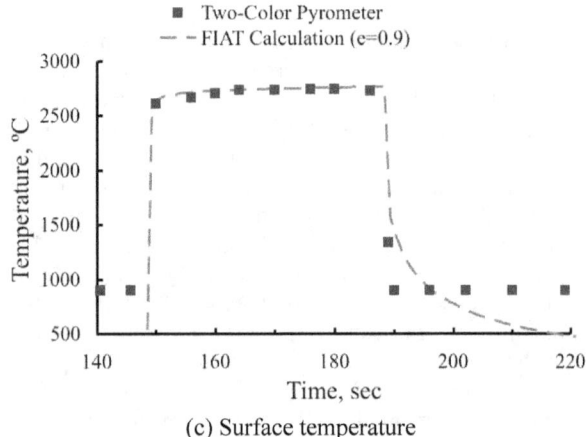

(c) Surface temperature

Fig. 8 Comparison of experimental and calculated thermal response for Model 3B. Heating rate=580 W/cm^2; stagnation pressure=0.45 atm; heating time=40 sec.

with the FIAT code are in reasonably good agreement with the experimental data except for 1) a mismatch in the prediction for the thermocouple closest to the surface (0.52 cm deep), 2) a faster temperature rise than measured for this depth, and 3) a failure to predict the bondline temperature response lag as shown in Fig. 6b.

Figures 7a, 7b, and 7c present a comparison of predicted in-depth, bondline, backface, and surface temperatures from the FIAT code with experimental measurements for Model 7B at a heating rate of 580 W/cm^2. In this case, the code predicts well the response of the thermocouple nearest the surface (0.52 cm deep), the peak in-depth temperatures at 1.016 cm and 1.755 cm depth, and the maximum bondline and backface temperatures. The measured surface temperature is about 300°C lower than the calculated level but is unaccountably lower than other pyrometer-measured temperatures at this same heating condition. Again, the calculated response does not accurately simulate the lag in temperature rise at the 1.02 cm and 1.76 cm deep locations or at the bondline.

The data from the test of Model 3B show similar results in Figs. 8a, 8b, and 8c. For this model, Fig. 8a shows that the thermocouple probe melted at about the temperature expected for platinum (1769°C). The FIAT calculation is seen to predict well the maximum temperatures measured at 1.33 and 2.07 cm depths, and the maximum of the two bondline temperatures, but the calculated response did not match the lag in the measured temperature rise seen in Figs. 8a and 8b. The calculated and experimental surface temperatures are seen to be in excellent agreement in this case for the entire test time (Fig. 8c). Another example presented for results at a heating level of 580 W/cm^2 is given in Fig. 9a, 9b, and 9c for Model 6A. This is the thinnest 10.16 cm diameter model tested with a pre-test thickness of 2.24 cm. The thermocouple probe closest to the surface (0.52 cm deep) melted at a temperature consistent with the melting point for platinum (Fig. 9a) as would be expected. The other response of the other in-depth thermocouple (1.43 cm deep) is matched by the computer prediction for maximum temperature (Fig. 9b). The calculated and measured surface temperatures are seen to agree well (Fig. 9c).

Fig. 10 shows results for Model 2 tested at a heating rate of 580 W/cm^2 for the longest test times and highest integrated heat load of all the 10.16 cm models. This model was heated for 86 seconds with a total heat load of 49.9 kJ/cm^2. Both of these two tests exceeded the total heat load value (36 kJ/cm^2) expected for the Stardust SRC entry with the nominal entry trajectory

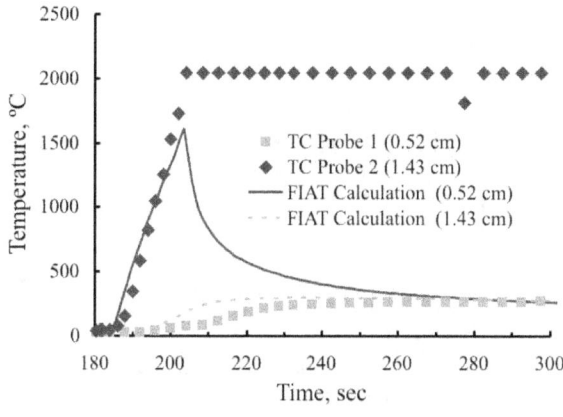

(a) In-depth temperatures

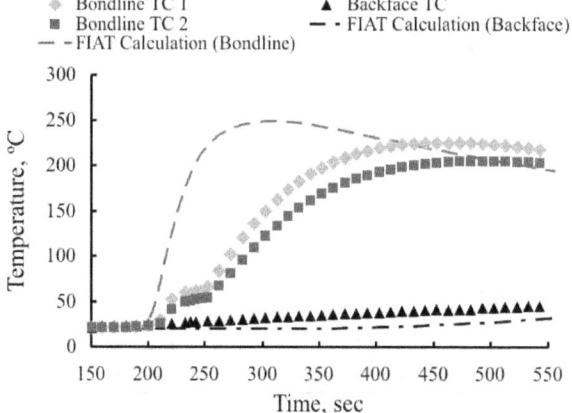

(b) Bondline and backface temperatures

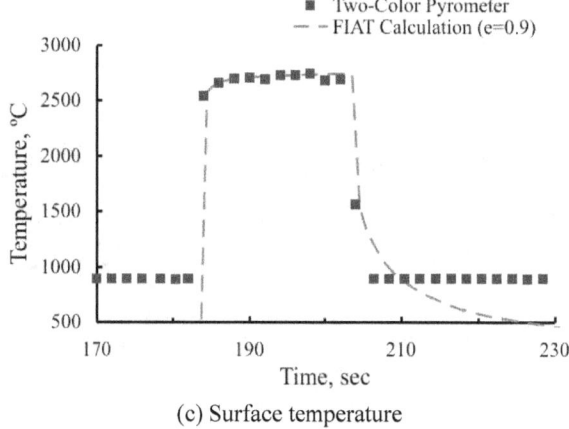

(c) Surface temperature

Fig. 9 Comparison of experimental and calculated thermal response for Model 6A. Heating rate=580 W/cm^2; stagnation pressure=0.45 atm; heating time=20 sec.

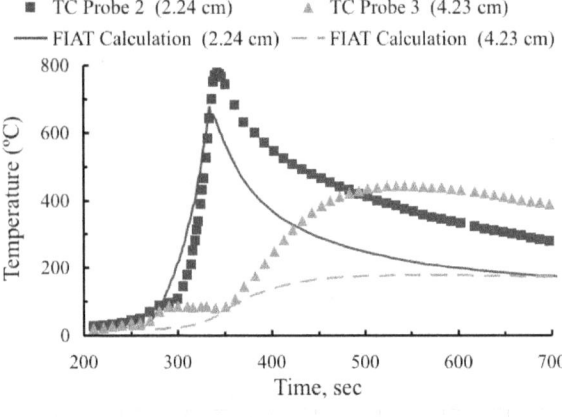

(a) In-depth temperatures

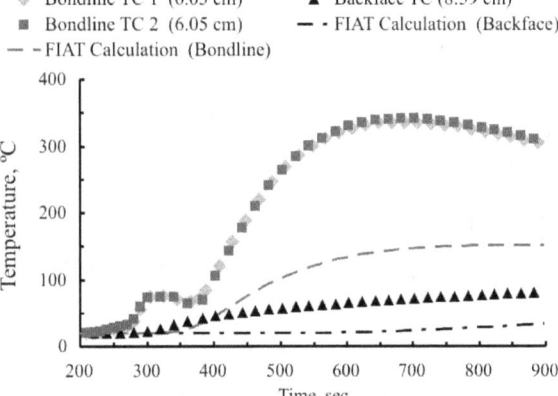

(b) Bondline and backface temperatures

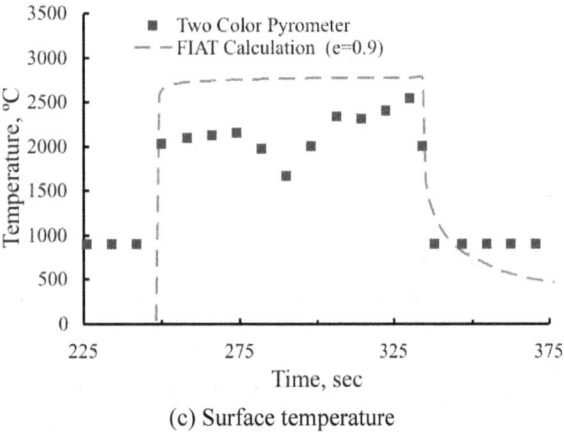

(c) Surface temperature

Fig. 10 Comparison of experimental and calculated thermal response for Model 2. Heating rate=580 W/cm^2; stagnation pressure=0.45 atm; heating time=86 sec.

but were somewhat less than the expected heat load of 55 kJ/cm^2 for an overshoot trajectory entry. This was one of the two thickest models tested with 6.04 cm of PICA backed by the 2.54 cm thick AETB layer. The thermocouple probes closest to the surface (0.89 cm deep) indicated failure from melting within the first 20 seconds of exposure with a response very similar to that shown for Model 3B and Model 6A (see Figs. 8a and 10a) and not shown here.

In Figs. 10a, 10b, and 10c, it is seen that the computer model results badly under-predict the in-depth, bondline, and backface measured temperatures in contrast to the much better agreement on thinner PICA models. The best explanation for this FIAT underprediction is that the assumption of one-dimensional ablation and heat conduction inherent in the FIAT model is not valid on these thick models with a large side wall area exposed to high heating levels. This conclusion is also supported by temperature rise differences between the computer predictions and the measured values. The more rapid onset of the measured in-depth temperature rise seen in Figs. 10a, and 10b is consistent with heat being conducted inward from sidewall heating. A post-test cross-section photo (Fig. 11) of one of these models after being cut into two halves clearly shows that considerable degradation had progressed from the model sides toward the center, thus invalidating the assumption of one-dimensional heat transfer assumed in FIAT calculations.

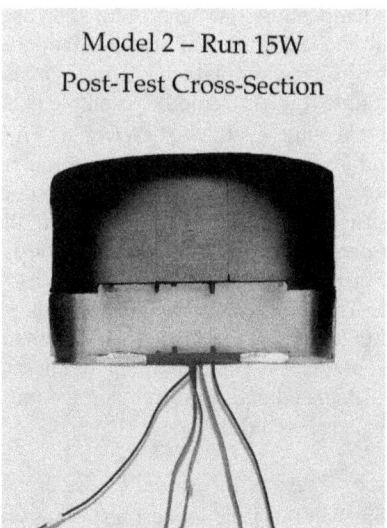

Fig. 11 Post-test photo of Model 2 cross-section that shows internal ablation resulting from sidewall heating

The measured in-depth temperature data characterized by an expected increase with time followed by a leveling off at temperatures between 0°C to 100°C to a constant or, in some cases, a decreasing temperature value (cf., Figs. 6b, 7b, 8b, 9b, 10a, 9b, 11b) has been observed in other heating tests of PICA and similar phenolic impregnated materials. For example, this same feature is evident in temperature data from arcjet tests during PICA development [14]. Similar features are seen in data from other heating tests of materials with phenolic resin impregnation dating back to at least 1968 but apparently have not been documented. An unidentified endothermic process within the PICA material can explain this behavior. Phase transition processes are known to cause similar effects on transient temperature data in other materials. It is clear that the FIAT code with the material properties and ablation chemical kinetics used for this study did not capture this behavior. The resolution of this modeling inconsistency is the subject of a separate investigation.

3. APPLICATION TO STARDUST FLIGHT HEAT SHIELD DESIGN

An objective of this investigation was to verify the Stardust SRC forebody heat shield design for Earth re-entry. The major design criterion for this vehicle heat shield was a maximum allowable bondline temperature of 250°C. The revised PICA properties derived from iterative adjustment to provide a best fit to data shown herein was used with the FIAT computer code to recalculate the surface recession, maximum temperatures, and design margins. Fig. 12 presents the results of this newer calculation and a comparison to the original design with the calculated bondline temperature as a function of spacecraft entry time plotted for both the baseline design trajectory heating rate (1200 W/cm^2) and for a 25% increase in heating rate (1500 W/cm^2). The result from the original calculation using the baseline Stardust properties also is shown. It is seen that the calculated maximum bondline temperatures for the cases of nominal design heating and of a 25% added margin are all well below the design maximum allowable temperature of 250°C. The recalculated maximum temperature of about 116°C is also less than that from the earlier calculation with baseline properties of 190°C. These results provide added confidence in the performance of PICA material for the Stardust heat shield design.

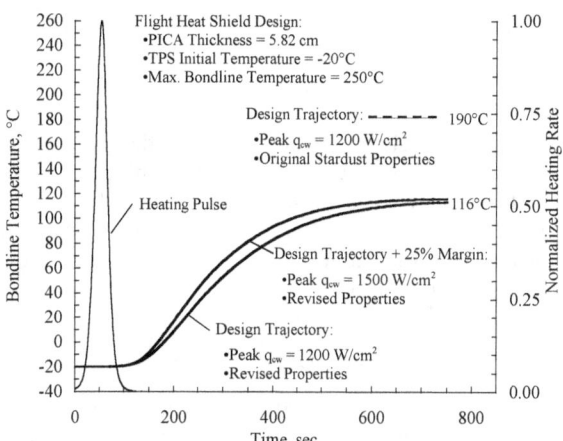

Figure 12 Comparison of Calculated bondline temperatures for Stardust heat shield design using baseline and revised model properties

4. Concluding Remarks

Extensive arcjet tests at conditions simulating the design Earth entry heating conditions for the Stardust Sample Return Capsule were conducted as part of this investigation to evaluate the heat shield design. The resulting data on ablative surface recession and internal temperature response were used to iteratively modify thermophysical properties for PICA material used in the FIAT computer code to satisfactorily predict the experiment response using surface recession rate and maximum internal temperatures as criteria. An apparent endothermic process at low temperatures during PICA ablation resulted in a delayed internal temperature rise that was not captured by computer code results using either the baseline or the revised properties. A separate study is underway to investigate this previously undocumented process. The predictive results using the FIAT code, however, were in reasonable agreement with measured surface recession and maximum internal temperature data so that the use of this code with the revised property set can predict with good confidence the performance of the actual Stardust heat shield design. It was concluded that the results of this study have validated the original Stardust PICA forebody heat shield design, and provided evidence for lower than previously predicted maximum temperatures at the adhesive bondline attaching the shield to the spacecraft structure. These results increase confidence in the heat shield design for the Stardust Sample Return Capsule.

Acknowledgments

This work was supported by NASA Ames Research Center under Contract NAS2-99092 with Eloret Corporation. The assistance of Bill Willcockson of the Lockheed Martin Company in providing data and material is gratefully acknowledged. The authors would like to thank James Arnold and Ethiraj Venkatapathy of NASA Ames for continued support and their valuable comments and suggestions.

References

1. Vellinga, J., et al, "Environmental Design Considerations for Stardust," Report 97ES-197, Lockheed Martin Company, Januaary 1997.
2. Tran, H., Johnson, C.E., Rasky, D.J., Hui, F.C., Hsu, M.-T., Chen, T., Chen, Y.-K., Paragas, D., and Kobayashi, L., "Phenolic Impregnated Carbon Ablators (PICA) as Thermal Protection Systems for Discovery Missions," NASA TM-110440, April 1997.
3. Tran H., Johnson, C.,Hsu, M-T., Smith, M., Dill, H., Chen-Jonsonn, A., "Qualification of the Forebody Heatshield of the Stardust's Sample Return Capsule," Paper 97-2482, 32nd AIAA Thermophysics Conference, Atlanta, Georgia, June 23-25, 1997.
4. Covington, M.A., Goldstein, H.E., Balboni, J.A., Terrazas-Salinas, I., Chen, Y.-K., Olejniczak, J., Martinez, E.R., Hienemann, J.M., "Analysis and Modeling of the Performance of a Low Density Carbon Phenolic Material for Atmospheric Entry Thermal Protection," NASA TM (to be published).
5. Winovich, W., and Carlson, W., "The 60 MW Interaction Heating Facility," 25th International Instrumentation Symposium, Anaheim, California, May 1979.
6. Olynick, D., Chen, Y.K., and Tauber, M., "Forebody TPS sizing with Radiation and Ablation for the Stardust Sample Return Capsule," Paper 97-2474, AIAA 32nd Thermophysics Conference, Atlanta, GA, June 23-25, 1997.
7. Zoby, E.V., "Empirical Stagnation-Point Heat Transfer Relation in Several Gas Mixtures at High Enthalpy Levels," NASA TN D-4799, June 1968.
8. Anon., *Standard Practice for Internal Temperature Measurements in Low-Conductivity Materials*, ASTM Standard E-377, December 1996.
9. Zoby, E.V. and Sullivan, E.M.,"Effects of Corner Radius on Stagnation-Point Velocity Gradients on Blunt Axisymmetric Bodies," NASA TN X-1067, March 1965.
10. Chen, Y.-K., and Milos, F.S., "Ablation and Thermal Response Program for Spacecraft Heatshield Analysis," Paper 98-0273, AIAA Aerospace Sciences Meeting & Exhibit, 36th, Reno, NV, Jan. 12-15, 1998.
11. "Final Report on Thermal Properties of Lightweight Charring Ablators," FMI EMTL Final Report No. 1648, Fiber Materials, Inc., July 1994.
12. Goldstein, H.E., et al, *J. Macromolecular Science-Chemistry*, A(34), PP.649-673, July 1969.
13. Anon., "User's Manual, Aerotherm Chemical Equilibrium Computer Program, (ACE81)," Acurex Report UM-81-11/ATD, Acurex Corporation, Aerotherm Division, Mt. View, California, August 1981.
14. "An Assessment of the Influence of Material Variables on the Ablation Response of PICA Type Materials," FMI Report FMI-Pm0-96-036, Fiber Materials, Inc. February 1996.

Development and Test Plans for the MSR EEV

Robert Dillman[1], Bernard Laub[2], Sotiris Kellas[3], Mark Schoenenberger[4]

[1] NASA Langley, MS 472, Hampton VA 23681, USA; Email: Robert.A.Dillman@nasa.gov
[2] NASA Ames, MS 234-1, Moffet Field CA 94035, USA; Email: Bernard.Laub@nasa.gov
[3] General Dynamics, MS 495, Hampton VA 23681, USA; Email: S.Kellas@larc.nasa.gov
[4] NASA Langley, MS 365, Hampton VA 23681, USA; Email: Mark.Schoenenberger-1@nasa.gov

ABSTRACT

The goal of the proposed Mars Sample Return mission is to bring samples from the surface of Mars back to Earth for thorough examination and analysis. The Earth Entry Vehicle is the passive entry body designed to protect the sample container from entry heating and deceleration loads during descent through the Earth's atmosphere to a recoverable location on the surface. This paper summarizes the entry vehicle design and outlines the subsystem development and testing currently planned in preparation for an entry vehicle flight test in 2010 and mission launch in 2013. Planned efforts are discussed for the areas of the thermal protection system, vehicle trajectory, aerodynamics and aerothermodynamics, impact energy absorption, structure and mechanisms, and the entry vehicle flight test.

1. INTRODUCTION

The overall Mars Sample Return (MSR) mission scenario [1] includes a Mars lander to place surface samples into a container with redundant seals, a small rocket to raise the container into low Mars orbit, and an orbiting spacecraft to capture this payload and insert it into the Earth Entry Vehicle (EEV). The Earth-return portion of the spacecraft then carries the EEV toward Earth on a near-miss trajectory; before passing Earth the EEV is released on an 11-12 km/s entry trajectory. The landing site has not been officially selected, but should include controlled ground- and air-space covering a large area of predominantly soft terrain, such as found at several military installations including the Utah Test and Training Range (UTTR) selected for Genesis and Stardust [2]. After landing and recovery, the EEV and the enclosed sample container are then transported to a dedicated sample handling facility, the design of which is under study by the JPL Mars Program.

As current plans do not call for sterilization of the samples before landing on Earth, containment of the returned materials is necessary for protection of the terrestrial environment. The NASA Planetary Protection Officer has established a draft containment assurance requirement calling for the probability of release of a Martian particle larger than 2.0 microns into Earth's biosphere to be less than 10^{-6}. This is orders of magnitude beyond the reliability requirements levied on any previous planetary entry system [3], and has driven many aspects of the EEV design [4]. For example, the original forward thermal protection system (TPS) used state-of-the-art low density materials; however, these lacked sufficient flight heritage to achieve the desired vehicle reliability, and were replaced with fully dense carbon-phenolic – while much heavier, carbon-phenolic has extensive flight history and well-understood performance. Similarly, the traditional parachute for terminal descent was removed; to reach 10^{-6} the vehicle needed to maintain sample containment even during a hard landing after parachute deployment failure, so the EEV design was made robust enough to tolerate non-parachute landing as the nominal case. Removal of the parachute also deleted the associated deployment mortar, which had its own set of failure modes that contributed to the total system risk. The probabilistic risk assessment (PRA) used to track the overall probability of loss of containment assurance [5] was a vitally important tool for making these design decisions.

Fig. 1. Solid Model of the MSR EEV

The current MSR EEV is a 0.9 m diameter blunt body with an entry mass of 42 kg, including 0.5 kg for the Mars samples, and multiple layers of containment and protection. The vehicle forebody is a 60° half-angle cone with a spherical nose; the aft side is concave, with a central hemispherical lid that latches in place after insertion of the sample container. The JPL-produced sample container fits in the center of the vehicle, inside a flexible containment vessel that is sealed in Mars orbit before launch toward Earth, and both of these components are designed to accommodate higher loads than those seen during landing on clay, sand, or soil. Wrapped around the containment vessel and sample container is a spherical impact energy absorber, which limits their deceleration load if the vehicle lands on a harder surface. This vehicle geometry is preliminary and subject to change, as we are years away from launch, but the development and test plans have been laid out based on this configuration.

2. THERMAL PROTECTION SYSTEM

Fully dense carbon-phenolic (CP) was chosen for the EEV forward TPS based on its extensive flight heritage; it has been through thousands of tests and used on hundreds of flights, for missile re-entry heat shields, for the Shuttle solid rocket nozzle throats, and for the Galileo and Pioneer Venus probe heat shields. However, the heritage manufacturing processes that were used to fabricate the heat shield nose caps for the interplanetary missions are not fully documented. The methods used for the tape-wrapped CP used on the conical flank of the vehicles are well known, but there are gaps in the process information for the chopped-molded CP used at the stagnation point.

Upcoming TPS efforts focus on recreating the missing steps of the chopped-molded heritage processes. In 2005 several different chopped-molded CP samples will be fabricated, using different combinations and variations of the available processes, to see which approach reproduces the heritage capabilities. Sample performance will be tested in the Ames arc jet facilities, and the mechanical and thermal properties will also be compared. Once the heritage chopped-molded CP has been reproduced, the methods used will be thoroughly documented for future use.

An aft TPS capability and heritage survey is also planned for 2005, along with the selection of a preferred aft TPS material and identification of any associated design and test requirements. Earlier MSR EEV designs carried a nominal 10 mm aft TPS thickness instead of completing the selection of a specific TPS, as funding for this survey was previously unavailable.

In 2006 we plan to perform several TPS thickness studies for the forward and aft heat shields, based on updates to the entry trajectory and Monte Carlo estimation of the worst case heat load. We also plan to create detailed designs of the various TPS joints and penetrations on the vehicle. These TPS joints include those between the nose and flank CP materials, the flank CP to the aft TPS, and the seam in the aft TPS where the lid opens and closes. The penetrations are all in the aft TPS, and include those needed for the EEV mechanical attachment to the parent spacecraft as well as ones for the electrical cable bundles carrying sensor info and survival heater power.

In 2007 we plan to conduct arc jet tests for each of the TPS joints and penetrations, and also for vehicle locations of interest such as the forward and aft stagnation point and the vehicle shoulder. Each configuration will be tested at least four times: at least two samples will be exposed to their expected peak heat flux plus margin to prove material survival, and at least two more will see half that level, which better matches the energy absorption into the bulk of the TPS. At both of these test levels, the test duration will be calculated to input the full entry heat load into the sample coupons. Some 60-plus TPS coupons will be fabricated, tested, inspected, and analyzed as part of this test series, at a cost of over a million dollars.

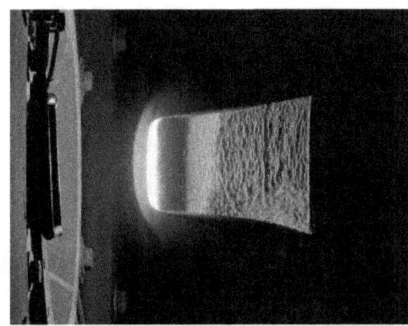

Fig. 2. TPS Arc Jet Testing

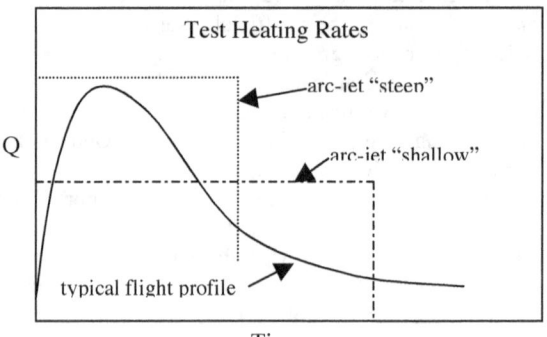

Fig. 3. Plot of Heat Flux (Q) v. Time

In 2008, a full scale engineering model of the TPS will be fabricated using the flight designs and heritage processes. After assembly to an engineering model of the EEV structure, it will undergo environmental testing and post-test inspection to look for any unexpected problems with the full-scale fabrication processes that were not evident on sample coupons.

Once detail designs using heritage materials have been tested and qualified for flight, TPS efforts will focus on preparation for the anticipated EEV flight test, discussed later in this paper. Redesign and TPS re-sizing to accommodate changes in mission requirements, vehicle configuration, and predicted heat load will likely continue until an eventual mission design freeze, but at a lower level of effort than the initial development tasks.

3. TRAJECTORY, AERO, AND AEROTHERMAL

There are no active systems on board the EEV during entry, descent, and landing, except for radio tracking beacons. An active guidance system was avoided due to the risks posed by potential failure modes; the vehicle relies instead on an accurate initial trajectory and the simple physical laws of ballistics, which still produce an acceptably small landing ellipse.

The concave aft shape of the EEV was sculpted to avoid the possibility of stable backwards orientations during re-entry. The vehicle is intended to be pointed nose-first at atmospheric interface, with a 2 rpm spin for stability, but possible failures of the spin-eject system on the parent spacecraft may lead to off-nominal entry conditions. Trajectory simulations of Earth entry performed using preliminary computational fluid dynamics and direct simulation Monte Carlo aero data show the vehicle to be self-reorienting from off-nominal entry states. Even from the extreme case of a backwards entry (180° angle of attack) with full spin-stabilization, the EEV pitches over to a forward orientation before the heat pulse.

Aerothermal calculations of the heat flux distribution around the EEV indicate that the coolest regions during entry will be the aft surfaces where the EEV lid joins the body [6], which simplifies the design of TPS joint and nearby mechanical penetrations. However, concerns about Mars dust reaching orbit with the sample container and possibly contaminating the outside of the EEV have led to an upcoming re-design. The aft body of the EEV will be re-examined to see if shape changes can raise the entry flux high enough to push the surface temperature past the 500°C sterilization level while maintaining the vehicle reorientation capability.

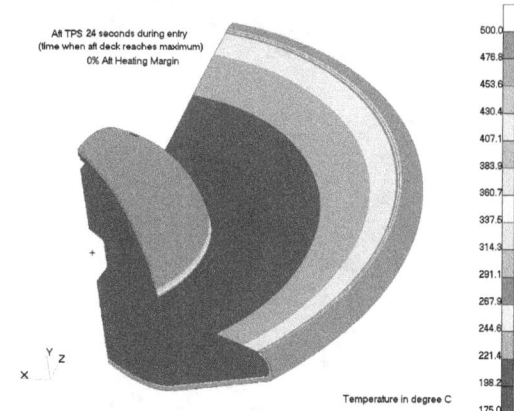

Fig. 4. EEV Entry Temperatures

Development plans for 2005 include updating heating and footprint calculations for a range of entry trajectories and the aerothermal calculations for the altered aft body shapes. In 2006 we plan to conduct several test runs in the Langley 20-inch Mach 6 air tunnel to anchor the aerothermal calculations for the most promising shapes, and some limited testing in larger facilities. Later in 2006 we plan to test the dynamic aero performance of the EEV, using a combination of ballistic range tests, spin tunnel tests, and drop tests of a full size EEV model. In 2007, we produce an updated aero database for the EEV and perform Monte Carlo trajectory runs across a range of possible entry conditions to confirm vehicle reorientation and evaluate the worst case entry conditions. Tasks in 2008-9 focus on updating the entry predictions as the vehicle design matures, and on providing analytic support for the EEV flight test.

4. IMPACT ENERGY ABSORBER

The EEV impact energy absorber has three main components: a relatively rigid inner shell, a crushable foam-filled cellular structure, and a tough outer shell for penetration resistance. In the nominal landing scenario, the vehicle will land in a well characterized region of mostly soft terrain, such as UTTR where the clay and sand are interrupted by only a few gravel roads and small concrete pads. For this case, the EEV's kinetic energy at impact is absorbed by deformation of the ground, as well as crushing and fracture of the vehicle structure and TPS. Full scale drop tests conducted at UTTR, using rigid, instrumented penetrometers and a rigid model of the EEV, showed a deceleration load of

1500 g's, well below the 2500 g requirement for preservation of the scientific value of the samples.

Fig. 5. UTTR Terrain

Fig. 6. Full Scale EEV Drop Model

For the off-nominal case of a hard surface landing, the impact energy absorber is designed to limit the loads at the interface to the sample container and containment vessel to less than 3500 g's [7]. In this case, some science degradation is expected, but sample containment is still maintained. The cell walls crush and tear to attenuate the impact loads, but the inner shell of the absorber remains intact. Full size impact tests at NASA Langley using a simulated sample container produced loads of under 3000 g's; a mechanical accelerator system was necessary to achieve the desired impact velocity of 41 m/s, which was higher than terminal velocity in the dense sea level air, but the impact absorbers repeatedly performed as intended.

Fig. 7. Langley Impact Dynamics Research Facility

Fig. 8. Impact Absorber, Post-Test

Development plans for 2005 call for comparison of the existing composite energy absorber to an alternate metallic design and investigation of fabrication methods applicable to the metallic absorber. This is in response to an ongoing study on the risk effects of switching from a composite structure to a metallic one, due to earlier difficulties in analyzing the composite structure for 10^{-6} reliability – if the vehicle structure is switched from composite to metallic, it may be beneficial to change the energy absorber as well. In 2006 we will perform impact speed crush tests on our flight materials to generate data necessary for design of the absorber; the current engineering development models used non-flight materials to reduce expenses and simplify fabrication. In 2007 we begin conducting tests again of the full size impact absorber, this time with flight materials; two tests are planned with empty impact hemispheres, and two more are planned carrying flight-like sample containers. In 2008 we plan to impact test the absorber inside a full-size model of the vehicle structure, to verify analysis of the absorber-structure interactions and the impact behavior of the structure. This data is necessary to properly size the energy absorber; we need to know how much of the structural mass breaks free on a hard surface impact, and how much stays attached and must be decelerated by the energy absorber. Later in 2008 we plan to impact test another absorber inside a flight-like structure, with simulated TPS mass and a flight-like sample container. These full-up tests are designed to confirm that the absorber works as intended when assembled with the flight hardware, since the impact absorber will not be included in the EEV flight test.

Test results will be used to verify impact analysis models which could then be used for further design adjustments, and/or design modifications in the event of significant changes to the mission requirements. For example, if the Mars sample mass increases significantly, the overall EEV diameter will grow, and the earlier drop tests onto the ground at UTTR may be invalidated.

5. STRUCTURE AND MECHANISMS

The EEV structure supports the heat shield and maintains the vehicle drag area to achieve the desired terminal descent velocity. The structure must survive the 130 g atmospheric interface deceleration, but is not required to survive landing. The vehicle structure will thus be designed and tested to the entry loads, plus margin, rather than to the higher levels associated with ground impact.

Components intended to operate after launch from Earth, such as lid placement sensors, the lid latches, and the (retractable) launch lock bolts, will be vibration tested to the expected launch loads and then tested to verify proper operation across their expected thermal range. Mechanical components that must survive through entry and descent, such as the EEV body vents, will be tested to the atmospheric deceleration levels. The lid latches, which hold the lid and body of the impact absorber together, will see additional testing as part of the impact absorber. All EEV components will also go through vacuum thermal cycle testing to verify compatibility with the expected environment.

Preliminary design and analysis of a metallic structure for the EEV is expected in 2005 as part of the comparison between the current composite design and a metallic one. In 2005 we begin design of the lid latches and the launch lock bolts; they are planned for completion in 2006, along with the mechanical TPS penetrations, body vents, and lid closure sensors. In 2007 we will fabricate engineering models of the lid latches and EEV launch locks, for development work and environmental testing. In 2008 the engineering model of the EEV structure will be fabricated for integration and environmental testing with the TPS engineering model. In 2009 the development plans call for fabrication and assembly of flight versions of the EEV structure and mechanisms for use in the flight test in 2010.

Structural and thermal analysis of the various mechanical components will be conducted throughout the design process, using the relevant environments from launch, deep space, Mars orbit, trajectory maneuvers, and Earth entry. Structural and thermal analysis of the full EEV assembly will also be required, in the several different configurations experienced during the proposed MSR mission. Finally, as with the other subsystems, some redesign of the mechanical components will likely be required due to changes in mission requirements.

Fig. 9. EEV Cross Section

6. EEV FLIGHT TEST

The MSR EEV flight test is intended to functionally test critical aspects of the vehicle design that cannot be fully duplicated in ground testing and analysis. The flight will demonstrate the integrated TPS performance in the actual, time-varying flight environment, with the TPS interacting with the surrounding flow and with the underlying structure; ground testing in arc jets simulates only one heat flux level per test, and cannot match all the environmental variables. This test will also verify the EEV aero and aerothermal performance, confirm that critical risk parameters are within their design limits, and demonstrate that there are no unknown system-level issues.

The flight test is intended to validate the nominal vehicle performance, and as such will duplicate the EEV entry trajectory and vehicle size, shape, mass, and materials. This will allow the flight test to match the mission's entry trajectory, entry heating, deceleration, terminal velocity, and nominal landing at the chosen site. Rather than conducting a flight test using one particular set of extreme conditions, this test is intended to validate the performance models used in the PRA so that they can be reliably used for repeated Monte Carlo runs.

Given the low probability of landing on a hard surface, the flight test is unlikely to prove the performance of the impact energy absorber. In order to provide more useful data from the components under stress during this test, the sample container and impact absorber will be replaced, for this test only, with a high-g data recorder, an inertial measurement unit to track the vehicle trajectory, numerous thermocouples, and several pressure sensors. The impact sphere and sample container will receive sufficient testing outside the flight test for the overall test program to cover the entire mission scenario.

However, the future of the proposed EEV flight test is still uncertain. A 2001 study by Sandia looking at relevant launch vehicles concluded that the flight test would cost roughly $30 million, mostly to buy a suitable launch vehicle. The high cost, as well as the possibility that a launch failure during this flight test could delay the launch of the MSR mission, has led to an effort to quantify the benefits of the flight test and to see if the same results can be achieved through expanded ground testing.

7. CHANGES AFTER FLIGHT TEST?

Assuming a successful flight test in 2010, there is debate about whether to allow changes to the EEV design before launch to Mars. Some concerns exist that changing the proven vehicle would invalidate the flight test heritage; however, minor changes should be allowable, as long as they do not require alteration of the analytical methods verified by the flight test.

The most serious requirements changes from the test flight to the interplanetary mission would be the increased mission time and the addition of planetary protection requirements. The mission duration may have limited effect on the space-rated materials used on the vehicle; the Galileo probe structure and heat shield, for example, flew through space for years before reaching Jupiter. Planetary protection, however, is a very significant change: the flight test has no extraordinary concerns about the presence of microorganisms, but the mission to Mars must deal with the possibility of Earth organisms contaminating the Mars samples, and as such will have to implement stringent hardware cleaning processes. Since the flight test vehicle is intended to be a duplicate of the Mars mission hardware, these cleaning processes will also have to be imposed on the flight test hardware.

8. CONCLUSION

Plans for the development and testing of the MSR EEV were outlined here, from present tasks through a flight demonstration and 2013 mission launch. It should be noted that all of these plans are preliminary works in progress which are expected to continue to change as the design matures and as requirements and funding constraints vary.

It should also be mentioned that there are several other MSR components under development at JPL which interact closely with the EEV subsystems discussed here, including the sample container, the flexible containment vessel, the spin-eject mechanism that releases the EEV from the parent spacecraft, and the micrometeoroid shield needed to protect the EEV during flight to and from Mars. Interface requirements for these and other MSR systems will need to be developed in the future, but are beyond the scope of this paper.

9. REFERENCES:

1. Gershman, R., Adams, M., Mattingly, R., Rohatgi, N., Corliss, J., Dillman, R., Fragola, J., and Minarick, J., *Planetary Protection for Mars Sample Return*, COSPAR-PTP1-0011-02, 34th COSPAR Scientific Assembly, October 2001.
2. Desai, P. N., Mitcheltree, R. A., and Cheatwood, F. M., *Sample Return Missions in the Coming Decade*, IAF Paper IAF-00-Q.2.03, October 2000.
3. NPG 8020.12B, *Planetary Protection Provisions for Robotic Extraterrestrial Missions*, April 1999.
4. Mitcheltree, R. A., Hughes, S. J., Dillman, R. A., and Teter, J. E., *An Earth Entry Vehicle for Returning Samples from Mars*, AAAF Paper ARVS-102, March 2001.
5. Fragola, J. R., Minarick, J. W., and Putney, B., *Mars Sample Return Probabilistic Risk Assessment Final Report*, Science Applications International Corporation Report, September 2002.
6. Amundsen, R. M., Dec, J. A., Mitcheltree, R. A., Lindell, M. C., and Dillman, R. A., *Preliminary Thermal Analysis of a Mars Sample Return Earth Entry Vehicle*, AIAA Paper 2000-2584, June 2000.
7. Kellas, S., and Mitcheltree, R. A., *Energy Absorber Design, Fabrication and Testing for a Passive Earth Entry Vehicle*, AIAA Paper 2002-1224, April 2002.

Validation of Afterbody Aeroheating Predictions for Planetary Probes: Status and Future Work

Michael J. Wright,[1] James L. Brown,[1] Krishnendu Sinha,[3] Graham V. Candler,[3] Frank S. Milos,[1] Dinesh K. Prabhu[2]

[1] NASA Ames Research Center, MS 230-2, Moffett Field, CA 95050
[2] ELORET Corp., 970 W. Fremont, Suite 8, Sunnyvale, CA 94087
[3] University of Minnesota, 107 Akerman Hall, Minneapolis, MN 55455

ABSTRACT

A review of the relevant flight conditions and physical models for planetary probe afterbody aeroheating calculations is given. Readily available sources of afterbody flight data and published attempts to computationally simulate those flights are summarized. A current status of the application of turbulence models to afterbody flows is presented. Finally, recommendations for additional analysis and testing that would reduce our uncertainties in our ability to accurately predict base heating levels are given.

1. INTRODUCTION

Uncertainty levels associated with aeroheating predictions for the design of the afterbody of planetary probes are typically assumed to be in the range of 200-300%, a level that can have a significant impact on Thermal Protection System (TPS) material selection and weight. This conservatism in the afterbody heat shield design will also shift the center of gravity aftward, which reduces the stability of the probe and in some circumstances may necessitate the addition of ballast in the nose. Current design practice for an afterbody heatshield assumes a laminar, fully catalytic, non-ablating surface. The predictions thus obtained are then augmented by a large factor of safety to account for turbulent transition, material response, and uncertainties in the baseline computations. A primary reason for this uncertainty is a sparsity of data for validation of our computational tools. Ground test data are usually complicated by sting interference effects. Little flight data exist, and recent attempts to propose dedicated flight experiments have failed to reach fruition. Therefore, it is important to thoroughly understand the limited flight data that are available to improve the design fidelity of the next generation of Earth and planetary entry vehicles and to assess the need for additional focused flight testing.

This paper will discuss four general topics. First, we review the relevant flow regimes and physical models for afterbody flows. Next, the paper surveys the available flight data for validating afterbody-heating predictions and reviews prior computational analyses of these data. Then, we conduct a brief survey of the state of the art in computing turbulent afterbody flowfields. Finally, we provide recommendations for areas of further work, and possible flight data that would aid in reducing the afterbody aeroheating design uncertainty.

2. FLOW REGIMES AND PHYSICAL MODELS

During entry, a planetary probe will pass from a free-molecular (collisionless) to a non-continuum and finally to a continuum flow regime. The transition between these regimes is usually determined by evaluation of the freestream Knudsen number $Kn_\infty = \lambda_\infty / D$, where λ_∞ is the mean free path and D is the body diameter. Free molecular flow is usually defined as the region where $Kn_\infty > 100$, while continuum flow is usually defined as the region where $Kn_\infty < 0.01$. This criterion is not accurate for separated base flows, because the local mean free path in the separation region can be much larger than that in the freestream. A more accurate determination can be made by using the density (ρ) gradient length local Knudsen number[1]

$$Kn_{GLL} = \frac{\lambda}{\rho}\left|\frac{d\rho}{dl}\right|_{max} \quad (1)$$

where λ and ρ are local values and the derivative is evaluated along the maximum gradient direction. Following the work of Boyd et al.,[1] we assume that continuum breakdown occurs when $Kn_{GLL} > 0.1$. This criterion results in a more useful determination of continuum breakdown in a separated flow, because while the local mean free path in the separation region can be quite large, the density gradient is usually small, which delays the onset of non-continuum effects. For afterbody flows the highest values of Kn_{GLL} are typically observed near the flow separation point due to large density gradients. Numerical solutions for free molecular and non-continuum flows are typically obtained using a Direct Simulation Monte Carlo (DSMC) methodology. An excellent review on the status of DSMC calculations for non-continuum wake flows was presented by Moss and Price.[2] However, for many problems of interest the majority of the aeroheating occurs in continuum flow, where Navier-Stokes based computational fluid dynamics (CFD) methods are applicable. The remainder of this paper will deal with the continuum flow regime.

When a probe enters a planetary atmosphere at high velocity, the resulting shock wave will thermally excite, dissociate, and possibly ionize the gas. In order to accurately model the resulting flowfield including the wake of the probe a non-equilibrium model is usually required.[3] Each chemical and thermal relaxation process has an associated characteristic time, and the rapid expansion of the flow into the wake will decrease the collision rate, which freezes the slower processes (such as vibrational relaxation) while the faster chemical relaxation processes continue at a finite rate. The details of the base flow structure and resulting heating rates can be very sensitive to the non-equilibrium state of the gas.[4] An excellent review of the thermodynamic and chemical-kinetic models for a non-equilibrium flowfield is given by Gnoffo et al.[5]

The afterbody flowfield will likely transition from a laminar to a turbulent flow during the entry. Wake transition begins in the far wake and travels upstream with increasing freestream Reynolds number (Re) until reaching the neck, where it is (temporarily) stopped by the adverse pressure gradient. In the base region transition begins in the separation shear layer. Lees[6] gives a transition correlation for the free shear layer in a two-dimensional or axisymmetric flow that is based on a local transition Reynolds number, defined as

$$\mathrm{Re}_{tr} = \rho_e u_e L / \mu_e \qquad (2)$$

where L is the running length of the shear layer from the separation point and the local density, velocity and viscosity are evaluated based on fluid properties at the outer edge of the shear layer. The critical transition Reynolds number is a function of the edge Mach number, and ranges from about 2×10^4 at Mach 2 to 5×10^6 at Mach 5. This criterion is based on free-flight data, but does not include effects of upstream ablation product gas injection, which could have a destabilizing effect on the shear layer and separated flow region.

Low Re wake flows are steady and dominated by a small number of large vortices. As the freestream Reynolds number increases the extent of separation increases as well and the vortex structure becomes more complex. Eventually the vortices begin to oscillate and the base flow becomes unsteady. Typically, the Reynolds number at which the flowfield becomes

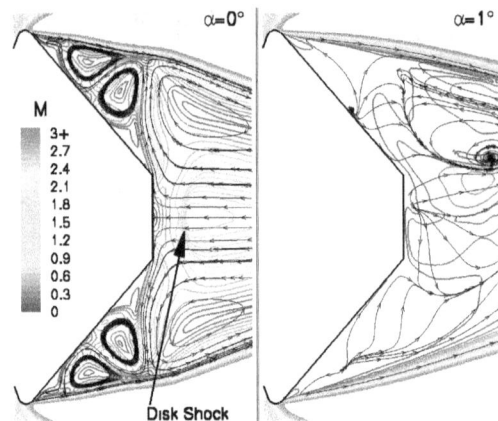

Fig. 1. Computed Mach number contours and streamlines in the symmetry plane of the Mars Pathfinder entry probe.

unsteady is near that at which transition to turbulence is predicted to begin. For many cases of interest, both events occur after the peak heating point on the trajectory. For these cases much of the heat pulse can be simulated assuming a continuum, laminar, steady flow.

Computational simulations of axisymmetric base flows with a flat base frequently show a disk shock in the near wake, caused when the reverse flow becomes supersonic along the rear stagnation line and a normal shock is required to slow the gas before impacting the body. The computed pressure and heat transfer on the flat base for these cases are much higher than would be the case if the disk shock were not present. Free flight experiments do not appear to show evidence of such a disk shock, although it would be difficult to see such a weak shock in a conventional Schlieren image. Clearly, since the presence of a disk shock has a significant influence on the predicted base heating, it is important to understand whether such a phenomenon is real or merely a computational artifact. One possibility is that the disk shock is a neutrally stable solution of the Navier-Stokes equations, which can occur only for an identically axisymmetric flowfield. Since no real flow is ever completely axisymmetric, this solution would rarely (if ever) occur in nature. In order to test this theory, two simulations were performed for Mars Pathfinder (Fig. 1). Freestream velocity is 6.6 km/s and density is 2.8×10^{-4} kg/m^3. The first solution was run assuming axisymmetric flow, and clearly shows the disk shock in the wake. The second was run as a three-dimensional flow at $\alpha = 1°$, and no disk shock is present for this case. In addition, the computed pressure and heat transfer on the flat base are a factor of three lower for the $\alpha = 1°$ solution. This result lends some support to the present hypothesis; however, the problem of disk shocks in axisymmetric base flows requires further study, including systematic comparison with experimental base heating and pressure data.

Finally, wake flows are sensitive to the details of the volume grid used in the CFD analysis. Therefore it is important to generate a grid that is well aligned to anticipated flow features. In particular, it is extremely important that the grid have sufficient points in the shoulder region to capture the rapid expansion and accurately predict the flow separation point and the angle of the resulting shear layer.[7] There must also be sufficient points in the separated flow region to resolve the vortical structure and the wake compression, or neck. At higher Reynolds numbers the wake will consist of multiple counter-rotating vortices that must be resolved. Care must also be taken to ensure that the grid completely encloses the subsonic portion of the wake, which can extend several body diameters downstream.

3. AVAILABLE FLIGHT DATA & PREVIOUS VALIDATION ATTEMPTS

Most relevant flight data for validation of afterbody aeroheating predictions was obtained during the Apollo program, although there are also limited data from other European and American entry probes. It is likely that Russian flight data also exist, although no references to any such data were located in the open literature. This section summarizes the available flight data, and discusses published attempts at post-flight analysis.

Table 1 Launch dates and entry conditions for Apollo program flight tests.

Flight	Launch Date	V (km/s)	α (deg)	γ (deg)	Refs
Fire-I	Apr. 14, 1964	11.56	0	-14.7	1,2,5
Fire-II	May 22, 1965	11.35	0	-14.7	3,4,14
AS-201	Feb. 26, 1966	7.67	20	-8.6	22
AS-202	Aug. 25, 1966	8.29	18	-3.5	22,25,27
Apollo 4	Nov. 9, 1967	10.73	25	-5.9	23,26,29
Apollo 6	Apr. 4, 1968	9.60	25	-6.9	23

3.1 Project Fire
Project Fire was an Apollo technology demonstrator program that resulted in two ballistic entry test flights, Fire-I[8-9] and Fire-II.[10-11] The primary objective of Project Fire was to understand the radiative heating environment of an Earth entry vehicle at Lunar return velocities, but the silica-phenolic afterbody was also instrumented with nine surface mounted thermocouples, one pressure sensor, and a radiometer. Table 1 shows the launch dates and entry conditions for the two Project Fire flight tests, and Figure 2 shows the vehicle

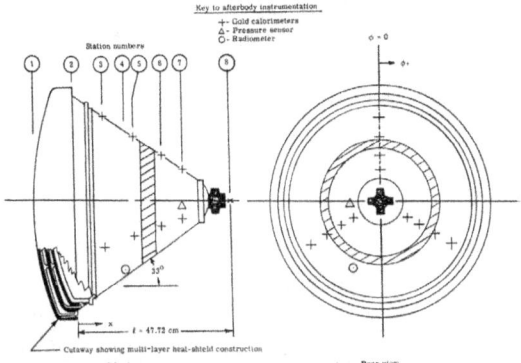

Fig. 2. Schematic of Fire reentry vehicle showing instrument placement (from [11]).

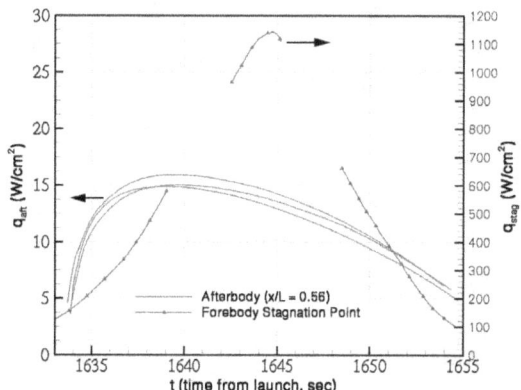

Fig. 3. Time histories of heat transfer measured during Fire-II flight on the conical frustum and at the forebody stagnation point (from [13]).

geometry and afterbody instrument placement. Unfortunately, the Fire-I probe experienced large angle of attack variations during entry, apparently due to a stage-separation anomaly in which the booster entered in front of the capsule, making the data from this flight difficult to interpret.[12] In contrast, the Fire-II entry was extremely successful. The vehicle maintained an angle of attack of less than 1° through the majority of the high heating portion of the entry, increasing to about 11° by the end of the experiment.[11] All afterbody instrumentation was functional during this flight, providing a valuable database of afterbody heating for a ballistic entry vehicle. Figure 3 shows the time histories of total heat transfer measured during flight at one afterbody station, and also at the forebody stagnation point. Peak afterbody heating at this location was about 1.5% of the peak stagnation point heating. The afterbody radiometer was determined to be functional, but did not measure any signal during the heating portion of the entry, indicating that radiative heating to the afterbody was negligible at these conditions.[11]

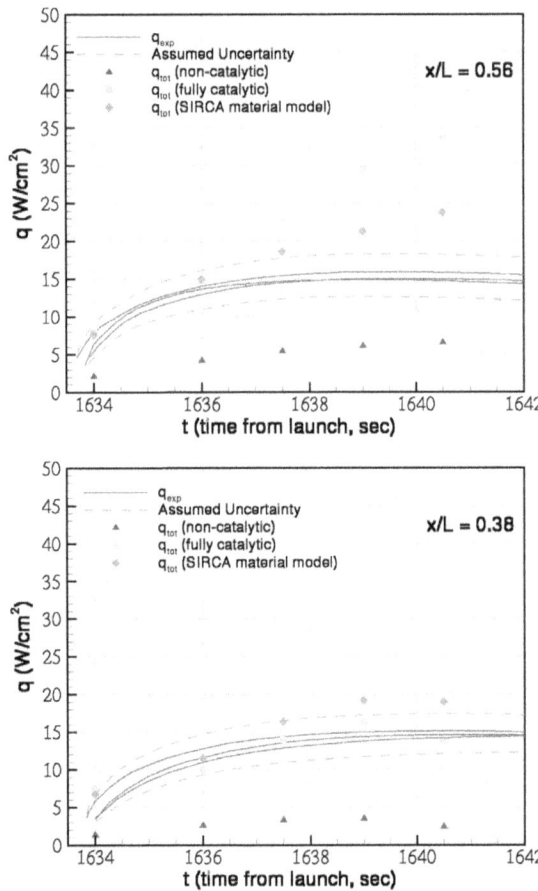

Fig. 4. Time histories of total heat transfer for Fire-II computed using several catalysis models on the afterbody as compared to flight data at two axial locations (from [13]).

The Fire-II afterbody flight data were analyzed in detail by Wright et al.[13] using a nonequilibrium Navier-Stokes code. The CFD results were computed assuming laminar flow, an assumption validated using the correlation of Lees.[6] A partially catalytic afterbody surface was assumed, with the catalytic efficiency of the afterbody TPS approximated using analogies to similar currently manufactured materials. Figure 4 shows the results of this analysis for two afterbody locations. The computations agreed with the flight data to within the experimental uncertainty over the early portion of the trajectory ($t < 1638$ s). The computations overpredicted the flight data later in the trajectory, especially on the rear of the body (larger x/L), but this result was attributed to TPS ablation, which was not modeled in the simulations.[13] The results demonstrated that modern CFD methods are capable of reproducing the flight data to within experimental accuracy as long as realistic surface boundary conditions are employed. A

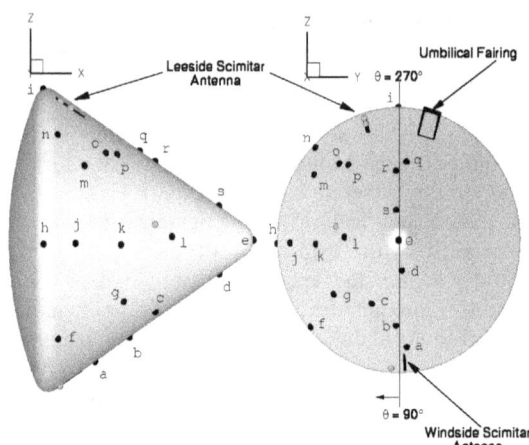

Fig. 5. Calorimeter locations on AS-201/202 afterbody. Orange symbols indicate inoperative instruments during AS-202 (from [19]).

more recent preliminary computation of the turbulent portion of the trajectory has also been published;[14] the results will be discussed in the following section.

3.2 Apollo Program

The Apollo program sponsored several dedicated flight tests to understand the heating environment of orbital and super-orbital entry probes. Once the design of the Apollo Command Module was determined, four flight tests were conducted which included forebody and afterbody instrumentation. The first two, AS-201 and AS-202, were conducted at orbital velocities, while the final two, Apollo 4 and Apollo 6, were conducted at super-orbital velocities representative of Lunar return. Table 1 shows the relevant entry parameters for these tests. The four flights together constitute the best database of flight afterbody heating data obtained to date. An onboard Inertial Measurement Unit (IMU) during the last three flights enabled an accurate trajectory reconstruction, and sounding rockets were used to reconstruct atmospheric properties.[15-16] The range of entry velocities and flight path angles during these flights were sufficient to span multiple flow regimes, from laminar to fully turbulent, and from minimal material response to strong pyrolysis injection and char formation. This range of conditions will permit a systematic study of the effects of turbulent transition and pyrolysis gas injection on turbulent heating levels.

The afterbody instrument package for AS-201 and AS-202 consisted of 23 surface-mounted calorimeters and 24 pressure transducers.[17] Calorimeter locations are shown in Fig. 5. Both flights were highly successful, with 16 of the calorimeters returning useful data on AS-201 and 19 on AS-202.[17] Pressure data were also obtained during the AS-201 flight, but the dynamic pressure during the AS-202 mission was too low for meaningful readings to be obtained on the afterbody. The afterbody heating rates for AS-201 were much higher than those for AS-202 due to the steeper entry angle, and therefore the heating for this mission was significantly affected by charring of the TPS.[18]

The afterbody heating data for AS-201 have not been investigated in detail using modern CFD methods. However, a recent paper analyzed the data for AS-202.[19] A total of 15 three-dimensional CFD solutions were run spanning the time from the onset of continuum flow until the separation region became unsteady. The surface was assumed to be fully catalytic, which was a reasonable assumption for the hydrocarbon-resin based Avcoat TPS material. The results were computed assuming laminar flow, validated using the correlation of Lees.[6] The computations generally agreed with the flight data to within the experimental uncertainty (±20%) for 15 of the 19 functional calorimeters.[19] The results at three calorimeter locations are shown in Fig. 6. The first (calorimeter "a") was in an attached flow region, the second ("m") was in separated flow, and the third ("j") was at a location where the flow separated and reattached during the entry. The heat pulse has two distinct lobes due to an atmospheric skip maneuver performed by the spacecraft during entry. Interestingly, both the flight data and the CFD results at calorimeter "j" clearly show the reattachment at $t = 4600$ s and separation at $t = 4800$ s, which indicates that the CFD solutions are not only accurately predicting the magnitude of the heating, but also the extent of separation. Relatively poor agreement was obtained for two calorimeters near the rear apex of the vehicle; the reason is not known at this time but it may be due to unmodeled details of the apex geometry. This work again demonstrated the ability of modern computational methods to accurately predict afterbody heating levels.

The Apollo 4 and 6 test flights were intended to qualify the entry system for Lunar return by entering at $\alpha=25°$ and a relative velocity of about 11 km/s. The actual entry velocity for Apollo 6 was only about 9.6 km/s due to a re-ignition failure in the upper stage.[20] The instrument package was modified for these flights, and consisted of 21 calorimeters, 10 pressure transducers, and 2 radiometers. The locations of the instrumentation are shown in Fig. 7. All 21 calorimeters provided useful data on each flight. Four of the calorimeters were placed near simulated protuberances and gaps in the flight vehicle; these data may be useful to validate the ability of modern CFD to predict local heating around geometrical singularities. The remaining calorimeters

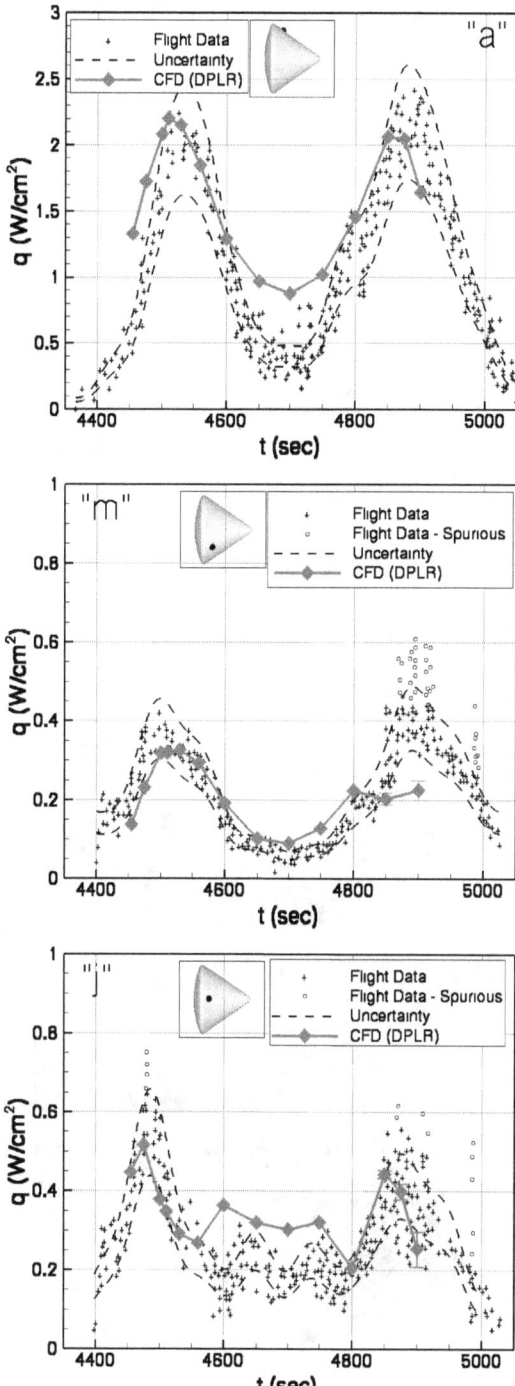

Fig. 6. Comparison of flight data and computed heat transfer for AS-202. Letters indicate calorimeter ID in Fig. 4 (from [19]).

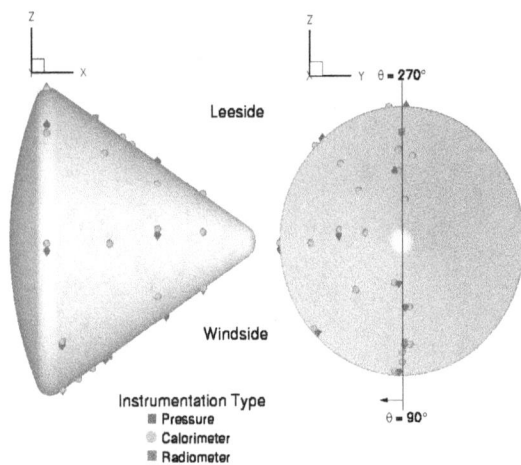

Fig. 7. Instrumentation locations on Apollo 4 and Apollo 6 conical afterbody.

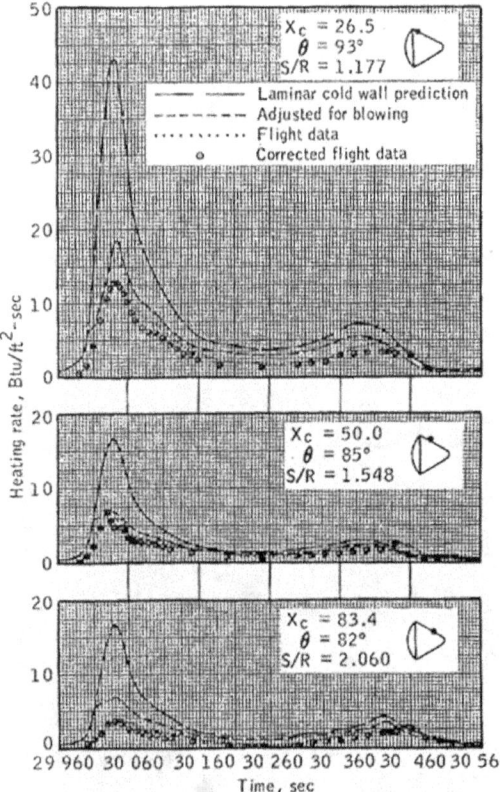

Fig. 8. Flight data and pre-flight engineering predictions from the Apollo 4 entry at three calorimeter locations on the attached flow portion of the afterbody (from [20]).

provided flight data of afterbody heat transfer on an ablating TPS material. The afterbody radiometers for both flights failed to detect a measurable signal,

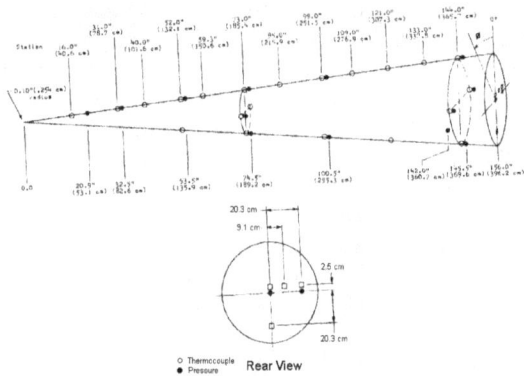

Fig. 9. Instrument locations on the Reentry F flight experiment (from [23]).

although they were determined to be functional. This result confirmed pre-flight predictions of negligible radiative heating,[21] and implies that there was zero afterbody radiative heating on the lower velocity AS-201 and AS-202 flights as well. There was little charring on the separated flow portion of the afterbody, and total heating levels were between 1-2% of stagnation point theory. The pressure and total heat transfer measured on the charred (attached flow) regions of the afterbody were corrected for wall blowing, but the resulting data were significantly lower than the preflight computations (see Fig 8). The level of underprediction was determined to be proportional to the forebody heating rate. Although a definitive reason for this effect has not been identified, it has been postulated that the cause was upstream blowing of ablation products into the boundary layer.[20,22] The afterbody heating data from these flights have yet to be looked at in detail with modern computational methods.

3.3 Other U.S. Flights with Afterbody Data

The Reentry F flight test was launched on April 27, 1968 from Wallops Island and entered at a relative velocity of 6 km/s on a ballistic trajectory.[23] The entry vehicle was a 3.92 m long 5° half-angle beryllium cone with a graphite nose tip, designed to provide transition and turbulent heat transfer data. The cone was instrumented with thermocouples and pressure sensors at 21 measurement stations, while the base had a total of 4 heat flux and 2 pressure sensors.[23] Instrument locations are shown in Fig. 9. Post-flight analysis of these data were somewhat complicated by thermal distortions, which resulted in a small effective angle of attack.[24] To date no comprehensive analysis of the base heating and pressure data given in [25] has been attempted with modern CFD tools.

The Viking program included two landers that entered the Martian atmosphere in July and September of 1976.

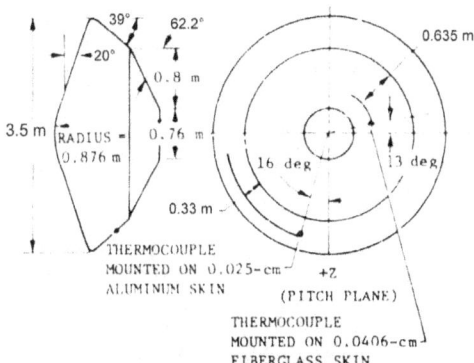

Fig. 10. Schematic of Viking entry probe showing sensor locations. (from [28]).

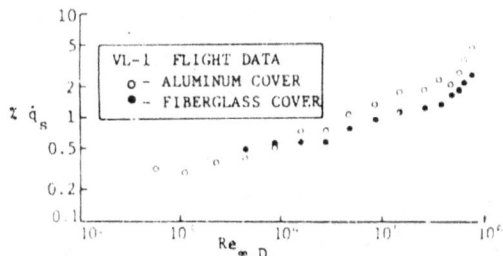

Fig. 11. Flight data from the afterbody temperature sensors on Viking 1. (from [27]).

Both probes were 70° sphere-cones which flew a lifting entry at a nominal angle of attack of 11°, and entered at a relative velocity of about 4.5 km/s.[26] Each probe included a base pressure sensor and two surface-mounted temperature sensors – one on the fiberglass inner cone and one on the aluminum skin of the outer cone.[27] A schematic of the Viking entry probe showing the temperature sensor locations is given in Fig. 10. Pre-flight analysis predicted afterbody heating to be 3% of the forebody stagnation point heating rate, but flight data indicated that the peak heating was actually about 4.2% of the stagnation value, as shown in Fig 11. The high heating levels, as well as the slope change observed in heating rate vs Reynolds number at $Re_D \sim 5 \times 10^5$, were believed to be evidence of turbulent transition on the base.[28] No attempt to reproduce these data with modern CFD techniques has been published.

The Galileo mission was launched October 18, 1989. The on-board 45° sphere-cone probe successfully entered the Jovian atmosphere on December 7, 1995 on a ballistic trajectory at a relative velocity of 47.4 km/s.[29] This probe survived the most severe heating environment ever experienced by a planetary entry capsule, with a peak ablating heat flux on the order of 30 kW/cm^2. Instrumentation consisted of 10 analog resistance ablation (ARAD) sensors on the forebody

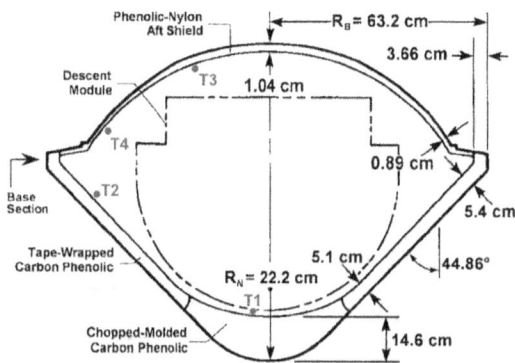

Fig. 12. Schematic of Galileo probe with thermometer locations marked (adapted from [30]).

and 4 resistance thermometers bonded to the structure beneath the carbon-phenolic TPS, as shown in Fig. 12. Both afterbody thermometers (T3 and T4 in Fig. 12) appeared to function normally. The only post-flight analysis of this data to date was performed by Milos et al.,[30] in which the thermometer response was simulated using a transient material response code coupled to a finite-element thermal analysis package. CFD analysis of the external flowfield was not performed. Instead, a triangular heat pulse was assumed with a total heat load based on engineering predictions. The results in [30] indicate that the post-flight analysis was not in good agreement with the afterbody flight data. Although it was possible to bound the flight data by varying the heat load and initial cold-soak structural temperature, Milos et al. were unable to reproduce the slope of the temperature increase. It remains to be seen whether a high-fidelity aerothermal analysis could improve the agreement with the flight data, although it should be noted that the Galileo flowfield is an extremely complex mix of optically thick radiation, strong ablation, and turbulent flow, and will present a significant challenge to the state of the art CFD methodology.

Mars Pathfinder was launched December 4, 1996 and successfully entered the Martian atmosphere on July 4, 1997 on a ballistic trajectory at a relative velocity of 7.5 km/s.[31] There was no surface-mounted instrumentation, but the aeroshell did contain nine thermocouples (TC) and three platinum resistance thermometers (PRT) at various depths in the TPS material as shown in Fig. 13. Of those on the afterbody, usable data were obtained from TC9, PRT1, and PRT3. Time histories of the temperature data at these locations are given in [32]. The only post-flight analysis of this data to date was performed by Milos et al.[32] In this analysis afterbody heating estimates were scaled from forebody CFD solutions. Using this assumption, Milos et al.[32] were able to reproduce the peak temperature at TC9, but not

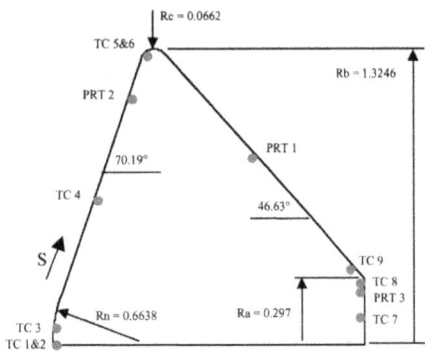

Fig. 13. Mars Pathfinder schematic showing instrument locations (from [32]).

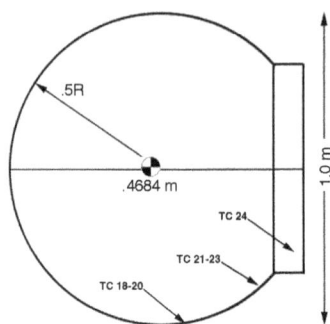

Fig. 14. Schematic of the MIRKA capsule with approximate location of aftbody thermocouples.

the time history of the temperature response. However, by assuming a "best-fit" heating profile that was longer in duration than the scaled profile they were able to demonstrate excellent agreement with the flight data. The assumed shape of the "best-fit" profile was in general agreement with pre-flight predictions,[33-34] but the heating rates required to match the data were considerably lower. To our knowledge no attempt has been made to reconcile this discrepancy.

3.2 European Flight Data

The MIRKA capsule was a German-led low-cost flight test flown as a piggyback payload on a Russian FOTON capsule. MIRKA successfully reentered the Earth's atmosphere on a ballistic trajectory at a velocity of 7.6 km/s on October 23, 1997.[35] The capsule, shown schematically in Fig. 14, was essentially a one meter diameter sphere with a flat base. The capsule was instrumented with 2 pyrometers, 3 rarified flow experiment (RAFLEX) pressure probes, and 25 thermocouples (TC) integrated into the TPS material at varying depths.[35] A total of seven TC's were on the afterbody. Several simulations of the MIRKA flight data have been published,[36-38] although most researchers have dealt only with the forebody flow. It was noted in [36] that the heat flux readings at TC15-17 were strongly influenced by hot pyrolysis gases injected upstream of that location, and it seems likely that those further downstream would also be affected.

The European Space Agency launched the Atmospheric Reentry Demonstrator (ARD) on October 12, 1998.[39] The probe reentered the Earth's atmosphere at a velocity of 7.5 km/s. ARD was a subscale Apollo-like capsule with a diameter of 2.8 m. The capsule afterbody was instrumented with 7 pressure sensors, 4 thermocouples, and 4 copper calorimeters.[40] In addition, the afterbody cone was coated with thermo-sensitive paint. Although the forebody thermocouples failed above about 800° C, those on the afterbody were functional throughout the entry.[40] However, the deduced heat transfer is not considered to be reliable.[40] Better results were obtained from the calorimeters, which provided heat transfer data throughout the entry.[40] Computational analysis of the ARD afterbody has been presented in [40]. While good agreement was obtained between the computations and flight data early in the trajectory, the CFD overpredicted the peak flight heating by as much as a factor of two. Possible reasons given for this discrepancy were delayed transition to turbulence or an inadequate gas chemistry model.[40-41]

4. TURBULENT FLOW SIMULATIONS

All of the previous computational results discussed in this paper dealt exclusively with laminar afterbody flows. However, a significant increase in heating rate occurs with the transition from laminar to turbulent flow. For windward acreage heating of a hypersonic entry vehicle, this enhancement can be a factor of 3-4. Similar enhancement of base heating rates is also possible. Incorporation of proper modeling of both transition and turbulence into the computational analysis will, thus, have a considerable impact on the aeroshell design. Unfortunately, there have been few published attempts to simulate turbulent afterbody flight data with CFD methods. This section briefly discusses the application of both traditional and state of the art turbulence models to afterbody flowfields.

Brown[42] recently calculated a variety of experimental flows in order to assess various existing turbulence models for use with real-gas Navier-Stokes simulations of hypersonic reentry vehicles. The test cases were selected based on the relevance of flow geometry and conditions and based on an assessment of the confidence in experimental results. Turbulence models assessed included compressibility-corrected versions of the Baldwin-Lomax model,[43] the one-equation Spalart-

Allmaras model,[44] the Wilcox two-equation k-ω model,[45] and Menter's two-equation SST k-ω model.[46]

One of the selected cases was the Hollis and Perkins[47] afterbody experiment of Mach 9.8 flow over a 70° sphere-cone. The experimental configuration is similar to the Mars Pathfinder spacecraft, although it was sting mounted, which changes the dynamics of wake closure and provides a path for upstream influence via the subsonic boundary layer on the sting. The nominal conditions for this experiment are air at $M_\infty = 9.8$, $T_\infty = 52.45$ K, $\rho_\infty = 0.00868$ kg/m^3, and a freestream Reynolds number of 9.2×10^4 based on diameter. The flow was assumed to be a perfect gas and an isothermal wall ($T_w = 300$ K) was specified.

Figure 15 shows Mach number contours from the SST turbulence model computation of the Hollis and Perkins experiment. The model surface is outlined in blue. Figure 16 shows the experimental heat transfer results along the model surface, along with the computed heat transfer for several of the turbulence models considered. The sting is included in the computations since the influence of the sting on the afterbody heat transfer results is likely to be significant. The flow is assumed to be laminar over the sphere-cone portion of the model. Separation occurs at the model shoulder ($s/R_b = 1$), and transition to turbulent flow is also specified to occur at this location. Reattachment occurs on the sting, with a recirculation zone washing most of the afterbody surface. The level of turbulence within the recirculation zone is obviously paramount to accuracy of the predictions for the afterbody heat transfer levels.

In Figure 16, the heat transfer results for the various turbulence models diverge considerably in the afterbody and sting reattachment regions. It is evident that the laminar and Baldwin-Lomax turbulent solutions under-predict heat transfer by as much as a factor of three in this region. The Spalart-Allmaras and k-ω models (not shown) similarly under-predicted the heat transfer in this region. Only the SST turbulence model accurately predicts the heat transfer over the entire afterbody and sting reattachment portions of this experiment, capturing the detailed variation on the afterbody frustum. Based on these results, as well as the other test cases chosen, Brown recommended the SST model for the computation of separated hypersonic flows.[42]

A particular shortcoming in afterbody turbulence model validation is the shortage of real-gas datasets. All of the datasets considered by Brown were amenable to treatment of the fluid as perfect-gas. Furthermore, although the Hollis and Perkins data are useful for

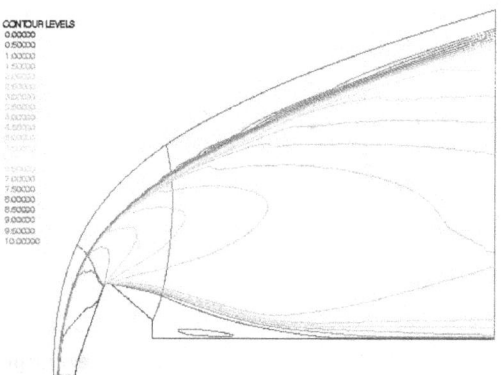

Fig. 15. Computed Mach Number contours for axisymmetric Mach 10, 70° Sphere-Cone of Hollis and Perkins (from [42]).

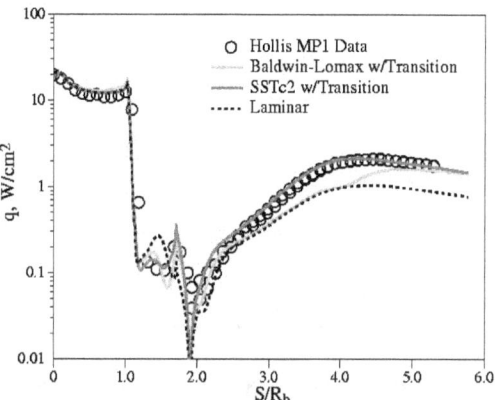

Fig. 16. Computed surface heat transfer for axisymmetric Mach 10, 70° Sphere-Cone of Hollis and Perkins (adapted from [42]).

afterbody heat transfer validation, the presence of the sting contaminates the assessment for heat transfer for the afterbody surfaces since closure of the recirculation zone is accomplished by attachment on the sting rather than with a free-flight wake closure.

Engineering prediction of turbulent flows relies heavily on Reynolds-averaged Navier-Stokes (RANS) simulations that compute the time-averaged flow field. However, RANS models can be inaccurate in high Reynolds number flows with large-scale separation. By comparison, detached eddy simulation (DES)[48] significantly improves predictions in massively separated flows by simulating the unsteady dynamics of the dominant length scales. DES methods have been shown to accurately predict the extent of the recirculation region and the base pressure in supersonic flows,[49] but such models have not been applied to hypersonic chemically reacting flowfields.

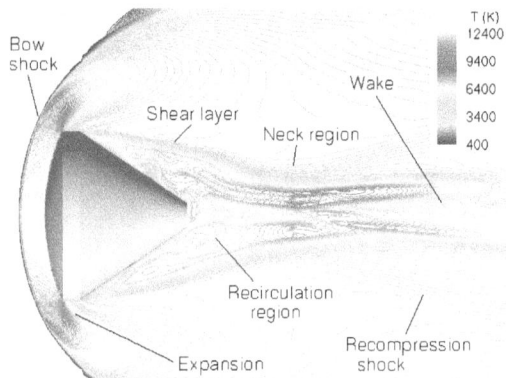

Fig. 17. Instantaneous pitch plane temperature contours around the Fire II vehicle computed with DES (from [14]).

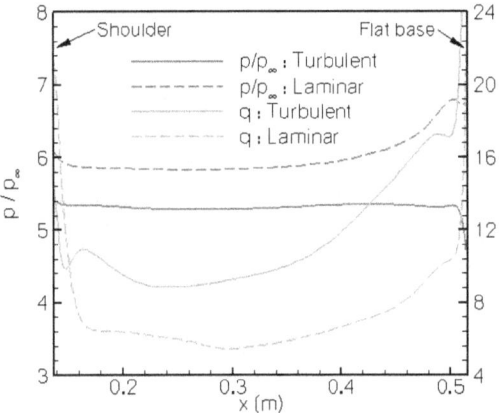

Fig. 18. Computed laminar and turbulent pressure and heat transfer on the afterbody of the Fire II vehicle (from [14]).

Recently Sinha et al.[14] used DES to study the flow field behind the Fire II flight vehicle at 35 km altitude, $M_\infty = 16$, and a freestream Reynolds number of 1.8×10^6 based on diameter. In these preliminary simulations the effect of non-equilibrium thermo-chemistry on the flowfield was neglected. The flowfield was simulated by solving the three-dimensional Favre-averaged Navier-Stokes equations. Turbulence was modeled using the Spalart-Allmaras one-equation turbulence model,[44] which was modified to operate in DES mode by introducing a new length scale d_{des}, defined as

$$d_{des} = \min(d, C_{des}\Delta) \qquad (3)$$

where d is the distance to the nearest wall, Δ is the largest dimension of the local grid cell, and C_{des} is an adjustable parameter calibrated by Shur.[48] This new length scale results in a form of turbulent eddy viscosity close to that of a large eddy simulation far from solid boundaries, and smoothly reverts back to the original RANS model near the wall.

Figure 17 shows the instantaneous temperature contours in the pitch plane of the Fire II vehicle. The main flow features are identified in the figure. The flow separates at the beginning of the conical afterbody and a large recirculation region forms behind the vehicle. The flow in this region is highly unsteady and three-dimensional, and is characterized by vortices and shear layers of varying intensity and length scales. The temperature in the wake is relatively high (6000 - 9000 K) except for a thin region close to vehicle wall. This corresponds to a laminar boundary layer on the afterbody. Some of the cold fluid from this region is swept away from the wall.

Time-averaged flow data were also computed by integrating the unsteady flow field over multiple characteristic times. The resulting flowfield was not entirely symmetric about the vehicle axis, as would be expected for an axisymmetric vehicle at zero angle of attack. The reasons for this asymmetry are currently under investigation.[14] Figure 18 shows the computed time averaged base pressure and heat transfer for the turbulent and baseline laminar computations. The pressure on the afterbody is lower in the turbulent simulation than the laminar case, whereas the turbulent heat transfer rate is higher than the laminar by 60-70%. The present results are preliminary, but are a promising first step in the application of sophisticated turbulence models to hypersonic base flows.

5. CONCLUSIONS AND RECOMMENDATIONS

The data obtained during Project Fire and the Apollo program provide an invaluable resource for the validation of modern computational tools for afterbody aeroheating. The six flight tests provide data spanning the entire range of Earth entry conditions, from axisymmetric to three-dimensional, non-continuum to continuum, laminar to turbulent, and non-ablating to fully ablating. The European ARD and MIRKA flights, together with the American Reentry F, are an additional valuable resource for Earth entry base heating. Two recent papers have looked at a portion of the Fire and Apollo data and have shown that modern computational methods appear to be fully capable of predicting afterbody heating to within the uncertainty of the flight data, at least for laminar flows without ablation. There has been less work done on understanding turbulent wake flows. Recent efforts using conventional and advanced turbulence models summarized herein are a good first step; however, more work needs to be done to fully understand the flight data. Data from Reentry F, a flat-based ballistic entry vehicle, may help to answer not only our ability to predict turbulent base heating, but also whether the disk shock phenomena is seen in flight. The Apollo 4 and 6 flight data include the effects of ablation and turbulence and will allow us to validate current methodologies in this environment. This work should be completed prior to recommending new flight testing so that we are better able to assess the gaps in our ability to predict Earth entry afterbody heating.

The state of affairs for other planetary destinations is not as good. Although code validation with Earth entry data certainly increases confidence in our ability to predict afterbody heating at other planets, differences in atmospheric composition and the associated chemical kinetics can only be fully resolved with in-situ flight data. Unfortunately, two thermo-couples on Viking and a single near-surface thermo-couple on Pathfinder are the only truly usable pieces of afterbody flight aeroheating data for any non-Earth entry. Worse, the recent trend has been a reduction or even elimination of heatshield instrumentation as a cost-saving or (perceived) risk-reduction measure. For example, there was no heat-shield instrumentation on either Mars Exploration Rover (MER) entry vehicle, the Stardust sample return capsule, or on the European Space Agency's Mars Beagle or Huygens Titan probes. Future planetary entry missions must include heatshield instrumentation in order to improve our understanding of these environments. The aftshell is often the safest place to incorporate instrumentation due to the low heating rates, and the results summarized in this paper give increased confidence in our ability to use such data effectively for code validation and improvement.

6. REFERENCES

[1] Boyd, I., Chen, G., and Candler, G., "Predicting Failure of the Continuum Fluid Equations in Transitional Hypersonic Flows," *Physics of Fluids*, Vol. 7, No. 1, 1995, pp. 210-219.

[2] Moss, J. and Price, J., "Review of Blunt Body Wake Flows at Hypersonic Low Density Conditions," AIAA Paper 96-1803, Jun. 1996.

[3] Gnoffo, P.A., "Planetary Entry Gas Dynamics," *Annual Review of Fluid Mechanics*, Vol. 31, 1999, pp. 459-494.

[4] Olynick, D., Chen, Y.-K., and Tauber, M., "Aerothermodynamics of the Stardust Sample Return Capsule," *Journal of Spacecraft and Rockets*, Vol. 36, No. 3, 1999, pp. 442-462.

[5] Gnoffo, P., Gupta, R., and Shinn, J., "Conservation Equations and Physical Models for Hypersonic Air Flows in Thermal and Chemical Nonequilibrium," NASA TP-2867, Feb. 1989.

[6] Lees, L., "Hypersonic Wakes and Trails," *AIAA Journal*, Vol. 2, No. 3, 1964, pp. 417-428.

[7] Gnoffo, P., Price, J., and Braun, R., "Computation of Near Wake Aerobrake Flowfields," *Journal of Spacecraft and Rockets*, Vol. 29, No. 2, 1992, pp. 182-189.

[8] Scallion, W.I. and Lewis, J.H., "Flight Parameters and Vehicle Performance for Project Fire Flight I," NASA TN D-2996, 1965.

[9] Slocumb, T.H., "Project Fire Flight I Heating and Pressure Measurements on the Reentry Vehicle Afterbody at a Velocity of 38,000 Feet Per Second," NASA TM X-1178, 1965.

[10] Lewis, J.H. and Scallion, W.I., "Flight Parameters and Vehicle Performance for Project Fire Flight II," NASA TN D-3569, 1966.

[11] Slocumb, T.H., "Project Fire Flight II Afterbody Temperatures and Pressures at 11.35 Kilometers Per Second," NASA TM X-1319, Dec. 1966.

[12] Woodbury, G., "Angle of Attack Analysis for Project Fire I Reentry Flight," NASA TN D-3366, 1966.

[13] Wright, M., Loomis, M., and Papadopoulos, P., "Aerothermal Analysis of the Project Fire II Afterbody Flow," *Journal of Thermophysics and Heat Transfer*, Vol. 17, No. 2, 2003, pp. 240-249.

[14] Sinha, K., Barnhardt, M., and Candler, G., "Detached Eddy Simulation of Hypersonic Base Flows with Application to Fire II Experiments," AIAA Paper 2004-2633, Jun. 2004.

[15] Hillje, E., "Entry Flight Aerodynamics from Apollo Mission AS-202," NASA TN D-4185, Oct. 1967.

[16] Hillje, E., "Entry Aerodynamics at Lunar Return Conditions Obtained from the Flight of Apollo 4," NASA TN D-5399, 1969.

[17] Lee, D., Bertin, J., and Goodrich, W., "Heat Transfer Rate and Pressure Measurements During Apollo Orbital Entries," NASA TN D-6028, Oct. 1970.

[18] Lee, D., "Apollo Experience Report: Aerothermodynamics Evaluation," NASA TN D-6843, Jun. 1972.

[19] Wright, M., Prabhu, D., and Martinez, E., "Analysis of Afterbody Heating Rates on Apollo Command Modules, Part 1: AS-202," AIAA Paper 2004-2456.

[20] Lee, D. and Goodrich, W., "Aerothermodynamic Environment of the Apollo Command Module During Superorbital Entry," NASA TN D-6792, Apr. 1972.

[21] Ried, R., Rochelle, W., and Milhoan, J., "Radiative Heating of the Apollo Command Module: Engineering Predictions and Flight Measurements," NASA TM X-58091, Apr. 1972.

[22] Lee, G., "Ablation Effects on the Apollo Afterbody Heat Transfer," *AIAA Journal*, Vol. 7, No. 8, 1969, pp. 1616-1618.

[23] Wright, R. and Zoby, E., "Flight Measurements of Boundary Layer Transition on a 5° Cone at a Mach Number of 20," NASA TM X-2253, May 1971.

[24] Alley, V. and Guillotte, R., "Postflight Analysis of Thermal Distortions of the Reentry F Spacecraft," NASA TM X-2250, May 1971.

[25] Dillon, J. and Carter, H., "Analysis of Base Pressure and Base Heating on a 5° Half Angle Cone in Free Flight Near Mach 20," NASA TM X-2468, Jan. 1972.

[26] Martin-Marietta Corp., "Viking Lander System, Primary Mission Performance Report," NASA CR-145148, Apr. 1977.

[27] Martin-Marietta Corp., "Entry Data Analysis for Viking 1 and 2," NASA CR-159388, Nov. 1976.

[28] Schmitt, D., "Base Heating on an Aerobraking Orbital Transfer Vehicle," AIAA Paper 83-0408.

[29] Givens, J., Nolte, L., and Pochettino, L., "Galileo Atmospheric Entry Probe System: Design, Development and Test," AIAA Paper 83-0098, Jan. 1983.

[30] Milos, F., Chen, Y.-K., Squire, T., and Brewer, R., "Analysis of Galileo Probe Heat Shield Ablation and Temperature Data," *Journal of Spacecraft and Rockets*, Vol. 36, No. 3, 1999, pp. 298-306.

[31] Spencer, D., Blanchard, R., Braun, R., Kallemeyn, P., and Thurman, S., "Mars Pathfinder Entry, Descent and Landing Reconstruction," *Journal of Spacecraft and Rockets*, Vol. 36, No. 3, 1999, pp. 357-366.

[32] Milos, F., Chen, Y.-K., Congdon, W., and Thornton, J., "Mars Pathfinder Entry Temperature Data, Aerothermal Heating, and Heatshield Material Response," *Journal of Spacecraft and Rockets*, Vol. 36, No. 3, 1999, pp. 380-391.

[33] Mitcheltree, R. and Gnoffo, P., "Wake Flow About Mars Pathfinder Entry Vehicle," *Journal of Spacecraft and Rockets*, Vol. 32, No. 5, 1995, pp. 771-776.

[34] Haas, B. and Venkatapathy, E., "Mars Pathfinder Computations Including Base Heating Predictions," AIAA Paper 95-2086, Jun. 1995.

[35] Schmitt, G., Pfeuffer, H., Kasper, R., Kleppe, F., Burkhardt, J., and Shottle, U., "The MIRKA Reentry Mission," IAF-98-V2.07, 49th International Astronautical Congress, Sep. 1998.

[36] Jahn, G., Schöttle, U., and Messerschmid, E., "Post-Flight Surface Heat Flux and Temperature Analysis of the MIRKA Reentry Capsule," *Proceedings of the 21st International Symposium on Space Technology*, Omiya, Japan, May 1998, pp. 532-537.

[37] Fertig, M. and Fruehauf. H., "Detailed Computation of the Aerothermodynamic Loads of the MIRKA Capsule," *3rd European Symposium on Aerothermodynamics*, Nov. 1998, pp. 703-710.

[38] Fruehauf, H., Fertig, M., and Kanne, S., "Validation of the Enhanced URANUS Nonequilibrium Navier-Stokes Code," *Journal of Spacecraft and Rockets*, Vol. 37, No. 2, 2000, pp. 218-223.

[39] Macret, J. and Leveugle, T., " The ARD Program: An Overview," AIAA Paper No. 99-4934, Jun. 1999.

[40] Tran, P. and Soler, J., "Atmospheric Reentry Demonstrator Post Flight Analysis: Aerothermal Environment," *Proceedings of the 2nd International Symposium on Atmospheric Reentry Vehicles*, Arcachon, France, Mar. 2001.

[41] Thirkettle, A., Steinkopf, M., and Joseph-Gabriel, E., "The Mission and Post-Flight Analysis of the Atmospheric Reentry Demonstrator," ESA Bulletin 109, Feb. 2002, pp. 56-63.

[42] Brown, J.L., "Turbulence Model Validation for Hypersonic Flows," AIAA Paper 2002-3308.

[43] Baldwin, B. and Lomax, H., "Thin Layer Approximation and Algebraic Model for Separated Turbulent Flows," AIAA Paper 78-257, Jan. 1978.

[44] Spalart, P. and Allmaras, S., "A One-Equation Turbulence Model for Aerodynamic Flows," AIAA Paper 92-0439, Jan. 1992.

[45] Wilcox, D., *Turbulence Modeling for CFD*, DCW Industries Inc., La Cañada, CA, 2nd Ed., 1998.

[46] Menter, F., "Two-Equation Eddy-Viscosity Turbulence Models for Engineering Applications," *AIAA Journal*, Vol. 32, No. 8, 1994, pp. 1598-1605.

[47] Hollis, B. and Perkins, J., "Comparisons of Experimental and Computational Aerothermodynamics of a 70° Sphere-Cone," AIAA Paper 96-1867, Jun. 1996.

[48] Shur, M., Spalart, P., Strelets, M., and Travin, A., "Detached Eddy Simulation of an Airfoil at High Angle of Attack," *4th International Symposium on Engineering Turbulence Modeling, Corsica*, 1999.

[49] Forsythe, J., Hoffman, K., and Squires, K., "Detached-Eddy Simulation with Compressibility Corrections Applied to a Supersonic Axisymmetric Base Flow," AIAA Paper 2002-0586, Jan. 2002.

Emerging Technologies

A SURVEY OF THE RAPIDLY EMERGING FIELD OF NANOTECHNOLOGY: POTENTIAL APPLICATIONS FOR SCIENTIFIC INSTRUMENTS AND TECHNOLOGIES FOR ATMOSPHERIC ENTRY PROBES

M. Meyyappan[1], J. O. Arnold[2]

[1]NASA Ames Center for Nanotechnology (NACNT), NASA Ames Research Center, Moffett Field, CA 94035 USA,
meyys@orbit.arc.nasa.gov
[2]NACNT, University Affiliated Research Center (UARC), Moffett Field, CA 94035, USA,
jarnold@mail.arc.nasa.gov

ABSTRACT

The field of Nanotechnology is well funded worldwide and innovations applicable to Solar System Exploration are emerging much more rapidly than thought possible just a few years ago. This presentation will survey recent innovations from nanotechnololgy with a focus on novel applications to atmospheric entry science and probe technology, in a fashion similar to that presented by Arnold and Venkatapathy [1] at the previous workshop forum at Lisbon Portugal, October 6-9, 2003.

Nanotechnology is a rapidly emerging field that builds systems, devices and materials from the bottom up – atom by atom – and in so doing provides them with novel and remarkable macro-scale performance. This technology has the potential to revolutionize space exploration by reducing mass and simultaneously increasing capability.

Thermal, Radiation, Impact Protective Shields: Atmospheric probes and humans on long duration deep space missions involved in Solar System Exploration must safely endure 3 significant hazards: (i) atmospheric entry; (ii) radiation; and (iii) micrometeorite or debris impact. Nanostructured materials could be developed to address all three hazards with a **single** protective shield, which would involve much less mass than a traditional approach. The concept can be ready in time for incorporation into NASA's Crew Exploration Vehicle, and possible entry probes to fly on the Jupiter Icy Moons

Orbiter (JIMO) mission.

Nanoelectronics: Future Exploration missions will require modular, reconfigurable electronics with performance at least comparable to that which exists in ground processors today, yet able to perform in harsh space environments despite very severe limitations on spacecraft resources. Nanotechnology will enable this, and revolutionize electronics in this century much as the integrated circuit did in the last.

X-ray tube for X-ray Diffraction and Fluorescence: An X-ray tube using carbon nanotubes has been developed that is substantially smaller and 10 times lighter than commercial X-ray tubes. The new technology will transform the study of planetary surfaces; permit lightweight, low power mass spectrographs; and facilitate habitat purification.

Nano Chemical Sensor: Chemical sensors using carbon nanotubes (CNT) and other nanostructures have been developed to detect volatiles such as water, ammonia, NOx, CO2, and hydrocarbons, enabling the use of extremely sensitive, light, and compact sensors.

High Thermal Conductivity Material: New nanotube based materials have been developed that will radically improve heat dissipation of high-performance computers and high power optical components by factors of 2X.

New Composite Materials that May Enhance Pressure Vessels for Atmospheric Probes of the Gas Giants: Studies underway suggest that Titanium/Fullerene composites may improve the capability of pressure vessels for probes like that used for the Galileo Probe mission to Jupiter.

Putting it all together: It appears that nanotechnology may be a key to enabling nanoprobes (1 $\leq$ 10 kg), helping realize planetary atmospheric scientists' desire for "multipile probes to multiple worlds"[1].

REFERENCE

1. J. O. Arnold[1] and E. Venkatapathy, *"Developments in Nanotechnology and Implications for Future Atmospheric Entry Probes.* International Workshop on Planetary Probe Atmospheric Entry and Descent Trajectory Analysis and Science, 6-9 October, 2003, Lisbon Portugal

Pico Reentry Probes:
Affordable Options for Reentry Measurements and Testing

William H. Ailor[1], **Vinod B. Kapoor**[2], **Gary A. Allen, Jr.**[3], **Ethiraj Venkatapathy**[4],
James O. Arnold[5] **and Daniel J. Rasky**[6]

[1] Director, Center for Orbital and Reentry Debris Studies, The Aerospace Corporation, 2350 E El Segundo Blvd, El Segundo, CA 90245, william.h.ailor@aero.org
[2] Senior Project Engineer, The Aerospace Corporation, 2350 E El Segundo Blvd., El Segundo, CA 90245, vinod.kapoor@aero.org
[3] Senior Research Scientist, Eloret Corporation
[4] Senior Staff Scientist and Planetary Exploration Technology Manager, NASA Ames Research Center, Moffett Field, CA 94035, ethiraj.venkatapathy-1@nasa.gov
[5] Senior Scientist, NASA Ames Center for Nanotechnology and Nanotechnology Area Manager, University Affiliated Research Center (UARC), Moffett Field, CA 94035, jarnold@arc.nasa.gov
[6] Senior Staff Scientist, NASA Ames Research Center, Moffett Field, CA 94035, drasky@arc.nasa.gov

ABSTRACT

It is generally very costly to perform in-space and atmospheric entry experiments. This paper presents a new platform - the Pico Reentry Probe (PREP) - that we believe will make targeted flight-tests and planetary atmospheric probe science missions considerably more affordable. Small, lightweight, self-contained, it is designed as a "launch and forget" system, suitable for experiments that require no ongoing communication with the ground. It contains a data recorder, battery, transmitter, and user-customized instrumentation. Data recorded during reentry or space operations is returned at end-of-mission via transmission to Iridium satellites (in the case of earth-based operations) or a similar orbiting communication system for planetary missions. This paper discusses possible applications of this concept for Earth and Martian atmospheric entry science. Two well-known heritage aerodynamic shapes are considered as candidates for PREP: the shape developed for the Planetary Atmospheric Experiment Test (PAET) and that for the Deep Space II Mars Probe.

1.0 INTRODUCTION

Space hardware reentering Earth's atmosphere faces a harsh heating and loads environment. In general, unprotected spacecraft hardware in this environment will melt, come apart, and disperse over a large area. While computer models can predict how such objects will respond to the environment and estimate the hazard to people and property on the ground from such events, very little debris has been recovered that can be used to calibrate these models. With the exception of the tragic loss of the Space Shuttle Columbia and its crew, over the last 44 years of space activities, it is estimated that fewer than 250 pieces have been recovered, and most of these were not examined. Columbia's heatshield protected most of that vehicle for a portion of the reentry, so many of the fragments recovered from it may not be representative of the reentry of an unprotected object.

The ideal approach for obtaining sufficient information to calibrate reentry breakup models would be to record information on attitudes, rates, temperatures, etc., as an object is actually reentering and breaking apart. Unfortunately, the environment prohibits reliable communications, so rather than transmit the data as it is recorded, it is better to retrieve it during the descent, when communication is feasible, or from ground recovery of the recorder after landing. Clearly, the data recorder must be protected from the entry heating and loads (it must have a heatshield and be designed to survive) and it must either be retrieved or broadcast its data. Since debris from a reentering satellite will be spread over hundreds of square miles anywhere on earth and possibly over water, recovery may be very difficult or impossible. If the survivable device is properly designed, it will separate from the host vehicle, follow its own trajectory, and reach a free fall state (terminal velocity, dropping straight down) at above 50,000 ft. The time to impact from this point depends on the trajectory, which in turn depends on the entry system shape and weight of the device, but typically ranges from 5 to 7 minutes.

These factors led to the design of a small, lightweight, survivable, self-contained device containing a data recorder, instrumentation, battery, and transmitter. The device, called a Reentry Breakup Recorder (REBR) and illustrated in Fig.1, will weigh less than 1 kg, has a heatshield, and will use internal sensors to measure and record attitude, rates, temperatures, and GPS location data. Some of these sensors could be externally mounted to the REBR device and would be demolished as reentry progresses. The recorded data would be "phoned home" during the free fall using the Iridium or similar orbiting communications satellites. There is no need to recover the device, although the GPS data could make that possible, in some cases. As illustrated in Fig.2, one or more of the devices would be "glued" to specific pieces of space hardware to record data during reentry of that hardware.

Design of REBR and its communications architecture has been ongoing for the last few years at The Aerospace Corporation. An initial design of the REBR payload (battery, basic sensors, data recorder, transmitter, and antennae) indicates a payload weight of about 0.35 kg. Preliminary testing, including a drop test from a high altitude balloon using Iridium as the communications pathway, has provided confidence in the overall concept. A major unknown has been the heatshield design and weight, and this paper highlights results of a study of these components.

While REBR was designed for reentry breakup research, a more general version of this device, which we call the Pico Reentry Probe (PREP), addresses the entry system design including the trajectory, the entry probe shape, and the thermal protection system (TPS). PREP is a modular concept and can be designed to conduct flight-testing of an integrated entry system, or flight qualification of subsystems such as TPS and innovative sensors and science instruments. It can also be used to perform low cost atmospheric science experiments.

Like REBR, PREP could be carried to orbit on a ride of opportunity, but would separate from the host vehicle prior to or early in the reentry. In the case of flight qualification of a TPS, PREP will be designed to record the TPS performance data during the high-speed portion of the flight and would send this information home during the terminal descent phase. The flight data will allow the TPS technologist to reconstruct the trajectory, compare the measured values with the predicted heating rates, and thereby establish the quality of the TPS performance. Since the proposed concept does not require ground tracking, targeting a reentry, or other operations normally entailed in atmospheric entry experiments, it is potentially a very inexpensive option for conducting some types of testing.

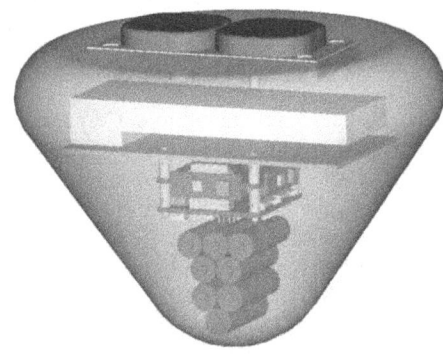

Fig. 1. The PREP design concept showing (from the bottom) batteries, data recorder, command and control board, transmitter, and antennae, all enclosed within the entry system with protective heat shield. Base diameter for PREP is 0.22 m (8.5 inches).

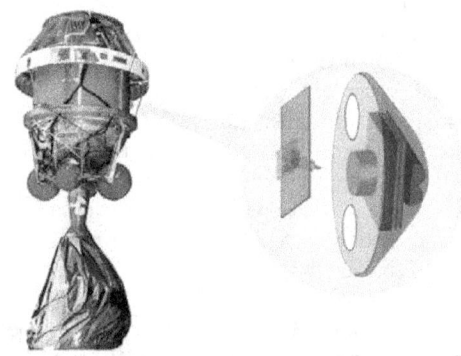

Fig. 2. The REBR or PREP would be attached to a launch stage or other hardware and carried to orbit. In the case of a reentry test-bed for TPS or sensor flight test and qualification, PREP will be part of an orbital system to be released on command.

2.0 OPERATIONAL CONCEPT (MISSION DESIGN)

Fig. 3 illustrates the mission concept for a typical reentry. PREP is dormant during the launch and operational lifetime of the host vehicle; it wakes up

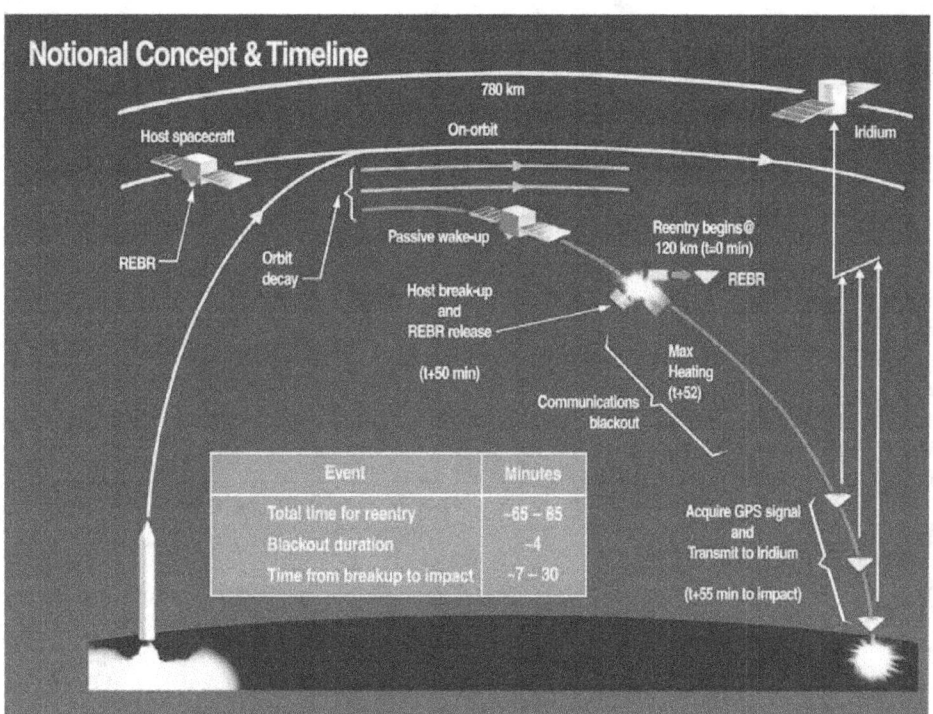

Fig. 3. Schematic of the mission scenario of REBR or PREP. The host launch stage or satellite would decay from orbit, reenter, and release the REBR or PREP.

and initializes itself as the host reenters the atmosphere. It acquires and stores data from its sensor suite during the reentry (and, in relevant cases, the breakup) of the host vehicle. A heat shield protects the PREP electronics from heating as reentry progresses, allowing it to survive while major parts of the host vehicle melt or ablate. It separates from the host vehicle at some point during the breakup process, and transmits the stored data through an overhead communication system (e.g. Iridium) prior to impact.

The times from the beginning of reentry to breakup and from breakup to impact will vary with each situation, but generally PREP has approximately 5 minutes to broadcast its data prior to impact. A host vehicle could carry several PREPs for redundancy and to record data specific to a particular location on the body or other area of interest. This feature may be of particular interest as a way to get in vivo information concerning satellite breakup.

One advantage of this concept is the cost of the hardware itself and, perhaps more importantly, of the infrastructure needed to conduct a test. A typical reentry test involves a specialized vehicle that can deorbit or insert the reentry vehicle into a specified location where radar, optical trackers and others wait to gather data. The cost of this infrastructure can be millions of dollars for each test. Since REBR and PREP take a ride of opportunity to space and use an existing communications network, very little infrastructure is required and much of the cost is eliminated.

The reentry heatshield is a key subsystem for these applications, and the TPS design requirements and material availability are discussed in the next section. Another key component is the communications system, discussed below in Section 4.3. Finally, this low-cost approach may enable new areas of atmospheric and material response investigations, and some potential applications of this system are proposed in this paper.

3.0 ENTRY SYSTEM DESIGN

Typical re-entry systems transport a payload package safely through the hypersonic/supersonic entry phases while protecting it from the external thermal and mechanical loads. If a soft landing is a requirement, then, in addition, the entry system will be designed to deploy a descent system (e.g., a parachute) to slow the vehicle. While it has been accomplished many times, deployment of a descent/parachute system is a complex and risky operation, requiring the ejection of the heat-shield and the backshell TPS prior to the deployment of the parachute at the right time during entry.

The design requirements for a PREP entry system are much simpler. The primary requirement is a system that allows sufficient time during the low-speed (M<4) phase to transmit the data to the orbiting communication system. Recovery of the data, not of the payload, is paramount. The subsystem requirements derived from this prime requirement drive the overall design towards a design solution with the least complexity, weight and cost. The system design is an iterative process and a brief description of the preliminary design process and the tools employed to guide the design is provided below.

The PREP design begins with an estimate for the payload mass and the volume. The current best estimate of the weight of the PREP payload (batteries, memory, command and control electronics, transmitter, and antennae) is about 350 grams. The objective is to keep this weight as low as possible to keep the overall entry system mass to a minimum, and to maximize the descent time for a given vehicle size. The largest data recorder/communication system component (see Fig. 1) can easily be accommodated with a maximum probe dimension of about 8.5 inches or 0.22 m diameter.

The next task is to select potential entry shapes. The shape, or Outer Mold Line (OML), determines the static-aerodynamic characteristics. The shape and the mass properties determine the static and dynamic stability (or orientation) of the probe during entry and descent. In keeping with the design philosophy of simplicity, we elected to evaluate the PAET and DS-II probe shapes shown in Figures 4 and 5.

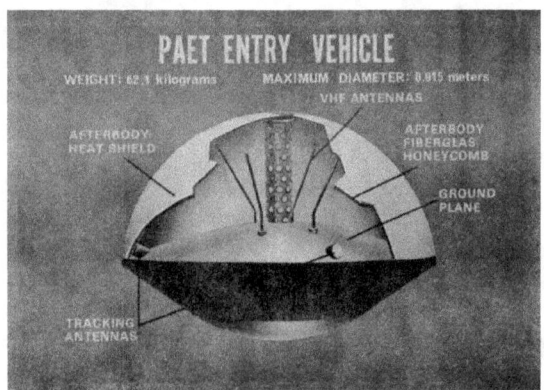

Fig. 4. Cut-away view of PAET Vehicle.

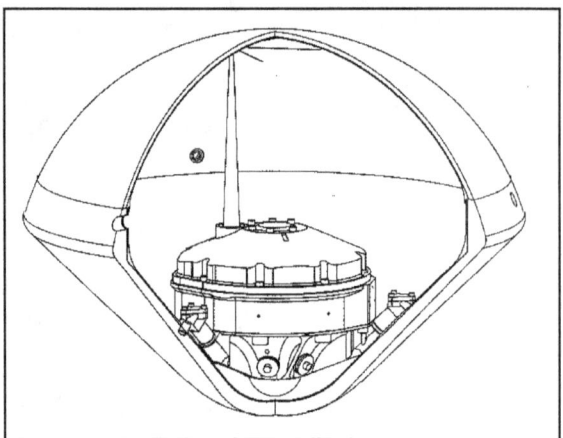

Fig. 5. DS-II Probe – Cut-away view of payload and the OML.

The well-studied aerodynamic characteristics of these two heritage aerodynamic shapes make them good candidates for PREP design. The PAET [1] flew in 1971; the OML of its heatshield is a simple sphere-cone with a cone half-angle of 55^0. The DS-II Mars Probe [2] flew to Mars in 1998, but was lost. Compared to PAET, the shape of the DS-II is slender. Its forebody is also a sphere cone with a conical frustum half-angle of 45 deg. DS-II shape has lower drag than PAET for a given base diameter. Considerable effort was spent during the DS-II design phase to determine the afterbody shape that assured not only stability but also the ability for the probe to orient itself correctly (forebody heat shield pointing in the direction of travel), independent of the orientation during orbital release from host spacecraft.

Given the aerodynamic characteristics, vehicle mass properties, and initial conditions (velocity, altitude, flight-path angle, and initial rates where appropriate), either three degrees of freedom (3- DOF) or 6-DOF simulations of the trajectory can be constructed. Higher fidelity 6-DOF Monte Carlo simulations can account for perturbation of a number of trajectory parameters or inputs and are often used in the detailed design phase to determine temperature limits. The 6-DOF simulations require both static and dynamic aerodynamic databases, detailed mass properties, and atmospheric properties. Such a level of analysis is beyond the scope of this paper. For the purposes of our current preliminary design, 3-DOF simulations are sufficient to determine the sub-system requirements and to determine whether the trajectory provides sufficient descent time for data transmission.

Once the velocity and altitude as a function of time are determined using 3-DOF simulations, the heating profile can be determined using well-known engineering formulas for the stagnation point, or using high fidelity CFD methods. The TPS material and the thickness required to keep the payload from over-heating are determined from the heating profile, using simulation methods for TPS response. It is a standard practice during preliminary design, especially with simple forebody shapes, to assume a TPS of constant thickness, based on its thickness at the stagnation point. Unless TPS mass fraction is significant, a constant thickness TPS is maintained to be conservative. For example, the Mars Pathfinder and Mars Exploration Rover heatshields used constant thickness TPS.

NASA Ames has spent considerable effort over the past decade to develop a fast, iterative, coupled trajectory and TPS sizing tool-kit known as TRAJ [3]. TRAJ has been validated using data from Pathfinder, PAET, Pioneer-Venus and other missions. TRAJ has been successfully applied to the TPS design of numerous missions including MER and MSL.

4.0 DESIGN ASSUMPTIONS FOR EARTH AND MARS MISSIONS

A total entry mass of 0.85 kg was assumed, including 0.35kg of payload. The design exercise is to assure that the entry probe aeroshell mass, which includes the mass of the TPS and structures, is less than 0.5 kg. The base-diameter of 0.22m (8.5 inches) was selected. The baseline TPS was Silicone Impregnated Reusable Ceramic Ablator (SIRCA) with a density of 260 kg/meter^3. If SIRCA is found not to meet the requirement, another TPS material, Phenolic Impregnated Ceramic Ablator (PICA), with a density of 228 kg/meter^3 and higher heat-flux performance is available. The internal structure was based on Pioneer-Venus structure: 0.0148-inch thick RTV-560 bondline with 0.125-inch thick sheet aluminum (2024 alloy). The TPS thickness is iterated until it meets the constraint that the bondline reach very close to 250°C; it is a function of the initial temperature of the TPS at the point of reentry (cold-soak temperature), and the best practice is to assume the cold-soak temperature to be 20 °C.

4.1 PREP Earth Mission Design Simulations and Key Results

In order to accommodate a range of missions, two entry angles (0^0 and -39^0) were selected to represent the extreme entry angle conditions for entry trajectory simulation. An entry angle of zero represents a slow orbital decay and entry, while an entry angle of -39^0 represents a very steep entry that the system might encounter after an abort, or to accommodate requirements for a flight test-bed for subsystems such as TPS or Instrumentation. The starting state vector and the input conditions are listed in Table 1.

Table 1. Input conditions assumed in simulating the trajectory for PREP Earth entry.

Altitude at entry, km	93.0
Radial distance at entry, km	6464 km
Inertial velocity at entry, km/s	6.88 km/sec
Inertial entry angle (gamma)	$-(0, 39^0)$
Inertial heading angle (psi)	108^0
Geocentric latitude	34.16^0
TPS temperature at entry	293 ^{0}K

The predicted trajectories and corresponding heat flux histories for the four cases (the two shapes, PAET and DS-II, each at two entry angles) are shown in Figs. 6 and 7, respectively. As mentioned earlier, in addition to predicting the trajectory, the TRAJ code also predicts the heat-flux profile and computes the TPS required at the stagnation point, based on a prescribed baseline TPS.

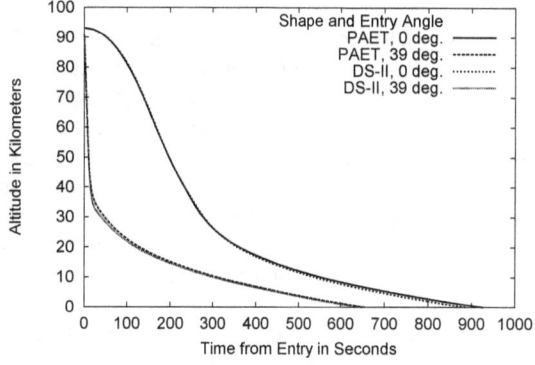

Fig. 6. The trajectory histories in terms of altitude vs. time are compared for the two entry flight path angles.

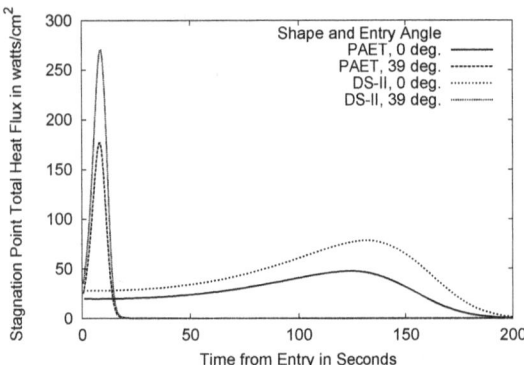

fig. 7. Predicted heat-flux histories for the PAET and DS-II shapes and for the limiting entry angles (0 deg., –39 deg.) are compared for sub-orbital entry conditions.

The DS-II and the PAET are very similar in terms of the altitude descent characteristics, both for the slow decay sub-orbital descent (zero entry angle) and for a steep (-39^0) entry angle descent. As expected, the most severe heat-flux is experienced by the steepest trajectory, but the highest heatload is experienced by the shallow trajectory. The TPS material selection depends on the peak heat-flux, whereas its thickness is determined by the shallow trajectory. SIRCA is a most efficient and lightweight solution for moderate heat-fluxes up to 180 W/cm^2, and is a very good choice for the PAET-shaped probe. On the other hand, the peak heat-flux experienced by the DS-II shape at the steep entry angle (-39^0) far exceeds the SIRCA performance limit. As a result, the TPS choice for the DS-II was PICA, which can withstand peak heat-fluxes up to 1200 W/cm^2. The maximum allowed stagnation pressure for PICA (1/2 atm) was not exceeded during the heat pulse.

The key results from the coupled trajectory and TPS sizing simulations are provided in Table 2. As noted, the TPS sizing is performed at the stagnation point and a constant TPS thickness is assumed all around the forebody in determining the weight and the mass fraction of the heatshield. The heatshield mass includes an aluminum skin (structures). The aft-shell TPS and structural mass were not computed, but it is typically a small percentage of the forebody mass, since the aft region heating is typically less than 4% of the stagnation heating. A PAET-shaped probe heat shield with a mass of about 0.16 kg or 19% of the total entry mass will be adequate to withstand the heating encountered in any trajectory between the two extreme entry angles studied here. The mass required for the heatshield of the DS-II shaped probe (estimated to be close to 67% of the entry mass, or 0.57kg) is unacceptably large. As a result, the PAET shape is preferred for earth entry applications.

In addition to the heat load, mechanical entry loads of up to 67 g's will be experienced, and the aeroshell and the payload must be designed to withstand such loads (the Pioneer-Venus Probe was designed to withstand up to 400 g's). They are not considered a problem for either the aeroshell or the payload.

The results clearly show that the PAET probe shape is preferable to the DS-II shape. The heating experienced by PAET shape is well within the capability of SIRCA. A mass fraction allocation of 20% for the heatshield, based on a 0.85 kg entry mass, is adequate to handle the expected range of entry angles. In the case of PAET, there is adequate mass margin available to accommodate aft-shell TPS and structures, as well as additional lightweight instrument or sensors, if necessary for the Earth applications.

4.2 PREP Mars Mission Simulations and Key Design Results

One of the potential applications of PREP involves science missions on Mars: a series of PREP devices containing sensors to detect atmospheric volatiles could be released from orbit periodically, to measure seasonal variations of the atmospheric composition [4]. This could possibly be performed by a very small, dedicated mass spectrograph or by the novel sensors from the emerging field of nanotechnology. The motivation to consider Pico-Probes for Mars is a result of the science goals outlined in the Decadal Survey [5], asking for revolutionary advances in nano/micro sensors that point to future miniaturized sensor detector systems and the need to find low-cost alternatives to current mission designs.

The input conditions assumed for the Mars mission scenario are similar to the Viking entry conditions. As mentioned above, the TRAJ code has been validated with data from Viking and Pathfinder, and is designed to simulate trajectories in various planetary systems. Once again we consider two shapes, the DS-II and the PAET, released from orbit with an inertial velocity of 4.6 km/s (Viking entry conditions). The entry angle is assumed to be -17^0.

The probe size, entry mass and payload mass are exactly the same as the Earth entry case discussed earlier. The baseline TPS is SIRCA and the initial cold-soak temperature of the heatshield is conservatively assumed to be 20^0 C.

Table 2. Results from the four cases (DS-II and PAET shapes at two different entry angles), obtained from the TRAJ code.

	DS-II		PAET	
Entry Angle	0^0	-39^0	0^0	-39^0
Ballistic Coefficient	31	31	27	27
Deceleration load, G	7.5	70	7.48	67
Stag. Point peak Heat Flux, W/cm^2	78	271	48	177
Stag. Point heat-load, joules/cm^2	8641	2204	5317.	1417
TPS material	PICA	PICA	SIRCA	SIRCA
TPS thickness, cm	4.85	1.05	1.39	0.29
Heatshield mass fraction, % of entry mass of 0.85 kg	67%	15%	19%	4%
Total mass, kg	0.85	0.85	0.85	0.85
Payload mass, kg	0.35	0.35	0.35	0.35
Heatshield mass, kg	0.57	0.12	0.16	0.033
Excess mass capacity to account for backshell, etc., kg	(-0.07)	0.38	0.34	0.467
Total EDL time, s	888	650	927	658
Transit time for communication, s	644	623	689	627

The key results from the simulations are shown in Figs. 8 and 9, and in Table 3. The results are very encouraging and the heat-flux, heat-load and the heatshield mass fraction are much smaller than the Earth entry cases and a mass fraction of less than 7% is adequate. Though the results are very encouraging, designing a Mars mission requires additional consideration. The lack of a GPS system requires 6-axis accelerometers on-board for trajectory reconstructions. Uplink to the orbiting satellite is a more demanding task and requires evaluation of orbital uplinks available during entry. One option is to use the spacecraft that releases the probe as the uplink station, and then the PREP mission would have to be designed to ensure adequate visibility between the probe and the spacecraft during the communications phase.

Future study of point designs of PREP or slightly larger nanocraft should be conducted to address the possibility of flying novel nano/micro volatile gas detectors and currently-evolving lightweight mass spectrographs to meet the New Frontiers Science mission objectives mentioned earlier.

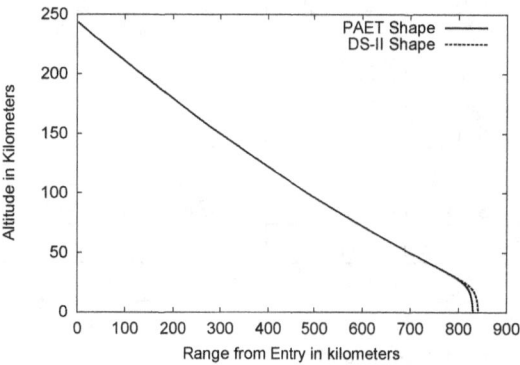

Fig. 8. PREP Mars entry trajectory of altitude vs. range shows the PAET and the DS-II shapes are very similar, and the requirements for communication uplink (or look angle) are not a discriminating factor in determining the shape.

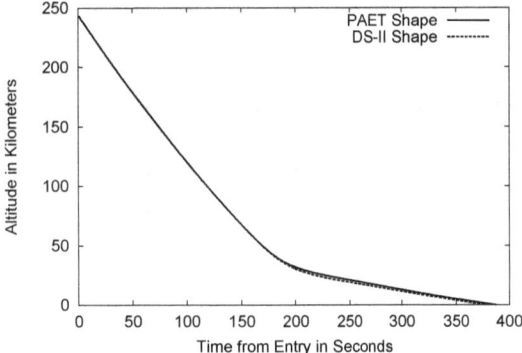

Fig. 9. PREP Mars entry trajectories for the PAET and the DS-II configurations are very similar - not a discriminating factor in the design.

Table 3. Key results for the PREP entry simulation for Mars entry from orbit. As a result of the low orbital entry speed and the thin Martian atmosphere, the TPS requirements are less severe than for Earth. SIRCA as the baseline material will provide adequate thermal protection for either of the two shapes. The required heatshield mass fraction is less than 7%.

	DS-II	PAET
Ballistic Coefficient kg/m^2	22	16
Deceleration load, g's	9.4	9.8
Stagnation point Heat Flux (max), W/cm^2	49.39	31.39
Stagnation point heat-load, J/cm^2	1977	1248
TPS thickness, cm	0.41	0.31
TPS mass, kg	0.0551	0.0358
TPS mass fraction, %	6.5	4.2
Total EDL time, s	378	387
Time for communication, s	107	112

4.3 Communications Architecture

A key component of the PREP design is the communication of data before impact, requiring that a receiver be above the reentry vehicle during its final free-fall. If PREP is returning to Earth, several options are available to receive and relay the broadcast data. These include commercial GEO systems (Astrolink, Spaceway, and Inmarsat), commercial LEO/MEO systems (Iridium, Globalstar, ICO, and Orbcomm), dedicated GEO (TDRSS), and dedicated aircraft (P3 Orion). After analysis of the requirements for coverage, availability, required power (Effective Isotropic Radiated Power), and cost, Iridium was chosen as the preferred candidate for PREP. The main consideration for Iridium was its full-time global coverage, coupled with its immediate availability: PREP would simply "phone home" at the end of its mission, with no advance scheduling required. On the basis of this selection, more detailed analysis of the Iridium constellation is required, including an investigation of the effect of the vehicle position and velocity on the communications link. A side benefit of the choice of Iridium is that the frequencies of the L1 carrier for GPS and of Iridium are within a few percent of each other, allowing the possibility of using a single antenna for both applications simultaneously, with appropriate filtering.

The initial concept for data transmission was the use of two omni-directional antennas to get full-sky coverage of the transmitted data. One of the associated drawbacks is the high power required to transmit continuously using both antennas, which would also entail high mass because of the additional hardware required, and data dropout if the vehicle is tumbling. To simplify the communications architecture and reduce the mass and power requirements, a design that uses a single antenna was selected. This required a method of assuring that the single antenna could point "up" toward the communications assets and will be achieved through center-of-gravity management: aeroshaping the heat shield to produce an aerodynamically stable freefall. While a parachute or streamer could be used, such a system would add to the complexity of the device, and every effort will be made to avoid use of such "active" systems. For the Mars mission experiments, the communications system will be engineered to interface with the available Mars communications infrastructure, using the existing Orbiter constellation at the time of the experiment. REBR and PREP hardware design is modular, with standard industry interfaces. The modems will be different for different applications and communications environments.

4.4 Safety

PREP uses the "fire and forget" concept—the probes can come down anywhere and complete their mission before impacting the ground. On ground impact, it is possible that one of the devices could strike an individual or cause property damage. The probability that a PREP device will injure an individual can be estimated on the basis of orbit inclination. Assuming a 1-ft² size, the highest casualty expectation would be on the order of 1x10⁻⁶ per device. This is well within the published DoD and NASA safety guidelines (1x10⁻⁴) for reentering hardware. Assuming a terminal ballistic coefficient of 31.1 kg/m², (DS/2 shape, worst case) the impact velocity at sea level is 22.3 m/sec. The likelihood of an individual being struck and injured by a PREP device is very small.

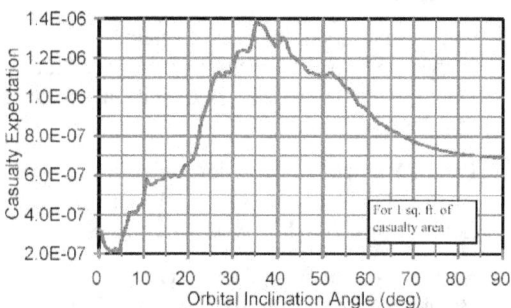

Fig. 10. The variation in casualty expectation as a function of orbit inclination is due to the population density under the orbit [6].

Safety related to orbiting satellites and manned systems such as the Space Shuttle and the International Space Station is not believed to be an issue with PREP, which is designed to operate in a reentry environment and would normally be attached to, or deployed from, space hardware that is going to reenter. As a result, it would not be left in orbit for extended periods as a stand-alone item, but would be attached to a larger, trackable object for much of its life.

4.5 Mission Applications

As a small, lightweight, instrumented reentry probe designed to collect information on heatshield performance, the upper atmosphere, or other information during an actual reentry, PREP will employ technology that enables other applications. For example, the PREP technology could be used for a "black box" for hardware designed to survive reentry [7], such as a single-stage-to-orbit vehicle. In this case, telemetry data would be recorded by the device, and in the event of a reentry accident, the data would be recovered from the device either by direct broadcast as discussed above or by recovery of the device, if GPS data as to its location is available.

PREP could be made available in kit form, allowing universities and researchers relatively inexpensive access to an environment that has been virtually inaccessible prior to this time. In this application, a researcher would use the kit for the basic data collection and communications functions, but would add instruments customized for the particular application.

5.0 SUMMARY AND CONCLUSIONS

Basic design of a small, lightweight, self-contained device to record data during reentry and breakup of space hardware and to broadcast the information to an orbiting communications system prior to ground impact has been ongoing for several years. These devices would collect data during actual breakup of space hardware during reentry, information that would be used to validate and calibrate reentry survival models critical to predicting reentry casualty risks and to help spacecraft manufacturers design space hardware that will respond in predictable and repeatable ways to the reentry environment. A major uncertainty in the design of this REBR has been specific information about the heatshield material and design. The preliminary studies described here indicate that a small, sub-1 kg reentry probe looks very promising. Two materials, SIRCA and PICA, and two shapes, from the DS-II and PAET probes, were considered in this study. Results show that the PAET probe shape using SIRCA is preferred for this application. The total weight of the REBR with this shape and material is 0.85 kg - less than the goal of 1 kg, and providing some margin for refining the shield design, modifying structures, adding sensors, and the like. The heatshield mass fraction for this case is approximately 20% of the total mass of the Recorder.

The novel mission design and communications approach used for the REBR could have benefits for other mission concepts: eliminating the requirement of a dedicated launch and reentry into an instrumented range could substantially lower the cost of testing new heat shield materials. The same basic mission design and communications architecture would be used for this application.

The heatshield design for a possible Mars atmospheric probe was also considered. Results for the SIRCA shield material are more favorable than those for a reentry into Earth's atmosphere, yielding a mass fraction for the heatshield of less than 7% of the total probe mass. It was noted that use of this concept on another planetary body requires careful mission design, to assure that a receiver for communications from the small probe is properly located during the probe's broadcast period.

Data collected by small probes of the type described here could provide actual physical evidence about the environments of other planets, could enable new ways of testing and evaluating heat shield materials by exposing them to an actual reentry environment, and could serve as inexpensive flight test vehicles for the validation of TPS engineering and science instrumentation. In such ways, these devices would help new technologies bridge the 'valley of death" from TRL 4 to 6, helping secure mission insertion by providing hard performance information to risk-adverse project managers.

Maturation of the concept of using small probes in the atmospheric entry environment may lead to improved models for estimating hazards associated with reentering space hardware. This work may also lead to some very interesting, and potentially very cost effective, science missions of the type advocated in the decadal report.

6.0 REFERENCES

1. Seiff, A., D. Reese, S. Sommer, D. Kirk, E. Whiting and H. Niemann, "PAET, An Entry Probe Experiment in the Earth's Atmosphere," ICARUS, 18, 525–563, 1973.

2. "Deep Space 2, Entry System (Aeroshell)," http://nmp.jpl.nasa.gov/ds2/tech/entry.html.

3. Allen G.A., P.J. Gage, E. Venkatapathy, D.R. Olynick and P.F. Wercinski, "A Web-Based Analysis System for Planetary Entry Vehicle Design," AIAA Paper No. 98-4826, Jun. 1998.

4. Arnold, J.O. and E. Venktapathy, "Developments in Nanotechnology and Implications for Future Atmospheric Entry Probes," International Workshop on Planetary Probe Atmospheric Entry and Descent Trajectory Analyses and Science, October 6-9, 2003, Lisbon, Portugal.

5. "New Frontiers in the Solar System: An Integrated Exploration Strategy," Space Studies Board, National Research Council, July 8, 2002, http://www.nap.edu.

6. Patera, R.P. and W.H. Ailor, "The Realities of Reentry Disposal," Advances in Astronautical Science, AAS 98-174, September 1998.

7. Kapoor, V. and W.H. Ailor, "The Reentry Breakup Recorder: A Black Box for Space Hardware," 17th Annual AIAA/USU Conference on Small Satellites, 2003.

NANOSTRUCTURED THERMAL PROTECTION SYSTEMS FOR SPACE EXPLORATION MISSIONS

SECOND INTERNATIONAL PROBE WORKSHOP
NASA AMES RESEARCH CENTER, MOFFETT FIELD, CA
AUGUST 23-26, 2004

J. O. Arnold[1], Y.K. Chen[2], T. Squire[2], D. Srivastava[1], G. Allen, Jr[3],
M. Stackpoole[3], H. E. Goldstein[4], E. Venkatapathy[2] and M. P. Loomis[2]

[1]NACNT, University Affiliated Research Center, Moffett Field, CA 94035, USA jarnold@mail.arc.nasa.gov, deepak@nas.nasa.gov
[2]NASA Ames Research Center, Moffett Field, CA 94035, ykchen@mail.arc.nasa.gov, tsquire@mail.arc.nasa.gov, evenkatapathy@mail.arc.nasa.gov, mloomis@mail.arc.nasa.gov
[3]Eloret Corporation, Ames Research Center, Moffett Field, CA 94035, gallen@mail.arc.nasa.gov, mstackpoole@mail.arc.nasa.gov
[4]Consultant, Ames Research Center, Moffett Field, CA 94305, hgoldstein@mail.arc.nasa.gov

ABSTRACT

Strong research and development programs in nanotechnology and Thermal Protection Systems (TPS) exist at NASA Ames. Conceptual studies have been undertaken to determine if new, nanostructured materials (composites of existing TPS materials and nanostructured composite fibers) could improve the performance of TPS. To this end, we have studied various candidate heatshields, some composed of existing TPS materials (with known material properties), to provide a baseline for comparison with others that are admixtures of such materials and a nanostructured material. In the latter case, some assumptions were made about the thermal conductivity and strength of the admixture, relative to the baseline TPS material. For the purposes of this study, we have made the conservative assumption that only a small fraction of the remarkable properties of carbon nanotubes (for example) will be realized in the material properties of the admixtures employing them. The heatshields studied included those for Sharp leading edges (appropriate to out-of-orbit entry and aero-maneuvering), probes, an out-of-orbit Apollo Command Module (as a surrogate for NASA's new Crew Exploration Vehicle [CEV]), a Mars Sample Return Vehicle and a large heat shield for Mars aerocapture missions. We report on these conceptual studies, which show that in some cases (not all), significant improvements in the TPS can be achieved through the use of nanostructured materials.

1. INTRODUCTION

Carbon nanotube (CNT)-based materials have the potential to revolutionize the design of future aerospace vehicles. As discussed in [1], CNTs exhibit Young's modulus of over 1 Tera Pascal, and tensile strength of about 200 Giga Pascal. They are about hundred times stronger than steel at 1/6th its weight. CNTs have thermal conductivities on the order of 3000 W/m°K in the axial direction - seven times higher than the thermal conductivity of copper. Perpendicular to this direction, the thermal conductivity is (0.25 W/m°K), essentially that of an insulator. Carbon nanotube materials can also be electrical conductors, semiconductors or insulators and can have piezoelectrical properties suitable for very high force activators. Composites made of CNTs and other nanostructured materials may benefit from these remarkable properties.

The low thermal conductivity in directions normal to the fiber and the high temperature stability when protected from oxidizing environments make them ideal candidates for both ablative and blanket-based heat shields. The high axial thermal conductivity of CNTs allows their use as passive heat pipes, transporting heat from hot spots on thermal shields to cooler areas, improving heat shield performance and reducing weight. Herein we analyze various cases of interest for future human and robotic space exploration missions, to evaluate benefits that could flow from incorporating nanostructured TPS materials. For our study, we have not assumed that the admixtures in which the nanomaterials are embedded will be endowed with the same remarkable material properties exhibited, for instance, in pure, single wall carbon nanotubes. Instead, we have made conservative assumptions about the level of enhancement that might reasonably be expected in the material properties of the admixtures, relative to those of baseline TPS materials.

2.0 VEHICLES ENTERING EARTH'S ATMOSPHERE

2.1 Vehicle with Sharp Leading Edges

The first application we considered was for vehicles with sharp leading edges (Fig. 1). As explained in

[2,3], such high Lift/Drag vehicles provide good aerodynamic performance, giving wide cross range capabilities, valuable for out-of-Earth-orbit missions. Such vehicles also offer value to aerogravity assist missions for solar exploration at Venus, for example, or for targeting the placement of nano- or micro-probes to Jupiter, cases where rapid deployment of assets such as orbiters and probes is needed.

Fig. 1 Proposed highly maneuverable hypervelocity, out-of-Earth-orbit vehicle [3], known as SHARP V-5.

Fig. 2 shows the 2-D coupled thermal/stress model of a wing leading edge that was developed [4] as a part of efforts conducted at Ames during the second Reusable Launch Vehicle Program. The UHTC is made of HfB_2 (20 volume percent SiC). This leading edge is attached to an SiC structured wing fixture. Research [4] demonstrated that a 2-D thermo-structural model can reliably predict 3-D performance for this configuration.

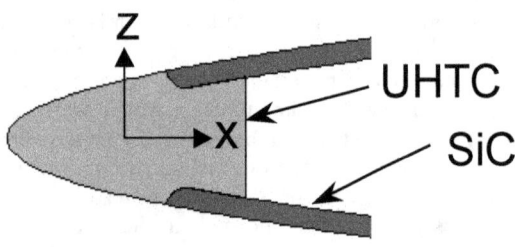

Fig 2. Representation of the nose of a Sharp leading edge vehicle.

Fig. 3 displays the flight conditions for the sharp leading edge depicted in Fig. 2. The initial time (t = 0) corresponds to the atmospheric pierce point altitude of 122 km. This trajectory was designed to minimize the integrated heat load. Heating was predicted with real-gas Computational Fluid Dynamics (CFD) codes. Note that the wing leading edge flew at an angle of attack of 35^0 for a little more than 20 minutes and then dropped rapidly to 15^0. Since the wing was at angle of attack, the heating is not symmetrical.

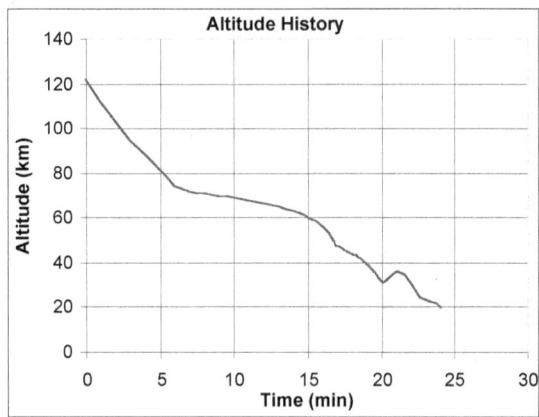

Fig. 3 (a) Altitude-time history.

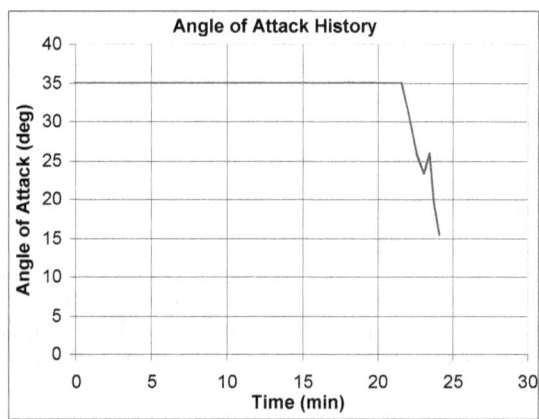

Fig. 3(b) Angle-of-attack history.

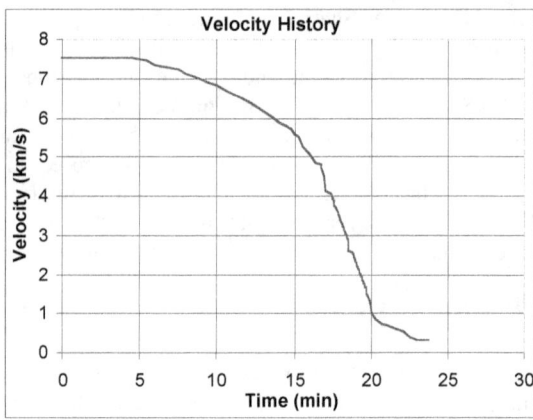

Fig. 3(c) Velocity History

In order to explore how nanotechnology could improve the performance of UHTC Sharp leading edges, two cases were considered. The first (nominal) case considered the UHTC HfB_2/SiC, using its established thermal conductivity and assuming it to be isotropic. The second case considered a UHTC/nanostructured admixture, assumed to have, in the X direction (Fig. 2), twice the thermal conductivity of HfB_2/SiC and in the Z direction, the same thermal conductivity as $HfB2/SiC$. The increased thermal conductivity in the X direction for such an admixture might be accomplished by fibers containing nanotubes (possibly CNTs) laid up with their axes in the X direction. The assumed value of twice the nominal thermal conductivity is far lower than that for pure CNTs, and allows for imperfect phonon transport in composite fibers. Fig. 4 shows the thermal conductivity for both cases as a function of temperature. It was assumed in both cases that there is no change in the isotropic mechanical properties of the UHTC.

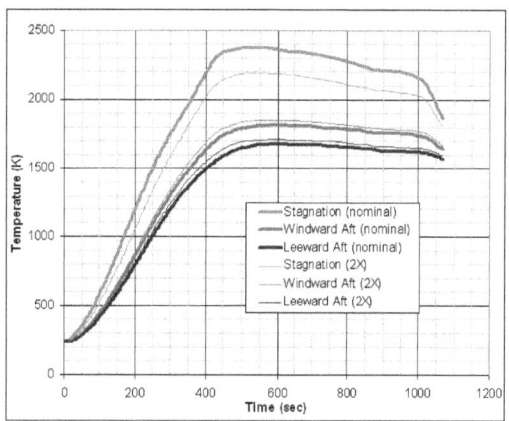

Fig. 5(a): Temperature profiles for locations circled in Fig. 5(b).

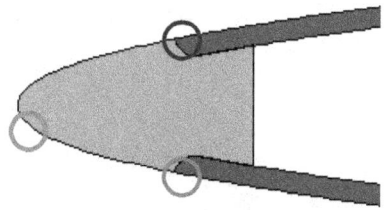

Fig. 5(b). Circles indicate three key locations on the sharp UHTC leading edge plotted in Fig. 5(a).

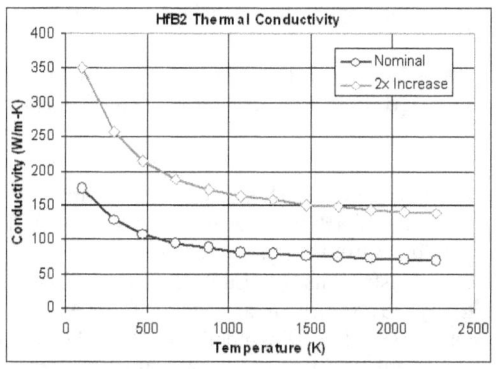

Fig. 4. Thermal conductivity as a function of temperature.

Fig. 5(a) shows the results of the thermal analysis for three key locations on the sharp leading edge, located by the circles in Fig. 5(b). The bold curves represent the nominal case, with the isotropic thermal conductivity, while the thin curves represent the case where this property is doubled in the X-direction.

The airfoil is at a positive angle of attack, so the stagnation point is on the lower edge, and the leeward attachment to the SiC structure runs cooler than that on windward side. It is important to note that the case with heterogeneous thermal conductivity results in a drop in the peak temperature at the stagnation point of 182 ^0K (328 ^0F). Such a reduction would make the UHTC less likely to oxidize and would increase the lifetime and reusability of the leading edge material. It is also noteworthy that the temperature (38^0K or 68^0F) at the UHTC/SiC attachment points is only slightly higher for case 2 than for the nominal case, suggesting that the assembly is viable for both cases.

We believe that the reason for all these changes in temperature is the 'passive heat pipe effect', arising from the increased thermal conductivity in the X-direction. With this effect, the equilibrium radiation from the sides of the UHTC effectively cools the stagnation point and the surrounding material.

Fig. 6 (a) is a plot of the principal stress history in the middle of the UHTC leading edge, the location of which is indicated by the circle in Fig. 6(b). Note that there is a very significant reduction in the tensile stress for the nanostructured material: doubling the thermal conductivity in the X-direction reduces the peak interior stress from 51.4 Mpa to 32.8 Mpa, a decrease of 36 percent.

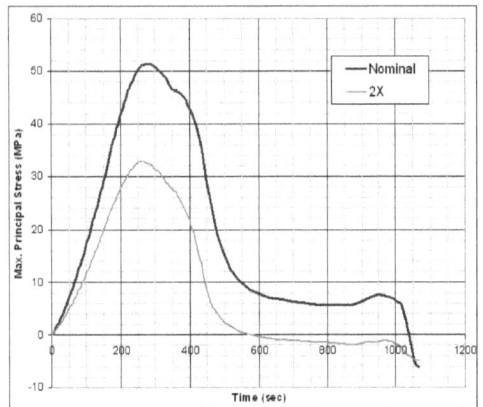

Fig. 6(a). Principal tensile stress in the UHTC leading edge versus time.

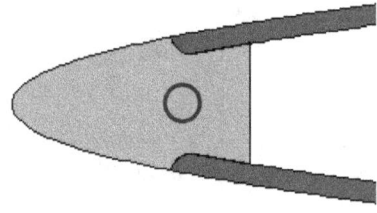

Fig. 6(b). Location of principal tensile stress in UHTC leading edge.

Future-generation materials for use on space transportation vehicles of the type described above require substantial improvements in material properties, leading to increased reliability and safety. UHTCs composed primarily of metal diborides are candidate materials for sharp leading edges on hypersonic re-entry vehicles. It is yet to be determined if they possess the properties necessary to withstand the extreme environments experienced at the leading edges during re-entry without undergoing some recession, oxidation or thermal shock. The design and processing of aligned nanotube-reinforced composites (composed of UHTC-type matrices with nanotube reinforcements) studied herein promises enhanced performance. It is hoped that, because of the extremely high thermal conductivity of the nanotubes, the overall thermal conductivity of the nanotube-reinforced UHTC system will show an increase over that of the baseline system, and yield increased performance. The conceptual study above offers considerable encouragement that the use of nanostructured materials will produce improvements.

If designed and processed correctly, we believe nanotube-reinforced UHTC composites could lead to a system with higher thermal conductivity than that found in the base UHTC material. These nanotube-reinforced composites could potentially dissipate heat away from localized regions of the leading edge and enable vehicles with leading edge geometries that provide improved performance and greater cross range. The drop in temperature at these localized regions may be sufficient to eliminate recession in these systems and allow for a reusable system. As part of this study, not shown here, there were indications that this approach may also yield improvements in other material properties, including toughness and thermal shock resistance, and so offer performance superior to that of current materials proposed for sharp leading edge applications. This approach may also yield composites that fail in a more graceful manner than current UHTC systems.

Initial samples comprising of aligned nanotubes in a refractory matrix have been processed. Preliminary mechanical properties and microscopy confirm that a preferred alignment has been achieved in them.

2.2 Out-of-Earth-orbit crewed vehicle (Apollo Command Module as a surrogate Crew Exploration Vehicle)

NASA's new vision for Space Exploration features the development of a Crew Exploration Vehicle (CEV) that would be used progressively for (1) out-of-orbit flight demonstrations, (2) Lunar Return missions and eventually, (3) return of astronauts from Mars. The precept of "spiral development" has been deemed appropriate for the CEV, giving rise to the opportunity to design a multi-use vehicle with a heatshield that is replaceable and able to be upgraded with a higher performance TPS when new missions demand it or when it becomes available. No concepts for the CEV have yet been developed by NASA's Office of Exploration (OExP), so we chose the Apollo Command Module as a surrogate CEV shape, and evaluated a nano-TPS for it.

Other studies ongoing [5] at Ames suggest that multi-functional approaches to the TPS for the CEV could result in significant mass savings. These concepts involve the use of a low-molecular-weight TPS, able to serve triple duty as Thermal, Radiation and Impact Protective Shields (TRIPS). This work suggests that fully dense Carbon Phenolic (CP) or lower density versions of CP would be good candidates as a starting points for TRIPS. Our study therefore concerns the potential benefits of using Carbon Phenolic as part of a nano-structured TPS for an Apollo Command Module shaped CEV.

At present, we have only analyzed the Apollo AS 202 out-of-orbit case. Wright, Prabhu and Martinez recently published [6] heat flux contours for this flight using a modern real-gas CDF code (DPLR). While their work was focused on afterbody flows, their forebody heat flux distribution is useful for our needs, and has been adopted herein. Fig. 7 shows the geometry of the Apollo vehicle used in [6].

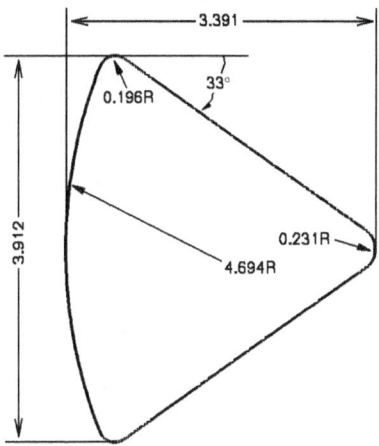

Fig. 7. Schematic drawing of the outer mold line of the AS-202 Apollo Command Module used in [6].

Fig. 8 (based on the work in [6], and provided to the present authors), displays the heat flux contours for Apollo AS 202 at the peak heating conditions, at a speed of 7.8 km/sec and an angle of attack of 18°. The stagnation point is on the lower (windward) edge of the figure.

Fig. 8. Heat flux contours for the Apollo AS 202 flight at peak heating conditions from [6]. The stagnation point is on the lower edge of the figure.

Fig. 9 displays the heat flux distribution along the plane of symmetry passing through the stagnation point for the Apollo AS 202 flight at the peak heating conditions specified above, and for the heat flux contours shown in Fig. 8. The peak heat flux at the stagnation point is slightly more than 100 W/cm^2 and falls to a value near 40 W/cm^2 on the leeward edge of the forebody.

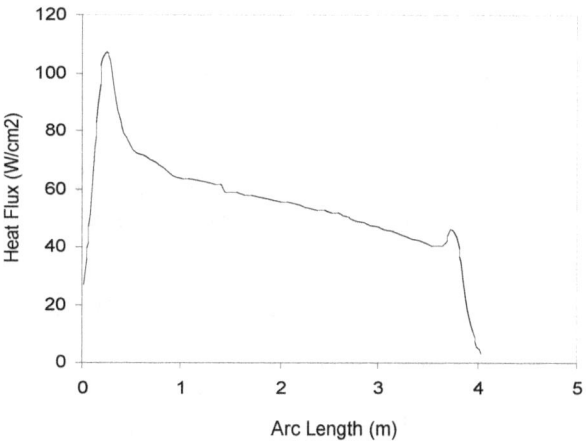

Fig. 9. Computed heat flux distribution in the plane of symmetry for the Apollo AS 202 flight at peak heating conditions from [6], corresponding to Fig 8. The origin of the streamline distance is the windward side edge or the forebody.

Fig. 10, predicted by the BLIMP program [7], displays the normalized stagnation point history of AS-202.

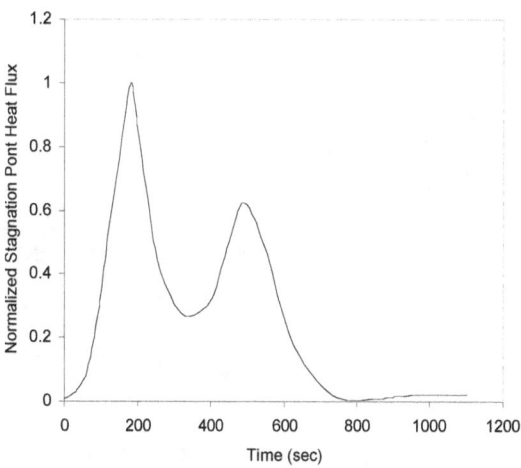

Fig. 10. Normalized stagnation point heating history for Apollo Command module for flight AS-202 computed with the code from [7].

Figs. 11 and 12 display the results of our calculations along the plane of symmetry, using the TITAN Code

[8]. The calculations are 2-D for the boundary conditions corresponding to the distributions in Figs 8 and 9.

Two cases were considered: (1) a heat shield made of fully dense Carbon Phenolic, whose properties are given in [9], and (2), a heatshield made of an admixture of fully dense CP and carbon nanotube composite fibers. For case 1, the thickness of the CP to sustain the bond line temperature at 250 °C was 4.65 cm. Fig. 11 shows the history of the stagnation point temperature while that for the bondline at the stagnation point is shown in Fig. 12. Splash down occurs at about 1100 seconds. Maintaining the bondline temperature at 250 °C or less at splashdown is the primary TPS requirement.

The material properties for the second case (95% CP + 5% nano-fiber by volume) are assumed to be identical to those for pure CP, except that the thermal conductivity of the admixture along the plane of symmetry is 500 W/m°K, higher than the value for pure fully dense CP at 0.55 W/m°K. As Fig.11 shows, the addition of carbon nanotube fibers has no effect on the stagnation point temperature history. There is a reduction of about 30°C in the bondline temperature at the stagnation point in case 2, compared to case 1. This reduction illustrates the passive heat pipe effect: a migration of heat from the hot stagnation point region to the cooler, downstream portion of the heat shield. It is estimated that this effect results is an overall saving of 5-10% in the heat shield mass.

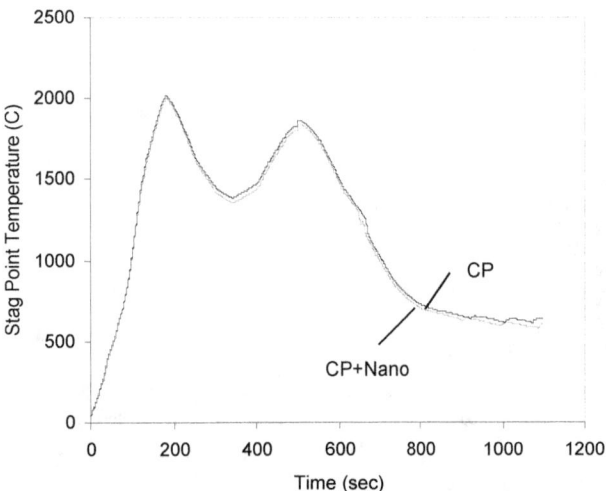

Fig. 11. TITAN [8] solutions for the stagnation point temperature history for the AS 202 trajectory and heating rate profile shown in Figs. 10 and 11.

While this effect is not as large as we had hoped for, it does illustrate that nanostructured TPS could be useful for improved designs for heat shields. One improvement could be in safety margins for bondline temperatures and another could be for specialized cooling of local "hot spots" like those areas where cavity heating may occur, such as the strut mounting locations commonly used for support of probes in launch stacks. As our colleagues at Ames extend their CFD study to the higher speed cases (e.g., Apollo 4 at Lunar Return speeds), we will continue our conceptual studies to look for benefits in these cases.

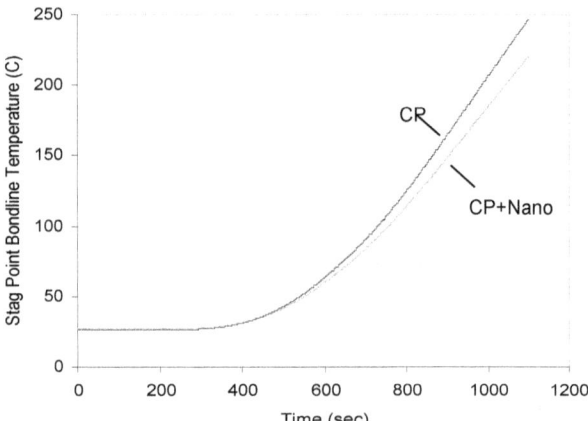

Fig. 12. TITAN solutions for the bondline temperature history for the AS 202 trajectory and heating rate profile shown in Figs 9 and 10. The prime TPS requirement is that the bondline temperature does not exceed 250 °C.

2.3 Sample Return Mission

Two NASA missions to return samples to Earth are currently underway: Genesis and Stardust [10,11]. Genesis will return (in September 2004) samples of solar wind material expelled from the Sun; Stardust will return ejecta from Comet Wild 2, in January 2006. Both spacecraft will enter the Earth's atmosphere at hyperbolic return speeds. These missions represent the beginning of a concerted effort to understand the Solar System by returning samples recovered from deep space to the Earth for detailed scientific study.

The entry system we chose to analyze is a generic sample return capsule (SRC) similar to the proposed Mars Sample Return (MSR) Earth Entry Vehicle (EEV) [12]. Because of the possibility of returning a biohazard from Mars, the MSR EEV TPS system is required to have a probability of failure of one in a million or less. CP was chosen for the TPS material on the MSR EEV because of its widespread use in ballistic missile nose tip applications and for the Pioneer-Venus and Galileo atmospheric probes [13, 14].

Fig. 13 depicts the entry of the MSR EEV, which will occur at speeds ranging from slightly above Lunar Return (11 km/sec) to as high as 13 km/sec, depending upon the trans Mars-Earth trajectory chosen. The sample is contained in an inner sphere, inside a crushable, insulated outer sphere behind the CP heat shield, as shown in the cut-away in Fig. 13.

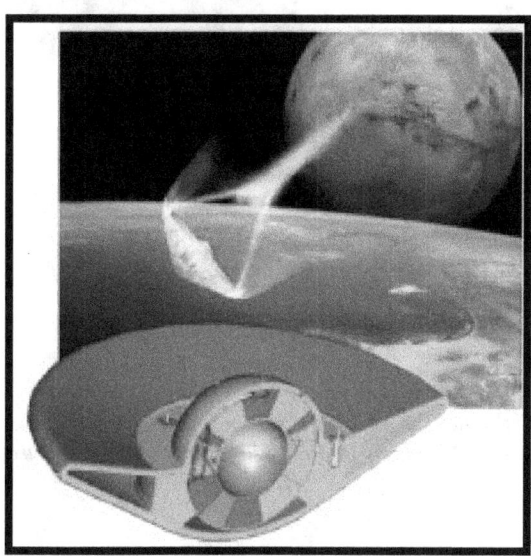

Fig. 13. Artist's concept of the MSR EEV. Courtesy of J. Corliss, Langley Research Center.

Since we saw a reduction of only 5-10% in heat shield mass during our study of nanostructured CP for the surrogate CEV study, we decided to explore nanostructured Reinforced Carbon-Carbon (RCC) - a thermostructural TPS - for a vehicle of the same general configuration as the MSR EEV. RCC has been quite successfully used on the wing leading edges of the Space Shuttle.

Fig. 14 depicts the cross-sectional view of a Sample Return Capsule (SRC) fitted with a 6.35 mm thick RCC forebody heat shield comprised of a 0.9 m base diameter, blunted 60^0 half-angle cone. The base diameter of the SRC RCC forebody is the same as that for the MSR EEV studied in [12]. We stress that the purpose here is to study options for a generic SRC, and not advocate an alternate MSR EEV vehicle.

As for the UHTC study in section 2.1, we carried out calculations of the thermal response to the Earth entry for two cases: (1) RCC with accepted values of its material properties provided by D. Curry of the Johnson Space Center [15], and (2) a nano-structured modification of RCC, whose thermal conductivity is assumed to be homogenously double that of RCC. Fig. 15 (a) depicts the heat flux as a function of time at the stagnation point of the RCC forebody. The entry speed was 11.5 km/sec and the heating was computed using TITAN, a state-of-the-art code [8]. Fig. 15 (b) displays the resulting temperature histories as a function of time for key locations at the centerline.

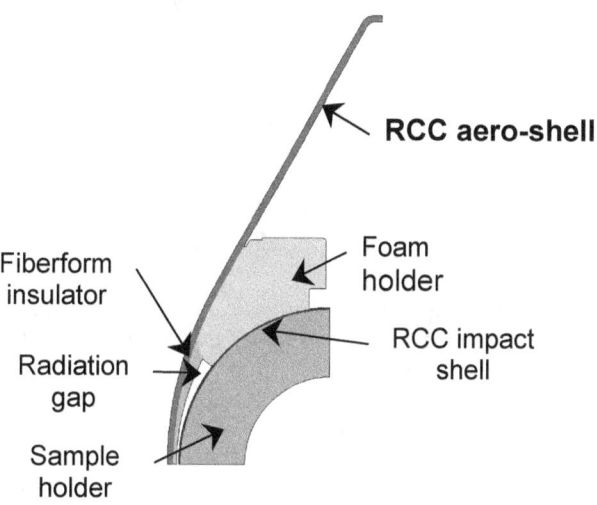

Fig. 14. Axisymmetric cross-section of an SRC, similar to the MSR EEV in [12], except that it is fitted with a Reinforced Carbon-Carbon heat shield. This is intended to be a generic SRC, not advocated here as a replacement of the MSR EEV.

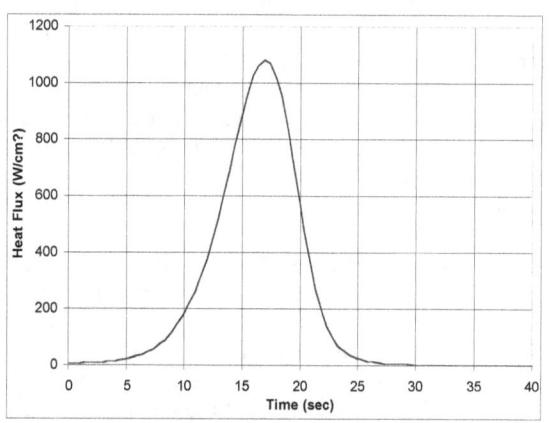

Fig. 15(a) Heat flux (W/cm^2) Versus Time (sec) for the SRC Entering the Earth's Atmosphere at 11.5 km/sec.

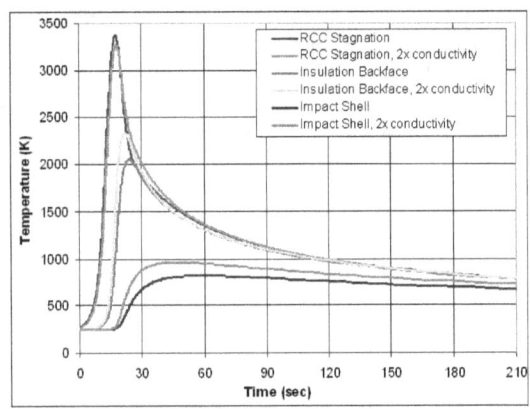

Fig. 15(b). Temperature history at the centerline of the SRC, resulting from the heating from the environment shown in Fig. 15 (b).

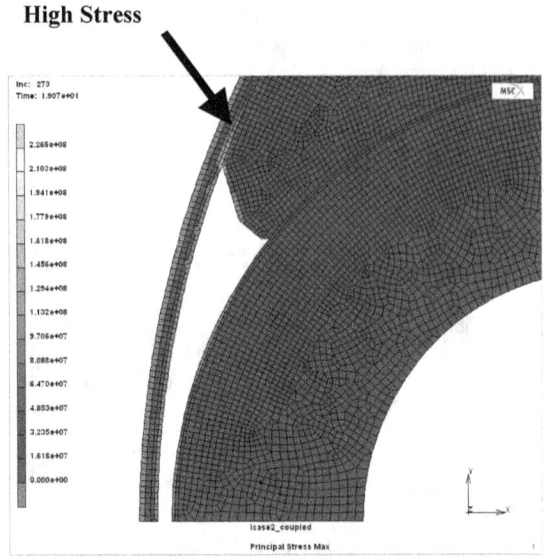

Figure 16. Stress diagram for axisymmetric SRC. Note that the high area of stress is at the attachment point where the forebody and underlying fiberform structure are joined.

The results of the calculations for cases (1) and (2) indicated that the peak tensile stress, located at the juncture of the RCC and the Fiberform insulator, was 228 Mpa, beyond the typical in-plane tensile strengths for RCC, which range from 50–300 MPa. Stress maps for both cases were similar, and it appeared to the authors that the attachment of a thin RCC aeroshell to the substructure could be a major design issue for the RCC forebody.

However, it should be pointed out that yet another property of nanostructured materials could remedy this problem. Estimates by one of us (D.S.), using standard micromechanics models, suggest that tensile strengths of 1 to 2 GPa could be achieved in a nanostructured RCC with an admixture of 90 percent RCC/10 percent carbon nanotube composite and, if so, the issue would be resolved.

It is significant to note that if one could achieve these results, the mass reduction of the nanostructured RCC SRC compared to the MSR EEV baseline, would be 14 kg, a 32 percent total entry mass savings. This mass savings can be understood by comparing the cross sectional views in Figs. 13 and 14. Note that much of the supporting structure for the CP in Fig. 13 is eliminated by use of an RCC forebody in Fig. 14, and this is the source of much of the mass savings.

We are aware that the "open" appearance of the SRC afterbody might cause concern from an aerodynamics/heating perspective, but private communications [16] suggest the CFD for this should be achievable with modern codes and hoped-for flight validation experiments.

3.0 MARS HUMAN AEROCAPTURE VEHICLE

NASA's new Space Exploration Program includes the goal of implementing a human Mars mission. Again, no mission planning for this mission is yet available from the OExP.

During the decade of the 1990's, Johnson Space Center led NASA's development of detailed Reference Missions for the Human Exploration of Mars [17]. These studies clearly showed that mass lifted into low Earth Orbit (LEO) is the principal metric to be minimized for affordable Human Mars Exploration Missions. Aerocapture, and subsequent out-of-orbit descent to the surface of Mars was identified as a "winner" for mass reduction regardless of the propulsion system used for the trans-Earth to Mars trajectory insertion (chemical, nuclear or solar electric). The studies pioneered the multifunctional use of structures as a mass-saving tool. For example, the shroud of the launch vehicle for Earth surface to LEO, containing the Mars exploration systems, doubled as the Mars aeroshell of the aerocapture/descent vehicle. Three of the current authors (J.A., Y.K.C and E.V.) participated in those studies along with others from Ames Research Center, and we adopt herein their results carried out for the JSC-led mission analysis.

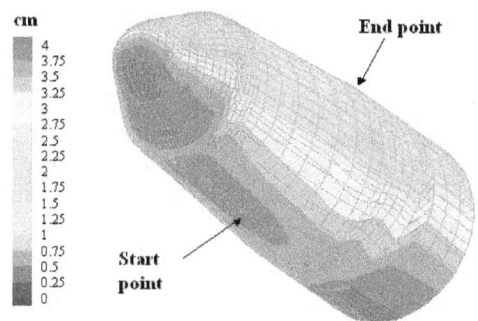

Fig. 17. Perspective view of the 28 meter long human Mars aerocapture vehicle from [17]. The bar indicates the local heat shield thickness in cm.

Fig. 17 shows a perspective of the Human Mars Aerocapture vehicle with a TPS thickness map, optimized for the vehicle encapsulated in a lightweight material, Silcone Impregnated Reusable Ceramic Ablator (SIRCA) [9]. The bar specifies TPS thickness in cm. Fig. 18 displays the stagnation point heat flux history for the aerocapture maneuver. As seen, the heat flux reaches almost 300 W/cm^2 during the maneuver. The primary constraint for the TPS sizing is that the bond line at any location does not exceed 250 ^{O}C. Fig. 19 displays the heating distribution along the streamline distance (represented by S), connecting the points labeled "start point" and "end point", respectively, in Fig 17. The reference length, L is 11.5 m and the start point is in the plane of symmetry.

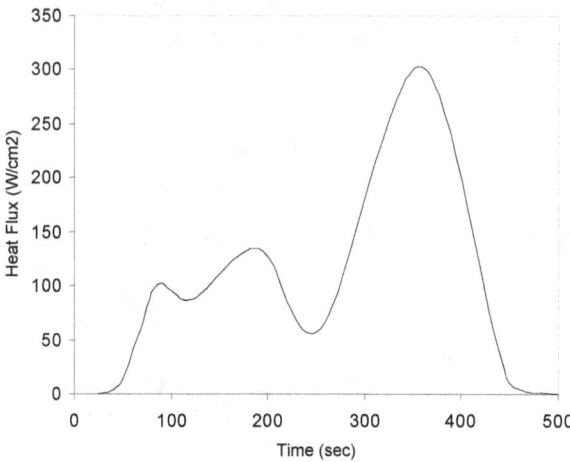

Fig. 18. Heat flux history during aerocapture at the stagnation point discussed above and in [17].

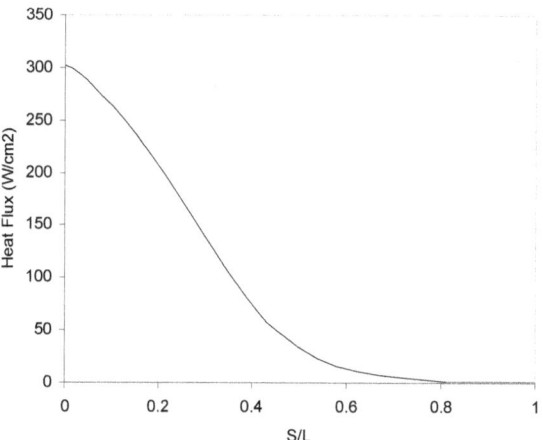

Fig. 19. Heating distributions along the streamline whose start and end points are shown in Fig. 17.

Fig. 20 is a plot of the percentage reduction in thickness from the SIRCA TPS thickness achieved by replacing the pure SIRCA with a 95% SIRCA-5% carbon nanotube admixture. The carbon nanotube fibers were assumed to be aligned along the streamline coordinate S/L. For pure SIRCA, the thermal conductivity at room temperature is 0.06 W/m^{O}K, while that for the admixture at room temperature is 480 W/m^{O}K.

This calculation was performed with the TITAN code [8] and the thermal conductivity is a function of temperature. As the plot shows, there are significant computed percentage reductions in local TPS thickness along the streamline on the windward (hot) side of the heat shield, but there is an <u>increase</u> in thickness on the windward (cool) afterbody region. The average thickness reduction over the streamline is only 0.98 percent, which is small. However, if there is a hot spot reasonably close to a minimum TPS thickness, the passive heat pipe effect might still be used to decrease the total TPS weight significantly. We note that if the carbon nanotube fibers were affixed to a heat sink, e.g. a conformal water tank for human life support, the TPS mass reductions could even be more significant. The carbon nanotube fibers could facilitate such a design, but the required system study is beyond the scope of the present paper.

SUMMARY AND CONCLUDING REMARKS

We have reported herein the results of conceptual studies to explore some of the benefits that might flow from the emerging field of nanotechnology, by improving TPS performance for missions of continuing interest to NASA. In particular, these studies are intended to suggest areas that might be appropriate for further study. We stress that the material properties of the nanostructured materials adopted in the studies are conjectural, but might be achieved through considerable research and development efforts.

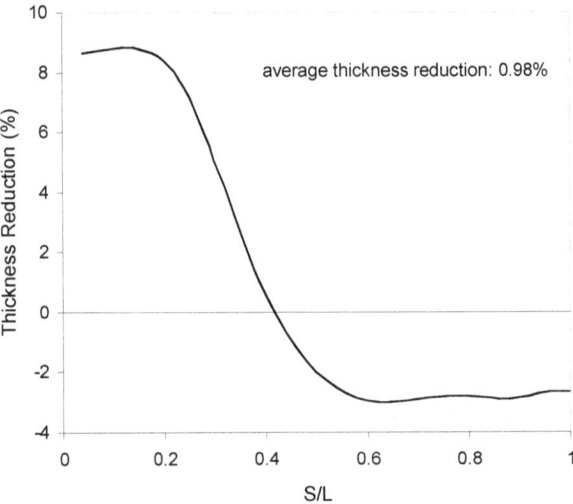

Fig. 20. Plot of predicted local TPS thickness reductions along a streamline connecting the start and endpoints specified in Fig. 17.

With these cautions noted, our conceptual studies indicate the following:

- UHTC properties may be modified by nanostructuring thermal conductivity, improving resistance to oxygen attack and thermal shock resistance. Such materials would find applications in aerogravity assist missions and for high L/D out-of-Earth-orbit vehicles, capable of greater cross range capabilities than vehicles with blunter leading edges.

- The passive heat pipe effect provided by nanostructured materials admixed with standard CP and SIRCA heat shield materials modestly improves TPS performance for an Apollo-type CEV and the Human Mars Aerocapture Vehicles. Passive heat piping, where aeroheating is delivered to heat sinks (e.g. water tanks on human vehicles), could provide good system design options, but an estimation of the benefits from this approach was beyond the scope of the present study.

- Our studies suggest that a significant portion (32 percent) of the total entry mass of a Sample Return vehicle could be achieved through the use of a nanostructured RCC thermostructural heat shield. This would be achieved by the increased tensile strength that may be afforded by carbon nanostructured materials.

ACKNOWLEDEGMENTS

The authors gratefully acknowledge that our efforts were enabled by funding from Ames Internal Research and Development program under the aegis of the Center's Strategic Research Council. J.A. acknowledges support under grant NAG2-1580 SC 20030034. J.A. and D.S acknowledge support under UARC Task Order T0.014.0.HP.IN, Contract NAS2-3144. The authors appreciate the efforts by Mary Gage in preparing the manuscript.

REFERENCES

[1] Meyyappan, M. and Srivastava, D., Handbook for Nanoscience, Engineering and Technology, Chaper18, CRC Press 2003, Editors: Goddard, W. III, and Brenner, D., Lyshevskik, S., and Iafrate, G.

[2] Arnold, J. O., Johnson, S. M. and Wercinski, P. F., "SHARP: NASA's Research and Development Activities in Ultra High Temperature Ceramic Nose Caps and Leading Edges for Future Space Transportation Vehicles", Paper IAF-01-V5.02, IAF Symposium, Toulouse France, October 1-5, 2001

[3] Reuther, J., Kinney, D., Smith, S., Kontinos, D., Saunders, D, and Gage, P., "A Reusable Space Vehicle Design Study Exploring Sharp Leading Edges, AIAA Conference Paper 2001-2884, June 2001.

[4] MSC.Marc Nonlinear Finite Element Solver, MSC.Software Corporation, 260 Sheridan Ave., Suite 309, Palo Alto, CA 94306 (www.marc.com).

[5] Loomis, M. P, and Arnold, J. O., "Thermal, Radiation and Impact Protective Shields (TRIPS) For Robotic and Human Space Exploration Missions", 2nd International Planetary Probe Workshop NASA Ames Research Center, Moffett Field, CA August 23-27, 2004.

[6] Wright, M. J., Prabhu, D. K., and Martinez, E. R., "Analysis of Afterbody Heating Rates on the Apollo Command Modules, Part I; AS-202". 37th AIAA Thermophysics Conference, Portland OR, AIAA Paper 2004-2456.

[7] Bartlett, E. P, Abett, M. J., Nicolet, W. E., and Moyer, C. B. "Improved Heat-Shield Design Procedures for Manned Entry Systems, Part II: Application to Apollo". Contract NAS9-9494, June 22, 1970, Aerotherm Corporation.

[8] Chen, Y.K., and Milos, F. S., "Two-Dimensional Implicit Ablation Thermal Response and Ablation Program for Charring Materils on Hypersonic Space Vehicles", Journal of Spcecraft and Rockets, Vol 38, No. 4, 2001, pp 473-481.

[9][NASA Ames Web Site for TPSX]:
http://tpsx.arc.nasa.gov/index.html

[10] NASA JPL Web Site for Genesis:
http://genesismission.jpl.nasa.gov

[11] NASA JPL Web Site for Stardust:
http://stzrdust.jpl.nasa.gov

[12] Mitcheltree, R., Hughes, S. R. Dillman and Teler, J., "An Earth Entry Vehicle for Returning Samples from Mars", AAAF paper ARVS-102, Second Atmospheric Reentry Vehicles and Systems, March 26-29, Arachon, France.

[13] Fimmel, R. O., Colin, L. and Burgess, E. NASA SP-461, "Pioneer Venus", 1983.

[14] Milos, F.S., "Galileo Probe Heat Shield Ablation Experiment," Journal of Spacecraft and Rockets, Vol. 34, No. 6, 1997, pp. 705-713 and Milos, F.S., Chen, Y.K., Squire, T. H. and Brewer, R.A., "Analysis of Galileo Probe Heatshield Ablation and Temperature Data," Journal of Spacecraft and Rockets, Vol. 36, No. 3, 1999, pp. 298-306.

[15] Curry, D. private communication, July 2004, Properties of Reinforced Carbon-Carbon.

[16] Wright, Michael J. Private Communication, July 2004.

[17] Bret, D., Editor: "Reference Mission 3.0 Addendum to the Human Exploration of Mars: The Reference Mission of the NASA Mars Exploration Study Team". JSC June, 1998.
http://ares.jsc.nasa.gov/HumanExplore/Exploration/EXLibrary/docs/MarsRef/contents.html

NASA AMES ARC JETS AND RANGE, CAPABILITIES FOR PLANETARY ENTRY

Ernest F. Fretter

Thermophysics Facilities Branch, NASA Ames Research Center, MS 229-4, Moffett Field, CA 94035, e-mail: Ernest.F.Fretter@nasa.gov

ABSTRACT

NASA is pursuing innovative technologies and concepts as part of America's Vision for Space Exploration. The rapidly emerging field of nanotechnology has led to new concepts for multipurpose shields to prevent catastrophic loss of vehicles and crew against the triple threats of aeroheating during atmospheric entry, radiation (Solar and galactic cosmic rays) and Micrometorid/Orbital Debris (MMOD) strikes. One proposed concept is the Thermal Radiation Impact Protection System (TRIPS) using carbon nanotubes, hydrogenated carbon nanotubes, and ceramic coatings as a multi-use TPS. The Thermophysics Facilities Branch of the Space Technology Division at NASA Ames Research Center provides testing services for the development and validation of the present and future concepts being developed by NASA and national and International research firms. The Branch operates two key facilities - the Range Complex and the Arc Jets. The Ranges include both the Ames Vertical Gun Range (AVGR) and the Hypervelocity Free Flight (HFF) gas guns best suited for MMOD investigations. Test coupons can be installed in the AVGR or HFF and subjected to particle impacts from glass or metal particles from micron to _ inch (6.35-mm) diameters and at velocities from 5 to 8 km/s. The facility can record high-speed data on film and provide damage assessment for analysis by the Principle Investigator or Ames personnel. Damaged articles can be installed in the Arc Jet facility for further testing to quantify the effects of damage on the heat shield's performance upon entry into atmospheric environments.

1.0 ARC JET COMPLEX

1.1 Mission

We provide ground-based high-enthalpy flow environments in support of experimental research and development activities in thermal protection materials, aerothermodynamics, vehicle structures, and hypersonics.

1.2 Heritage

The Ames Arc Jet Complex has a rich heritage of over 40 years in TPS development for every NASA Space Transportation and Planetary program, including Apollo, Space Shuttle, Viking, Pioneer-Venus, Galileo, Mars Pathfinder, Stardust, X-33, X-34, SHARP-B1 and B2, X-37 and MER-A and B. With this early TPS history came a long heritage in the development of arc jets. These facilities are used to simulate the exit and entry heating that occurs for locations on the body where the flow is brought to rest (stagnation point or nose cap, wing leading edges and on other TPS areas of the space craft). Exposures have been run from a few minutes to over an hour, from one exposure to multiple exposures of the same sample, in order to understand the TPS materials' response to a hot gas flow environment representative of real hyperthermal environments.

The Ames Arc Jet Complex is a key enabler of the three major areas of interest to TPS development: selection, validation, and qualification.

1.3 Facilities

The Ames Arc Jet Complex has seven available test bays located in two separate laboratory buildings. At the present time, four bays contain Arc Jet units of differing configurations that are serviced by common facility support equipment. This support equipment includes two D.C. power supplies, a steam ejector driven vacuum system, a water-cooling system, high-pressure gas systems, data acquisition system, and other auxiliary systems. The magnitude and capacity of these systems is the primary reason why the Ames Arc Jet Complex is unique in the aerospace testing world. The largest power supply can deliver 75 MW for a 30 minute duration or 150 MW for a 15 second duration. This power capacity, in combination with a high-volume 5-stage steam ejector vacuum system, enable facility operations that can match high-altitude atmospheric flight on relatively large size test objects. The arc heaters themselves are of either the Ames designed segmented constricted type or the Hüls design. When combined with a variety of nozzles of both conical and semielliptical cross sections, the resulting facility capabilities offer wide versatility for testing both large flat-surface test objects as well as stagnation flow models that are fully immersed in the test stream.

1.3.1 Aerodynamic Heating Facility (AHF)

The AHF can operate with either a 20-MW constricted arc heater or a Hüls arc heater. The constricted heater operates at pressures from 1 to 9 atm and enthalpy levels from 11 to 33 MJ/kg while the Hüls heater operates at pressure from 1 to 40 atm and enthalpies from 3.5 to 9.5 MJ/kg. Either heater can be coupled with a family of conical nozzles with exit diameters ranging from 76 to 914 mm. A large add-air mixing plenum allows for very low enthalpies for ascent heating simulations. A five-arm fully programmable model insertion system provides exposures of up to five test samples during a single run. Table 1 summarizes testing features for the constricted heater.

- Air or Nitrogen gases
- 20-MW Ames-designed constrictor arc heater or 12 MW Hüels arc heater.
- Nozzles from 3 to 36" exit diameter (76 to 914 mm)
- Samples sizes up to 8" diameter (203mm) or 26 x 26" (660 by 660 mm) wedge configuration
- Pressures from 0.005 to .125 atm (with Hüels heater in excess of 5 atm)
- Heat fluxes from less than 1 on a wedge to over 300 W/cm2 on a 4" dia hemisphere
- 5-arm fully programmable model insertion system

1.3.2 Interaction Heating Facility (IHF)

The IHF is equipped with a 60-MW constricted heater that operates at pressures from 1 to 9 atm and enthalpy levels from 7 to 47 MJ/kg (3000 to 20000 Btu/lb). The facility is designed to operate with interchangeable conical nozzles with exit diameters ranging from 152 mm (6") to 1 m (41"). When the heater is coupled with the semielliptical nozzle, the test stream is suitable for testing flat panels of up to 610 x 610mm (24" by 24") in simulated hypersonic boundary layer flow environments.

Testing features include:

- 60-MW Ames-designed constrictor arc heater
- Nozzle exit sizes from 152mm to > 1m (6" to 41")
- Stagnation, free jet wedge, or flat panel with semielliptic nozzle
- Stagnation pressures from 0.01 to over 1 atm
- Heat fluxes from 5 to >6000 kW/m^2
- Enthalpies from 7 to 47 MJ/kg (3000 to 20,000 Btu/lb)
- Power supply capable of delivering 75 MW for 30 minutes or 150 MW for a 15 second duration

1.3.3 Panel Test Facility (PTF)

The PTF facility operates with a 20-MW constricted heater that is coupled with a semielliptical nozzle. The heater operates at pressures from 1 to 9 atm and enthalpy levels from 7 to 35 MJ/kg (3000 to 15000 Btu/lb). The test stream generated is suitable for the simulation of boundary layer heating environments on flat panel samples of approximately 355 by 355mm (14" by 14"). However, it is possible to test sample sizes of 406 by 406mm (16" by 16").

The testing features include:

- 20-MW Ames-designed constrictor arc heater
- Semielliptic nozzle
- Test samples up to 355 by 355 mm (14" by 14")
- –4 deg to +8 deg inclinations of the surface of the test sample
- Run durations up to 30 minutes possible
- Cold wall (fully catalytic) heat flux from 6 to 340 kW/m^2 (0.5 to 30 Btu/ft^{2}s)
- Surface pressures from 66 to 4700 Pa (.0006 to .05 atm)

1.3.4 Turbulent Flow Duct (2x9)

Turbulent Flow Duct (2x9) is a supersonic duct used to study highly active, turbulent, two-dimensional fluid flows over a flat surface. The duct is rectangular and can accommodate models 203mm wide by up to 508mm (8" to 20") of any desired depth. The duct operates at surface pressures from 0.02 to 0.15 atm and of shear stresses from 5 to 70 kg/m^2. A Hüls arc heater operating at enthalpy levels from 3 to 9 MJ/kg produces the flow.

Testing features include:

- Air or nitrogen gases
- Linde (Hüls) free-length arc heater (12-MW)

- Test samples of 203mm high by 508mm long (8" by 20")
- Surface pressures from 0.02 to 0.15 atm
- Cold wall heat fluxes from 20 to 700 kW/m^2 (2 to 60 Btu/ft^2-s)
- Enthalpy range from 3 to 9 MJ/kg (1300 to 4000 Btu/lb)

1.3.5 Developmental Arc Jet Facility (DAF)

The Development Arcjet Facility, (DAF), at the NASA Ames Arc Jet Complex is designed for the following potential uses (1) high life-cycle testing of TPS for 2nd and 3rd generation RLV, (reusable launch vehicles) (hours in duration); (2) TPS materials testing in simulated high enthalpy Earth atmospheric entry environments (40 to 100 MJ/kg); (3) quick-turnaround thermal protection materials tests in all planetary atmospheric gases, (e.g. carbon dioxide, hydrogen-helium, nitrogen/methane, and argon); (4) test bed for new arc jet diagnostic instrumentation; and (5) chemical vapor deposition (CVD) and nano-technology materials experiments (e.g. diamond film, carbon nanotube production).

An existing one-inch diameter segmented arc heater serves as the DAF plasma generator. The test cell is supported by existing systems including dc power supply, high pressure cooling water, gas delivery, and vacuum pumping. The unique features of this facility are its small scale, ease of use, low cost, and low power requirement compared with the rest of the Arc Jet Complex

Capabilities

- Multiple gases or gas mixtures (N$_2$, O$_2$, Air, CO$_2$, Ar, H$_2$, He, CH$_4$) for simulations of a wide range of planetary entry profiles
- 3-MW Aerotherm™ segmented arc heater
- Stagnation or free wedge configurations
- Multiple model insertion system, up to 10 positions available

2.0 RANGE COMPLEX

2.1 Range Heritage

NASA Ames has a long tradition in leadership with the use of ballistic ranges and shock tubes for the purpose of studying the physics and phenomena associated with hypervelocity flight. The Range Complex has provided critical testing in support of many of NASA's Space Transportation and Planetary Programs including Mercury, Gemini, Apollo, Shuttle, Viking, Pioneer Venus, Galileo, Cassini, Stardust, Mars Odyssey, Mars Exploration Rovers, Mars Science Laboratory, International Space Station, National Aerospace Plane and X-37.

Cutting-edge areas of research run the gamut from aerodynamics, to impact physics, to flow-field structure and chemistry. This legacy of testing began in the NACA era of the 1940's. Today it continues to provide unique, critical and mission enabling support of the Nation's programs for planetary geology and geophysics; exobiology; solar system origins; earth atmospheric entry, planetary entry, and aerobraking vehicles; and various vehicle configurations for supersonic and hypersonic flight.

2.2 Overview

The Test Complex currently consists of three ranges: the Ames Vertical Gun Range (AVGR), the Hypervelocity Free Flight (HFF) Facilities and the Electric Arc Shock Tube (EAST).

2.2.1 Vertical Gun Range

The Ames Vertical Gun Range (AVGR) was designed to conduct scientific studies of lunar impact processes in support of the Apollo missions. In 1979, it was established as a National Facility, funded through the Planetary Geology and Geophysics Program. In 1995, increased science needs across various discipline boundaries resulted in joint core funding by three different science programs at NASA Headquarters (Planetary Geology and Geophysics, Exobiology, and Solar System Origins). In addition, the AVGR provides programmatic support for various proposed and ongoing planetary missions (i.e. Stardust, Deep Impact).

Utilizing its 0.30 cal light-gas gun and powder gun, the AVGR can launch projectiles to velocities ranging from 0.5 to nearly 7 km/sec. By varying the gun's angle of elevation with respect to the target vacuum chamber, impact angles from 0° to 90° with respect to the gravitational vector are possible. This unique feature is extremely important when examining crater formation processes.

The types of projectiles that can be launched include spheres, cylinders, irregular shapes, and clusters of many small particles. The projectiles can be metallic (i.e. aluminum, copper, iron), mineral (i.e. quartz, basalt), or glass (i.e. pyrex, soda-lime). For example, soda-lime spheres can be launched individually for sizes ranging from 1.5 to 6.4mm (1/16 to 1/4 inch) in diameter; in groups of three for sizes ranging from 0.2 to 1.2mm; or as a cluster of many particles for sizes ranging from 2 to 200-μm.

The target chamber is roughly 2.5 meters in diameter and height and can accommodate a wide variety of targets and mounting fixtures. The chamber can maintain vacuum levels below 0.03 torr, or can be back filled with various gases to simulate different planetary atmospheres. Impact events are typically recorded using high-speed video or film.

2.2.2 *Hypervelocity Free-Flight Facility*

The Hypervelocity Free-Flight (HFF) Range currently comprises two active facilities: The Aerodynamic Facility (HFFAF) and the Gun Development Facility (HFFGDF). The HFFAF is a combined Ballistic Range and Shock-tube Driven Wind Tunnel. The primary purpose of this facility is to examine the aerodynamic characteristics and flow-field structural details of free-flying aeroballistic models. The HFFAF has a test section that is equipped with 16 shadowgraph-imaging stations. Each station can be used to capture an orthogonal pair of images of a hypervelocity model in flight. These images combined with the recorded flight time history can be used to obtain critical aerodynamic parameters such as lift, drag, static and dynamic stability, flow characteristics, and pitching moment coefficients. For very high Mach number (i.e. M> 25) simulations, models can be launched into a counter-flowing gas stream generated by the shock tube. The HFFAF is the Agency's only aeroballistic capability, and is the only ballistic range in the nation that is capable of testing in atmospheres other than air. The facility can also be configured for hypervelocity impact testing and shock tunnel testing.

The HFFGDF is used for gun performance enhancement studies, and occasional impact testing. The Facility utilizes the same arsenal of light-gas and powder guns as the HFFAF to accelerate particles ranging in size from 3.2mm to 25.4mm (1/8 to 1 inch) diameter to velocities ranging from 0.5 to 8.5 km/s (1,500 to 28,000 ft/s). Both facilities support three of NASA's strategic Enterprises: Aerospace Technology, Human Exploration and Development of Space, and Space Science. Most of the research effort to date has centered on Earth atmosphere entry configurations (Mercury, Gemini, Apollo, and Shuttle), planetary entry designs (Viking, Pioneer Venus, Galileo and MSL), and aerobraking (AFE) configurations. The facility has also been used for scramjet propulsion studies (NASP) and meteoroid/orbital debris impact studies (Space Station, and RLV).

2.2.3 *Electric Arc Shock Tube*

The Electric Arc Shock Tube Facility is used to investigate the effects of radiation and ionization that occur during very high velocity atmospheric entries. In addition, the EAST can also provide air-blast simulations requiring the strongest possible shock generation in air at an initial pressure loading of 1 atmosphere or greater. The facility has three separate driver configurations. Depending on test requirements, the driver can be connected to a diaphragm station of either a 102mm (4 inch) or a 610mm (24 inch) shock tube. The high-pressure 102mm shock tube can also drive a 762mm (30 inch) shock tunnel. Energy for the drivers is supplied by a 1.25-Mj-capacitor storage system. It can be charged to a preset energy level at either a 0- to 40-kV mode (1530 µF) or a 0- to 20-kV mode (6120 µF). Voltage, capacitance and arc-driver components are selected to meet, as effectively as possible, the test objectives of a given program.

3.0 SUMMARY

The Arc Jets and Range Complex directly support NASA's three main goals: to understand and protect our home planet, to explore the universe and search for life, and to inspire the next generation of explorers. These Facilities also support the three enabling goals of ensuring provision of space access, extending the duration and boundaries of human space flights, and enabling revolutionary capabilities through new technology.

Ames Research Center's Arc Jets & Ballistic Range Complex forges fruitful partnerships with organizations (government, industry and academia) that need to completely, accurately and efficiently test concepts that use innovative techniques, materials and/or design ideas.

We provide a wide variety of hyperthermal and hypervelocity test conditions to examine the aerothermodynamics and flow field characteristics of entry (Earth or other planetary atmospheres) and hypersonic vehicles, simulate meteor or asteroid impacts on a planet or moon surface, and to simulate micrometeoroid impacts on a spacecraft.

Our goal: to provide the Nation with unique, critical and mission enabling testing capabilities by conducting low-cost "flight tests" in ground based facilities. The Arc Jets and Range Complex has a remarkable, comprehensive suite of highly adaptable world-class test hardware. When combined with our staff's extensive expertise and wide range of test experiences, we offer a unique set of testing possibilities.

THE INSTRUMENTED FRISBEE[(R)] AS A PROTOTYPE FOR PLANETARY ENTRY PROBES

Ralph D Lorenz[(1,2)]

[(1)]*Lunar and Planetary Lab, University of Arizona, Tucson, AZ 85721-0092, USA email: rlorenz@lpl.arizona.edu*
[(2)]*Planetary Science Research Institute, The Open University, Milton Keynes, UK*

ABSTRACT

A Frisbee has been equipped with sensors, batteries and microcontrollers for data acquisition to record its translational accelerations and attitude motion. The experiments explore the capabilities and limitations of sensors on a rapidly-rotating platform moving in air, and illustrate several of the complex gyrodynamic aspects of frisbee flight. The experiments constitute an instructive exercise in aerospace vehicle systems integration and in attitude reconstruction.

1. INTRODUCTION

A remarkable feature of the flight of spin-stabilised disc-wings (most familiar in a recreational form such as the 'Frisbee', a trade-mark of Wham-O, Inc.) is that a wide range of nonmonotonic flight behaviours may be generated by manipulating the launch conditions (launch speed and elevation, angle of attack, and spin rate.) These are a result of the combined gyrodynamic and aerodynamic properties of the disc [e.g. 1,2].

The aerodynamic properties of flying discs have been studied in wind tunnel tests with a disc mounted on a motorized [3]. This work (see also [4]) has shown how the pitch moment coefficient, as well as the roll and sideforce coefficients, vary as the advance ratio increases to around unity. Note that the lift and drag coefficients are relatively insensitive to spin.

Iin recreational applications, a frisbee may routinely fly at angles of attack up to 90° in a conventional throw, and indeed up to -90° in throws such as the 'hammer' where the disc is thrown at a slightly negative angle of attack, to roll onto its back and descend near-vertically in an inverted attitude.

In order to explore these flight regimes and to investigate the capabilities of on-board measurements, we have undertaken experiments with free-flying discs equipped with on-board sensors and data acquisition equipment.

An additional motivation for these tests is the evaluation of attitude and trajectory determination methods for planetary probes. These, such as the Mars Pathfinder probe or the Huygens probe to Titan, are spun for attitude stability. Their dynamics are inferred from the data telemetered from a small number of sensors such as gyros and accelerometers to determine the vehicle's flight path, and to infer the density of the atmosphere from the aerodynamic deceleration in flight. Since the measured deceleration relates directly to both the atmospheric density and to the drag coefficient, it is vital for accurate atmospheric measurements that the drag coefficients as a function of angle of attack be known, and the angle of attack history be known.

Figure 1. Test site and launch configuration. Note the high angle of attack of the disc.

Disc-wings have also been proposed as an architecture [4] for Unmanned Aerial Vehicles (UAVs) - onboard sensing will be necessary to implement guidance and control on such vehicles.

2. INITIAL EXPERIMENTS

A commercial (175g 'Wham-O Competition Frisbee') disk was obtained for tests. Initial experiments used an Analog Devices ADXL202 two-axis accelerometer, with its pulse-width modulated digital data sampled and stored by a Parallax Inc., Basic Stamp 2 microcontroller (BS2IC). ~6mA. At 6mA this could be powered by small Lithium 'button' cells (model CR2032). The two accelerometer axes were mounted along and orthogonal to the axis of the disk and were calibrated simply by holding the disc with the sensitive axes along against the local gravity vector.

The microcontroller was programmed in Basic to sample and store the two accelerometer axes, as an 8-bit number (representing from +2 to -2g) into the 2K onboard EEPROM at about 65 sample pairs/sec. We have used similar equipment to measure swing dynamics of parachute-borne instrument packages [5].

The microcontroller and accelerometer were attached with silicone adhesive to the underside of the frisbee (see fig.2), with the accelerometer mounted close to the center of the disk. The batteries were similarly mounted close to the other items to minimize any displacement of the center of mass or change in moments of inertia. The equipment had a mass of about 28g, giving a flying weight of 204g. Further particulars of the construction, and code available for download, have been published in the electronics literature [6].) More serious weight-reduction efforts, e.g. use of surface-mount components, could reduce the instrumentation mass by factor of 2-3, although at the expense of considerable labor.

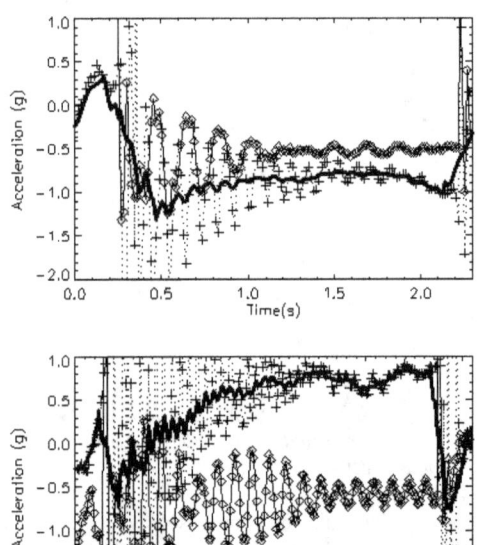

Figure 2. Results from 1st generation disc. Radial (solid/diamonds) and axial (dotted/crosses with smoothed thick line) acceleration. Top - conventional level throw. Note the offset in radial acceleration (due to centripetal acceleration off-center) and the spin modulation. Radial accelerometer shows damped 2x/period modulation due to nutation. Bottom - 'Hammer' throw, where disc is thrown in vertical orientation and rolls onto its back due to the large pitch moment at negative angles of attack -- hence axial signal of ~+1g during later part of flight. Note radial signal vanishes at ~1s – see fig 13.

Flow disturbance was minimized by fairing the equipment with adhesive tape, to present a smooth profile. Since wind-tunnel tests show that the pressure on the underside of a disk is modest and uniform, it is believed that the instrumentation's perturbation to the aerodynamic characteristics is minimal.

3. SECOND AND THIRD GENERATION TESTS

A second, more elaborate set of instrumentation was developed in early 2004 (figure 2). This used a more powerful microcontroller. The Netmedia Inc., BX-24 microcontroller is a device with 32,000 bytes of EEPROM for program and data storage, with 8 on-board 10-bit analog-to-digital converters, also programmable in a Basic programming language.
The BX-24 required a power source with higher current than CR2032 button cells could provide. We used a string of 6 1/3-AAA Nickel Metal Hydride cells giving 7.2V . These cells were mounted along the inside of the rim of the disc.

Components were mounted using a glue gun, usually in holes or recesses cut with rotary tools both to maximize the security of their attachment (discs tend to be made of high-density polyethylene or polypropylene to which adhesives do not bond well), to minimize the projection of components into the airflow and to minimize the change in weight and moment of inertia introduced by their installation.

Figure 3. 2nd Generation Frisbee showing equipment mounted on underside.

This second-generation disc employed ADXL202 accelerometers as before (this case mounted on the rim of the disc, rather than projecting up from its centre.) Two sun sensors, photodiodes yielding a current with a roughly cosine response to sun angle were mounted, one roughly flat on the upper side of the disc, the other on a steeply angled part of the rim. During early

Tucson summer, the sky is persistently clear and thus sun sensors yield excellent signals.

The magnetometers are small, cylindrical fluxgate units (FGM-1) sold for mobile robotics. They provide a 5V square wave pulse with a period proportional to the field along the sensitive axis. One was mounted tangentially inside the rim of the disc; the other was mounted radially, canted at 45 degrees to the disc axis.

Sensors tried on the 2nd generation disc also include a Sharp GP12A02 infrared range sensor (this projects a spot of IR onto a scene, and uses a position-sensitive detector to measure the spot position and thus determine the distance to the reflecting surface.) This sensor gives an analog voltage for distances between about 20cm and 80cm. Although some interpretable data was obtained, the somewhat uneven sensor output and low update rate (new distances are determined only 30 times per second) eroded the sensor's usefulness.

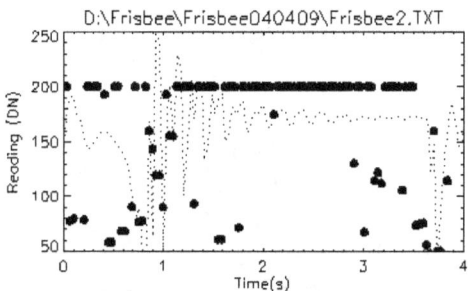

Figure 4. Radial acceleration (dotted line) and IR data (circles). Usable signal is only obtained at the beginning and end of the flight.

A small sonar unit (Devantech SRF-04) was modified to transmit an ultrasound pulse across the underside of the disc, in the hope of obtaining a spin-modulated time-of-flight measurement that might therefore indicate the mean airspeed across the underside of the disc. This was generally noisy, although did sometimes yield a perceptably spin-modulated signal.

Finally, a Fujiwara XP150 pressure sensor was mounted on the curved rim. This is a differential pressure sensor with built-in temperature compensation and amplification giving a 0-4.25Voutput range for pressure differences of 150 Pa. For this application, the sensor output was amplified by a further factor of 30 in the hope of measuring directly the suction peak on the upper leading edge of the disc. A differential measurement is somewhat unsatisfactory in that the 'reference' pressure is that inside the lip on the underside of the disc, which itself changes during the rotation. Again, some notably spin-modulated signals were recovered, but not reliably.

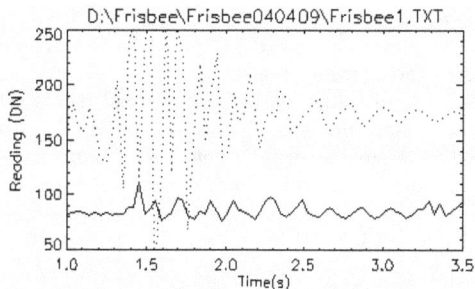

Figure 5. Axial acceleration (dotted line) and speed-of-sound 'anemometer' : spin modulation suggests there is some flight speed information in the data.

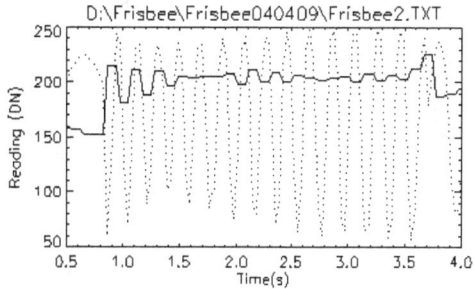

Figure 6. Magnetometer data (dotted line) showing constant amplitude and thus attitude over the last 1.5s of flight. Solid line is pressure sensor output - strongest immediately after release, and notably spin-modulated, as expected.

This 2nd generation testbed confirmed that the attitude sensors and the BX24 and battery installation were workable, even though they brought the instrumented disc mass up to 260g. It was noted that the large number of sensors, confronted with the fixed sampling throughput of the microcontroller, gave sampling rates slightly lower than ideal.

Thus the platform was modified into the 3rd generation disc, with which the principal results of this paper were obtained. This used fundamentally the same architecture, except with a second BX24 microcontroller running in parallel to double the data acquisition capability. A second FGM-1 magnetometer was added, as well as a second accelerometer. This two-axis device, an ADXL210, was mounted flat on the disc center. This device has a sensitivity somewhat low for flight measurements (it is identical to the ADXL202 but with a range of +/-10g) but permits the recording of the accelerations during the throw.

The sonar was modified with a down-looking sonar used as an acoustic altimeter. The device itself (SRF-08) is the same as the SRF-04 used in the time-of-flight experiment, but with the incorporation of a dedicated microcontroller to perform timing. This allows the master processor (the BX24) to perform other duties during the altimetry meassurement. Whereas the time-of-flight pulse took under a millisecond to cross the disc, in the altimetry application the waiting time would be prohibitive - at typical heights of 3m, the sound pulse takes some 18ms to return.

4. SENSOR CALIBRATION

Some analog signal processing was performed in hardware (e.g. resistor network on the sun sensors, amplifier for the pressure sensor.) Further processing was performed in software to yield an 8-bit number for each sensor to facilitate storage as a single byte.

The attitude sensors (sun sensors and magnetometers) were calibrated by mounting the frisbee on a motorized alt-azimuth mount (that of a Meade LX200 20cm telescope). Although the orientations were simply set manually with the motors turned off, this approach yielded a very stable, smooth mount for making measurements at a selection of attitudes. After aligning level and due north, the calibration attitudes were simply read off from the telescope setting circles. A 'lazy susan', a bearing table for presenting condiments in a kitchen, was attached to the mount and could be spun by hand.

Figure 7. Calibration arrangement.

Results are shown in figure 3. The signal processing was designed to yield an approximately cosine response with regard to sun (or field) direction. Between the peak and the trough of the reading, the orientation measurements may have a sensitivity of around 1 DN per degree (all measurements are recorded and reported as 8 bit integer Data Numbers DN.)

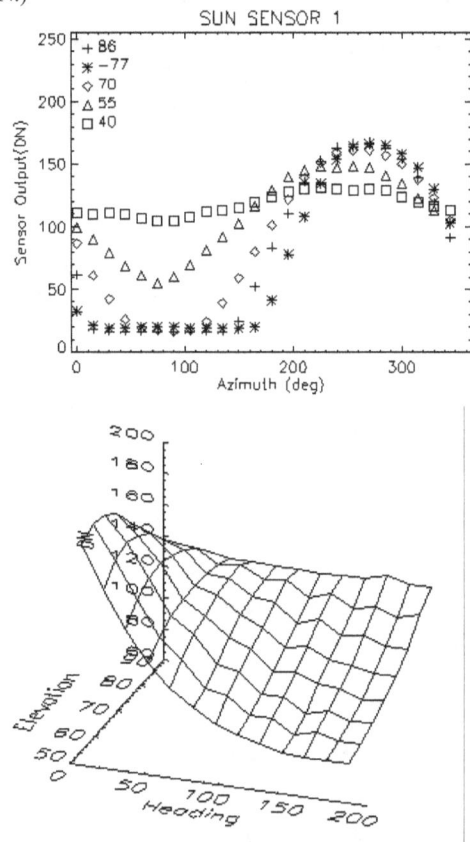

Figure 8. Example sensor calibration data. (a) shows sun sensor 1 output, with the disc attached to the telescope mount and not rotating. Various symbols correspond to different elevation angles (90 degrees is vertical) and the telescope (and spin axis) is rotated to different azimuths. (b) carpet plot of the minimum Magnetometer 1 output as the disc spins with the spin axis at various elevations and spin axis azimuths

5. RANGE INSTRUMENTATION

To determine aerodynamic coefficients, the flight conditions (specifically flight speed and flight path angle) must be known. These are difficult to obtain with on-board measurements, since GPS measurements are not practicable on such short flights, and so additional instrumentation is needed.

A video record from a conventional camcorder was digitized after the tests, and X-Y positions of the disc in the image plane were recorded for each frame to yield a record of the translational motion of the disc for most of its flight, in particular providing a launch speed and angle constraint. The image coordinates were converted into physical distance using red cones placed in the field as ficucial markers every 2.5m. We note the work of Hummel [7,8], who has conducted free-flight measurements using a dedicated high-speed video system (120 and 200 frames per second), deriving aerodynamic coefficients (assumed to be smoothly-varying functions of angle of attack) from the trajectory data alone.

6. FLIGHT RESULTS

The video trajectory data for one flight (#4) is shown in figure 4. The characteristic 'airfoil' trajectory is observed, with a near-linear shallow upwards ramp before the disc has slowed appreciably, then a slow and steepening descent. The video record was obtained looking north, while the disc was thrown in an ESE direction. From the prespective of the thrower (not apparent in this video record) the disc curved slightly to the left (north) towards the end of its flight. Eastward distance was 22m, with a ~6m slide to the N.

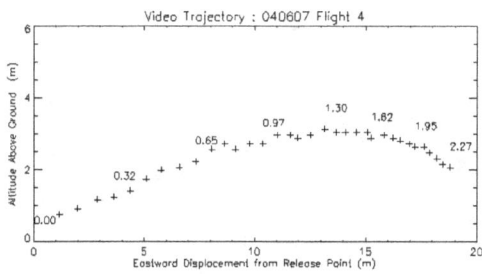

Figure 9. Frisbee trajectory from video record. (a) X-Y position, with time in seconds indicated along the trajectory.

Figure 10 shows the attitude sensor and accelerometer data from the same throw. Spin modulation on is seen on both magnetometers and sun sensors at ~6.5 Hz. Modulation envelope varies due to the slow precession of the spin axis during flight.

The accelerometers are over-ranged (span is +/-2g) at launch and impact. The radial accelerometer is spin-modulated as expected about its zero value of 170DN, though with some twice-per-rotation nutation signal just after launch. Hummel [8] has similarly observed nutation in the first second or so of frisbee flight - the nutation is caused by the angular momentum vector imparted by the thrower being misaligned with the axis of maximum moment of inertia. Energy dissipation, either by flexing of the disc, or by aerodynamic forces, tend to damp this nutation quickly. The Axial accelerometer shows spin modulation (coning) due to sensor misalignment with principal axis, about a near-constant flight signal of ~1.3g.

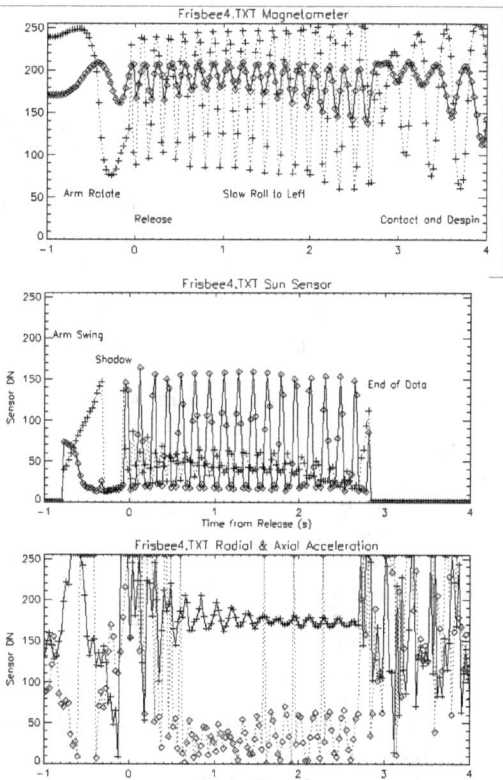

Figure 10. Data from dynamics sensors - same flight as figure 4. Magnetometer record (a) shows spin modulation on magnetometers 1 (dotted line/crosses) and 2 (solid line/diamonds) at ~6.5 Hz. Modulation envelope varies due to tilt of spin axis during flight. Sun sensors (b) show similar effects. Accelerometers (c) are over-ranged (span is +/-2g) at launch and impact. Radial accelerometer (solid line/crosses) is spin-modulated as expected about its zero value of 170DN, though with some twice-per-rotation nutation signal just after launch. Axial accelerometer (dotted line/diamonds, with zero level ~110) shows spin modulation (coning) due to sensor misalignment with principal axis, about a near-constant flight signal of ~1.3g.

Figure 6 zooms in on these parts of the acceleration signals, with a magnetometer overlain to indicate the spin phase. The accelerometers are particularly sensitive to nutation and coning effects, which are well-known in spacecraft engineering. Nutation is commonly observed in spinning satellite deployments due to slight tip-off errors on separation - Spencer et al. [9] show accelerometer data from the Mars Pathfinder entry probe with a nutation signal (although they incorrectly label it as 'coning'), and a nutation was generated on the Giotto spacecraft [10] by a dust impact during its close approach to comet Halley in 1986.

7. ATTITUDE RECONSTRUCTION

The determination of the spin axis and phase of a rotating vehicle is of course a standard problem in spacecraft attitude dynamics, and a combination of sun sensor and magnetometer is often used (e.g. [11,12]) A spacecraft is likely to maintain a constant spin axis and rate over periods of weeks, and a variety of on-board filtering approaches are often used to estimate the attitude. A major difference here is that the spin axis of a frisbee in flight is being precessed rapidly.

Full exploitation of the data from frisbee flights with a fixed (rather than evolving and experimental) sensor configuration may be best accomplished with a forward model of the flight and attitude dynamics, which computes explicitly the expected signal from each of the sensors. By performing Euler rotations, the dot product of the magnetometers with the field vector, and that of the sun sensors with the solar vector, can be calculated at every instant of flight, and a sensor model applied to derive the corresponding stored data number. Expected acceleration signals can similarly be computed. Launch conditions and a model of the aerodynamic coefficients can be adjusted such that the model time series of the various sensors match those recorded.

For the present proof of concept, we employ a heuristic approach, wherein the envelope of the spin-modulated sensor data is used to define, by manual best fit to the data such as that in figure 3, the spin axis direction.

In this exercise, tests were performed quite deliberately in the morning around 8am, when the sun was sufficiently high above the horizon to give a good signal, but was still well in the East. Early afternoon tests would have suffered not only from more convectively unstable conditions with stronger winds, but the sun direction and magnetic field direction would then be approximately collinear, making the attitude determination degenerate.

Since mechanical construction was performed with hand tools and adhesives, without precision alignment fixtures, sensor orientation was determined by post-hoc measurement. As an example, sun sensor 2 was nominally mounted flat in the plane of the disc but after the mounting adhesive had a perceptable, but difficult to measure, inclination to the plane. Modest spin modulation is therefore evident in its flight data (e.g. at one point in flight 6, the reading varies 20-45 ; later the variation is 25-70). Both of these ranges (other data show the sensor to report approximately 20+130

$\cos(\Theta)$ where Θ is the angle between the sensor normal and the sun) are consistent with a mounting offset of 6 degrees.

For the flight shown in figures 4 and 5, the sensor data around 1s after release was best fit with an elevation of 70 degrees (i.e. spin axis was 20 degrees from vertical) and an azimuth of 120 degrees (i.e. the spin axis was tilted towards WNW). If the disc were flying horizontally in an ESE direction, it would therefore have a 20 degree angle of attack (see figure 8 for the relationship of the body sensor axes to lift and drag, and the definition of angle of attack and flight path angle.)

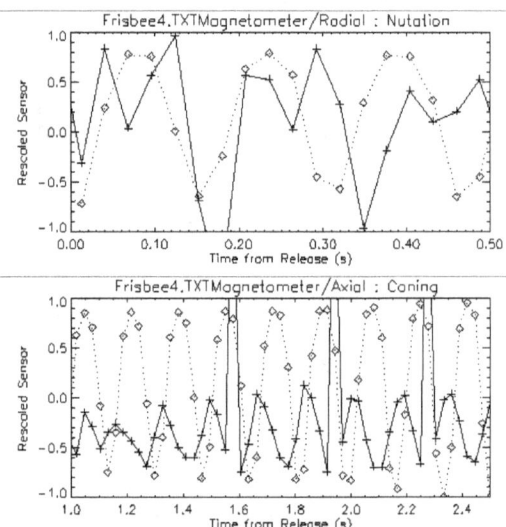

Figure 11. Nonideal dynamic signatures in the acceleration records (solid line/crosses), with magnetometer for spin-phase reference (dashed line/diamonds). (a) shows a nutation signature (~twice per spin period) in the radial acceleration early in the flight - compare with figure 5 later in the flight where this signature is damped out. (b) shows coning (once per spin period) in the axial accelerometer, indicating that the accelerometer is not truly aligned with the spin axis. The axial accelerometer also shows nutation early in the flight.

At ~2.5s, shortly before impact, the attitude has precessed due to the disc's pitch moment. The best-fit attitude has an elevation of 60 degrees and a heading of 150 degrees. This corresponds to a precession of about 14 degrees in total.

8. ACCELERATIONS

The acceleration recorded by the radial accelerometer is by design spin-modulated. The mean value is nominally zero, although sensor positioning and orientation may introduce a centripetal acceleration component, with an amplitude related to the near-constant spin rate. While the modulation envelope is defined by the lift and drag forces, the phase of the peaks and troughs in the signal relative to the peaks and troughs in the attitude sensor record can be used to infer the projection of the flight direction in the spin plane, a useful parameter in guiding a disc wing via spin-phased actuators such as flaps.

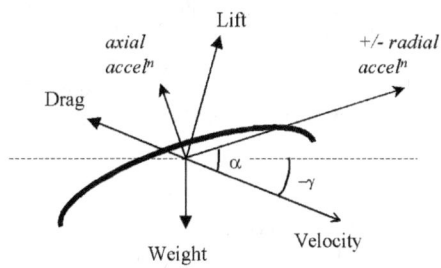

α = Angle of Attack
γ = Flight Path Angle

Figure 12. Forces on flying disc, and their measurement. Measurements are body-fixed, ideally along the spin axis and orthogonal to it (i.e. radial).

Note that the radial signal amplitude and phase is affected not only by the magnitude of the lift and drag, but also by the attitude (see figure 12) which controls how these two forces are projected into the spin plane and thus the radial sensing direction.

Figure 13 shows the radial accelerometer record towards the end of flight, with magnetometer 1 overlain as a phase reference. It can be seen that the acceleration peaks before the magnetometer, indicating a particular heading. However, within 4 spin periods (~0.6s) the phase reverses, with the acceleration peaking after the magnetometer. This does not indicate a 180 degree turn in flight, but rather the changing dominance of the two terms that contribute to the radial acceleration $(Dcos(\alpha)-Lsin(\alpha))/M$, where M is the mass of the disc (0.26kg) and other terms are defined in figure 12. Using the coefficients from [3] the net radial force coefficient $C_dcos(\alpha)-C_lsin(\alpha)$ can be calculated, and is shown in figure 13.

It is seen that this function has a maximum value at a modest angle of attack (~8 degrees, coincidentally close to the angle at which a disc has zero pitch moment, i.e. flies in trim.) Between 20 and 35 degrees, the function decreases rapidly and becomes negative. It is this change of sign that is responsible for the change in phase of the radial acceleration signal.

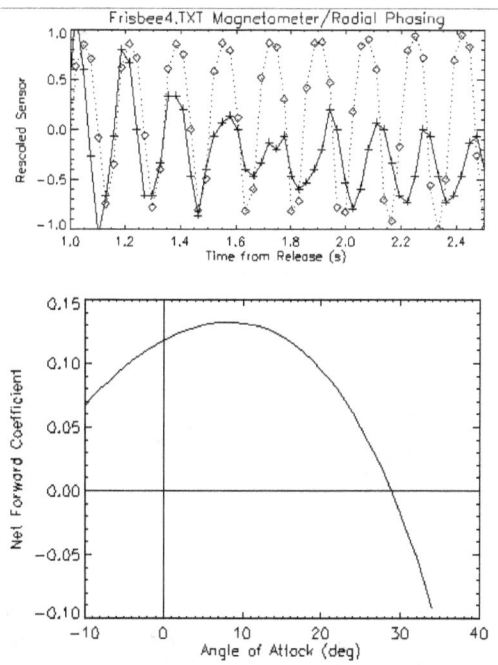

Figure 13. (a) The radial acceleration record (solid line/crosses) against the magnetometer record (dotted line/diamonds). Note that the acceleration initially peaks before the magnetometer, a phasing that could be used to determine the direction of flight. Within 4 spin periods (<1s) the phasing reverses, due to the swiftly-increasing angle of attack (b) The net forward radial component of lift and drag, $C_dcos(\alpha)-C_lsin(\alpha)$, using a quadratic and linear fit to Cd and Cl : at just under 30°, the net component is zero. Beyond 30°, the component is negative, and hence the phase of measured radial acceleration with respect to the forward direction is reversed.

The function is zero, corresponding to a vanishing of the radial signal, at a critical angle of attack of around 28 degrees. In principle, one can deduce from the radial acceleration record when the disc encounters this critical angle (though of course the value will depend on the actual variation of the coefficients with angle of attack - the plot shown is from only the linear and parabolic fits.)

9. AERODYNAMIC COEFFICIENTS

Knowing the disc mass properties, the flight speed and the attitude as a function of time, the aerodynamic coefficients can be produced – lift and drag from instantaneous accelerations resolved into the lift and drag directions., Moment coefficient measurement requires differencing between successive attitude determinations e.g. (fig 14.) to determine precession rates.

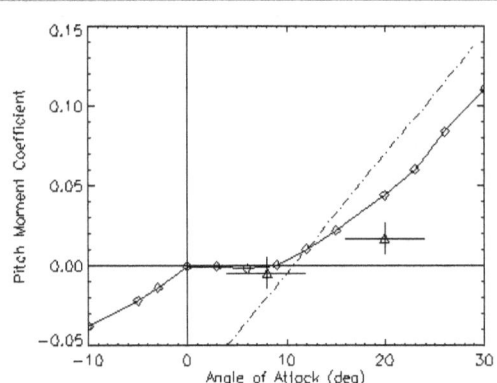

Figure 14. Pitch moment coefficient (diamond/solid line from [4], dot-dash line from [7]). Triangles from flight data in this paper.

10. CONCLUSIONS AND FUTURE WORK

This work has demonstrated that useful flight data on frisbee dynamics can be obtained with onboard instrumentation at rather modest effort and expense. Additional range instrumentation is required to document the flight velocity in order to recover aerodynamic coefficients : a conventional video camera has proven adequate.

A combination of magnetometer and sun sensor data can adequately constrain the attitude of the disc in flight. Sensor mounting offers some tradeoffs. Precise spin-axis alignment is difficult, especially since the equipment modifies the mass properties of the disc, whereas data reduction is simplest if both sensors are aligned with the spin axis. On the other hand, sensors mounted in the spin plane can provide spin-phase information. It has been easiest to mount sensors at an intermediate orientation, and recover the attitude by empirical fitting of sensor readings to data from known orientations.

Accelerometer data has generally been of good quality, although care sensor mounting position and orientation can improve the signal by suppressing coning and nutation sensitivity. Nonetheless, relatively simple algorithms can determine when a sensor is pointed in the forward direction during steady flight - allowing the actuation of control surfaces at a specific spin phase for manoeuvring. Similarly, the acceleration peaks relative to the attitude references can permit on-board determination of heading. A combination of these two may permit autonomous guided flight of disc-wings. An interesting recreational possibility might be a frisbee that homes in on, or attempts to avoid, a person.

Combined attitude information, accelerations, and speed documentation from the video record, allow recovery of aerodynamic coefficients. Our measurement of lift, drag and pitch moment coefficients are in agreement with published wind-tunnel measurements. The free-flight technique lends itself to application in conditions (e.g. high-α) that are challenging for wind-tunnels.

The ultrasonic sensors (speed of sound, and range) have not given encouraging results to date, perhaps as a result of turbulent conditions around the transducers. The infrared ranger gave similarly erratic results, perhaps largely due to its modest refresh rate (~25Hz) or poor performance in very bright conditions.

ACKNOWLEDGEMENTS

Jessica Dooley is thanked for assistance with the experiments and for digitizing the video record. The work was partly supported by Cassini.

REFERENCES

1. Bloomfield, L. A., The Flight of the Frisbee, Scientific American, p.132 April 1999
2. Lorenz R.D., Flying Saucers, New Scientist, 40-41, 19 June 2004
3. Potts, J.R. and W.J. Crowther, Frisbee Aerodynamics, AIAA-2002-3150, 20th AIAA Applied Aerodynamics Conference and Exhibit, 24-26 June 2002, St. Louis, Missouri.
4. Potts, J. R. and W. J. Crowther, Disc-Wing UAV: A Feasibility Study in Aerodynamics and Control, CEAS-AARC-2002.
5. Dooley, J. M. and R. D. Lorenz, A Miniature Probe-Parachute Dynamics Testbed, in Proceedings of the Workshop on Planetary Entry and Descent Trajectory Reconstruction and Science, Lisbon, Portugal October 2003 (published as ESA SP-544, European Space Agency, Noordwjik, the Netherlands)
6. Lorenz, R. D., Frisbee Black Box, Nuts and Volts Vol.25 No.2 (February 2004) pp.52-55
7. Hummel, S. and M. Hubbard, Identification of Frisbee Aerodynamic Coefficients using Flight Data, 4th International Conference on the Engineering of Sport, Kyoto, Japan, September 2002.
8. Hummel, S., Frisbee Flight Simulation and Throw Biomechanics, MS Thesis, UC Davis, 2003
9. Spencer J et al, Mars Pathfinder Entry and Descent Analysis, Journal of Spacecraft and Rockets, 1999.
10. Paetzold, M., Bird, M. K., Volland, H. 1991. GIOTTO-Halley encounter - When was the large nutation generated?. Astronomy and Astrophysics 244, L17-L20

*THERMAL, RADIATION AND IMPACT PROTECTIVE SHIELDS (TRIPS) FOR ROBOTIC AND HUMAN SPACE EXPLORATION MISSIONS

SECOND INTERNATIONAL PROBE WORKSHOP
AUGUST 23-26, NASA AMES RESEARCH CENTER, MOFFETT FIELD CA

M.P. Loomis[1] and J. O. Arnold[2]

[1]NASA Ames Center for Nanotechnology (NACNT) Moffett Field, CA 94035, USA mloomis@mail.arc.nasa.gov
[2]NACNT, University Affiliated Research Center, Moffett Field, CA 94035, USA jarnold@mail.arc.nasa.gov

ABSTRACT

New concepts for protective shields for NASA's Crew Exploration Vehicles (CEVs) and planetary probes offer improved mission safety and affordability. Hazards include radiation from cosmic rays and solar particle events, hypervelocity impacts from orbital debris/micrometeorites, and the extreme heating environment experienced during entry into planetary atmospheres. The traditional approach for the design of protection systems for these hazards has been to create single-function shields, i.e. ablative and blanket-based heat shields for thermal protection systems (TPS), polymer or other low-molecular-weight materials for radiation shields, and multilayer, Whipple-type shields for protection from hypervelocity impacts. This paper introduces an approach for the development of a <u>single</u>, multifunctional protective shield, employing nanotechnology-based materials, to serve simultaneously as a TPS, an impact shield and as the first line of defense against radiation. The approach is first to choose low molecular weight ablative TPS materials, (existing and planned for development) and add functionalized carbon nanotubes. Together they provide both thermal and radiation (TR) shielding. Next, impact protection (IP) is furnished through a tough skin, consisting of hard, ceramic outer layers (to fracture the impactor) and sublayers of tough, nanostructured fabrics to contain the debris cloud from the impactor before it can penetrate the spacecraft's interior.

1. INTRODUCTION

NASA's new vision for Space Exploration calls for a sustained and affordable robotic and human program for the exploration of space beyond low Earth orbit. The human and robotic vehicles involved in these missions must survive long-duration exposure to radiation from Solar Particle Events (SPE), Galactic Cosmic Rays (GCR) and micrometeorites. In many instances, the vehicles - planetary entry probes, sample return capsules and NASA's new Crew Exploration Vehicle (CEV) - must also survive very harsh aerothermal heating during hypervelocity, atmospheric maneuvers. High launch costs continue to motivate significant weight reduction for such vehicles, driving a need for multi-functional materials that perform structural roles, while providing shielding against these harsh space environments.

A new generation of strong, lightweight materials, able to fill this need, is emerging from the developing field of nanotechnology. The fabrication approach to these materials is from the bottom up, so materials of the future can be designed for multiple functions when the material properties for one function are suitable for another. For example, materials comprised of elements with low atomic weight, such as hydrogen and carbon, make good radiation shields because less secondary radiation is produced in collisions with high-speed cosmic rays and solar particles. It also happens that carbonaceous materials filled with low-molecular-weight pyrolizing materials such as carbon phenolic make good ablative heat shields for missions involving high aeroconvective entry heating. In regions of lower heat flux, flexible blankets filled with fibrous insulating materials are used for thermal protection. Similarly, Whipple-type shields, (for protection from hypervelocity impact) have inner layers of tough, fibrous materials to slow down and contain shield penetrants. Clearly, many of these shielding materials have common elemental constituents and similar associated properties, allowing the possibility for one material to perform several functions.

Carbon nanotubes (CNTs) have many properties that make them ideal candidates as the basic building block for multifunctional materials. Their high strength, toughness and low weight make them ideal materials as fibers for impact shields. Their low molecular weight, ability to be functionalized with hydrogen, and ability to form lightweight composites with materials such as polyethylene, make them ideal for use as radiation shields. Their low thermal conductivity in directions normal to the fiber, and high temperature stability when protected from oxidizing environments, make them apt for both ablative and blanket-based heat shields. The high axial thermal conductivity of CNTs allows their use as passive heat pipes to transport heat from hot spots on thermal shields to cooler areas, resulting in lighter, thinner heat shields.

Here, we discuss our approach to conduct research to integrate, redesign and re-engineer heat shields, radiation shields, and impact shields, using nanotechnology-based multifunctional materials. The goal is to develop a single shield with significant weight savings, increased functionality and improved safety and affordability for NASA's next-generation space exploration vehicles.

2.0 FUTURE MISSIONS BENEFITING FROM TRIPS

2.1 Robotic Missions

The NRC New Frontiers Decadal Report [1] envisions missions to the outer planets with multiple atmospheric probes. The new Jupiter Icy Moons Orbiter (JIMO) mission plan [2] involves very long (8-10 years) interplanetary transit voyages to these regions of the solar system. The use of this transportation system for the deployment of atmospheric probes is being discussed. These long duration flights will result in large Total Integrated Doses (TID) of radiation, and involve a greater risk of micrometeorite strikes, providing a technology "pull" for TRIPS technology for robotic missions. TRIPS will enhance more conventional missions and shorter duration robotic missions involving atmospheric entry (Mars, Venus and sample return{s}). The outer ceramic micrometeorite shield on TRIPS would help prevent heat shield erosion in the event of a robotic probe having to enter the Mars atmosphere during a dust storm.

2.2. Human Lunar Missions

Mature concepts of the CEV to be employed on the planned new Lunar missions are not available. However, it is clear that the transit times to/from the moon are short (3 days) and that the re-entry environment will be similar to Apollo (11 Km/sec entry speed and peak heating rates near 400 W/cm^2), if the geometry and mass are similar to that for the Apollo Earth Return Vehicle (Apollo Command Module). Peak entry heating for Apollo was ten times that on the Space Shuttle wing leading edge, therefore ablative heat shield systems/materials (Apollo used an ablator called AVCOAT 5026, which is no longer available) will be required, and entry heating is a serious hazard. Integrated radiation fluxes for normal sun activity are small during the short transit times, but strong solar flares could occur, and it is desirable that the transit vehicle offer protection from them. Micrometeorite/Orbital Debris (MMOD) impacts are possible for Lunar missions. The longer the CEV stays in Lunar orbit or on the Moon, the greater the risk of a micrometeorite strike, and TRIPS would therefore reduce risk of loss of vehicle and crew and be mass efficient for human Lunar Missions.

2.3 Human Mars Missions

During the 1990's, the Johnson Space Center led NASA's development of detailed Reference Missions for the Human Exploration of Mars [3,4]. These studies clearly showed that mass lifted into low Earth Orbit (LEO) is the principal metric to be minimized for affordable Human Mars Exploration Missions. Aerocapture, and subsequent out-of-orbit descent to the surface of Mars was identified as a "winner" for mass reduction, regardless of the propulsion system used (chemical, nuclear or solar-electric) for trans Earth to Mars trajectory insertion, and the vehicles that perform these maneuvers require protective shields. These studies pioneered the notion of multifunctional structures as a mass-saving tool. For example, the Earth surface to Low Earth Orbit (LEO) launch shroud, containing the Mars exploration systems, doubled as the Mars aerocapture/descent vehicle aeroshell. Clearly, additional mass reduction in the aerocapture/descent systems could be achieved through a single system providing protection against multiple threats, but these benefits have not yet been quantified by systems analysis. In this case, the outer ceramic micrometeorite shield on TRIPS would help prevent heat shield erosion in the event of a crewed aerocapture/descent vehicle having to maneuver in the Mars atmosphere during a dust storm.

Fig. 1. Nuclear Thermal Rocket/Mars Aerocapture Vehicle leaving Low Earth Orbit. The Apollo shaped cap on the front served as the Earth Return Vehicle in the mission study [3,4].

3.0 THERMAL PROTECTION SYSTEMS HERITAGE AND DEVELOPMENT

NASA Ames has a 40 + year heritage in developing tools to predict aeroconvective heating environments for entry vehicles and Thermal Protection Systems (TPS) to allow safe entry, descent and landing. This heritage stretches back to Apollo, and leaders in the vehicle's heat shield development are still active at Ames. Ames played a central role in the development of the tile and blanket TPS employed on the Space Shuttle and the carbon phenolic ablative heat shield for the Galileo entry probe.

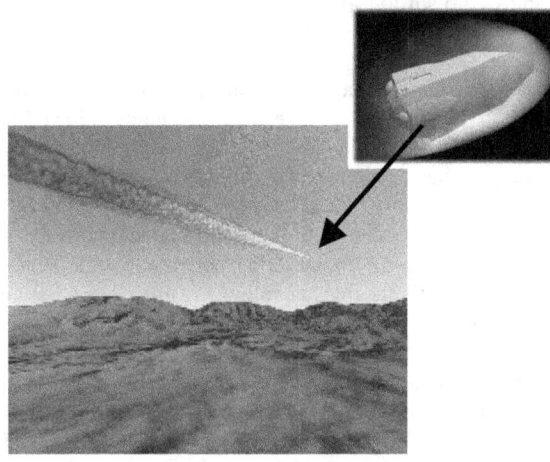

Fig. 2. Aerocapture maneuver where aerodynamic deceleration mass-effectively replaces the need for retro-rockets for insertion into Mars orbit. Aerocapture speeds at Mars range from 7 - 8.5 km/sec and the braking would occur over a ground track covering about 1/3 of the circumference of the Red planet. The aerocapture vehicle was sized to be 28 meters long.

More recently, new ablative materials developed at Ames have been adopted by Agency missions: Silicon Impregnated Ceramic Reusable Ablator (SIRCA) was flown on the afterbody of the entry vehicles for both the Mars Pathfinder Mission and, most recently, the MER missions. SIRCA was sized for the Human Mars Aerocapture vehicle [3,4] and considered to be a viable candidate, as was the commonly used ablator SLA 561-V, developed by Lockheed-Martin Astronautics. Another ablator developed at Ames, Phenolic Impregnated Carbon Ablator (PICA), is suitable for very high heat fluxes (up to about 1,200 W/cm^2). This very lightweight ablator enabled the Stardust Discovery Mission and will protect the Earth Return Capsule during its 12.7 Km/sec re-entry in January, 2006.

New, mid-density ablative heat shield materials, appropriate for use on crewed Moon and Mars missions, need to be developed, and our work on TRIPS will be associated with such an effort. At present, it appears that a new mid-density material will have its roots in PICA and the fully-dense carbon phenolic heritage.

4.0 TRIPS TECHNOLOGY DEVELOPMENT APPROACH

The concept being proposed here (Fig. 3) is to undertake a steady, evolutionary technology development approach, with the long term goal of developing nano-based materials for use in aerocapture and entry vehicles employed in robotic and human exploration missions. The materials would constitute a *single-shield* system, capable of simultaneous protection against aerodynamic atmospheric heating, solar and cosmic radiation and micrometeorite/orbital debris strikes.

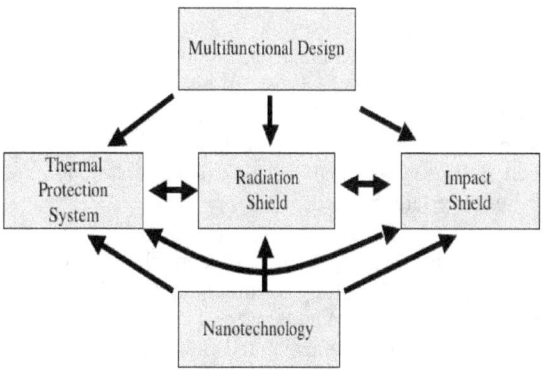

Fig. 3. Key concept of the Thermal, Radiation, Impact Protective Shield (TRIPS) development approach.

In the following sections, we discuss how nano-based materials can be employed independently in shields against the aforementioned hazards, and then discuss the commonality of the materials and concepts for a single shield protecting against multiple hazards.

4.1 Concept: Nanotechnology-Based Shield for Solar Particles and Cosmic Rays

Lightweight materials such as hydrogen, lithium and boron make better radiation shields than those made of high atomic weight systems, since less secondary radiation is produced during the collision process with high-speed cosmic rays and solar particles. This is in contrast to X-rays and gamma rays, which are better shielded by heavy materials.

While not as effective as hydrogen, carbon is also an effective radiation shield. Carbon chain polymers such as polyethylene or polystyrene contain a significant fraction of hydrogen and are often used in radiation shielding. For baseline NASA radiation shielding comparisons, polyethylene is the standard material.

Polymer-carbon nanotube composites have the potential to improve radiation shielding performance if the nanotubes can be functionalized, or filled with significant amounts of hydrogen, lithium or boron. Given the high strength of carbon nanotubes, these properties may enable a multifunctional material with high strength and high radiation shielding capability to be fabricated.

4.2 Approaches for attaching lightweight atomic species in carbon nanotubes and making fibers

Many approaches have been developed recently to functionalize and or fill carbon nanotubes with a variety of materials. Perhaps the largest interest comes from the fuel cell industry, where there is potential for a huge market for reversible hydrogen storage. These techniques are based either on the use of high pressure, electrochemical methods, or filling by capillary action as the nanotube is formed. The results have been somewhat disappointing, especially in terms of hydrogen storage, where early claims of large storage capability were later refuted.

Bauschlicher [5] has used rigorous methods of computational nanotechnology to understand the bonding of hydrogen to carbon nanotubes and Fig. 4 was provided by him. Jaffe [6] has estimated the maximum atomic hydrogenation of carbon nanotubes and storage of H_2 within them would lead to a maximum mass fraction of hydrogen at about 10 percent. We would seek to reach this limit for radiation shielding, provided that other properties of interest for the nanostructured TRIPS material such tensile strength or thermal conductivity were not inappropriately compromised by carbon-carbon bond stretching by the hydrogenation.

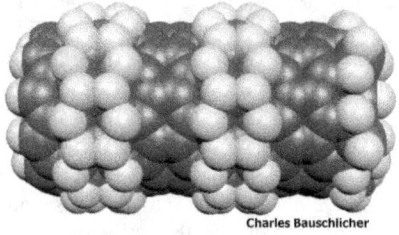

Fig.4. Hydrogenated carbon nanotube

A proposal [7] by our colleagues at NASA Goddard to fill carbon nanotubes with $LiBH_4$ shows significant promise.

Another method of functionalization is ion implantation. This is a technique common in the electronics industry, but has not received as much attention as the other methods for filling nanotubes, since the technique is not reversible. For NASA applications in radiation shielding, reversibility is not an issue, since the desire is to have the hydrogen a permanent part of the material. Furthermore, the method is compatible with either pre- or post-processing of carbon nanotube polymer composites. This may be advantageous, because many groups (such as the U.S. Army Research Labs) are working on the development of high strength carbon nanotube polymer composites for other applications such as bulletproof vests. Post-processing the best composites developed within or outside of NASA using hydrogen ion implantation may be an efficient use of resources.

In the ion implantation technique, ions are implanted directly into the composite with the energy selected for penetration through the film thickness. For a given energy, the distribution of the ions in the material is roughly Gaussian. Varying the energy can provide a more uniform distribution. Minimal damage to the composite during implantation will result if the film or fiber is relatively thin, enabling the use of low ion energy beams which will not break the carbon-carbon bonds and still penetrate through the proper depth. Once implanted, the hydrogen may functionalize or form covalent bonds with the interior or exterior of the nanotube, as well as form molecular hydrogen, inside the tubes or in the intersticies, as shown in Fig. 5.

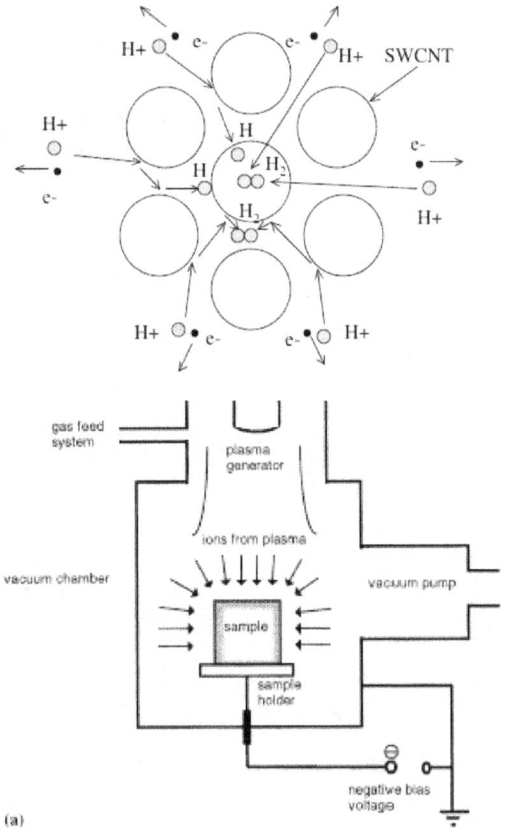

Fig. 5. Schematic of apparatus and process for plasma immersion ion implantation.

We have identified two methods of ion implantation that may be useful in this application. The first uses a commercially available ion gun, which simply accelerates and implants directly into the sample. The second method is plasma immersion or plasma source ion implantation [8,9]. In this technique, the sample is immersed in a low temperature plasma chamber filled with hydrogen or other species. When the sample is biased to a negative voltage, electrons are driven away and ions are accelerated towards the sample and become implanted. It can be shown that if the energy of the ions is kept below 70 eV, the ions will penetrate and dope thin samples but not dislocate carbon atoms from the nanotube lattice, thus retaining the high strength characteristics of the fibers. The method has many advantages over beam-line ion implantation, including high dose rate and uniform coverage.

4.3 Expertise for manufacture of fibers

Several groups have made remarkable progress recently in producing high strength carbon nanotube fibers. We highlight two here to demonstrate the progress which can be leveraged for NASA's purposes and TRIPS developments:

(1) Researchers have developed [10] a procedure for spinning composite carbon nanotube fibers that are tougher than spider silk and any other natural or synthetic organic fiber reported so far. The new fibers are being used to make supercapacitors and to weave textiles. To prepare the fibers, Ray H. Baughman, Alan B. Dalton and their coworkers at UTD and at Trinity College Dublin use single-walled nanotubes synthesized from CO and a surfactant (lithium dodecyl sulfate) in a coagulation-based spinning process. The process produces nanotube-polyvinyl alcohol gel fibers that the group converts to 100-meter-long nanotube composite fibers roughly 50 μm in diameter. On the basis of strength tests, the Texas researchers report that their nanotube product can be drawn into fibers that exhibit twice the stiffness and strength and 20 times the toughness (ability to absorb mechanical energy without breaking) of steel wire of the same weight and length. The fiber toughness is more than four times that of spider silk and 17 times greater than Kevlar fibers used in bullet-proof vests.

(2) Pasquali and Smalley have reported [11] that a sulfuric acid-based superacid makes an excellent medium for dispersing single-walled carbon nanotubes (SWNTs) at concentrations that are useful for industrial processes. They also found that the acids coat SWNTs with a layer of protons. This discovery enabled them to process the dispersion into the first continuous fibers of aligned, pristine SWNTs. Fibers like these might be used to make ultralight, ultrastrong materials with remarkable electronic, thermal, and mechanical properties. This phenomenon allows the team to overcome the tubes tendency to clump together, and they can make solutions composed of up to 10 percent SWNTs by weight – ten times more concentrated than any previously prepared dispersions. At these high concentrations, the SWNTs self-align in a liquid-crystalline phase, similar to the polymer used for making Kevlar. More dilute dispersions employ hard-to-remove detergents and polymer additives and are considered impractical for industrial purposes.

4.4 Modeling

Modeling will be used to predict radiation shielding effectiveness in order to better guide the development and experimental efforts. Based on our collaborations with NASA Goddard, we have decided to use the GEANT4/MULASSIS suite of codes for applications in this area.

GEANT4 is a toolkit for the simulation of the passage of particles through matter. Its application areas include high-energy physics and nuclear experiments, medical, accelerator and space physics studies. GEANT4 exploits advanced software engineering techniques and object oriented technology, to achieve the transparency of the physics implementation and hence provide the possibility of validating the physics results. The GEANT4 software was developed by RD44, a world-wide collaboration of about 100 scientists participating in more than ten experiments in Europe, Russia, Japan, Canada and the United States. A description of the code, which can run on a Windows-based laptop, can be found in [12].

The MUlti-LAyered Shielding SImulation Software (MULASSIS) is a Monte Carlo simulation-based tool for dose and particle fluence analysis associated with the use of radiation shields. Users can define the shielding and detector geometry as planar or spherical layers, with the material in each layer defined by its density and elemental/isotopic composition. Incident particles can be any GEANT4 particles, including protons, neutrons, electrons, gammas, alphas and light ions. There is a wide choice for their initial energy and angular distribution. A description of the code can be found in [13].

4.5 Facilities for testing of radiation-shielding capabilities of materials

Ground testing of shielding materials for space radiation is routinely done at proton and heavy ion accelerator facilities, two of which are found in California. A good review of this topic can be found in the article by Miller [14].

The Crocker Nuclear Laboratory at UC Davis houses a medium-energy particle accelerator, the Davis 76-inch isochronous cyclotron, with associated facilities, and scientific and technical personnel. NASA, The Naval Research Laboratory, JPL and Lawrence Livermore National Laboratory have all used the CNL Cyclotron to support their research in various areas of radiation effects produced by solar and cosmic radiation. The facility can be used to produce protons from 1 to 70 MeV

Loma Linda University Medical Center's Proton Therapy Center has the world's smallest variable-energy proton synchrotron. The accelerator has a range of energies between 40 and 250 MeV. It is designed to deliver a sufficient beam of particle energy to reach the deep localized solid tumors in patients. When not in use for patient treatment, the facility is available for biophysics and radiobiology experiments. In 1994 NASA and LLUMC officials signed a Memorandum of Agreement to study ways to protect astronauts from radiation in space. LLU and NASA scientists are using the University's proton laboratory to simulate cosmic and solar radiation encountered by astronauts, plants, animals, and supportive hardware.

The $34-million NASA Space Radiation Laboratory (NSRL) at Brookhaven is one of the few places in the world that can simulate the harsh cosmic and solar radiation environment found in space. The facility, opened in 2003, employs beams of heavy ions extracted from Brookhaven's Booster accelerator, the best in the United States for radiobiology studies. The NASA Space Radiation Laboratory features its own beam line dedicated to radiobiology research, as well as state-of-the-art specimen-preparation areas. These beams simulate the high-energy, high-charge (HZE) components of galactic cosmic rays that constitute the biologically most significant component of space radiation.

5.0 IMPACT SHIELD CONCEPT AND INTEGRATION WITH THERMAL/ RADIATION SHIELD

5.1 Impact Shield Concept for TRIPS

Christiansen has developed [15] a strategy for incorporation of MMOD shielding into a flexible, deployable concept for the Transhab, which we would adopt and propose to modify for TRIPS; it is said [15] to be the most capable MMOD shield yet developed. Christiansen's approach is depicted in Fig. 6 [15, p.54]. Christiansen reports that his 8 cm thick MMOD shield can prevent back-wall penetration of a 3.6 mm diameter aluminum sphere that strikes the front Nextel layer at an incidence angle of 45° and a velocity of 5.8 km/sec. The test article had no material filling the voids between each layer of the shield.

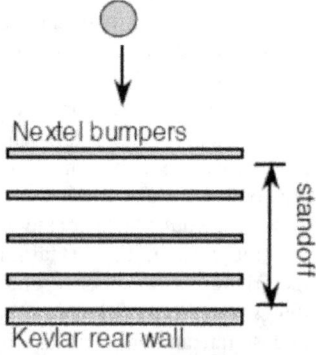

Fig. 6. Adopted from [15 p.54], depicting a MMOD shield developed for the Transhab.

Each Nextel layer in Christiansen's MMOD Transhab shield provides a shock to the penetrating threat, breaking it into smaller debris particles the deeper it goes. The Kevlar layers use the stopping power of this material (used for bullet-proof vests) to arrest the debris and prevent it from penetrating the Transhab interior.

As previously discussed, CNT nanostructured composite fibers have been developed [10] that are 17 times tougher than those from which Kevlar fabric is woven. We would adopt a fabric for TRIPS woven from the CNT nanostructured fibers, with weave spacing to be determined by ballistic range tests. We would also

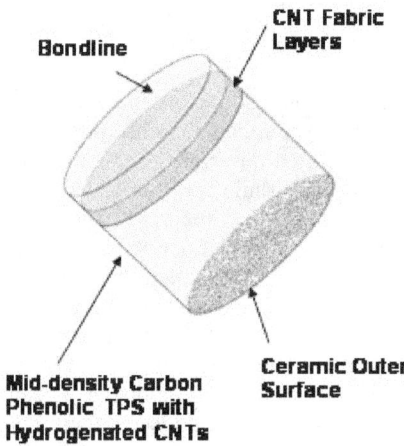

Fig. 7. Conceptual TRIPS Lay-Up. The number of ceramic layers and CNT fabric layers below the outer mold line and their spacing is determined from ballistic range tests.

consider Nextel or other suitable materials, perhaps more capable as a heat shield, to provide the multiple shocking layers, whose function is to progressively break penetrating debris into smaller pieces

Finally, we note that Christiansen, [15, p. 74] first suggested the value of carbon nanotube fibers in MMOD shields, holding particular promise for the intermediate and rear wall materials.

5.2 Impact Shield Modeling

Christiansen's work [15] provides an approach to develop Ballistic Limit Equations (BLE's) that are conservative in their prediction of the amount of shielding necessary for design purposes to protect against debris penetration for a given MMOD threat. These equations use appropriate materials constants and are derived from hypervelocity impact testing conducted in a ballistic range. Further, the approach includes a rationale for the use of BLE's for impact velocities beyond those tested in the ballistic range. Our plans also take account of recommendations of the Columbia Accident Investigation Board [16], that improved physics-based codes should be developed to predict damage to spacecraft by debris.

5.3 End-to-End impact and arc jet testing

Following the work reported in [17], simulated MMOD particles, traveling at hypervelocities, would be fired into test articles in the ballistic range, to develop understanding of impact shield performance and to validate ballistic limit equations (BLEs). Our work will mimic mission profiles in which a presumed MMOD strike would occur prior to the vehicle executing atmospheric maneuvers, when thermal protection is required. Subsequent to the ballistic range testing, the MMOD-damaged test article will be exposed to aeroconvective heating in an arc jet, simulating the aerocapture/entry heating. From these results, databases to define safety limits for MMOD-damaged TRIPS can be derived, as for the Space Shuttle [17].

6.0 INITIAL EVALUATION OF TRIPS MASS SAVINGS

6.1 Approach and limitations

The analysis in this section is limited to the TPS and radiation shielding aspects of TRIPS and is intended only as an initial evaluation of the benefits of our concept. We chose to evaluate an Apollo shape and mass, since it can be considered as a first approximation to a CEV that might be used for out-of-Earth Orbit, Lunar return and Mars return missions. We envision a CEV with an upgradable, replaceable heat shield that could be developed in a "spiral" approach to meet the increasingly higher entry severity (Earth orbit, Earth return, Mars return) and in a fashion that allows for TRIPS research and development. We selected Carbon Phenolic for the TPS material for three reasons: its heritage (military, Pioneer-Venus [18] and Galileo [19]); the low atomic weights of its constituents - carbon + Phenolic (C_6H_6O); and the rule-of-thumb that low atomic weight materials are superior for radiation shields.

6.2. Thermal – conceptual TPS sizing

Our colleague, Dr. Gary Allen, provided the analysis herein with a code that can perform trajectory, engineering aerothermodynamics, TPS sizing and mass estimation for a uniform thickness heat shield. His calculations were validated against Apollo Command Module test flight data.

Fig. 8 shows three trajectories, plotting altitude versus range in km. AS 202 is a rather high speed, out-of Earth orbit, Apollo test flight. AS 501 for Apollo 4 was a rather long flight, and is the closest we have come to demonstrating aerocapture through the use of roll modulation to execute a lifting maneuver with a blunt body. The dotted curve represents a Mars return mission, using the Apollo Command Module to perform aerocapture at 12.5 km/sec to a 700 km altitude.

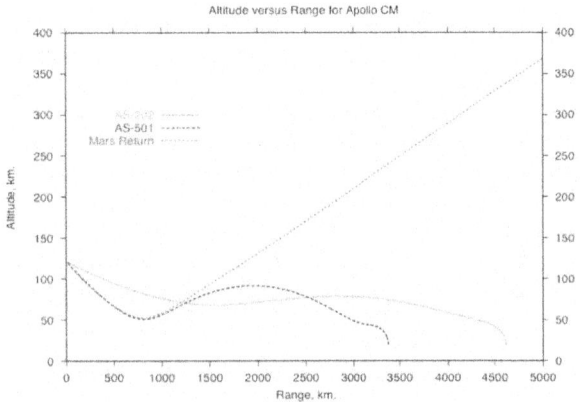

Fig. 8. Trajectories for the Apollo Command Module for the current study

Fig. 9 is the companion chart, showing total heat fluxes (convective plus hot gas radiation). Table 1 summarizes the results of Allen's calculations. Note the very low recession rates predicted in the last two columns of the Table: thickness (sized for the stagnation point) and heat shield recession from ablation. The bondline temperature is the sizing constraint and in each case was chosen to be 250 ^{0}C, with no margin beyond this level (zero bondline temperature margin).

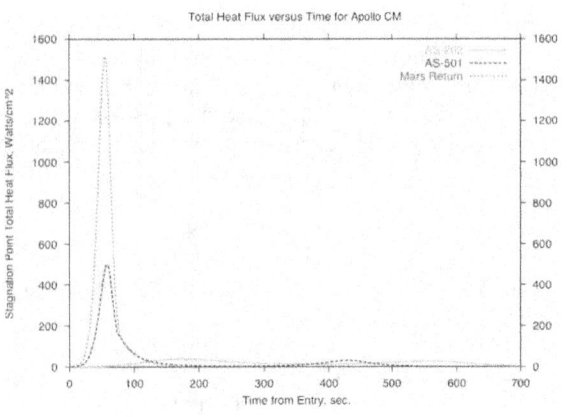

Fig. 9. The peak heat fluxes for the three trajectories in Fig.8 are 39, 521 and 1500 W/cm^2, respectively. The heat fluxes include both convective q_c and hot gas shock layer radiation q_r. The subscript max means maximum values.

Vehicle Design	Vel	$q_{c\text{-max}}$	$q_{r\text{-max}}$	Heat Load	Carbon-Phenolic	
					Thick.	Recess.
	km/s	W/cm^2	W/cm^2	J/cm^2	cm	cm
AS-202	8.7	39	0.00	12678	3.86	0.05
AS-501	11.2	185	336	21590	4.05	0.05
Mars R	12.5	241	1283	42480	4.46	0.12

Table 1. Summary of Apollo Command Module TPS Sizing Calculations.

6.3 Radiation shielding

Fig. 10, taken from [20], plots the 5 cm depth dose equivalent (rem/yr) versus the absorber aerial weight (g/cm^2). The plotted doses are the sum of the (nearly constant) galactic cosmic ray flux and the solar flux, held constant at the solar minimum. The plot does not account for the event of Coronal Mass Ejections (CMEs). The plot does illustrate that the best of the selected shield materials, in order of efficiency, are liquid hydrogen, liquid methane, polyethylene and graphite, depicting the rule-of-thumb that lower atomic weight materials are better for radiation shielding. The horizontal line at 50 rem/yr locates the 1999 recommended maximum allowable annual depth-dose [20] for astronauts working in low Earth orbit.

As noted in [20], crews on interplanetary missions must have safe haven from CMEs and suggestions for shielding range from 5 – 20 g/cm^2 aluminum equivalent. From Fig. 10, it can be seen that 10 gm/cm^2 of graphite and polyethylene are respectively, equivalent to and slightly better shielding materials than 20 cm of

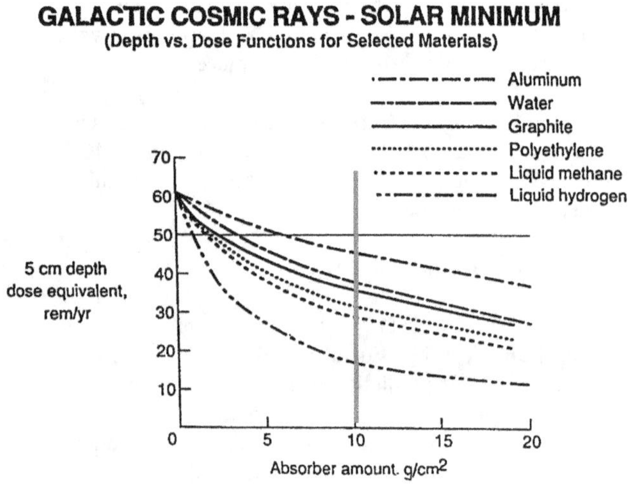

Fig. 10. Five cm depth (in tissue) dose equivalent in rem/yr vs absorber amount, g/cm^2

aluminum. As located by the vertical bar at 10 gm/cm^2, the materials trade space for TRIPS, using a composite of Carbon Phenolic and hydrogenated CNT's lies between polyethlylene and graphite, correlating to the 20 gm/cm^2 upper bound aluminum equivalent [20].

6.4 Impact shielding

This initial analysis of our concept does not include evaluation of impact shielding.

6.5 CONCLUSIONS FROM INITIAL EVALUATION OF TRIPS

The Carbon Phenolic thickness of the TPS for the Mars return case shown in Table 1 is 4.46 cm. The aerial weight of the heat shield is the product of the heat shield thickness and its density: (4.46 cm)x(1.5gm/cm^3) = 6.7 g/cm^2. From the discussion in section 6.3, the aerial weight of the radiation shield is 10 g/cm^2. Assuming that Carbon Phenolic would be approximately as effective as graphite/polyethylene as a radiation shield for this first-cut analysis, we see that the dual-use TPS/radiation shield approach provides TPS **and** about 70 per cent of the upper range of the suggested radiation shielding, encouraging us to develop TRIPS technology. For a mid-density Carbon Phenolic TPS, the fraction of radiation shielding would be less, perhaps 20-40 percent.

As NASA improves its understanding of the biological effects of radiation on deep space missions, it is hoped that less radiation shielding will be required. In this event, the concept of TRIPS will become even more important.

6.4 Self-healing TRIPS

Polyethylene was tested as a potential heat shield material in the early days of thermal protection materials development. To the best of our knowledge, it has not been flown as a TPS material, because it liquifies in the ablation process. It is possible that a layer of polyethylene, placed between the bond line of the vehicle structure and the TRIPS, would melt and fill a hole caused by MMOD, perhaps preventing enlargement of the hole by cavity heating. The resulting system would amount to a self-healing TRIPS.

7.0 SUMMARY

We have presented an approach to develop a single protective shield that can protect robotic and human space transportation vehicles from the triple threat associated with deep space missions: thermal (aerothermodynamic entry heating), radiation and MMOD strikes.

A simple study has shown that using a fully dense Carbon Phenolic TPS for a Mars return capsule also provides about 70 percent of the needed radiation shielding for astronaut health. Use of mid-density Carbon Phenolic TPS would reduce this fraction to 20-40 percent.

We believe that emerging nanotechnologies will enable the development of TRIPS, leading to safer, more affordable space exploration.

ACKNOWLEDGEMENTS

The authors acknowledge helpful discussions regarding the TRIPS concept with Messrs Bernard Laub and Howard Goldstein. We acknowledge Dr. Gary Allen's work in the TPS sizing calculations. We also acknowledge support from NASA Ames Internal IR & D funding from the center's Strategic Research Council. We greatly appreciate Mary Gage's help in preparing our manuscript. J.O. Arnold acknowledges support under NASA grants NAG2-1580 SC 20030034 and NAS2-03144 TO.018.0.HP.ASN.

REFERENCES

[1] Space Studies Board, National Research Council, "New Frontiers in the Solar System: An Integrated Exploration Strategy," July 8, 2002: Web site: http://www.nap.edu.
[2] Jupiter Icy Moons Orbiter, Web site: http://www.jpl.nasa.gov/jimo, and http://www.jpl.nasa.gov/jimomissions.cfu
[3] Drake, B. Editor: "Reference Mission 3.0 Addendum to the Human Exploration of Mars: The Reference Mission of the NASA Mars Exploration Study Team". JSC June, 1998.
[4] Ref Mars Mission Web Site can be found at http://ares.jsc.nasa.gov/HumanExplore/Exploration/EXLibrary/docs/MarsRef/contents.html
[5] Bauschlicher, C. W., ``High coverages of hydrogen on a (10,0) carbon nanotube" Nano Lett.}{1}{223 (2001). }
[6] Jaffe, R.L (NASA Ames), Private communication, July 2004.
[7] Benavides, J. (NASA GSFC), Private communication.
[8] Anders, A., "Fundamentals of Pulsed Plasmas for Material Processing," Surface and Coatings Technology, 2003.
[9] Ensinger, W. "Semiconductor Processing by Plasma Immersion Ion Implantation," Materials Science and Engineering A253,1998.
[10] Baughman, R. H. and Dalton, A. B., CHEMICAL & Engineering News, June 16, 2003

http://pubs.acs.org/cen/topstory/8124/8124notw8.html. and Nature, 423, 703 (2003).

[11] Pasquali, M. and Smalley, R. CHEMICAL & Engineering News, December 15, 2003
http://pubs.acs.org/cen/topstory/8150/8150notw7.html.
[Macromolecules] Web site:
http://dx.doi.org/10.1021/ma0352328

[12] Website: http://wwwasd.web.cern.ch/wwwasd/geant4/geant4.html

[13] Web site:
http://reat.space.qinetiq.com/mulassis/mulassis.html

[14] Miller, J., "Proton and Heavy Ion Accelerator Facilities for Space Radiation Research," Gravitational and Space Biology Bulletin 16(2), June 2003 which can be found on line at.
http://pubs.acs.org/cen/topstory/8124/8124notw8.html

[15] Christiansen, E. L., "Meteoroid/Debris Shielding" NASA Johnson Space Center, TP-2003-210788, August 2003.

[16] Columbia Accident Investigation Board Report, August 2003.

[17] Curry, D. M., Pham, V. T., Norman, I., and Chao, D. C., "Oxidation of Hypervelocity Impacted Reinforced Carbon-Carbon", NASA/TP-2000-209760, March 2000.

[18] Fimmel, L. Colin, L., and Burgess, E. "Pioneer Venus" NASA S-P 461.

[19] Milos, F.S., "Galileo Probe Heat Shield Ablation Experiment," Journal of Spacecraft and Rockets, Vol. 34, No. 6, 1997, pp. 705-713 and Milos, F.S., Chen, Y.K., Squire, T. H. and Brewer, R.A., "Analysis of Galileo Probe Heatshield Ablation and Temperature Data," Journal of Spacecraft and Rockets, Vol. 36, No. 3, 1999, pp. 298-306.

[20] Stanford, M. and Jones, J. A., "Space Radiation Concerns for Manned Exploration" Acta Astronautica, Vol. 45, Issue 1, July 1999, pages 39-47.

Honored Historians

FROM H.G. WELLS TO UNMANNED PLANETARY EXPLORATION

John W. Boyd[(1)]

[(1)]*NASA Ames Research Center, Moffett Field CA, USA, Email: John.W.Boyd@nasa.gov*

The possibility of planetary exploration has been a dream of the human race since Galileo discovered the moons of Jupiter in 1610. Visual sightings of bodies entering Earth's atmosphere have been made by Earth's inhabitants over the centuries.

Over time, the many meteor showers (Leonid, Perseid) have provided dramatic evidence of the intense heat generated by a body entering Earth's atmosphere at hypervelocity speeds. More recently (in 1908), few viewed the Tunguska meteor that impacted in Siberia, but the destructive power on the countryside was awesome.

As an aside, Edward Teller (a physicist and friend), who was born in 1908, used to tell stories of how he and several of his Hungarian colleagues (Von Karmen, Von Neumann, Szilard) came to Earth on this meteor. Many years ago Edward also wore a tie with the letters "ET" on it, long before the movie. Then he claimed to be a Martian.

I have often wondered if Harvey Allen (born in 1910), developer of the blunt body that made entry probes possible, was somehow affected by that event, or by Teller.

I want to mention one more event before we talk about Allen and his impact on planetary probes. From the accounts of H.G. Wells' War of the Worlds written in 1898, the Martians had already solved the entry problem in the last century. From accounts, it was a 30-meter diameter cylinder with a circular nose.

Fig. 1. Mars to Earth entry probe.

There was no indication of the heat shield material except to say that it flaked off when touched.

I would like to take a more serious view and talk about a man to whom all of us in the planetary probe world owe much. H. Julian Allen's work made possible the safe return of all of our Mercury, Gemini and Apollo astronauts as well as the successful entry of our planetary entry probes.

Fig. 2. H.J. Allen.

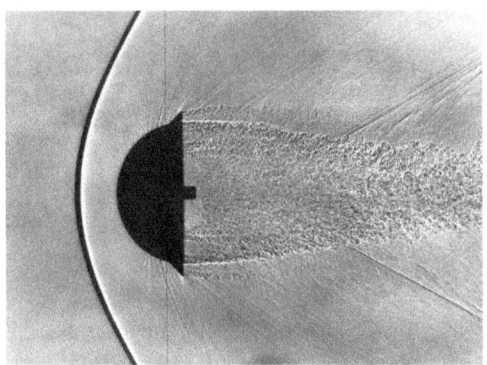

Fig 3. Blunt/Cone

He was the originator of the concept of bluntness as an aerodynamic technique for greatly reducing the severe heating problem of spacecraft entering the atmosphere. His concept also revolutionized the basic design of ballistic missiles, for which he received the Air Force Association Science Trophy.

This audience does not need a tutorial on the rationale for a blunt body. I will just say that Harvey knew that the kinetic energy lost by a missile, or a warhead, as it enters the earth's atmosphere is totally converted into heat. The heat comes from two sources, and, significantly, is generated in two places: inside and outside the boundary layer. The heat appearing outside the boundary layer is generated by shock-wave compression. Of the heat generated in the boundary layer, some results from compression but much more arises from viscous shear or skin friction. The heat generated by the shock wave outside the boundary layer is for the most part well removed from the body and cannot reach it by convection through the insulating blanket of boundary layer. The answer seemed quite obvious to Harvey – make the nose blunt in order to strengthen the bow shock wave.

One of Harvey's earliest scientific contributions of significance to the aviation industry was the development of a general theory of subsonic airfoils. The so-called low-drag airfoils, used on aircraft such as the Mustang Fighter in World War II, were improved considerably through this general theory. Beginning in the mid-1940s, his technical leadership was a driving force behind the development of the various high-speed wind tunnels, hypervelocity ranges and arc jets at the NASA Ames Research Center, which now represent one of the country's primary resources used in the development of advanced aircraft and spacecraft. These national facilities, in a sense, are a monument to his career.

In addition to being a distinguished scientist and engineer himself, Harvey was also an inspiring leader. A whole generation of aeronautical engineers was guided and inspired by him at the Ames Research Center. He served in a number of leadership positions at Ames, capping his career by a term as Center Director from 1965 to 1969.

I worked very closely as his colleague and assistant for about twenty years. When Ames was established in 1939/40 at this site selected by Charles Lindbergh, Harvey was working at Langley and demanded they transfer him or he would resign – they called people who transferred from Langley to Ames the "Mayflower Society."

They did transfer him and Ames flourished because of it. I have no doubt that this man was a genius. He was a true original with startling scientific insight and the uncanny ability to forge paths into entirely new areas and to inspire others to follow him.

He also had a lifelong fascination with airships though he knew they were impractical. In fact, he could have been considered an expert on the subject of airships. At noontimes he told many stories of the various near accidents of airships. The most famous was when he was at Mines Field in Los Angeles and witnessed the takeoff of the Graf Zeppelin as piloted by Hugo Eckhart. Evidently the airship was having difficulty lifting off – so much so that the crew was madly throwing excess weight, including lettuce and other foods, off the airship. Harvey would tell this story and relate its absurdity quite vividly. We had some fun and unusual lunchtime conversations with Harvey.

Allen became Ames' second Director in 1965, and used the position to maintain Ames' preeminence in basic and applied research. He had little use for the political game-playing of Washington, preferring to send subordinates to the endless meetings at Headquarters, and worked hard to keep Ames' essential character as a research center. He continued his well-known habits of casually popping into the offices of his fellow researchers to discuss interesting ideas, thus making sure that he was continually aware of every aspect of Ames' work. Not content with the paper-pushing duties of a senior administrator and always a scientist at heart, Harvey continued his own research during his tenure as Director, while constantly encouraging and nurturing new scientific and engineering talent. He trusted younger researchers to follow their own ideas wherever they led, without the imposition of arbitrary restraints and conditions from above.

As mentioned earlier, under Allen's direction there began the development of hypervelocity ranges and arc jets to confirm and expand on his theories. Over a period of several decades, Ames engineers designed and built a stable of hypervelocity ranges and arc jets that simulated entry conditions, from Earth orbital velocities to speeds in excess of Earth escape speed.

The hypervelocity ranges were critical for understanding the aerodynamics of the Mercury, Gemini, and Apollo entry capsules. The ranges also served this purpose for the atmospheric entry probes that followed. The arc jets have been critical for the TPS development of all of NASA's entry vehicles.

In 1958 when NACA evolved into NASA, the staff whose research was mainly focused on aircraft began looking toward space. From what little we knew at the time of the atmospheric composition of our two closest neighbors, Mars and Venus, they had atmospheres composed primarily of CO_2.

About that time we had a visit from a famous astronomer from Pic du Midi Observatory in the Pyrenees Mountains (Zdenek Kopal). He assured us that CO_2 was the predominant gas in the atmosphere of Mars and Venus. With Harvey's support, in 1959 several of us proposed a project to look at the effects on the drag and stability of several shapes in CO_2.

Using one of the hypervelocity ranges we conducted tests, and the results published in TMX642 showed the stability of some of the shapes was considerably different than in air, and we were off in an exciting new direction.

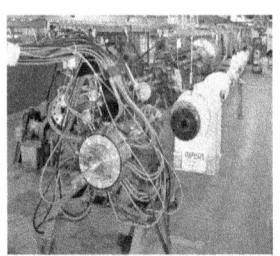

Electric Arc Shock-Tube Facility

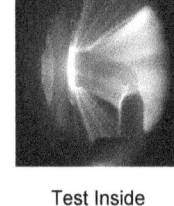

Test Inside an Arc Jet

Pressurized Ballistic Range

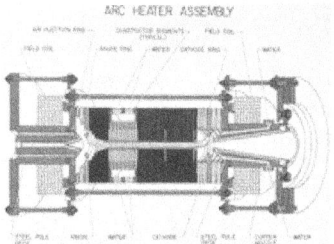

Arc Jet Heater

Vertical Gun Range

Hypervelocity Free Flight Range

Figure 4: Ranges/Arc Jets.

I believe Ames' first serious effort in Entry probes was the PAET Project, which was tested in our ballistic ranges and launched in June 1971 with a successful reentry in the earth's atmosphere, and we will speak of it later. I would like to look briefly at a wide range of these vehicles developed by this country and others.

The first of these was FIRE II, which was a technology demonstrator for Apollo launched in 1965 with an entry velocity of over 11 km/sec. This was followed by the early unmanned Apollos, PAET, Viking, Venus probes, etc. Most recently we have had the strikingly successful MER rovers, Spirit and Opportunity.

We are also looking forward to several upcoming entry events:

- Genesis – to collect solar wind particles and return them to Earth; Earth return, Sept. 2004;
- Huygens: to explore Saturn's moon, Titan; Jan. 2005;
- Stardust: to collect comet material; Earth return, Jan. 2006.

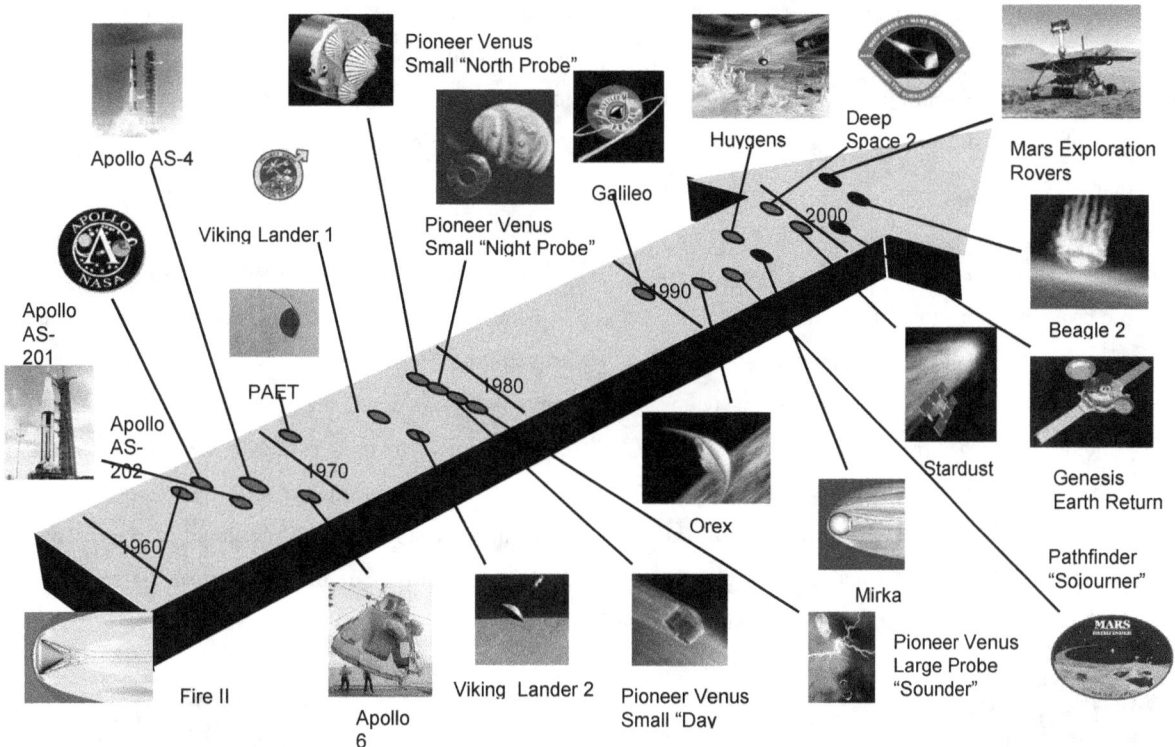

Figure 5. Entry vehicles.

Following in the same spirit of innovation exhibited by Harvey Allen, his colleague Al Seiff proposed in 1963 that small probes be sent to Mars and Venus to determine the structure and composition of their then relatively unknown atmospheres. Al's concept was to measure the structure (density, temperature and mean molecular weight) of the unknown planetary atmospheres from the aerodynamic response of a blunt-bodied vehicle throughout its entry from hypersonic to subsonic flight. His concept of "inverting" the entry physics "problem" into a tool for planetary science was brilliantly demonstrated by the Planetary Entry Experiment Test (PAET).

Fig. 6. Al Seiff (right) with his Project manager for PAET, Dave Reese.

Next, we see a photograph of the forebody of the PAET vehicle showing various instrument windows and ports. It had a beryllium nose cap and an ablator on the flank of the forebody. The inset shows a shadowgraph of a PAET model in flight in the Ames Ballistic Range.

Along the right of the figure are artists' concepts of the three probe missions to Mars, Venus and Jupiter. Al was the PI on the atmospheric structure investigations for all three, and he and his teams clearly met and exceeded the goals set in 1963. Thus we see that the seminal PAET demonstration led to humankind's vastly increased knowledge of the atmospheres of Mars, Venus and Jupiter via the Viking, Pioneer-Venus, and Galileo probe missions.

Having known Al and his work, I can say that he was a wonderful man who literally "touched the stars" with his intellect and instruments. Al was an inspiration to those he led and they generally had great fun doing their work. His example and mentoring gave rise to a considerable number of persons who went on to achieve much in space science, space technology and NASA management.

In conclusion I would like to share with you some artifacts that Harvey Allen gave me for safekeeping thirty-five years ago. They are authentic heat shield plugs taken from Glenn's 1962 Mercury capsule, White's 1965 Gemini capsule, and Apollo 6's unmanned capsule in 1968, presented to Harvey by JSC in recognition of his solution to the re-entry heating problem.

Fig 7. PAET, Viking, Pioneer-Venus and the Galileo Probe

LESSONS FROM THE PIONEER VENUS PROGRAM

Steven D. Dorfman

Boeing Satellite Systems, P.O. Box 92919, Los Angeles, CA 90009-2919

Good evening. It is indeed a pleasure to address a group of scientists and engineers who share my enthusiasm for space exploration. When Bernie Bienstock, tonight's host from Boeing, first asked me to be a keynote speaker tonight he suggested I tell war stories about Pioneer Venus. Well, for a veteran like me, that's a tempting offer. And it's a temptation I don't intend to resist. So my remarks tonight will be more of a personal Pioneer Venus memoir then a program history. As the Hughes Program Manager for the Pioneer Venus project, I gained a perspective on what it takes to design, build, test, launch and fly planetary missions.

Before beginning, let me extract a few lessons from my experience in the planetary exploration program and as program manager of the Pioneer Venus Program. I feel it is the job of experienced veterans to convey to younger members of the scientific and engineering community the problems and issues that were encountered, and how we resolved them.

First, I learned the importance of a continuous, relatively low cost robotic exploration program to complement large scale expensive robotic and manned missions. Funding such programs is an unnatural act for Congress, especially today with the emphasis on manned space flight. I believe it is the responsibility of groups like you to give substance to the idea and be effective advocates for these types of missions.

Second, I learned the importance of mission success. Those of you building instruments or spacecraft should not be beguiled by "faster, better, cheaper." Those ideals only come after quality, the paramount requirement. Your efforts will be quickly forgotten if focus on schedule, cost and innovative ideas does not result in mission success. Keep your eyes on the goal.

A key part of any mission should be an extensive test program. Theory and engineering will indeed produce an impressive spacecraft, but unless it can be proven to operate per specification in the required environment, you may not achieve success. Testing innovation, to simulate the representative environments, is required.

Third, I learned from Pioneer Venus how satisfying it is to be involved in the excitement of the Planetary Programs. I've been associated with many programs over my long career, but none has been more satisfying than my six years heading up the Pioneer Venus program for Hughes.

My planetary exploration experience began in the early 70's with the Outer Planets Grand Tour. At that time, the alignment of the outer planets presented a unique opportunity to fly a single spacecraft to Jupiter, Saturn, Uranus, Neptune and Pluto. In fact, this opportunity was singled out by President Nixon as a major national objective, in much the same way President Bush has announced the missions to the Moon, Mars and beyond as part of his Vision for Space Exploration in February, 2004. In response to Nixon's announcement, JPL planned to begin a major contract procurement. At Hughes I was asked to form a team to compete for this contract and I succeeded in enlisting our best and brightest. As the planning began, and the cost estimates developed, NASA rapidly determined that the Grand Tour was too expensive. As a result, Congress balked and refused to fund the effort.

In an effort to salvage this exciting mission, JPL decided to build the spacecraft in-house as the Mariner Jupiter Saturn (MJS) Program. Thus my Hughes team found itself without an objective. Ultimately the MJS program morphed into the very successful Voyager program that flew two spacecraft to all the outer planets except Pluto and returned extraordinary science data and spectacular photos, testimony to JPL's technical and budgetary creativity.

It was during this period that the idea of the Pioneer Venus program developed. Richard Goody and Mike McElroy of Harvard, Tom Donahue of Michigan, Don Hunten of Arizona and other scientists sold the concept of a series of exploration missions to Venus with comparative planetary atmosphere science as its centerpiece and a launch every few years. The mission was assigned to the Ames team who had been so successful with the Pioneer Program over many years, most recently a flyby of Jupiter, beating JPL to the punch with an amazingly inexpensive mission.

Ames had historically relied on TRW, now NGST, as their contractor. Even though Ames wanted to continue with TRW, since they had an excellent track record, NASA insisted on a competition for the Pioneer Venus program. We had a good team already assembled and the competition for the contract was fierce, but in the end Hughes was awarded the contract. Ironically the major difference between Hughes and TRW was our superior approach to science integration. Since we believed it was our weak point we worked extra hard on the science objectives and overachieved!

The Pioneer Venus program we won was really two missions; an Orbiter to orbit Venus for one Venusian Day (about 9 earth months) and a Multiprobe Bus. The later spacecraft consisted of a probe carrier that, in turn, transported one Large probe and three Small Probes to Venus and released them on a ballistic trajectory for descent through the Venusian atmosphere. As an additional incentive to sell the program to Congress, NASA defined the Pioneer Venus program as a "management experiment" designed to develop and test ways to reduce the cost of planetary programs by streamlining programmatics. If this sounds familiar, think "faster, better cheaper."

All the appropriate executives, including the heads of NASA, Ames and Hughes, agreed with this noble objective for the Pioneer Venus Program, with one very important exception, Ames Pioneer Program Manger Charlie Hall. Charlie and his team had been very successful over the years by paying scrupulous attention to detail. He was committed to making sure the Pioneer Venus program was a technical success, and he was not about to allow Hughes or any other hardware-provider to cut corners in order to save money. Thus Charlie's philosophy was in direct conflict with the management's experiment objective.

We began the Pioneer Venus contract in late 1974 with a planned launch of the Orbiter in May 1978 and the Multiprobe in August 1978. Because we had four years, we thought there was plenty of time. As it turned out, we barely made the launch dates.

The Orbiter was relatively straightforward, compared to the Multiprobe Bus and Probes that had to survive descent through the harsh Venusian atmosphere. To help overcome our many Multiprobe problems we formed a strong global team. The GE reentry team in Philadelphia, experienced in designing vehicles to enter the earth's atmosphere, was assigned the responsibility for the Probe entry system, including protective heat shielding and parachute design to extract the science-laden Large Probe pressure vessel and control its descent through the Venusian clouds. Since the Probes had to remain stable as they descended through the Venus atmosphere, we used the aerodynamic expertise at the Hughes Missile Division, NASA's Ames Research Center and the Langley Research Center. Since the pressure at the surface of Venus was equivalent to an ocean depth of 3300 feet, we went to the Navy's David Taylor Research Center for their deepsea expertise. To test the pressure vessel at the high pressure and temperatures anticipated at Venus we went to the only facility capable of simulating the Venus surface environment, the Southwest Research Institute in San Antonio, Texas. We had dozens of subcontractors all over the world.

As we developed our design, we began an extensive program to validate the ability of our Probe hardware to withstand the Venus environment. During this testing, we encountered numerous problems, mostly associated with adapting earth-based hardware to operate in the anticipated Venus environment. For example, the Large Probe pressure vessel imploded with a very loud bang the first time we tested its ability to withstand the high pressure and temperature on the Venusian surface. We had to go back and redesign, increasing the pressure vessel wall thickness. In addition, during the first tests of the parachute system, our parachute system ripped apart and had to be redesigned. Finally, at the aptly named test range in Truth or Consequences, New Mexico, we successfully demonstrated the parachute design by dropping it from a helium filled balloon at 100,000 feet.

The first time we tested the Small Probe's ability to withstand the hot temperature of Venus we found the interior overheated. Although the Large Probe thermal control was successful with an internal nitrogen atmosphere, this technique did not allow the Small Probe thermal design to close. After much experimentation, we determined that nitrogen gas was too conductive for the Small Probe and needed to be replaced with xenon, a much heavier inert gas.

Design of the probe penetrations and windows was an equally challenging problem, especially for the Large Probe Infrared Radiometer. The only material that would meet the requirements of this instrument at the high temperatures and pressures was natural diamond. We went to the South African diamond company, DeBeers, to find a diamond large enough for our needs. They came up with a 200 carat diamond which was then polished to a _ inch window. The company that processed the diamond was so excited about being part of planetary exploration that they issued an impressive brochure explaining how the window was produced. As difficult and expensive as that process was, it was even more challenging and expensive to determine out how to attach this window to the titanium pressure vessel. We finally developed a brazing process that provided a leak-proof seal.

Resolving these problems, and many more like them, put us behind schedule. By the end of the program, our teams were working 24/7 in order to accomplish the final assembly, integration and test of our multiple spacecraft. The outcome was a real cliffhanger but at the end, with an all-out effort, we launched the Orbiter in May 1978 and the Multiprobe in August 1978, exactly on the original schedule set 6 years earlier

Even after launch there were problems. Our onboard computer on both the Orbiter and Multiprobe Bus experienced single event upsets from high-energy particles. We redesigned the firing sequence of the Orbiter solid motor many times, to find the optimal programming sequence. It was fired by an autonomous onboard timer, with no direct control from the ground stations. If the firing occurred too early, the Orbiter would miss Venus altogether. Too late, and the Orbiter would enter the Venusian atmosphere. In addition, the Orbiter developed a small nutation which could have affected the solid rocket motor firing. Thus we needed to spin-up the Orbiter to make it more stable. In the end, we resolved these last problems and on December 4, 1978 the Orbiter solid rocket motor fired and injected the spacecraft into Venus orbit – exactly as planned.

And five days later, on December 9th, five spacecraft, including the Multiprobe Bus, the Large Probe and 3 Small Probes, approached Venus. They had separated 3 weeks earlier to target 5 different landing sites on Venus. In order to conserve battery power, the one Large and three Small Probes that communicated directly with Earth did not begin transmitting until approximately 20 minutes prior to entry, presumably enough time for the DSN to acquire the transmissions from all probes. Since the probes had been dormant for 4 months, we had no idea of their health status or for that matter whether they were still working. Furthermore, with the need to acquire four probes in a few minutes, there was limited room for DSN error. Needless to say, we were a very anxious group in mission control. The specter of 6 years' work going down the drain was quite real. As you can imagine, there were loud cheers in mission control and around the world as cheerful Aussie operators at the DSN station in Tidbinbilla announced "probe acquired" shortly after the onboard timer turned the probes on. The probes transmitted scientific and engineering data for the one-hour descent to the surface. Even though there was no plan to survive landing, one Small Probe continued to transmit data from the surface for over another hour before being consumed by the heat at the surface.

In the post-mission press conference, I was honest in saying I was delighted that everything worked the first time we tried it at Venus, since almost nothing worked the first time we tested it on Earth. Both the Orbiter and Probe missions were very successful. Most of the desired Probe data was returned and the Orbiter, required to operate for one Venus day (9 months), continued to operate for over 10 years.

The management experiment was a limited success. Charlie Hall was correct in his vision that mission assurance was more important than squeezing the last dollar out of costs. Nevertheless, with our efforts to complete the program efficiently, the total spacecraft cost for the two missions was approximately $100M for 6 spacecraft or about $300M in today's dollars. NASA got good value. After much personal conflict with Charlie over how the program was to be executed, we eventually became good friends with the common objective of a successful outcome. Our final award fee was close to 100%.

Subsequently the Planetary Program fell on bad times, partially due to NASA's focus on the Space Shuttle and International Space Station. Planetary launches declined to once per decade instead of one every few years, with programs like Galileo or Magellan absorbing most of the limited funds. The original plan, to continue visiting Venus with Probes, was abandoned. In fact, the next probe to enter a planetary atmosphere was the Galileo probe, 17 years later.

My company, Hughes, reduced its efforts on planetary exploration and redirected the focus on communication satellites, becoming the world leader in this field. Several members of the Pioneer Venus team went on to develop the satellite direct-to- home service, DirecTV, which became a huge financial success for Hughes.

I'm pleased to note that there is now a renaissance in the Planetary Programs. Under Dan Goldin's leadership, the Discovery Program has resurrected lower cost, smaller programs like the Mars lander, Sojourner, which was enormously successful and captured the public's imagination. Today, there are two robots exploring the surface of Mars, the Genesis spacecraft is on its way to Earth, the Casini spacecraft is now orbiting Saturn and the Huygens Probe is due to enter the Titan atmospheric early next year. Maybe it's even time to return to Venus!

Thank you for inviting me to speak. Good luck to all of you. You are in an interesting field at an interesting time.

My thanks to Bernie Bienstock for inviting me to the Probe Conference and helping me prepare this paper.

Finally, preparing this paper reminded me how much of a privilege it was to work with an outstanding team at Hughes, Ames, JPL, our subcontractors and the scientific community. I salute them all. They made Pioneer Venus a success.

List of Attendees

Attendees

Firstname	Lastname	Affiliation	email
Doug	Abraham	JPL	Douglas.S.Abraham@jpl.nasa.gov
John	Abrams	Stanford University	jlabrams@stanford.edu
Elena	Adams	U of Michigan	eya@umich.edu
William	Ailor	The Aerospace Corporation	William.H.Ailor@aero.org
Rebecca	Allen Diamond	ELORET	rdiamond@mail.arc.nasa.gov
Marla	Arcadi	ELORET	marcadi@mail.arc.nasa.gov
Jim	Arnold	UARC	jarnold@mail.arc.nasa.gov
Sami	Asmar	JPL	sami.asmar@jpl.nasa.gov
David	Atkinson	U of Idaho	atkinson@ece.uidaho.edu
Parviz	Bahrami	NASA ARC	parviz.a.bahrami@nasa.gov
Tibor	Balint	JPL/Caltech	tibor.balint@jpl.nasa.gov
Gilles	Beaufils	Consultant	legillou@club-internet.fr
Sarah	Beckman	NASA ARC	Sarah.E.Beckman@nasa.gov
Reta	Beebe	New Mexico State U	rbeebe@NMSU.Edu
Carlo	Bettanini	U of Padova	carlo.bettanini@unipd.it
Juan	Betts	The Aerospace Corporation	juan.f.betts@aero.org
Bernard	Bienstock	The Boeing Company	bernard.bienstock@boeing.com
Thierry	Blancquaert	ESA/ESTEC	thierry.blancquaert@esa.int
Scott	Bolton	JPL	Scott.J.Bolton@jpl.nasa.gov
Jean-Marc	Bouilly	EADS-ST	jean-marc.bouilly@space.eads.net
Robert	Calloway	NASA LaRC	Robert.L.Calloway@nasa.gov
Silvia	Calzada	ESA Keplerlaan	Silvia.Calzada@esa.int
Sebastiano	Caristia	CIRA	s.caristia@cira.it
Jackeline	Carpio	SJSU, Mech & Aero Eng	SJSUSpartanEngineer@yahoo.com
Richard	Centner	ELORET	rcentner@mail.arc.nasa.gov
Gary	Chapman	AerospaceComputing Inc.	GTChapman@att.net
Eric	Chassefiere	CNRS/UPMC	Eric.Chassefiere@aero.jussieu.fr
F. Neil	Cheatwood	NASA LaRC	F.M.Cheatwood@LaRC.NASA.gov
Jason	Cheng	SJSU, Mech & Aero Eng	jasoncheng@sbcglobal.net
Anthony	Colaprete	NASA ARC	anthony.colaprete-1@nasa.gov
Giacomo	Colombatti	U of Padova	giacomo.colombatti@unipd.it
William	Congdon	Applied Research Associates Inc.	bcongdon@msn.com
Charles	Cornelison	NASA ARC	charles.j.cornelison@nasa.gov
Alan	Covington	ELORET	acovington@mail.arc.nasa.gov
Sylvia	Cox	NASA ARC	sacox@mail.arc.nasa.gov
James	Cutts	JPL	James.A.Cutts@jpl.nasa.gov
Sandra	Dashora	NASA ARC	sdashora@mail.arc.nasa.gov
Christopher	Dateo	ELORET	cdateo@mail.arc.nasa.gov
Linda	Del Castillo	JPL	Linda.DelCastillo@jpl.nasa.gov
Prasun	Desai	NASA LaRC	prasun.n.desai@nasa.gov
Robert	Dillman	NASA LaRC	Robert.A.Dillman@nasa.gov
Nik	Djordjevic	Lockheed Martin	nik.djordjevic@lmco.com
Jessica	Dooley	LPL, University of Arizona	dooleyj@email.arizona.edu
Matthew	D'Ortenzio	NASA ARC	mdortenzio@mail.arc.nasa.gov
Robert	Dougherty	Lockheed Martin	robert.j.dougherty@lmco.com
Chris	D'Souza	C S Draper Laboratory	dsouza@jsc.draper.com
Dan	Dumbacher	NASA HQ	dan.dumbacher@nasa.gov
Tom	Edwards	NASA ARC	Thomas.A.Edwards@nasa.gov
Donald	Ellerby	NASA ARC	Donald.T.Ellerby@nasa.gov
Larry	Esposito	LASP	larry.esposito@lasp.colorado.edu

Jay	Feldman	ELORET	jfeldman@mail.arc.nasa.gov
Francesca	Ferri	CISAS-U of Padova	francesca.ferri@unipd.it
Skip	Fletcher	NASA ARC	Leroy.S.Fletcher@nasa.gov
Francois	Forget	LMD/IPSL/CNRS	Francois.Forget@lmd.jussieu.fr
Robert	Frampton	The Boeing Company	robert.v.frampton@boeing.com
Scott	Francis	Georgia Institute of Technology	gte221w@mail.gatech.edu
Ernest	Fretter	NASA ARC	Ernest.F.Fretter@nasa.gov
Peter	Gage	ELORET	pgage@mail.arc.nasa.gov
Ken	Galal	NASA ARC	kgalal@mail.arc.nasa.gov
Amit	Ganguli	Stanford University	aganguli@stanford.edu
Matt	Gasch	ELORET	mgasch@mail.arc.nasa.gov
Paulo	Gil	Inst Superior Tecnico, DEM-SMA	p.gil@dem.ist.utl.pt
Howard	Goldstein	RIACS/USRA/ARC	hgoldstein@mail.arc.nasa.gov
Sergey	Gorbunov	ELORET	sgorbunov@mail.arc.nasa.gov
Dana	Gould	NASA LaRC	Dana.C.Gould@nasa.gov
Sean	Gourley	SJSU, Mech & Aero Eng	sean.gourley@balliol.oxrod.ac.uk
Mike	Green	NASA ARC	mgreen@mail.arc.nasa.gov
Rob	Grover	JPL	Myron.R.Grover-III-116692@jpl.nasa.gov
Michael	Gubert	NASA MSFC	Michael.K.Gubert@msfc.nasa.gov
Karen	Gundy-Burlet	NASA ARC	gundy@email.arc.nasa.gov
Michael	Gusman	ELORET	mgusman@mail.arc.nasa.gov
Samad	Hayati	JPL	Samad.Hayati@jpl.nasa.gov
T	Hightower	NASA ARC	T.M.Hightower@nasa.gov
Howard	Houben	NASA ARC	houben@humbabe.arc.nasa.gov
Scott	Hubbard	NASA ARC	G.Scott.Hubbard@nasa.gov
Frank	Hui	NASA ARC	frank.c.hui@nasa.gov
Winifred	Huo	NASA ARC	Winifred.M.Huo@nasa.gov
Andy	Ingersoll	Caltech	api@gps.caltech.edu
Ed	Irby	NASA ARC	Edward.B.Irby@nasa.gov
Bonnie	James	NASA MSFC	Bonnie.F.James@nasa.gov
Roger	Johnson	NASA LaRC	r.k.johnson@nasa.gov
Sylvia	Johnson	NASA ARC	Sylvia.M.Johnson@nasa.gov
Carl	Justus	NASA MSFC	jere.justus@msfc.nasa.gov
Vinod	Kapoor	Aerospace Corp	Vinod.Kapoor@aero.org
Bobby	Kazeminejad	Austrian Academy of Sciences	Bobby.Kazeminejad@rssd.esa.int
Devin	Kipp	Georgia Institute of Technology	gtg486n@mail.gatech.edu
Dean	Kontinos	NASA ARC	dean.a.kontinos@nasa.gov
Bernard	Laub	NASA ARC	Bernard.Laub@nasa.gov
Denis	Lebleu	Alcatel Space	denis.lebleu@space.alcatel.fr
Jean-Pierre	Lebreton	ESA/ESTEC	jean-pierre.lebreton@esa.int
Wayne	Lee	JPL	wayne.j.lee-100938@jpl.nasa.gov
Jason	Leong	Lockheed Martin	jason.leong@lmco.com
Mary Kae	Lockwood	NASA LaRC	m.k.lockwood@larc.nasa.gov
Ralph	Lorenz	University of Arizona	rlorenz@lpl.arizona.edu
David	Lozier	NASA ARC	dlozier@mail.arc.nasa.gov
Evans	Lyne	U of Tennessee	jelyne@utk.edu
Daniel	Lyons	JPL	Daniel.T.Lyons@jpl.nasa.gov
Paul	Mahaffy	NASA GSFC	Paul.R.Mahaffy@nasa.gov
Ed	Martinez	NASA ARC	ermartinez@mail.arc.nasa.gov
Joe	Marvin	ELORET	jmarvin@mail.arc.nasa.gov
Jim	Masciarelli	Ball Aerospace & Technologies Corp	jmasciar@ball.com
Ryan	McDaniel	NASA ARC	Ryan.D.McDaniel@nasa.gov
Lori	McNeill	ELORET	lmcneill@mail.arc.nasa.gov
Angus	McRonald	Global Aerospace Corporation	Angus.McRonald@jpl.nasa.gov

Unmeel	Mehta	NASA ARC	unmeel.b.mehta@nasa.gov
Ching	Meng	John Hopkins University	ching.meng@jhuapl.edu
David	Meyers	ELORET	dmeyers@mail.arc.nasa.gov
Meyya	Meyyappan	NASA ARC	meyya@orbit.arc.nasa.gov
Kevin	Miller	Ball Aerospace & Technologies Corp	kmiller@ball.com
Frank	Milos	NASA ARC	Frank.S.Milos@nasa.gov
Michelle	Munk	NASA MSFC	Michelle.M.Munk@nasa.gov
Quoc	Ngo	NASA ARC	qngo@mail.arc.nasa.gov
Tomomi	Oishi	NASA ARC	toishi@mail.arc.nasa.gov
Joseph	Olejniczak	NASA ARC	jolejniczak@mail.arc.nasa.gov
Ricardo	Olivares	NASA ARC	Ricardo.A.Olivares@nasa.gov
Brandon	Owens	Stanford University	owensbd@stanford.edu
Periklis	Papadopoulos	SJSU / ELORET	ppapado1@email.sjsu.edu
Chul	Park	ELORET	cpark@mail.arc.nasa.gov
Amisha	Patel	SJSU, Mech & Aero Eng	aeroslackers333@yahoo.com
Robert	Peck	Gunn High School	krwnpeck@pacbell.net
Leora	Peltz	The Boeing Company	leora.peltz@boeing.com
Craig	Peterson	JPL	Craig.Peterson@jpl.nasa.gov
Nathan	Poffenbarger	SJSU, Mech & Aero Eng	nathanpoffenbarger@sbcglobal.net
Dinesh	Prabhu	ELORET	dprabhu@mail.arc.nasa.gov
Hardeep	Purewal	Stanford University	hpurewal1@aol.com
Zack	Putnam	Georgia Institute of Technology	gte482u@mail.gatech.edu
Ajith	Rajan	Oregon State University	ajith@orst.edu
Philip	Ramsey	U of Tennessee	pramsey@utk.edu
Anna	Rapo	SJSU, Mech & Aero Eng	hopeasilma7@yahoo.com
Daniel	Rasky	NASA ARC	Daniel.J.Rasky@nasa.gov
Daniel	Reda	NASA ARC	Daniel.C.Reda@NASA.gov
Keith	Reiley	The Boeing Company	barbara.I.torrico@boeing.com
Richard	Reinert	Ball Aerospace & Technologies Corp	rreinert@ball.com
Yehia	Rizk	NASA ARC	yrizk@mail.arc.nasa.gov
James	Robinson	NASA HQ	james.r.robinson-1@nasa.gov
Laura	Rogers	Caltech	lrogers@its.caltech.edu
Reuben	Rohrschneider	Georgia Institute of Technology	gte087x@mail.gatech.edu
Minakshi	Sant	NASA ARC	msant@mail.arc.nasa.gov
Keisuke	Sawada	Tohoku University	sawada@cfd.mech.tohoku.ac.jp
Keith	Schreck	SJSU, Mech & Aero Eng	keithsspace_aiaa@yahoo.com
Gerhard	Schwehm	European Space Agency	gerhard.schwehm@esa.int
David	Schwenke	NASA ARC	schwenke@nas.nasa.gov
Surendra	Sharma	NASA ARC	Surendra.P.Sharma@nasa.gov
Prabhakar	Shatdarshanam	SJSU, Mech & Aero Eng	prasub@yahoo.com
Hsiao-Wei	Shieu	SJSU, Mech & Aero Eng	Alex_ae@sbcglobal.net
Victor	Shum	SJSU, Mech & Aero Eng	yshum12@pacbell.net
Mikhail	Simakov	Institute of Cytology	exobio@mail.cytspb.rssi.ru
Kristina	Skokova	ELORET	kskokova@mail.arc.nasa.gov
Charles	Smith	NASA ARC	casmith@mail.arc.nasa.gov
Alan	Somers	Caltech	somers@its.caltech.edu
Jarvis	Songer	Lockheed Martin	jarvis.t.songer@lmco.com
Michael	Souder	Stanford University	kungfu@stanford.edu
Thomas	Spilker	JPL	Thomas.R.Spilker@jpl.nasa.gov
Mark	Spiwak	The Boeing Company	mark.a.spiwak@boeing.com
Thomas	Squire	NASA ARC	Thomas.H.Squire@nasa.gov
Deepak	Srivastava	UARC/Ctr for Nanotechnology	deepak@nas.nasa.gov
James	Stallcop	NASA ARC	James.R.Stallcop@nasa.gov
Ryan	Stephan	NASA LaRC	Ryan.A.Stephan@nasa.gov

Michael	Tauber	ELORET	mtauber@earthlink.net
Jeff	Umland	JPL	Jeffrey.W.Umland-103434@jpl.nasa.gov
John	Underwood	Vorticity Ltd	John.Underwood@vorticity-systems.com
Marcel	Van den Berg	ESA/ESTEC	mvdberg@rssd.esa.int
Benjamin	Vancrayenest	CNES/von Karman Inst Fluid Dynamics	vancraye@vki.ac.be
George	Vekinis	Ntl Ctr Scientific Rsch "Demokritos"	gvekinis@ims.demokritos.gr
Matthew	Velante	SJSU, Mech & Aero Eng	yank_in_cali@yahoo.com
Ethiraj	Venkatapathy	NASA ARC	evenkatapathy@mail.arc.nasa.gov
Kiran	Vermuri	SJSU, Mech & Aero Eng	kvemuri@email.sjsu.edu
Rowen	Vishwa	SJSU, Mech & Aero Eng	Aeronut8@aol.com
Dunyou	Wang	ELORET	dwang@mail.arc.nasa.gov
Paul	Wercinski	NASA ARC	Paul.F.Wercinski@nasa.gov
Rob	Wilbur	NASA ARC	rwilbur@mail.arc.nasa.gov
Michael	Wilder	NASA ARC	michael.c.wilder@nasa.gov
Bill	Willcockson	Lockheed Martin	william.h.willcockson@lmco.com
Adriaan	Window	University of Queensland	adriaan@abwindow.net
Paul	Withers	Boston Un, Center for Space Physics	withers@bu.edu
Ah-San	Wong	University of Michigan	aswong@umich.edu
Michael	Wright	NASA ARC	Michael.J.Wright@nasa.gov
Leslie	Yates	AerospaceComputing Inc.	lyates@aerospacecomputing.com
Boris	Yendler	Lockheed Martin	boris.yendler@lmco.com
Larry	Young	NASA ARC	Larry.A.Young@nasa.gov
Richard	Young	NASA ARC	Richard.E.Young@nasa.gov
Gabriel	Zavala-Diaz	NASA ARC	gzdiaz@mail.arc.nasa.gov
Wayne	Zimmerman	JPL	wayne.f.zimmerman@jpl.nasa.gov

www.ingramcontent.com/pod-product-compliance
Lightning Source LLC
Chambersburg PA
CBHW081717170526
45167CB00009B/3607